# *An Introduction to* TOXICOGENOMICS

# *An Introduction to* TOXICOGENOMICS

*Edited by*

**Michael E. Burczynski**

CRC Press is an imprint of the
Taylor & Francis Group, an **informa** business

The Editor's photograph courtesy of Srinivas Maganti.

CRC Press
Taylor & Francis Group
6000 Broken Sound Parkway NW, Suite 300
Boca Raton, FL 33487-2742

First issued in paperback 2019



ISBN-13: 978-0-8493-1334-9 (hbk)
ISBN-13: 978-0-367-39530-8 (pbk)
Library of Congress Card Number 2002041296



**Library of Congress Cataloging-in-Publication Data**

An introduction to toxicogenomics / edited by Michael E. Burczynski.
p. cm.
Includes bibliographical references and index.
ISBN 0-8493-1334-1 (alk. paper)
1. Genetic toxicology. 2. Genomics. 3. Proteomics. 4. Gene expression. I. Burczynski, Michael E.

RA1224.3 .I586 2003
616'.042—dc21

2002041296
CIP

**Visit the Taylor & Francis Web site at**
**http://www.taylorandfrancis.com**

**and the CRC Press Web site at**
**http://www.crcpress.com**

# Dedication

*This textbook is dedicated to my parents, Michael and Claire, who always encouraged me to follow in their footsteps and mimic their own unrivaled pursuit of perfection.*

# Preface

*An Introduction to Toxicogenomics* is intended to serve as a primary source of information for a wide audience, ranging from undergraduates to established scientists and clinicians in the field of toxicology. Since the advent of cDNA microarrays, oligonucleotide array technology, and gene chip analyses in the last decade, genomics has revolutionized the entire field of biomedical research. Toxicologists immediately recognized the impact that the new subdiscipline of toxicogenomics could have on the study of drug toxicity and rapidly embraced this technology as one of the bright potential futures of toxicological analysis. This textbook is thus an attempt to consolidate the concepts underlying this new field and introduce them to the very population of scientists who will use toxicogenomic approaches in their upcoming research endeavors.

The chapters in Section 1, Fundamentals of Expression Profiling Analysis, were written in such a manner as to introduce the reader (toxicologist or non-toxicologist) to the basic principles of microarray analysis. The first chapter serves as an overview to the actual physical platforms for the analyses themselves: the various microarrays and gene chips available and the caveats and methods currently encountered in preparing these platforms for use. Chapter 2 investigates and discusses expression profiling results generated by the two main platforms available to researchers: cDNA vs. oligonucleotide arrays. Chapter 3 is dedicated to the handling and quality control of microarray data, and covers essential topics in data analysis such as normalization, standard curves, spike-in controls, and techniques for data reduction. In the last chapter of this introductory section, Chapter 4 serves as an introduction to the complex world of cluster analysis by providing examples and specific insights into principal components analysis, hierarchical clustering methods, *k*-means clustering algorithms, self-organizing maps, and other clustering approaches.

Section 2 is a segue from a general discussion of microarray analysis fundamentals to the application of expression profiling techniques to the specific field of toxicology. Chapter 5 introduces general concepts in toxicogenomic studies and also distinguishes between the two fundamental types of toxicogenomic analyses in the field: mechanistic vs. predictive toxicogenomics. Section 3 describes toxicogenomic findings in two of the main model systems employed in toxicological research today. Chapter 6 details the toxicogenomic evaluation of transcriptional responses in rat liver following administration of well-studied toxicologically relevant compounds *in vivo*, while Chapter 7 investigates similar responses, as well as temporal changes in expression, in primary cultures of rat hepatocytes. Comparing and contrasting the findings reported in these two chapters reveal important similarities and differences in expression between these two model systems.

In Section 4, the field of mechanistic toxicogenomics is explored and reviewed. Chapter 8 provides insights into the transcriptional networks affected by the well-known toxicant dioxin (TCDD) via its effects on the aryl hydrocarbon receptor and the xenobiotic response element (XRE). Chapter 9 summarizes mechanistic explorations of phase II enzyme induction by detailing expression profiles evoked by the antioxidant *t*-butylhydroquinone, which works via the antioxidant response element (ARE). In addition to the well-characterized induction of xenobiotic and antioxidant enzymes by tBHQ, many other functionally annotated groups of genes have been found, providing a shining example of how mechanistic toxicogenomic studies generate exciting new leads for the characterization of novel roles for proteins in cells and tissues. In Chapter 10, many excellent toxicogenomic studies published to date are summarized and reviewed, demonstrating the degree to which researchers around the world have begun to employ expression profiling studies in the field of toxicology.

In Section 5, what some might call the "holy grail" of toxicogenomics (that is, the field of predictive toxicogenomics) is presented and discussed. Chapter 11 provides a basic description of toxicogenomic databases and the considerations that accompany these database designs, whether for small, contained studies or for open-ended, in-house construction. Chapter 12 subsequently outlines the expansive and comprehensive approach to predictive toxicogenomic analysis described by authors from GeneLogic, Inc. (Gaithersburg, MD). Their database ToxExpress™ is described in detail, and they provide fundamental examples of applied toxicogenomics in their program, along with impressive results. Chapter 13 describes the first specific application of unsupervised hierarchical clustering approaches to the problem of class discovery in the field of toxicogenomics, and Chapter 14 describes applications of supervised learning methods to the problem of class prediction in the field of toxicogenomics. The conclusion of Chapter 14 introduces the new and promising field of clinical pharmacogenomics/toxicogenomics and summarizes important concepts and aspects in a field that, while largely unproven to date, nonetheless possesses a near limitless potential to impact and benefit human health in the future.

The final chapter in the final section focuses on the future of toxicogenomics and asks several questions: What can we expect to become the norm? What new complementary technologies are rising on the horizon? Will expression profiling data be incorporated into federal regulatory reviews? If methods of prediction are successful, will they weigh as heavily in the decision-making process as carcinogenesis studies and other toxicity screens? These and many other uncertainties define the future of toxicogenomics as a field.

An *Introduction to Toxicogenomics* has gathered many of those scientists at the forefront of this field in order to present a comprehensive and faithful representation of this new subdiscipline. It is my hope that all of the exciting potential of toxicogenomic analysis — from mechanistic studies at the academic laboratory bench, to massive predictive efforts in the pharmaceutical industry, to clinical diagnostics becoming available at the bedside — is realized in the near future. The continued efforts of scientists in this field and those entering this field in the coming years will certainly determine whether toxicogenomics remains a possibility, or a reality.

# Acknowledgments

I first and foremost thank all of the authors contained herein, without whom this book could not have been written. Their expertise and their willingness to work together to create a textbook on this exciting topic was unparalleled and is gratefully acknowledged.

I also must thank various individuals at CRC Press who made this process as enjoyable as it was demanding. Specifically, Kristina Rosello, Erika Dery, Gerry Jaffe, and Stephen Zollo provided me with much-appreciated humor and guiding lights along each step of the way.

I thank the Mississippi College Chemistry Department and in particular Dr. Ed Valente for introducing me to medicinal chemistry. I thank Dr. Greg Möller and Dr. Jerry Exon at the University of Idaho for introducing me to the field of toxicology. I also thank the Pharmacology Department at the University of Pennsylvania and my thesis advisor Dr. Trevor Penning for the excellent training I received in pharmacology and toxicology.

I thank Mark Johnson, who gave me a start in the field of toxicogenomics during my postdoctoral work at Johnson & Johnson; William Trepicchio, who introduced me to the field of clinical pharmacogenomics/toxicogenomics at the Genetics Institute; and, finally, Andrew Dorner, who has continued to provide me with guidance in these pursuits at Wyeth Research.

I also thank the members in my own Clinical Pharmacogenomics Laboratory, as well as my colleagues at Wyeth Research who continually help me grow as a scientist each day: specifically Jen Stover, Natalie Twine, Judy Oestreicher, Maryann Whitley, Andrew Hill, William Mounts, Fred Immerman, Andrew Strahs, Mike Agostino, and Donna Slonim. I also extend appreciation to Vasu Maganti, my co-worker and faithful partner in the prestigious Wyeth Golf League at Rolling Green.

I would be remiss not to mention other members of my family who supported me throughout this process — specifically, Andy S. Mells and Heather S. Tinks, my little brother and sister.

I also thank my agents Jay Poynor and Erica Orloff, who graciously released me from work on my novels and allowed me to focus on this textbook for the better part of a year. I appreciate you both more than words can say.

Finally, I thank Jennifer — whose loving support, patience, and understanding are simply boundless.

# The Editor

**Dr. Michael Burczynski** received his Ph.D. in Pharmacology from the University of Pennsylvania and is currently an investigator in the Division of Molecular Medicine at Wyeth Research in Cambridge, MA. Some of his most recent articles have appeared in the scientific journals *Toxicological Sciences*, *Cancer Research*, *Biochemistry*, and *The Journal of Biological Chemistry*. He received a Bristol Myers Squibb Young Investigator Award in 1999 from the American Association of Cancer Research for his work on the role of human aldo-keto reductases in oxidative stress. He is a member of the American Association of Cancer Research, the Society of Toxicology, and several other professional organizations. He has resided in various locales, including Idaho, London, Philadelphia, New York, and Boston. His interests include classical piano, hockey, golfing, falconry, and flyfishing with his father in the Catskill Mountains in upstate New York. He is also a novelist and writes under the pen name Michael St. Clair. He is currently at work finishing up his first two novels: a biomedical thriller entitled *The Apocalypse Gene* and a serial killer mystery entitled *The Pain Resurrector*.

# The Contributors

**Thomas K. Baker**
Department of Lead Optimization Toxicology
Lilly Research Laboratories
A Division of Eli Lilly and Company
Greenfield, Indiana

**Michael E. Burczynski**
Division of Molecular Medicine
Department of Genomics
Wyeth Research
Cambridge, Massachusetts

**Mark A. Carfagna**
Department of Nonclinical Safety Assessment
Lilly Research Laboratories
A Division of Eli Lilly and Company
Greenfield, Indiana

**Arthur L. Castle**
Department of Toxicology
Gene Logic, Inc.
Gaithersburg, Maryland

**Robert T. Dunn II**
Senior Research Scientist
Investigative Toxicology
Pharmacia Corporation
Kalamazoo, Michigan

**David Gerhold**
Safety Assessment Department
Merck Research Laboratories
West Point, Pennsylvania

**Marnie A. Higgins**
Department of Lead Optimization Toxicology
Lilly Research Laboratories
A Division of Eli Lilly and Company
Greenfield, Indiana

**Andrew Hill**
Expression Profiling Informatics
Wyeth Research
Cambridge, Massachusetts

**Youping Huang**
Biometrics Research
Wyeth Research
Princeton, New Jersey

**Frederick Immermann**
Biometrics Research
Wyeth Research
Pearl River, New York

**Jeffrey A. Johnson**
School of Pharmacy
Environmental Toxicology Center
Waisman Center, Center for Neuroscience
University of Wisconsin
Madison, Wisconsin

**J. Kevin Kerzee**
Regulatory and Applied Toxicology
Baxter Healthcare
Round Lake, Illinois

**Kyle L. Kolaja**
Investigative Toxicology
Pharmacia Corporation
Skokie, Illinois

**Jiang Li**
School of Pharmacy
University of Wisconsin
Madison, Wisconsin

**Jennifer L. Marlowe**
Center for Environmental Genetics
Department of Environmental Health
University of Cincinnati Medical Center
Cincinnati, Ohio

**Donna L. Mendrick**
Department of Toxicology
Gene Logic, Inc.
Gaithersburg, Maryland

**Michael S. Orr**
Department of Toxicology
Gene Logic, Inc.
Gaithersburg, Maryland

**Mark W. Porter**
Department of Toxicology
Gene Logic, Inc.
Gaithersburg, Maryland

**Alvaro Puga**
Center for Environmental Genetics
Department of Environmental Health
University of Cincinnati Medical Center
Cincinnati, Ohio

**John C. Rockett**
Gamete and Early Embryo Research Branch
Reproductive Toxicology Division
National Health and Environmental Effects Research Laboratory
Office of Research and Development
U.S. Environmental Protection Agency
Research Triangle Park, North Carolina

**Thomas Rushmore**
Drug Metabolism Department
Merck Research Laboratories
West Point, Pennsylvania

**Timothy P. Ryan**
Department of Lead Optimization Toxicology
Lilly Research Laboratories
A Division of Eli Lilly and Company
Greenfield, Indiana

**Julia Scheel**
Axaron Bioscience AG
Toxicology Program
Heidelberg, Germany

**Fiona Spence**
Safety Assessment
GlaxoSmithKline
The Frythe, Welwyn, United Kingdom

**Thorsten Storck**
Toxicology Program
Axaron Bioscience AG
Heidelberg, Germany

**Craig R. Tomlinson**
Center for Environmental Genetics
Department of Environmental Health
University of Cincinnati Medical Center
Cincinnati, Ohio

**Roger G. Ulrich**
Rosetta Inpharmatics
Merck Research Laboratories
Kirkland, Washington

**Marie-Charlotte von Brevern**
Toxicology Program
Axaron Bioscience AG
Heidelberg, Germany

**Jeffrey F. Waring**
Department of Cellular and Molecular Toxicology
Abbott Laboratories
Abbott Park, Illinois

**Maryann Whitley**
Expression Profiling Informatics
Wyeth Research
Cambridge, Massachusetts

**Sophie Wildsmith**
Safety Assessment
GlaxoSmithKline
The Frythe, Welwyn, United Kingdom

**Jian Xu**
Safety Assessment Department
Merck Research Laboratories
West Point, Pennsylvania

# Table of Contents

## SECTION 4 Mechanistic Toxicogenomics



## SECTION 5 Predictive Toxicogenomics



## SECTION 6 The Future of Toxicogenomics

# *Section 1*

## *Fundamentals of Expression Profiling Analysis*

# 1 Preparation and Utilization of Microarrays

*Sophie Wildsmith and Fiona Spence*

## CONTENTS



## 1.1 INTRODUCTION

Toxicogenomics describes the use of novel genomics techniques to investigate the adverse effects of exogenous agents.[1] Alterations in mRNA levels are often the earliest detectable cellular events initiated in response to a potential toxin, hence the interest in developing techniques to measure differentially expressed genes via mRNA. Various methods of transcript profiling have been described previously, with the most well-established being the northern blot. This technique has now been superseded by technologies that allow the simultaneous analysis of multiple genes, for example differential display[2] and serial analysis of gene expression (SAGE).[3] The focus of this

0-8493-1334-1/03/$0.00+$1.50

chapter is microarrays, one of the latest technologies to be used in the parallel monitoring of gene expression, otherwise known as the field of genomics. For scientists, it is a technology that has transformed the millennium into what could aptly be described as the "omic" era. According to Granjeaud et al.,[4] microarrays have become the preferred method for large-scale gene expression measurement.

Two major platform technologies are currently used for analysis of gene expression: cDNA microarrays and oligonucleotide arrays. cDNA microarrays, as developed in the laboratories of microarray pioneers such as Brown and Schena,[5] consist of a large number of genes deposited on a glass slide used for a multiplex reaction. Nucleic acid (usually DNA) is spotted, in a grid arrangement, onto a solid support such as glass slides or filter membranes. The microarrays serve as hybridization targets for cDNA made from tissue or cell lysates. The RNA from the sample is reverse transcribed, with simultaneous incorporation of label, and the resulting cDNA provides a signal where it binds complementary DNA. In this chapter, the spotted DNA on the slide is referred to as the *target*, while the labeled cDNA sample is called the *probe*.

The alternative technology available is the oligonucleotide array, often referred to as an *oligo chip*. Two methods are currently available for manufacturing oligo chips. In one method, presynthesized olignucleotides are immobilized on a support matrix; in a second variation, the oligonucleotides are synthesized directly on the support.[6] Affymetrix (Santa Clara, CA) has commercialized the latter procedure by combining photolithography based on the masking process for silicon chip manufacture[7] with combinatorial chemistry technology. The end product is a high-density oligonucleotide array (more than 250,000 oligonucleotide spots per $cm^2$).[8] Advantages of this technology include improved chip-to-chip reproducibility and specificity. These chips are expensive but can be purchased off the shelf with optimized protocols, which reduces the labor involved in producing and optimizing an in-house system and provides some assurance with regard to quality control; however, the elevated cost means that this platform is not well suited to academic or large-scale use.

## 1.2 OPTIONS FOR MICROARRAY PLATFORMS

The two main choices available to the potential user are to purchase microarrays from an external source or to fabricate them in-house. The former option has the advantage of reduced labor and capital investment. However, the cost of purchasing microarrays can be prohibitive, and problems can arise regarding subsequent intellectual property or royalties derived from experiments using genes on these arrays. Setting up an in-house microarray production facility has, until recently, been regarded as a major undertaking; however, instrument companies are now making benchtop spotters and supplying protocols, training and advice that enable competent scientists to produce their own microarrays within a few months. Although a large, skilled labor force is required for all of the steps of the process, there are many advantages of an in-house facility. These are: (1) *reduced cost of microarrays*, as the production of large numbers of chips ultimately results in significant savings; (2) *flexibility*, as a single facility can produce various "boutique" microarrays for several different groups; (3) *retention of intellectual property*, which is useful especially when the institution owns patents on novel genes of interest; and (4) *control over quality*, as quality controls can be instituted in conjunction with deposition of control targets on the chips.

### 1.2.1 Commercially Available Microarrays

Microarrays are currently available from a variety of sources: commercial companies, service providers, or academic institutions. A list of these is given in Table 1.1. The important issues to consider when purchasing microarrays from a vendor include:

- *Cost.* Calculations should be based on the number of microarrays required to cover the entire genome or that portion of interest to the researcher. It is important to note that the reproducibility of the system should also be factored in, because more variability in a particular platform will mean more replicate microarrays to achieve statistical significance and thus a higher cost.

- *Quality.* Quality is linked to reproducibility. Selecting only established companies and products is recommended; however, new products can often be tested on a trial arrangement, during which period the vendor may reduce costs and provide additional technical support.
- *Potential modifications/discontinuations.* Some vendors change their products frequently, even if only to add more genes. It is important to realize that such modifications and/or discontinuations may make it impossible to repeat experiments or continue an experiment at a later date. This problem is best circumvented by advance planning and buying sufficiently large batch sizes for each experiment.
- *Experience required.* Some vendor's protocols may be extensive and require previous experience in radioactive or fluorescence labeling.
- *Target DNA requirements.* When providing a service company with DNA for spotting, the amount required by the company may vary considerably. This is due to the various types of spotters commonly used.
- *Transport.* Microarrays may be damaged in shipping, due to changes in temperature and humidity or by impact. In our laboratory, we have experienced problems with DNA spots becoming dislodged from the glass during transport.

**TABLE 1.1**
**Obtaining DNA Microarrays from External Sources: A Selection of Resources**

| Manufacturer/Service Provider | Location | Product | Features |
|---|---|---|---|
| Agilent Technologies | Palo Alto, CA | Ready-to-go microarrays or contract printing | Oligonucleotides printed using ink-jet printers or cDNA microarrays |
| BD Biosciences Clontech | Palo Alto, CA | Human, mouse, and rat arrays available; custom arrays made to order | Glass, plastic, or nylon format |
| Genetix, Ltd. | Christchurch, Dorset | Microarray services | Includes hybridizations and scanning |
| Genome Systems, Inc. | St. Louis, MO | Gene Discovery Array | — |
| Genometrix, Inc. | The Woodlands, TX | Universal Arrays™, Risk Tox | — |
| Genomic Solutions | Ann Arbor, MI | Preprinted microarrays | — |
| Memorec | Koln, Germany | Microarray services | Based on the company's PIQOR technology |
| Mergen | San Leandro, CA | Standard and custom oligo arrays | Provides all aspects of microarraying as a service, from spotting to image processing |
| Microarrays, Inc. (microarrays.com) | Vanderbilt University, Nashville, TN | Custom contact printing | Slide lots of 160 |
| MWG Biotech | Ebersberg, Germany | Predesigned oligo arrays and custom arrays | 50-mer oligos |
| Perkin Elmer Life Sciences | Boston, MA | MICROMAX™ Human cDNA mMicroarray System I | — |

### 1.2.2 "Homemade" cDNA Microarrays

Before embarking on in-house production of microarrays (see Figure 1.1), it is necessary to consider the intended function of the microarray and thus the style and size of the finished product. Two main approaches may be considered: (1) include as many known genes as possible for the system in question, or (2) attempt to discern from the literature or previous in-house results which genes are the most relevant for the application. The former approach could be considered a "semi-open" system, which offers greater potential to obtain novel information on many genes. For this application, large microarrays would be appropriate, and the absolute accuracy of the sequences on the microarray would not constitute a major issue. An early example of this approach was the investigation of yeast gene expression changes with metabolic state, where DeRisi et al. monitored the expression of virtually every gene of *Saccharomyces cerevisae*.[9]

In contrast, the second approach is a completely closed system that will provide results based only on strictly predetermined genes, which may already be well characterized. These microarrays are not representative of the entire genome. The advantage of smaller size and reduced complexity is that these features facilitate the task of making a high-specificity, high-quality microarray for

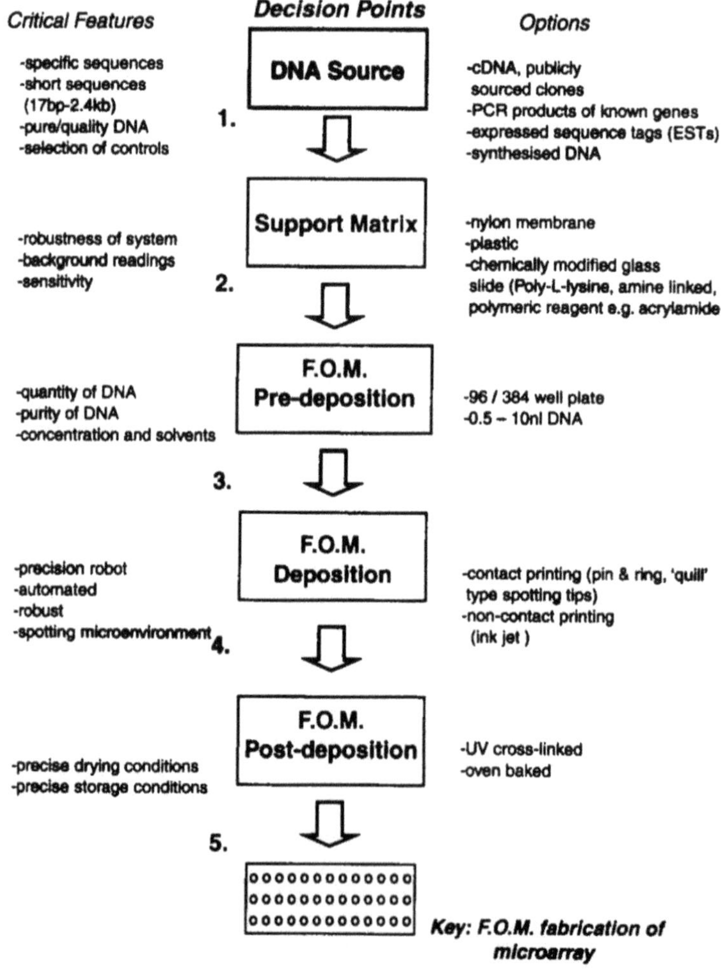


FIGURE 1.1 Decision points in the microarray fabrication process.

quantitative use. This option is useful for focusing specifically on a mechanism of action or for comparative studies where reproducibility and precision are important.

#### 1.2.2.1 Selecting Support Materials

The first steps in creating a microarray are to decide upon the supporting matrix to be used and the source of the genes or DNA that will be arrayed. The types of support materials currently available include glass, plastic, and nylon. This decision should be considered in conjunction with the purchase cost and capabilities of the microarray hardware. Some deposition robots may be optimized for specific materials. The type of detection system will often dictate the support material, for example, radioactivity is most commonly used with nylon or plastic arrays.

#### 1.2.2.2 Sourcing Clones

Target DNA can be obtained from a variety of sources. Many laboratories have a ready source of their own clones, which can be used to generate PCR products that are then deposited on the array. Numerous public sources of clones are also available, such as the Human Genome Mapping Project (HGMP) (http://www.hgmp.mrc.ac.uk/), I.M.A.G.E. consortium,[10] or Research Genetics (Huntsville, AL) (see Bowtell[12] for information on sources of clones). An advantage of this route is that a large number of clones can be obtained relatively inexpensively. Using a large pool of clones, rather than specifically choosing individual genes, has the advantage of reducing selection bias and may result in the discovery of novel gene expression patterns. Drawbacks of this approach include the sometimes questionable quality of sequences, various species availability, and the bias in large pools toward highly expressed genes. We have found a number of I.M.A.G.E. consortium clones to contain inserts with poly(A) stretches and repetitive sequences. The clone inserts are also highly variable in length. Using these clones without subcloning or reprocessing can result in nonspecific hybridizations and wildly different hybridization kinetics for genes with different target lengths.

In terms of availability, it is easier to find genes in the public domain that are highly abundant, commonly expressed in normal tissues/cells and therefore easier to clone. Rare, low-abundance transcripts, or those thought to be important or patentable in developing therapeutics, are more difficult to source.

Aside from finding commercial sources of clones, another approach is to generate clones in-house that are appropriate to the area of interest. This is proving to be increasingly popular, especially when combined with an open gene hunting method. Time and expense must be considered, but investment in quality at this early stage pays dividends later in terms of confidence in results.

Microarrays are more specific when they are developed for a single species. Obtaining clones of human genes is relatively easy, as so much of the genome is already available. Finding clones of other species proves more difficult. Furthermore, although a number of genes share a high degree of homology across species, some genes may be very different. Good examples of these are the glutathione transferase genes that consist of different subunits in rat than in human. In addition, the untranslated region of genes in rat is often considerably longer than in human. This is particularly problematic if using a cDNA approach, from the poly(A) end of the mRNA, as the reverse transcription enzyme may never reach the sequence in rat that corresponds to that in human.

The method used for generating labeled cDNA will determine the region of the genes on which they are deposited. For reverse transcription methodology using poly(T) as a primer, sequences should be chosen closest to the 3′ end of the gene.[9] Likewise, Heller et al.[11] selected sequences proximal to the 3′ end of the gene and selected for areas of least similarity to related and repetitive sequences.

When using gene-specific primers for generating DNA, obviously the region of the gene that is targeted should be the region that is most specific to that gene. It is helpful if all the genes of interest are in the same type of plasmid or in plasmids with the same primer sequences. This means that the same primer pairs can be used for all of the polymerase chain reactions (PCRs) for the simple purpose of reducing pipetting effort and errors.

It is worthwhile checking all sequences for overlaps in homology, as this can result in a dilution of signal through competition. Closely related gene families are likely to cross-hybridize. Heller et al.[11] found cross-hybridization between genes with 70 to 90% sequence homology and also between genes with short regions of identity over the length of the target. They minimized this problem by designing targets specific to gene family members.

Clontech (Palo Alto, CA) is an example of a company that has made considerable bioinformatics investments when designing their arrays. The exact sequence of each clone is known, and repetitive elements and poly(A) tracts have been avoided. Their cDNA labeling/amplification protocol uses specific primers so that sequences can be selected that are unique to the genes of interest. In Chapter 2, the phenomenon of cross-hybridization is further addressed in a demonstration of the results of a side-by-side comparison of data generated on cDNA microarrays and oligonucleotide arrays from the same samples.

#### 1.2.2.3 Preparation of DNA for Deposition

Microarrays are usually made by printing oligonucleotides or a section of the plasmid. The latter approach enables the relevant part of the cloned plasmid DNA to be amplified by PCR and then spotted down in a purified and concentrated form. PCR is usually carried out in a 96-well plate format — while faster throughput can be achieved using 384-well plates, in practice the smaller wells often do not provide sufficient DNA for deposition. The DNA products are usually purified using a high-throughput method such as that available from Qiagen (Dusseldorf, Germany). Product purity is usually verified using an A260/280 absorbance measurement and gel electrophoresis. Multiple PCR products (i.e., nonspecific amplification of DNA) result in heterogeneous DNA being deposited onto the chip, which obviously leads to nonspecific hybridizations and ultimately meaningless data. Thus, purification of all products is important in order to remove unwanted contaminants from the PCR reaction.

#### 1.2.2.4 Preparation of Oligonucleotides for Deposition

An alternative approach to using PCR products from clones is to generate or commission the synthesis of oligonucleotides for array deposition. These can be devised based on the sequences of genes of interest, without requiring the physical clone to be in hand. This approach reduces work load, storage space, and cataloging. Generally, oligonucleotides of 30 to 80 nucleotides are used. Lengthy, high-purity oligonucleotides can be costly to purchase because the synthesis yields are lower for longer molecules. Several companies, such as MWG Biotech (Ebersberg, Germany), provide synthesis of high-quality, salt-free, long nucleic acids that have been quality checked via MALDI-TOF mass spectrometry. An added benefit is that concentrated oligonucleotides are easy to store and are delivered purified, quantitated, and ready to dilute for spotting.

#### 1.2.2.5 Deposition of DNA

##### *1.2.2.5.1 Robotics*

Deposition can be categorized into contact or noncontact printing. Contact printing involves using a robot to mechanically "spot" down nanoliter droplets of DNA in solution.[12] Several spotting robots are available on the market (see Table 1.2), with a variety of spotting tip designs, including split (channeled) pins, flat-ended pins, and pin-and-ring technology. An example spotter is shown in Figure 1.2. To date, spotting robots have been the most practical technology for in-house use,

**TABLE 1.2**
**A Selection of Currently Available Robotic Spotting Systems**

| Company | Product | Models | Features |
|---|---|---|---|
| Affymetrix (acquired from Genetic Microsystems; http://biogem.ucsd.edu/BiogemGMSSpotter) | Pin-and-ring spotter | GMS417 | Useful for spotting various viscosities; capacity: 42 slides, 3 microplates; enclosed in a cabinet |
| Amersham Biosciences, Bucks, U.K. (acquired from Molecular Dynamics; www.amershambiosciences.com) | Fourth-generation spotter with spring-loaded capillary pens suitable for depositing different viscosity solutions | Lucidea™ | Capacity: 75 slides; 12 × 384 well plates; spot size: approx. 140–160 µm; humidity and temperature control in chamber; software with sample tracking and barcode-reading utility |
| BioRad, Hercules, CA (acquired from Virtek; www.biorad.com) | Macroarrays (quills) or microarrays (solid pins) | VersArray ChipWriter Pro | Capacity: 126 slides; 32 plates, with stacker as standard; spot size: 90–105 µm |
| | | VersArray ChipWriter compact system | Capacity: 20 slides; spot size: 100 µm; sample management software |
| Gene Machines, San Carlos, CA (www.genemachines.com) | Solid substrate arrayer that accommodates a variety of print heads and pin types, including solid and split pins from Telechem and Majer Precision. | OmniGrid® | Capacity: 100 slides; potential for integrated robotic server arm for plate stacking (72 plates); spot size: 100–200 µm ± 2.5 µm; HEPA-filtered air; humidity-controlled enclosure; sample tracker software |
| | | OmniGrid® Accent | Small benchtop version of the above, with a lower capacity of 50 slides and 3 plates |
| Genetix, Hamps, U.K. (www.genetix.co.uk) | Macroarraying for gridding onto membranes and other multiple uses (e.g., colony picking) | Q-Bot | Capacity: 2 × 36 plate holders as standard; sterile, filtered environment; sample tracking; barcoding |
| | Solid- pin microarrayer | QArray | Capacity: 84 slides, 5 microplates as standard, 70 optional; positive-pressure HEPA air filtration and humidity control; microarray and library software |
| | | QArray lite | Capacity: 90 slides, 5 microplates; HEPA filtration; humidity control; software |
| | | QArray mini | Capacity: 54 slides, 5 microplates; small footprint; enclosed system with humidity control |

*(continued)*

**TABLE 1.2 (CONTINUED)**
**A Selection of Currently Available Robotic Spotting Systems**

| Company | Product | Models | Features |
|---|---|---|---|
| Genomics Solutions, Ann Arbor, MI (www.genomicsolutions.com) | Solid dip-and-print titanium pins for sample conservation | GeneTAC™ RA1 | User-configurable number of slides (24–72) and microplates (3–11); over 13,000 spots per hour; computer interface; MicroSys, PixSys, and ProSys have humidity control |
| | Cartesian quill pen or non-contact ink-jet printers | PA (PinArray™) series uses Telechem quill pins; SQ (synQUAD™) series uses inkjet-style dispensing | For slides or membranes; capacities of 10, 50, or 100 slides and 1–100 microplates, depending on model; spot size: 75–200 µm; Clontracking software |
| PerkinElmer Life Sciences, Boston, MA (acquired from Packard Biosciences or GSI Luminomics; www.perkinelmer.com) | Contact spotting using flexible printhead that allows pin subsets to be used; requires Telechem quill pins | SpotArray™ 24 | Capacity: 20 slides, 4 plates; options include humidity control/barcoding |
| | | SpotArray™ 72 | Capacity: 68 slides, 4 plates; positive pressure cabinet with humidity control and HEPA filter; rapid spotting of 1536 samples in duplicate in 1 hour |
| | Noncontact piezoelectric printing | SpotArray™ Enterprise | Flexible system for sub-nanoliter printing on porous or nonporous, two- or three-dimensional substrates |
| | | BioChip Arrayer | Sub-microliter dispensing; spot size: 180 µm ± 15 µm; cabinet with ionized and filtered air; integrated camera for verifying dispensing, glass slide holders |

although noncontact methods such as bubble-jet[13] and ink-jet[14] printing are becoming more accessible and affordable (for example, via PerkinElmer Life Sciences; Boston, MA). These have the potential for accurate and even spotting, although issues regarding cross-contamination between different DNA samples still must be addressed.

Deposition of the PCR products or clones can be on glass, nylon, or other supports and can be performed by numerous methods — for example, noncovalent attachment via poly-L-lysine-coated glass slides[15] or covalent attachment, such as the silyl chemistry used by Schena.[16] Typically, 0.5 to 10 nl of DNA is deposited in a spot 100 to 150 µm in diameter and at a distance 200 to 250 µm from neighboring spots. The exact dimensions and quality of spots depend on the type of robot and the settings used. The type of pen tip will affect spot quality; for instance, those with quills may clog easily with viscous solutions or dust.[17] The type of pen tip will also affect the target DNA

FIGURE 1.2 The GeneTac RA1 spotter. (Photograph courtesy of Genomics Solutions; Ann Arbor, MI.) (See color insert following page 112.)

volume required; for example, the pin-and-ring systems hold a reservoir of DNA and are less thrifty than solid-pin systems.

#### 1.2.2.5.2 DNA Concentration and Length

It is imperative that the DNA is pure and deposited in excess. Ideally, all PCR products should be of similar concentration/molarity (approximately 500 ng/μl for glass[18]) and size, in order to achieve similar reaction kinetics for all hybridizations. The ideal length deposited is often debated, but Heller et al.[11] found no significant difference in hybridization signal for products ranging from 0.2 to 1.2 kb. Some advantages may be gained by spotting single-stranded PCR products (for example, preventing self-hybridization). Watson et al.[18] identified around a twofold increase in signal when using single-stranded, rather than double-stranded DNA, but they emphasized the necessity of identifying the correct strand for deposition when using single-stranded spotting. Some work has been conducted on the optimal length of spotted oligonucleotides.[19,20] Qiagen and BD Biosciences use oligomers of 70 and 80 nucleotides respectively, suggesting that this is approximately the optimal length.

#### 1.2.2.5.3 Microenvironmental Conditions for Deposition

The microenvironment used for microarraying is also significant; a number of researchers use high-efficiency particulate air (HEPA) filters (to reduce airborne contaminants) and humidity-controlled chambers.[17] Humidity determines the rate of evaporation of water from the arrayed spots. Rapidly dried spots may be uneven, with the majority of DNA being located in the center, whereas slow drying may result in creeping and spot spreading. Spot quality, as defined by a perfect circular shape with an even density of DNA, is important. Irregular shapes and uneven signals, caused by variable rates of evaporation and leading to typical "doughnut" or "inverse doughnut" appearances of the spots following hybridization, can cause significant problems for analysis, as reproducibility between spots of different replicates is essential for interpretation.

Deposition methods have been published[17,21] and some vendors provide protocols. A number of user groups have been established for the discussion of optimized protocols, for example Amersham Biotech's (Amersham, Bucks, U.K.) Microarray Technology Access Programme (MTAP). While individual laboratories independently strive for the best signal and lowest background readings on glass slides, a number of slide vendors are addressing these problems with new technologies such as orientated deposition[21] and using metal-coated slides.

#### 1.2.2.6 Postdeposition Processing

Following printing, it is suggested that slides should be left for 24 hours at room temperature to permit thorough drying of the DNA. The deposited DNA is then immobilized, usually by ultraviolet irradiation. Alternatively, spotted slides can be oven-baked at approximately 80°C for 2 to 4 hours to enhance immobilization. Once immobilized, the remaining postdeposition processing will be dependent, to a large extent, on the spotting process that has been used. Using silylated slides, Schena et al.[22] rehydrated the spotted DNA before rinsing with sodium borohydride and ethanol. When using glass, it is common to wash the slides with solvents to remove any contaminants such as grease and any loosely attached DNA. This is followed by a boiling step to denature the DNA.

## 1.3 USING MICROARRAYS

### 1.3.1 RNA Preparation

A number of issues regarding RNA and its preparation are worthy of consideration. Some of the main ones include cell heterogeneity, tissue extraction, and RNA integrity. The RNA source may be from cell culture or from tissue samples, including tissue banks and biopsy specimens. In general, preparation from cell culture is often considered easier and more reproducible. Using tissue samples for expression profiling results in more variable RNA quality, especially when measuring the effect of drugs or toxicants in target tissues where ongoing related or unrelated pathological processes can interfere with mRNA expression. In addition, the type of tissue itself may have a profound effect on the extraction process and the resultant RNA quality; for example, in our experience, mRNA extractions from liver are of much higher quality than those obtained from stomach. The integrity of the extracted RNA is crucial in being able to reverse transcribe reasonable (500-bp) lengths of cDNA from it, regardless of the platform used. Tissues of potential interest for gene expression analysis should therefore be snap-frozen in liquid nitrogen immediately after harvesting, and stored at –80°C until processed further.

### 1.3.2 Sample Labeling

For cDNA microarrays, a single round of transcription is used in order to generate a labeled cDNA probe from the sample RNA. Fluorescent or radioactively labeled probes can be made from either total RNA or purified mRNA. Duggan et al.[23] suggest 50 to 200 µg of total RNA per slide or 2 to 5 µg of poly(A) mRNA. The quantity of RNA required may place a limitation on experiments, especially if tissue is scarce or only one cell type has been isolated (for example, by laser capture microdissection [LCM]). This problem can be circumvented for fluorescence by using a probe preparation step that incorporates a PCR step rather than reverse transcription alone; however, because standard PCR is not linear, this can lead to problems of quantitation. It may also selectively amplify some genes leading to nonrepresentative expression profiles; hence, other, non-PCR-based methods of amplifying the nucleic material may be preferred.[24]

Another way to amplify signal is to use signal-amplification methods, such as the Tyramide signal amplification technology (PerkinElmer Life Sciences) or the dendrimer technology (Genisphere; Oakland, NJ). Alternatively, greater sensitivity can be obtained by using radioactive labeling; P33-labeled cDNA on filter arrays requires only 50 ng of total RNA per experiment.[23]

In our experience, the labeling step is the primary cause of variation across experiments.[25] The complex biological reaction is prone to error from differences in quality and quantities of the constituent ingredients. This reaction step is also subject to user-induced variations in probe labeling. Examples include commercially available transcriptional enzymes stored in glycerol and thus difficult to pipette accurately. Enzymes such as transcriptases are fragile and have a short half-life at room temperature, and their activity can be substantially reduced by high temperatures and frothing caused by over-zealous pipetting.[26]

Methods for sample labeling are dependent upon both the slide type and the detection equipment used. Radioactivity is often used with nylon membrane arrays, whereas fluorescent labeling is generally used with glass. Incorporating fluorescent label during the reverse transcription reaction may have an effect on the efficiency of the enzyme, potentially leading to truncated transcripts. If only one nucleotide is labeled, then a transcriptional bias may be present.

Dual-label hybridization is a technique often used to compensate for differences in spotted genes from array to array. Two samples are labeled with paired fluorophores that are competitively cohybridized to the same slide. Cy3 and Cy5 are the most commonly used fluorophores,[9] primarily for historical reasons, although other combinations include fluorescein and lissamine[5] and Cy3 and rhodamine.[11]

Following preparation, labeled samples are purified in order to remove contaminating fluorescent nucleotides or debris such as cellular protein, lipid, and carbohydrates, all of which can cause fluorescence-contaminating particulate matter.[23] Purification is usually carried out using filter spin columns such as Qiaquick (Qiagen) or gel chromatography columns such as Biospin 6 (BioRad; Hercules, CA). Occasionally, an additional ethanol precipitation is performed.[15]

### 1.3.3 Hybridization

This step may also give rise to significant variation depending upon the support and the chemistries used for deposition. The surfaces with deposited DNA are easily damaged at this stage. In particular, membranes may be abraded, resulting in uneven and high backgrounds.

#### 1.3.3.1 Incubation and Automation

Glass microarrays are often hybridized by spotting a small volume of sample (for example, 20 μl) onto the microarray and then carefully dropping a coverslip onto it. This has the effect of spreading the solution over the entire slide while eliminating air. Sealant may then be applied around the periphery of the coverslip in order to prevent dehydration of the solution. Problems with this method include seepage of the sealant under the slide (causing high backgrounds) or incorporation of air bubbles.

Alternative approaches are to use humidity chambers for the incubation step (thus obviating the need for sealant) or to use hybridization (sealed) chambers. TeleChem International, Inc. (Sunnyvale, CA) and Clontech manufacture hybridization chambers to reduce evaporative loss of samples from the slides. Amersham Biotech (Piscataway, NJ) has developed a hybridization station with agitation of sample solution. This mixing is intended to provide even coverage and hence an even hybridization. Likewise, PerkinElmer Life Science and GeneMachines (San Carlos, CA) also have automated hybridization stations. In our own laboratory, we have improved replicate reproducibility by 40% by using chambers designed in-house that enable free flow of the hybridization solution over the microarray. These are either rotated in a hybridization oven or connected to a proprietary automated system that carries out washing and drying steps.

#### 1.3.3.2 Stringency of Hybridization

The speed, extent, and specificity of hybridization is dependent on the stringency of the hybridization solutions, which is a function of the salt concentration and temperature. The most commonly used hybridization solution is sodium citrate buffer and saline (SSC) with the addition of detergent, although others are commercially available.[27,28] A number of researchers use additives in the hybridization solution in order to reduce backgrounds. These include Denhardt's reagent, sheared salmon sperm, Cot1DNA, tRNA, and poly(A). To improve binding at low copy number, formamide, dextran sulfate, or polyethylene glycol can also be used. The time and temperature for hybridization are a function of the hybridization solution and the complexity and length of the sample DNA. The optimum (i.e., maximum rate) hybridization temperature should be about 20 to 25°C below the melting temperature ($T_m$). The $T_m$ can be approximated using the following equation derived from solution hybridization kinetics.[29]

$$T_m = 81.5°C - 16.6(\log 10M) + 0.41(\%G + C) - 0.63(\% \text{ formamide concentration}) - 600/L$$

where M is the monovalent cation concentration, (%G + C) is the percentage of corresponding nucleotides in the probe, and L is the number of nucleotides in the DNA hybrid. When no formamide is used in the hybridization, that part of the equation is ignored.

It is important to note that fluorescent dyes may reduce the $T_m$ of the probe and it is best to determine the optimum hybridization temperature experimentally by increasing the stringency of the hybridization and washes until specific binding is obtained. The deposition chemistry should also be considered in calculating hybridization time. For example, the surface of silane-coated glass may deteriorate after prolonged incubation (>10 hours) at temperatures above 50°C.

#### 1.3.3.3 Posthybridization Washes

Following hybridization, the microarrays are subjected to a series of washes in order to remove unbound, labeled probe and nonspecifically bound sequences. The wash solution is usually of similar composition to the hybridization solution. More stringent conditions are often applied during the final washes, either by increasing the temperature or lowering the ionic strength of the buffer. Washing should take place under conditions of salt concentration and temperature that are equivalent to between 5 and 20°C below the $T_m$.[29] The microarrays are then dried, either by an air jet or by rapid centrifugation. Slow drying may result in streaking that interferes with image analysis.

### 1.3.4 Image Capture and Analysis

As discussed earlier (see sample labeling), the type of detection equipment used will determine the labeling method. Film or phosphorimaging plates (Molecular Dynamics, Inc.; Sunnyvale, CA) will be used with radioactivity, whereas optical systems are used for fluorescently labeled samples. Over the past two years, the number of commercially available microarray scanning instruments has increased considerably, along with a concomitant decrease in price. Although the number of wavelengths available is still somewhat limited, rapid, high-resolution scanners are now available from as little as $50,000. Factors to be considered when buying a scanner include compatibility with spotter slide format, throughput, technical support, and the ability to access the output data.

Some detection equipment is supplied with vendor software for analysis of microarrays (e.g., QuantArray from PerkinElmer). Independent software specifically designed for the market may have superior properties and additional capabilities. For example, Imaging Research, Inc. (Ontario, Canada) makes ArrayVision™ for image analysis, together with ArrayStat™, which has a number of useful statistical functions such as calculating the number of replicates necessary to be certain of a gene change of a given magnitude. BioDiscovery, Inc. (Los Angeles, CA) sells a semiautomated image-analysis program (ImaGene™) and a package called Genesight™ that incorporates data visualization, statistical analysis (including multivariate analysis), and data-mining capabilities. Some free image-analysis software is available on the Internet, such as NIH Image (developed at the U.S. National Institutes of Health[22]) and ScanAlyse 2 (from Stanford University; http://rana.standford.edu/software). Important criteria for image-analysis software include speed, ease of use, automation (especially of spot finding), and the ability to distinguish artifact from real signal. When scanning the slides, each microarray may be scanned several times, depending on the number of fluorophores being used for labeling. It is important to establish that this repetitive laser scanning process causes minimal photobleaching of the fluorophores.

## 1.4 FUTURE DEVELOPMENTS

Microarrays are becoming recognized as a powerful tool for examining global gene expression. In 1999, Affymetrix and homemade platforms dominated the array market with 43% and 24% of the market share, respectively.[30] The largest players in cDNA manufacture were Phase 1 (9%) and Incyte (3%); these two companies have subsequently ceased selling microarrays, and the market has become increasingly fragmented. Although these companies cite "other interests" as the reasons for changing strategy, it is possible that high development costs at such an early stage of market entry have made it difficult to recoup these costs by passing them on to the chip consumers. Despite the technical difficulties and labor involved, it seems that the prohibitive price of commercial microarrays, plus skepticism regarding quality, means researchers will still want to make their own.

The global method of gene expression profiling defined by microarray technology is now beginning to be supported by the measurement of proteins (proteomics) and metabolites (metabonomics) in drug development and profiling disease. Complementary technologies such as real-time PCR and protein profiling using surface-enhanced laser desorption ionization (SELDI-TOF) (Ciphergen; Fremont, CA)[31] can be implemented in parallel, and this holistic approach enhances identification of novel biomarkers and our understanding of the molecular mechanisms of disease.

In the forthcoming years, microarray technology is set to continue developing at a phenomenal pace. Automation and integration with other systems are likely to be the major focus of expansion. Under the pressures of cost, throughput, and efficiency issues, there will be a constant drive towards further miniaturization and the concept of a "lab on a chip."

## ACKNOWLEDGMENTS

Adapted from *Molecular Pathology*, 54(1), 8–16, 2001 with kind permission from the BMJ Publishing Group.

## REFERENCES

1. Aardema, M.J. and MacGregor, J.T., Toxicology and genetic toxicology in the new era of "toxicogenomics": impact of "-omics" technologies, *Mutat. Res.*, 499(1), 13, 2002.
2. Liang, P. and Pardee, A.B., Differential display of human eukaryotic messenger RNA by means of the polymerase chain reaction, *Science*, 257, 967, 1992.
3. Velculesu, V.E., Zhang, L., Vogelstein, B., and Kinzler, K.W, Serial analysis of gene expression, *Science* 270, 484, 1995.
4. Granjeaud, S., Bertucci, F., and Jordan, B.R., Expression profiling: DNA arrays in many guises, *Bioessays*, 21, 781, 1999.
5. Schena, M., Shalon, D., Davis, R.W., and Brown P.O. Quantitative monitoring of gene expression patterns with a complementary DNA microarray, *Science*, 270, 467, 1995.
6. Ginot, F., Oligonucleotide micro-arrays for identification of unknown mutations: how far from reality?, *Hum. Mut.*, 10(1), 1, 1997.

7. Lipshutz, R.J., Fodor, S.P., Gingeras, T.R., and Lockhart, D.J., High density synthetic oligonucleotide arrays, *Suppl. Nat. Genet.*, 21(suppl. 1), 20, 1999.
8. van Hal, N.L., Vorst, O., van Houwelingen, A.M., Kok, E.J., Peijnenburg, A., Aharoni, A., van Tunen, A.J., and Keijer, J., The application of DNA microarrays in gene expression analysis, *J. Biotechnol.*, 78(3), 271, 2000.
9. De Risi, J.L., Vishwanath, R.I., and Brown, P.O., Exploring the metabolic and genetic control of gene expression on a genomic scale, *Science*, 278, 680, 1997.
10. Lennon, G., Auffray, C., Polymeropoulos, M., and Soares, M.B. The Image Consortium: an integrated molecular analysis of genomes and their expression, *Genomics*, 33, 151, 1996 (I.M.A.G.E. consortium, http://www.bio.llnl.gov/bbrp/image/image.html).
11. Heller, R.A., Schena, M., Chai, A., Shalon, D., Bedilion T., Gilmore, J., Woolley, D.E., and Davis, R.W., Discovery and analysis of inflammatory disease-related genes using cDNA microarrays, *Proc. Natl. Acad. Sci. USA*, 94(6), 2150, 1997.
12. Bowtell, D.L., Options available — from start to finish — for obtaining expression data by microarray, *Nat. Genet. Suppl.*, 21, 25, 1999.
13. Okamoto, R., Suzuki, T., and Yamamoto, N., Microarray fabrication with covalent attachment of DNA using bubble jet technology, *Nat. Biotechnol.*, 18, 438, 2000.
14. Lemmo, A.V., Rose, D.J., and Tisone, T.C., Inkjet dispensing technology: applications in drug discovery, *Curr. Opin. Biotechnol.*, 9, 615, 1998.
15. DeRisi, J., Penland, L., Brown, P.O., Bittner, M.L., Meltzer, P.S., Ray, M., Chen, Y., Su, Y.A., and Trent, J.M., Use of cDNA microarray to analyse gene expression patterns in human cancer, *Nat. Genet.*, 14, 457, 1996.
16. Schena, M., Genome analysis with gene expression microarrays, *Bioassays*, 18(5), 427, 1996.
17. Cheung, V.G., Morley, M., Aguilar, F., Massimi, A., Kucherlapati, R., and Childs, G., Making and reading microarrays, *Nat. Genet. Suppl.*, 21, 15, 1999.
18. Watson, A., Mazumder, A., Stewart, M., and Balasubramanian, S., Technology for microarray analysis of gene expression, *Curr. Opin. Biotechnol.*, 9(6), 609, 1998.
19. Hughes, T.R. Mao, M., Jones, A.R., Burchard, J., Marton, M.J., Shannon, K.W., Lefkowitz, S.M., Ziman, M., Schelter, J.M., Meyer, M.R., Kobayashi, S., Davis, C., Dai, H., He, Y.D., Stephaniants, S.B., Cavet, G., Walker, W.L., West, A., Coffey, E., Shoemaker, D.D., Stoughton, R., Blanchard, A.P., Friend, S.H., and Linsley, P.S., Expression profiling using microarrays fabricated by an ink-jet oligonucleotide synthesizer, *Nat. Biotechnol.*, 19(4), 342, 2001.
20. Beaucage, S.L., Strategies in the preparation of DNA oligonucleotide arrays for diagnostic applications, *Curr. Med. Chem.*, 8(10), 1213, 2001.
21. http://cmgm.stanford.edu/pbrown.
22. Schena, M., Shalon, D., Heller, R., Chai, A., Brown, P.O., and Davis, R.W., Parallel human genome analysis: microarray-based expression monitoring of 1000 genes, *Proc. Natl. Acad. Sci. USA*, 93, 10614, 1996.
23. Duggan, D.J., Bittner, M., Chen, Y., Meltzer, P., and Trent, J.M., Expression profiling using cDNA microarrays, *Nat. Genet. Suppl.*, 21, 10, 1999.
24. Van Gelder, R.N., Von Zastrow, M.E., Yool, A., Dement, W.C., Barchas, J.D., and Eberwine, J.H., Amplified RNA synthesized from limited quantities of heterogeneous cDNA, *Proc. Natl. Acad. Sci. USA*, 87, 1663, 1990.
25. Wildsmith, S.E., Archer, G.E., Winkley, A.J., Lane, P.W., and Bugelski, P.J., Maximisation of signal derived from cDNA microarrays, *Biotechniques*, 30(1), 203, 2001.
26. Palmer T., in *Understanding Enzymes*, Wiseman, A., Ed., John Wiley & Sons, New York, p. 342.
27. TeleChem, http://www.hooked.net/~telechem.
28. Clontech, http://www.clontech.com.
29. Hames, B.D. and Higgins, S.J., Eds., *Gene Probes 2: A Practical Approach*, IRL Press, Oxford, 1995.
30. BioInsights, Redwood City, CA.
31. Merchant, M. and Weinberger, S.R., Recent advancements in surface-enhanced laser desorption/ionization time-of-flight mass spectrometry, *Electrophoresis*, 21, 1164, 2000.

# 2 Comparative Studies Using cDNA vs. Oligonucleotide Arrays

*Jiang Li and Jeffrey A. Johnson*

## CONTENTS



## 2.1 INTRODUCTION

The recent popularity of microarray technology can be attributed to its successful application to a wide range of topics including toxicity profiling (Toxicogenomics), drug screening and identification (pharmacogenomics), genomic characterization of disease, and other areas of recruiting "omics" technologies.[2,6,7,22,23,29] More than 2000 papers involving microarray analyses have been published, mostly within the last 2 years (Figure 2.1A). As discussed in the previous chapter, which focused mainly on cDNA microarrays, arrays exist in a variety of forms and can be classified based on any number of attributes, including length of target sequence (long cDNAs or oligonucleotides), commercial or custom-made, glass- or membrane-based, and spotted or *in situ* synthesized. Because researchers are interested in the application of microarrays in their systems, many commercial suppliers have made standard microarrays (Figure 2.1B) and analysis packages available. Commercial arrays vary in the number of genes they are able to screen, with some being capable of global screening and others capable of only specific, functional screening. For example, the Affymetrix human U95Av2 GeneChip® allows for detection of 9670 genes/expressed sequence tag (EST) clusters on one array, while the Incyte human UniGem V2.0 allows for detection of 8556 genes/EST clusters on one array.[36,38] In contrast, BD Clontech provides small-scale arrays that are particularly designed for toxicology, apoptosis, and cancer research.[37] At present, the two popular array systems

0-8493-1334-1/03/$0.00+$1.50

A.

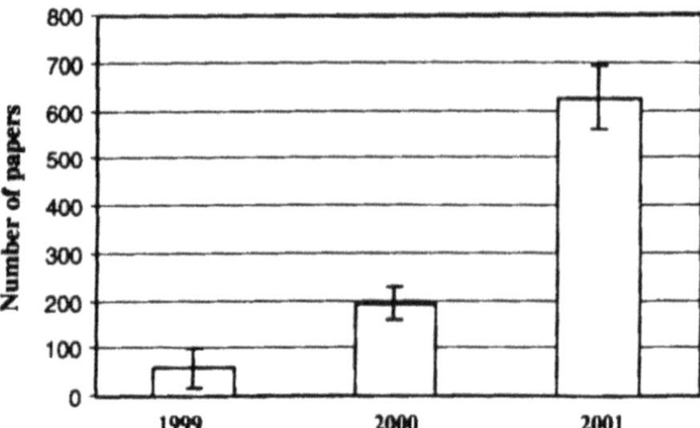


B.

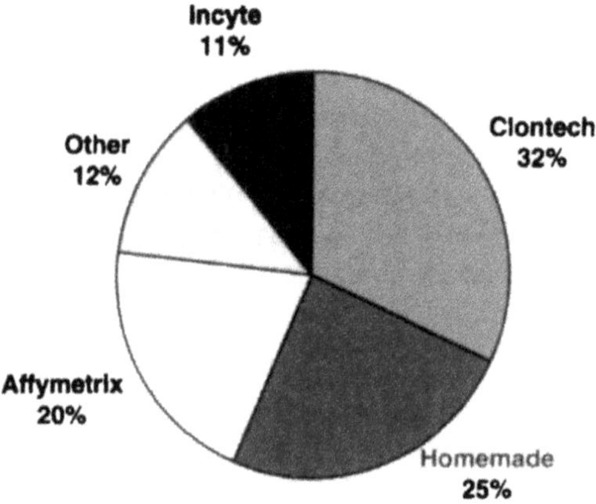


**FIGURE 2.1** Original articles associated with the term *microarray*. (A) The number of articles containing microarray data from 1999 to 2001. (B) Types of microarrays used from 1999 to 2001. Homemade arrays are custom spotted by the investigator and are usually the long cDNA variety. "Other" is a grouping of commercial array companies including Research Genetics, Sigma GenoSys, Super Array, Rosetta Inpharmatics, Takara, R&D Systems, NEN Life Sciences, and Phase-1 Molecular Toxicology. (From Li, J. et al., *Toxicol. Sci.*, 69, 383, 2002. With permission.)

are photolithographically synthesized (also called *in situ* synthesized) oligonucleotide microarrays and spotted long cDNA microarrays, the most commonly used, high-throughput array technologies in toxicogenomics.[7,9,22,24] The platforms differ in array manufacturing and design, array hybridization, scanning, and data handling.[6,11,22,27,30,31]

## 2.2 OVERVIEW OF THE cDNA MICROARRAY AND OLIGONUCLEOTIDE ARRAY PLATFORMS

On the oligonucleotide array, a given gene is currently represented by 14 to 20 different 25-mer oligonucleotides serving as unique, sequence-specific detectors. An additional control element on these arrays is the use of mismatch (MM) control oligonucleotides that are identical to their perfect match (PM) partners except for a single base difference in a central position. Presence of the mismatched oligonucleotide presumably enables cross-hybridization and local background to be estimated and subtracted from the PM signal. Eukaryotic mRNA is converted to biotinylated cRNA from T7-$(dT)_{24}$-primed cDNA. Each sample is then hybridized to a separate array. Transcript levels are calculated by correlation to cRNA spikes of known concentration added to the hybridization mixture. Differences in mRNA levels between samples are determined by comparison between any two hybridization patterns produced on separate arrays of the same type. Results are determined according to intensity levels normalized to the overall background for the chip.

As discussed in Chapter 1, there are two major types of cDNA microarrays: membrane-based arrays with porous surfaces such as nylon, initially developed by BD Clontech,[37] and chemically coated glass-based arrays, which are currently used more widely for commercial or custom-based purposes.[26] In both cases, thousands of cDNA fragments are robotically deposited on the substrate. The membrane-based microarrays are hybridized with $^{32}P$- or $^{33}P$-labeled cDNA targets, but the glass-based microarrays are hybridized with fluorescent-dye-labeled cDNA targets. After hybridization, the radioactive or fluorescent signal intensities are measured using a phosphorimager or laser scanner, respectively.[22] On the Incyte cDNA microarray platform, genes are generally represented by unique polymerase chain reaction (PCR)-derived cDNA probes, which are several hundred base pairs in length. These probes are then physically deposited on a two-dimensional grid etched onto a chemically modified glass slide. Aliquots from two purified mRNA samples are separately reverse transcribed using primer sets labeled with two different fluorophores (generally Cy3 and Cy5), and the resulting dye-labeled cDNA populations are used to probe the target elements in a competitive hybridization reaction. After hybridization, the glass slide is analyzed in a two-channel fluorescence scanner. The relative amounts of a particular gene transcript in the two samples are determined by measuring the signal intensities detected for both fluorophores and by calculating signal ratios.

A key assumption when employing microarrays as a tool to profile gene expression changes is that the quantified signal intensities are in linear relation to the real expression levels of the corresponding genes. Accumulated evidence indicates that data generated from oligonucleotide microarrays displays a good correlation to SAGE (serial analysis of gene expression), real-time RT-PCR, and northern blots.[10,12,19] In the cDNA array, the nonlinearities were revealed by serial dilution experiments.[13,26] Because measured expression levels are usually reported as a ratio of the signal from a target mRNA sample relative to one from a co-hybridized reference mRNA sample, this method complicates direct comparison with results from other technologies, as the reported ratios depend on the chosen mRNA reference. Thus, a high difference between gene expression levels of the pair-wise samples does not necessarily imply a high ratio.[18] Additionally, microarray users analyzing the same image using different software packages may formulate significantly different conclusions about levels of differential expression.[35]

### 2.2.1 Reproducibility of the Two Array Systems

The inherent variability of microarray data can often lead to false-positive signals, thereby concealing genuine biological changes. This is especially true for genes with low basal expression levels. Thus, independent of the platform used to generate the data, one must first assess the reproducibility of the microarray data. Two sources of randomly generated error associated with the reproducibility of microarray data are systematic and experimental errors.[20] While both oligonucleotide arrays and cDNA microarray platforms experience the challenges of experimental and systematic error, the extent of their influences may be quite different. Experimental error is a result of experimental design. This type of error reflects the variations in treatment, cell density, culture medium in cell lines, tissue regions harvested (reproducibility of the dissection), genetic background, and diet in animal models. In contrast, systematic error is generated during the analytical phase of microarray analysis. False-positive signals can be introduced at multiple steps: chip manufacture, possibly contributing to chip-to-chip variance; preparation of cRNA or cDNA for microarray analysis; hybridization or washing steps; and global scaling and normalization of overall signal intensities between chips. Methods of quality control and normalization approaches for controlling systematic error are covered in detail in the next chapter.

### 2.2.2 Oligonucleotide Array Advantages and Disadvantages

While synthesis of oligonucleotides by photolithography offers the advantage of abolishing the need to hydrolyze the oligonucleotide from its synthetic support and reattach it to the microarray, this approach does not allow for independent confirmation of the fidelity of synthesis. Because of this, and because this approach does not allow purification of the oligonucleotide prior to attachment to the microarray, oligonucleotide chip manufacturing can lead to internal errors, and batch-to-batch variance may also contribute to data bias, as separate samples are hybridized to separate chips when using oligonucleotide arrays.

### 2.2.3 cDNA Microarray Advantages and Disadvantages

In terms of sample labeling, the two-color approach offers several advantages. Notably, hybridization of two samples is performed on the same slide, eliminating the possibility of different spot morphologies, probe amounts, or hybridization inconsistencies that could alter the ratio. Second, the photo multiplier tube (PMT) voltages can be adjusted in different channels to equalize intensity values on each of the slides. Finally, the coefficients of variation (CVs) in ratios are typically lower than the CVs from raw hybridization signals. The two-color approach does have some disadvantages, though.[25,32] For example, different fluorescence-labeled nucleotides may be incorporated with varying efficiency, altering the ratio as a result of enzymatic contributions rather than transcript abundance.[14] Next, multiple experiment comparisons are not possible without replicating the reference sample, which, in some cases, may be difficult to obtain. Spectral overlap between dyes can also complicate interpretation of algorithms used in analysis. Finally, executing signal amplification schemes in two colors is more complex than in single-color because multiple haptens are required. Generally, because the hybridization difference between the different samples has been eliminated, dual-dye systems produce less variable data when compared with single-dye systems.[16]

IncyteGenomics[33] reported the highly reproducible performance of its cDNA microarray technology platform. The study design consisted of two independent evaluations with 70 arrays from two different lots and used three human tissue sources as samples: placenta, brain, and heart. Overall signal response was linear over three orders of magnitude, and the sensitivity for each element was estimated to be 2 pg mRNA. The calculated CV for differential expression for all nondifferentiated elements was 12 to 14% across the entire signal range and did not vary with array batch or tissue source. The minimum detectable fold change for differential expression was 1.4. Accuracy, in terms of bias, was less than 1 part in 10,000 for all nondifferentiated elements.

Kamb and Ramaswami[15] investigated the general properties of data generated from Affymetrix and Incyte arrays, and the intensities from two independent experiments using the same control RNA were plotted. In both cases, the data fit a straight line with a slope approximately equal to one and the intercept near zero. These results suggest that the data produced by both platforms were well behaved. The lack of obvious skewing or bias in the expression measurements implied the generation of a highly reproducible dataset; however, it does not necessarily follow that the data generated by the two platforms have any correlation with the data from well-recognized technologies such as SAGE or real-time PCR.

## 2.3 ISSUES AFFECTING THE SPECIFICITY OF MICROARRAY ANALYSES

In terms of microarray analyses, because both platforms may generate high-quality data it seems contradictory to point out the nonlinearity and incomparability of the data generated from microarray analysis and that from conventional gene expression technologies. To clarify this point, another important issue, specificity, should be considered during microarray analysis. Three major factors may significantly influence specificity of the microarray performance.

### 2.3.1 The Length and G+C Content of Probe

For the Incyte long cDNA array, every clone was selected from the UniGem database, which contains sequences that have been verified as being representative of the desired gene. The sequence of probes varied from 500 to 5000 base pairs, averaging 1000 bp.[38] One of the problems associated with having such a large range of target sequence is that it becomes difficult to control the hybridization efficiencies of various cDNA probes (see Chapter 1). Because the hybridization conditions are based on the length of nucleic acid fragments and the compositions of G+C and A+T (or A+U), hybridization efficiency can vary widely when long cDNA sequences are used as probes.

In the case of Affymetrix's oligonucleotide array, the 25-oligomer probes are designed to uniquely represent the desired cognate gene through blasting the GenBank database, thus minimizing cross-hybridization between similar sequences.[17] Using oligonucleotides also makes it easier to design and select probes from the same or different genes with similar G+C content and putative melting temperature. Other criteria for the selection of probes include unique sequence, minimum secondary structures, and 3′ terminal sequence selection, all of which serve to ensure greater specificity in the probe–target binding.[36] Even though the oligonucleotide probes have already been screened for comparable length and G+C content, there still exists a large variance in the single intensity of different probe set for the same gene. Therefore, the specificity of hybridization may depend on the region of ORF (open reading frame) chosen as a probe.

### 2.3.2 Cross-Hybridization with Related or Overlapping Genes

Cross-hybridization is an important limitation for the cDNA-based microarrays. As we know, different genes may have common domains and some degree of sequence identity or homology with genes from different species. Many individual genes are part of gene families, and in many cases, these genes have a great degree of sequence identity that can only be distinguished from each other through the design of site-specific hybridization probes.

Evertsz and coworkers[5] from IncyteGenomics gave an excellent example of the hybridization cross-reactivity within homologous gene families on glass cDNA microarrays (Incyte platform). They pointed out that the potential for hybridization cross-reactivity needs to be considered when interpreting the results. They attempted to establish guidelines for interpreting microarray results in which spotted cDNA targets were known or suspected to share moderate to high sequence homology when the arrays were applied for whole genome screening. A model array representing four distinct functional classes (families) — chemokines, cytochrome P-450 isozymes, G proteins, and proteases — was used for these experiments. The cDNA clones selected for this array exhibited pairwise sequence identities ranging from 55 to 100%, as determined by a homology scoring algorithm (LALIGN). As expected, hybridization signals were highest at the matching target elements. Targets containing less than 80% sequence identity relative to the hybridization probe sequences showed cross-reactivities ranging from 0.6 to 12%. Targets containing greater than 80% identity showed higher cross-reactivities (26–57%). These cross-reactive signals were analyzed for statistical correlation with the length of sequence overlap, percent sequence identity, and homology score determined by LALIGN. The results suggest that homologous regions of genes should be avoided when selecting clones for cDNA microarrays; they also indicate that cross-hybridizations could be conducted on microarrays where expressed gene sequences between the two species share 80% sequence identity or higher.

Numerous investigations have demonstrated that oligonucleotide arrays show a greater dynamic range in the reported differential expression ratio, both in general as well as in the case of members of gene families.[10,12,19,21] One possibility for this difference is the lack of cross-hybridization with oligonucleotide arrays. Data from the Affymetrix platform indicate 25-mer probes have the ability to distinguish between genes with up to 90% sequence homology. Therefore, if only one member of the gene family is differentially regulated under certain conditions, the ratio reported by oligonucleotides designed toward that particular gene may reflect the change in expression more accurately; whereas, the cDNA element may reflect a weighted average of changes in differential expression for that gene and others to which it cross-hybridizes, which may or may not demonstrate changes in expression.

### 2.3.3 The Potential Impact of RNA Splicing on Interpretation

An increasing number of alternatively spliced gene transcripts have also been identified, demonstrating that cDNAs derived from mRNA transcripts can also share exact identity over a portion of their sequences.[28] Many organisms employ alternative RNA splicing of genes that cannot be accurately measured using full-length, ORF-based, cDNA microarrays. As suggested previously with regard to cross-hybridization, oligonucleotide probe design also could theoretically overcome this difficulty.

## 2.4 COMPARATIVE EXPERIMENTAL STUDY USING THE TWO POPULAR ARRAY SYSTEMS

While high-throughout technologies are very efficient, they are also quite expensive. Because different laboratories may use different platforms to profile the same genes or similar treatments, combining expression measurements across microarray technologies or mutually transforming data across a variety of microarray databases would potentially reduce the need to duplicate experiments. Thus, many biostatisticians have tried to determine how to compare the data generated by different platforms by developing algorithms or software that could help microarray users share their databases.[1,3] Unfortunately, at the present it seems unlikely that these data can be normalized or transformed to a common standardized index of gene expression, or other intermediate value.

Chuo and colleagues[18] reported a poor correlation of mRNA measurements from two large-scale microarray studies analyzing gene expression in multiple cancer cell lines. The research was based on the publicly available data measured with Stanford-type cDNA microarrays and Affymetrix oligonucleotide microarrays. This research represented the first comparative study on the large-scale analysis of data generated from spotted cDNA arrays and microarrays with synthesized oligonucleotides, the two most commonly used microarray platforms. In general, corresponding measurements from the two platforms showed poor correlation. Clusters of genes and cell lines were discordant between the two technologies, suggesting that different biological relationships would be detected in gene expression analyses on the two platforms and emphasizing the fact that GC content, sequence length, average signal intensity, and an estimator of cross-hybridization (BLAST score, less than 1E-10 as a threshold) were found to be associated with the degree of correlation. This suggests gene-specific or, more correctly, probe-specific, factors influence measurements differently in the two platforms, implying a poor prognosis for broad utilization of gene expression measurements across platforms. However, without an independent third source of data, the authors were not in a position to determine which of these two platforms was the better one.

Our recent study was designed to assess the reliability of datasets generated by the two platforms from the same samples.[21] We had already established a model for upregulation of a cluster of antioxidant responsive element (ARE)-driven genes in the IMR-32 human neuroblastoma cell line by treatment with *tert*-butylhydroquinone (tBHQ; 10 μM) for 8 hr and 24 hr.[20] HuGene FL (Affymetrix), U95Av2 (Affymetrix), and UniGem V2.0 (IncyteGenomics) were chosen for a comparative study on 8- and 24-hr samples. The Affymetrix data generated from U95Av2 chips demonstrated that the mRNA of 218 (2.3% of total clones) genes was increased after 8 hr of tBHQ treatment. This list included most of the known ARE-driven genes, and nine genes selected for further analysis showed high consistency by RT-PCR. IncyteGenomics identified an increase in four genes and no decrease. These same four genes were also called by the Affymetrix microarray. A summary of gene expression data generated by the two platforms is shown in Table 2.1. The sensitivity (fluorescence intensity) and specificity (fold) were very different for selected genes when comparing the two platforms. Cross-hybridization was shown to partially contribute to the discrepancies in the data generated by the two platforms. According to our results, the data generated from oligonucleotide microarrays are more reliable for evaluating changes in gene expression than data from long cDNA microarrays. This chapter suggests some fundamental disparities in the results generated from the two global screening platforms.

Yuen and co-workers[34] compared the accuracy of microarray measurements obtained with an oligonucleotide array (GeneChip, Affymetrix) to a laboratory-developed cDNA array by assaying test RNA samples from an experiment using a paradigm known to regulate many genes measured on both arrays. The custom array contains 956 clones selected primarily from an NIA 15K library or purchased from Research Genetics; however, information about the length and sequence of the clones was not provided in detail. The dual-color labeling method was used to incorporate Cy3 or Cy5 into the cDNA generated from 20 μg of total RNA. These investigators selected 47 genes represented on both arrays, including both known regulated and nonregulated transcripts, and established reference relative expression measurements for these genes in the test RNA samples using quantitative real-time PCR (QRT-PCR) assays. The validity of the reproducible (average CV, 11.8%) QRT-PCR measurements was established through application of a new mathematical model. The performance of both array platforms in identifying regulated and nonregulated genes was identical. With either platform, 16 of 17 definitely regulated genes were correctly identified, and one definitely nonregulated transcript was falsely identified as regulated. Accuracy of the fold-change measurements obtained with each platform was assessed by determining measurement bias. Both platforms consistently underestimated the relative changes in mRNA expression between experimental and control samples. The bias observed with cDNA arrays was predictable for fold changes <250-fold by QRT-PCR and could be corrected. The bias observed with the commercial oligonucleotide arrays was less predictable and calibration was not possible. Following calibration, fold-change measurements generated by custom cDNA arrays were more accurate than those obtained by commercial oligonucleotide array.

In summary, these comparative studies indicate that the interchange of data across different platforms seems to be impossible. More comparative studies across these different platforms are still required to form a consensus opinion as to which platform is likely to generate more accurate microarray data.

## 2.5 FUTURE POSSIBILITIES

Microarray analysis has transformed from an interesting idea into a core technology in just a few short years. The recent trend of institutions bringing these tools to core facilities should further open the way for academic researchers to use this technology. While only a few of the academic core facilities surveyed offered access to the Affymetrix GeneChips, a much larger number offered cDNA array technology. Also, based on the number of publications associated with microarray technology, nearly 75% of papers used spotted cDNA arrays as a means of profiling the gene expression patterns since 1999. In addition, the wide use of the Affymetrix microarray system has been limited due to the more involved sample processing steps. The investigator is generally responsible for isolating RNA, creating double-stranded cDNA incorporating a T7 promoter, performing *in vitro* transcription (IVT) to get labeled cRNA, fragmenting the cRNA, and providing

**TABLE 2.1**
**Summary Statistics of Gene Expression Data Generated by Two Microarray Systems**

| | Incyte UniGem V 2.0 | | | Affymetrix HuGene FL | | | | | Affymetrix U95Av2 | | | | |
|---|---|---|---|---|---|---|---|---|---|---|---|---|---|
| | | 8 h ($n = 5$) | 24 h ($n = 1$) | | 8 h ($n = 3$) | | 24 h ($n = 3$) | | | 8 h ($n = 3$) | | 24 h ($n = 3$) | |
| Gene Name | Accession No. | Fold | Fold | | Fold | Call | Fold | Call | Accession No. | Fold | Call | Fold | Call |
| Thioredoxin reductase | D88687 | 2.04 | 2.70 | X91247 | 1.95 | I | 2.47 | I | X91247 | 2.56 | I | 2.20 | I |
| NAD(P)H: quinone oxidoreductase | M81600 | 1.04 | 1.20 | J03934 | 10.6 | I | 14.8 | I | M81600 | 5.40 | I | 5.45 | I |
| Neurofilament heavy | X15306 | 1.26 | 1.20 | X15306 | 2.5 | I | 1.97 | I | X15306 | 2.27 | I | 1.52 | NC |
| Transketolase | AW006207 | 1.30 | 1.20 | L12711 | 1.2 | NC | 2.20 | I | L12711 | 0.85 | NC | 1.63 | I |
| Breast cancer cytosolic NADP(+)-dependent malic enzyme | N/A | N/A | N/A | U43944 | 2.10 | I | 5.10 | I | U43944 | 3.26 | I | 2.16 | I |
| Malic enzyme 1 NADP(+)-dependent, cytosolic | NM_00239 | 1.72 | 1.60 | N/A | N/A | N/A | N/A | N/A | AL049699 | 11.06 | I | 5.98 | I |
| Aldo-keto reductase family 1, member C4 (AKR1C4) | S68287 | 1.60 | 4.60 | U05861 | 4.85 | I | 5.13 | I | U05861 | 3.57 | I | 17.78 | I |
| Aldo-keto reductase family 1, member C1 (AKR1C1) | M86609 | 1.10 | 1.80 | D17793 | 6.35 | I | 4.13 | I | D17793 (AKR1C3) | 13.0 | I | 16.2 | I |
| Ferritin, light chain | BE301211 | 1.14 | 2.30 | M11147 | 1.35 | NC | 2.33 | I | AL031670 | –0.41 | NC | 1.66 | I |
| Ferritin, heavy chain | AI816415 | 1.46 | 1.70 | L20941 | 1.35 | NC | 2.57 | I | L20941 | 0.97 | NC | 2.15 | I |
| Heme oxygenase 1 | Z82244 | 1.14 | 1.3 | Z82244 | 4.55 | I | 1.70 | NC | Z82244 | 7.38 | I | 2.74 | NC |
| γ-Glutamylcysteine ligase, regulatory subunit | NM_00206 | 1.04 | 1.30 | L35546 | 7.50 | I | 4.47 | I | L35546 | 4.99 | I | 3.74 | I |
| γ-Glutamylcysteine ligase, catalytic subunit | AL033397 | 1.20 | 1.26 | M90656 | 1.40 | NC | 1.38 | NC | M90656 | 1.33 | I | 0.94 | I |
| Glutathione reductase | N/A | N/A | N/A | X15722 | 2.2 | MI | 3.73 | I | X15722 | 4.96 | I | 1.97 | NC |

*Note:* Data analysis was performed across all pair-matched tBHQ and vehicle-treated groups. The fold and difference calls (or change calls) in Affymetrix HuGene FL and U95Av2 chips were calculated by Microarray Suite 4.1 and 5.0, respectively, and the fold changes were presented as means. Each transcript in the comparison analysis had five possible difference call (or change call) outcomes: (1) increase (I); (2) marginal increase (MI); (3) decrease (D); (4) marginal decrease (MD); (5) no change (NC). The folds in Incyte Unigem V.2.0 arrays were calculated by the Incyte algorithms and the results were presented as means. N/A, the gene probe is not available in that array. (From Li, J. et al., *Toxicol. Sci.*, 69, 383, 2002. With permission.)

quality control (QC) documentation of these steps. There are also additional costs in that Affymetrix recommends running a test chip to verify the quality of samples.

The oligonucleotide approach for the analysis of gene expression has also been criticized for a lack of sensitivity.[17,41] Because a single capture probe is not always sufficient to distinguish the expression of a particular gene, the use of multiple capture probes to represent a single gene in the Affymetrix array system is intended to avoid the problem of cross-hybridization as well as to increase the sensitivity. Thus, one of the likely causes of increased specificity in oligonucleotide arrays as compared with long cDNA arrays is a decrease in cross-hybridization with highly homologous genes.

Although it is true that shorter oligonucleotides (15- to 20-mers) promise more specific binding with probes, some papers recently indicated that 50- or 70-mers yield good sensitivity while maintaining the excellent specificity of shorter sequences.[17] When Operon Techologies[41] compared the sensitivities of 35-, 50-, 70-, and 90-mers for detecting highly expressed genes and genes expressed at moderate or low levels, they found that the 70-mers performed best. However, the results from MWG Biotech showed that 50-mer oligonucleotide arrays provided high specificity and excellent sensitivity, while 70-mers only increased formation of secondary structure.[17,40]

Ramakrishnan and colleagues[25] recently introduced the Motorola platform, which is based on a cross-linked polyacrylamide substrate that is photo-cross-linked to a glass slide and has specific functional groups to which the 5′ end of an oligonucleotide (30-mer) is attached via a hexylamine linker. They point out several features of this platform that may rival other commercialized platforms. First, a three-dimensional surface (the polyacrylamide matrix) has been shown to have a higher specific hybridization signal compared to glass. Second, high-quality, presynthesized oligonucleotides were used in contrast to *in situ* synthesis protocols that can generate amounts of deficient sequences depending upon the stepwise coupling efficiency and the length of the oligonucleotide. The probe present on the final commercial chip has been empirically tested prior to its attachment on the array. Third, single-color detection has been used rather than dual-color for ratio-based reporting of expression changes. Finally, automation of the target-preparation using the Qiagen BioRobot (Model 9604 or 3000) procedure shows high reproducibility and low variability as measured by hybridization to microarrays showing a linear and quantitative amplification probe hybridization signals.[4,39] Thus, oligonucleotide spotted arrays may be an optional choice for global screening. However, comparative studies between this platform with others also are necessary.

## 2.6 SUMMARY

Increasing evidence strongly implies that oligonucleotide-based arrays are more reliable for global screening compared to long cDNA array at the present time. Oligonucleotide arrays give a more accurate and comprehensive representation of gene expression profiles that in turn ensures with a higher probability that further research work will proceed down the correct path. Clearly, the recent efforts to genetically profile toxic compounds and promising pharmaceuticals require the best datasets possible for application to drug discovery. In terms of the quality-to-price ratio, new systems such as oligonucleotide spotted arrays are expected to find a place in the market. What the future holds, however, remains to be determined in this ever-evolving field.

## REFERENCES

1. Aach, J., Rindone, W., and Church, G.M., Systematic management and analysis of yeast gene expression data, *Genome Res.*, 10, 431, 2000.
2. Aardema, M.J. and MacGregor, J.T., Toxicology and genetic toxicology in the new era of "toxicogenomics": impact of "-omics" technologies, *Mutat. Res.*, 499, 13, 2002.

3. Dangond F., Chips around the world: proceedings from the *Nature Genetics* microarray meeting, *Physiol. Genomics*, 2, 53, 2002.
4. Dorris, D.R. et al., A highly reproducible, linear, and automated sample preparation method for DNA microarrays, *Genome Res.*, 12, 976, 2002.
5. Evertsz, E.M. et al., Hybridization cross-reactivity within homologous gene families on glass cDNA microarrays, *Biotechniques*, 31, 1182, 2001.
6. Greenberg, S.A., DNA microarray gene expression analysis technology and its application to neurological disorders, *Neurology* 57, 755, 2001.
7. Hamadeh, H.K. et al., Discovery in toxicology: mediation by gene expression array technology, *J. Biochem. Mol. Toxicol.*, 15, 231, 2001.
8. Hamadeh, H.K. et al., Gene expression analysis reveals chemical-specific profiles, *Toxicol. Sci.*, 67, 219, 2002.
9. Harkin, D.P., Uncovering functionally relevant signaling pathways using microarray-based expression profiling, *Oncologist*, 5, 501, 2000.
10. Harkin, D.P. et al., Induction of GADD45 and JNK/SAPK-dependent apoptosis following inducible expression of BRCA1, *Cell*, 97, 575, 1999.
11. Harrington, C.A., Rosenow, C., and Retief, J., Monitoring gene expression using DNA microarrays, *Curr. Opin. Microbiol.*, 3, 285, 2000.
12. Ishii, M. et al., Direct comparison of GeneChip and SAGE on the quantitative accuracy in transcript profiling analysis, *Genomics*, 68, 136, 2000.
13. Iyer, V.R. et al., The transcriptional program in the response of human fibroblasts to serum, *Science*, 283, 83, 1999.
14. Jia, M.H. et al., Global expression profiling of yeast treated with an inhibitor of amino acid biosynthesis, sulfometuron methyl, *Physiol. Genomics*, 3, 83, 2000.
15. Kamb, A. and Ramaswami, M., A simple method for statistical analysis of intensity differences in microarray-derived gene expression data, *BMC Biotechnol.*, 1, 8, 2000.
16. Kane, M.D., Aligning experimental design with bioinformatics analysis to meet discovery research objectives, *Cytometry*, 47, 50, 2002.
17. Kane, M.D. et al., Assessment of the sensitivity and specificity of oligonucleotide (50mer) microarrays, *Nucleic Acids Res.*, 28, 4552, 2000.
18. Chuo, W.P. et al., Analysis of matched mRNA measurements from two different microarray technologies, *Bioinformatics*, 18, 405, 2002.
19. Lee, S.B. et al., The Wilms tumor suppressor WT1 encodes a transcriptional activator of amphiregulin, *Cell*, 98, 663, 1999.
20. Li, J. and Johnson, J.A., Time-dependent changes in ARE-driven gene expression by use of a noise-filtering process for microarray data, *Physiol. Genomics*, 9, 137, 2002.
21. Li, J., Pankratz, M., and Johnson, J.A., Differential gene expression patterns revealed by oligonucleotide versus long cDNA arrays, *Toxicol. Sci.*, 69, 383, 2002.
22. Marcotte, E.R., Srivastava, L.K., and Quirion, R., DNA microarrays in neuropsychopharmacology, *Trends Pharmacol. Sci.*, 22, 426, 2001.
23. Morgan, K.T., Gene expression analysis reveals chemical-specific profiles, *Toxicol. Sci.*, 67, 155, 2002.
24. Pasinetti, G.M., Use of cDNA microarray in the search for molecular markers involved in the onset of Alzheimer's disease dementia, *J. Neurosci. Res.*, 65, 471, 2001.
25. Ramakrishnan, R. et al., An assessment of Motorola CodeLink microarray performance for gene expression profiling applications, *Nucleic Acids Res.*, 30, e30, 2002.
26. Ramdas, L. et al., Sources of nonlinearity in cDNA microarray expression measurements, *Genome Biol.*, 2, research0047, 2001.
27. Schulze, A. and Downward, J., Navigating gene expression using microarrays — a technology review, *Nat. Cell. Biol.*, 3, E190, 2001.
28. Schweighoffer, F. et al., Qualitative gene profiling: a novel tool in genomics and in pharmacogenomics that deciphers messenger RNA isoforms diversity, *Pharmacogenomics*, 1, 187, 2000.
29. Strausberg, R.L. and Austin, M.J., Functional genomics: technological challenges and opportunities, *Physiol. Genomics*, 1, 25, 1999.
30. van Hal, N.L. et al., The application of DNA microarrays in gene expression analysis, *J. Biotechnol.*, 78, 271, 2000.

31. Watson, S.J. and Akil, H., Gene chips and arrays revealed: a primer on their power and their uses, *Biol. Psych.*, 45, 533, 1999.
32. Weinstein, J.N. et al., The bioinformatics of microarray gene expression profiling, *Cytometry*, 47, 46, 2002.
33. Yue, H. et al., An evaluation of the performance of cDNA microarrays for detecting changes in global mRNA expression, *Nucleic Acids Res.*, 29, E41, 2001.
34. Yuen, T. et al., Accuracy and calibration of commercial oligonucleotide and custom cDNA microarrays, *Nucleic Acids Res.*, 30, e48, 2002.
35. Zapala, M.A. et al., Software and methods for oligonucleotide and cDNA array data analysis, *Genome Biol.*, 3, SOFTWARE0001, 2002.
36. www.affymetrix.com (Affymetrix).
37. www.clontech.com (BD Clontech).
38. www.incyte.com (IncyteGenomics).
39. www.motorola.com/lifesciences (Motorola Life Sciences).
40. www.mwgbiotech.com (MWG Biotech).
41. www.westburg.nl (Operon).

# 3 Quality Control of Expression Profiling Data

*Andrew Hill and Maryann Whitley*

## CONTENTS



## 3.1 INTRODUCTION

Nucleic acid-based microarrays were first used for large-scale parallel analysis of gene expression about 7 years ago.[1,2] Since that time, gene expression microarray techniques have continued to be developed by many investigators who have recognized the unprecedented potential of array

0-8493-1334-1/03/$0.00+$1.50

techniques for investigating many biological systems. In such a dynamic field of inquiry, it is not surprising that many of the basic techniques are still evolving, and the utility of specific microarray-related methods is greatly debated. However, if there is one point that most practitioners agree upon, it is that the quality of current microarray data is not guaranteed. In such an environment, quality control (QC) of array methods and data is crucial to successful use of the technology.

First, because microarray expression profiling techniques are still developing, some of the problems are not yet well understood, and pitfalls for the individual investigator are not necessarily obvious. For example, RNA isolation and preparation of target materials can be highly variable processes, and one's choices of methods can have strong effects on array data. Once arrays have been hybridized, differentiating a poor result from a good one is not always straightforward. In addition, following microarray data generation, a multitude of downstream data analysis methods can have dramatic impacts on results.

A second impetus for strong QC methods arises from the importance of maximizing consistency among expression profiling results from different investigators, laboratories, or even technologies. Expression profiling microarrays are expensive, experiments can be time consuming, and resulting datasets can be very large. Thus investigators have recognized the value of electronically integrating gene expression data from diverse sources in large, easily queried expression databases. The potential power of such databases for biological discovery is great, but it can only be realized if high-quality, comparable data can be generated by many independent investigators. Widespread application of common QC methods is crucial to achieving comparable data, and this has been recognized in community database initiatives such as MIAME.[3]

Quality control for array data may consist of several steps. First, one should try to prevent quality problems by optimizing sample preparation and hybridization and should be prepared to monitor quality with appropriate control reagents; this is the focus of Section 3.2. Second, one needs to detect poor-quality data when they occur; methods for doing this are discussed in Section 3.3. A third step for quality control involves selecting a data reduction method appropriate for the data and ensuring the consistency of the final reduced data, topics that are discussed in Section 3.4.

The majority of the authors' experience has been with the Affymetrix GeneChip™ oligonucleotide array platform (Affymetrix; Santa Clara, CA). For this reason, most of this chapter describes QC methods that can be applied directly to this platform. Nevertheless, we stress that the concepts described later can, and should, be applied to other common platforms. Briefly, Affymetrix GeneChip® arrays typically consist of ~500,000 distinct 25-mer oligonucleotides, synthesized by photolithography on a glass substrate. On most designs, 10 to 20 distinct 25-mers are tiled on the array to monitor each mRNA. Each of these complementary 25-mers (called *perfect match* [PM] oligos) is accompanied by a corresponding *mismatch* (MM) oligo that has the same sequence but for a single base change at the central position. MM oligos are used as controls for nonspecific binding in the Affymetrix GeneChip data reduction methodology.[2] In the remainder of the chapter, the term *probe* will be used to refer to a PM or MM oligonucleotide, and the term *target* will be used to indicate material that is hybridized to an array.

## 3.2 THE STARTING POINT: QUALITY CONTROL OF THE TARGET MATERIAL AND ARRAYS

### 3.2.1 Quality and Amount of cRNA Target

All expression profiling experiments start with the isolation of total or polyadenylated RNA from a biological sample. Two typical methods are available, depending on the type of array to be used. For spotted cDNA microarrays, the RNA is reverse transcribed (RT) into labeled complementary DNA (cDNA); for oligonucleotide arrays, unlabeled cDNA is purified and used as the template for *in vitro* transcription (IVT) back into labeled cRNA. The labeled target is then applied to the array. Description of the many methods for isolating, reverse transcribing, and labeling samples is beyond the scope of this chapter, and good guidelines are available elsewhere.[4] The focus of this section, therefore, is on the components of the process that can alter the quality of the target.

Several aspects of the RNA amplification and labeling process can greatly influence the representation of the initial mRNA in the final labeled pool, and these should be given careful consideration when choosing a protocol for target preparation. The choice of random primers or oligo-dT in the cDNA synthesis steps, the specificity and processivity of the enzymes used in the reverse transcriptase and T7 RNA polymerase reactions, and the method chosen for purifying the cDNA and cRNA will all influence the average length and/or 3′ bias of the final product. Different types of arrays will have inherently different biases in the probes, and, if the labeled target does not contain the complementary sequences to the probe sequences tiled on the array, false negatives will result.

In addition to the representation of messages, the use of different isolation, purification, RT, and IVT methods can also have an effect on the overall yield of the target preparation process and, more importantly, can alter the purity of the target cRNA. Variation in purity is especially problematic, because it means that target preparations that appear to contain the same mass of cRNA by ultraviolet absorbance methods can, in fact, contain quite different masses of labeled, hybridizable material. Comparability of the array readouts from such targets is then compromised because of the limited dynamic range of the arrays (see Section 3.2.2.2).

To limit variation resulting from target preparation methods, it is advisable to adhere to the following minimal guidelines: (1) assess RNA degradation prior to amplification and labeling, (2) process all samples in an experiment with identical protocols, and (3) quantitate the target material as accurately and consistently as possible. Finally, when validating or developing new protocols, especially for high-throughput screening (HTS) applications, the inclusion of no-template controls is necessary to assess the amount of carryover and other sources of impurities.

In typical single-round amplification reactions, conversion rates of total RNA to cRNA are on the order of 1 to 10, so 2 to 10 μg of total RNA is often required to achieve 10 μg of labeled cRNA target. Rehybridization of single target preparations to multiple arrays can be used to extend the use of small samples. In our hands, cRNA hybridization solutions for oligonucleotide arrays can typically be reused three times. Variance among repeated hybridizations of a single hybridization solution is usually significantly less than the variance introduced by other biological and technical sources of noise. If sample material is limiting for a particular application, then higher levels of RNA amplification can be achieved by multiple rounds of linear amplification. For example, starting with as little as 10 ng of total RNA, two rounds of linear amplification can yield ~10 μg of labeled cRNA target.[5] However, protocols that provide higher amplification are technically more difficult and also carry the risk of larger representational biases in the amplified product, so they should be used only when truly required.

### 3.2.2 Array Quality

#### 3.2.2.1 Array Variation

Variation in array quality occurs in various forms, regardless of the particular array platform that is used. For example, quality problems for oligonucleotide arrays may arise from incomplete oligo synthesis, spatial defects due to photolithography problems, or physical defects such as scratches. For spotted cDNA arrays, size and morphology of spots are commonly problematic. All these variations will tend to degrade the fidelity of array data. One serious drawback of most arrays is that none of the above-mentioned defects can be detected prior to actually using the array (posthybridization QC is addressed in Section 3.3), and preventative measures are very limited. To reduce overall variation due to array manufacture, arrays used in a particular experiment should be members of the same fabrication lot when possible, as inter-lot quality variation can in some cases be large. In addition, for "homemade" spotted cDNA arrays, preventative measures must include careful control of all aspects of the clone picking and array spotting processes; a description of these processes is far beyond the scope of this chapter, but many sources of good information are available in the literature.

#### 3.2.2.2 Dynamic Range of the Array: Sensitivity and Saturation

In most biological systems, the vast majority of messages are typically present at relatively low levels, below ~1 message in 100,000 (i.e., less than 10 parts per million [ppm]), while a small number of ubiquitous mRNAs can exceed 1 message in 1000 (1000 ppm). The particulars of the distribution can be expected to be different in different organisms or in different cell types. In general, we can say that most mRNAs are distributed across an abundance range of at least approximately three orders of magnitude, with a pronounced bias toward low-abundance messages. For the purposes of this chapter, we define ***dynamic range*** as the range of message abundance that can be robustly detected in the target, ranging from the lowest abundance message that produces a hybridization signal just above background levels up to the highest abundance message that saturates the response of the array.

Dilution and spike-in studies demonstrate two facts about the effective dynamic range of current oligo arrays. Spiking studies on oligonucleotide arrays suggest that the dynamic range of the readout is at best three orders of magnitude, and response over this range is not fully linear.[6] Spotted cDNA arrays also suffer from limited linear dynamic range.[7] Dilution studies demonstrate that the range of transcript abundances that can be reliably quantitated is dependent on the total mass of target material that is applied to the array. Thus, for example, the PM probe features for an abundant message can be in the linear range of response when a small mass of target is included in the hybridization, but the same features can saturate if the total mass of target in the hybridization is increased. Figure 3.1 shows data from a dilution study using *Caenorhabditis elegans* target that demonstrate the dependence of the effective dynamic range of the readout upon the mass of target applied to the array.

Due to the dependence of dynamic range on the total mass of target, proper matching of target mass to array is clearly crucial in order to maximize the fidelity of the readout. For the example shown in Figure 3.1, the use of 2 μg of *C. elegans* target (quantitated by ultraviolet absorbance) resulted in a large number of messages being detected (called ***present***), but comparison of this data to the result when 0.1 μg of target was used shows that high-abundance messages were saturating when 2 μg of target was used. On the other hand, reduction of the target material to 0.1 μg reduced nonlinearities among the high-abundance readouts, but at the cost of a large loss of sensitivity to detect rare messages. In this particular case, the best compromise was found by applying 0.5 to 1 μg of target material, which led to near-maximal numbers of genes being detected, while limiting saturation of the most ubiquitous messages. Note that, for *C. elegans*, the determined target range is an order of magnitude lower than the array manufacturer's generic recommendation[8] of at least 10 μg.

### 3.2.3 Housekeeping and Spiked Messages for QC

Internal controls of some kind are crucial to determine if an array experiment has failed and, if so, at what stage the failure took place. Both endogenous and spiked transcripts can be useful. Table 3.1 summarizes several housekeeping, polyadenylated spike, and cRNA spike messages that we have used for QC purposes on oligonucleotide arrays.

#### 3.2.3.1 Housekeeping Genes

Housekeeping genes are generally defined as mRNAs that are highly and invariably expressed in all samples in any given experiment. Because they are endogenous to the samples and already present in the target, they are the best representation of the quality of the sample. Commonly used examples are genes such as β-actin, glyceraldehyde-3-phosphate dehydrogenase (GAPDH), or

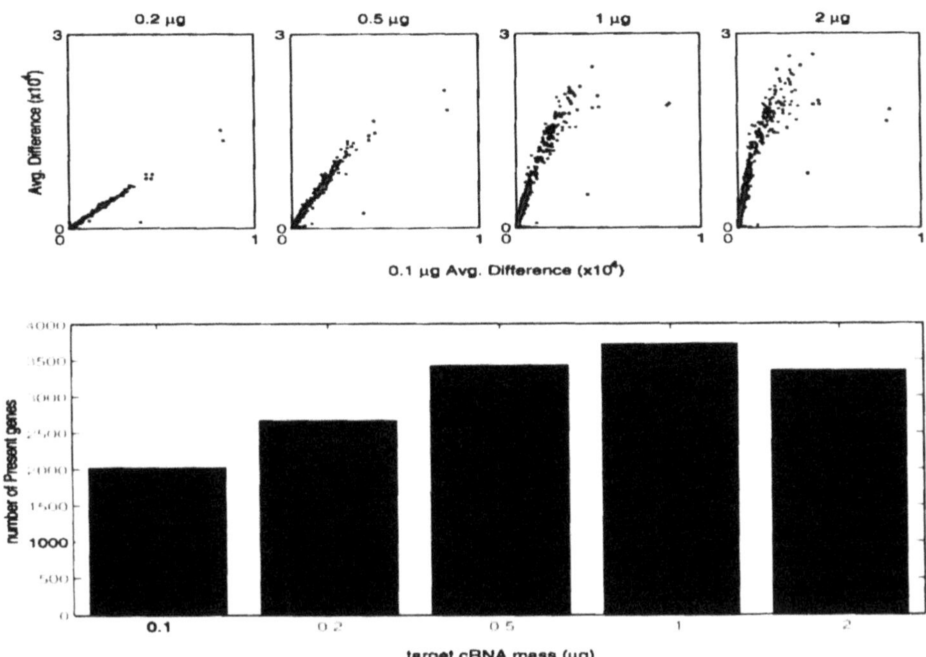


**FIGURE 3.1** Dynamic range of array readouts. A single *C. elegans* cRNA preparation was serially diluted, cRNA spikes were added, and each dilution was applied to two or three replicate arrays. (Bottom panel) The number of genes called present by the Affymetrix MAS 4.0 software was proportional to target mass, reaching maximal levels by 0.5–2 µg. (Top panels) Pairwise scatter plots of mean readouts for 0.2-, 0.5-, 1-, and 2-µg target hybridizations against the 0.1-µg target hybridization. As target mass increases, hybridization intensity increases, but readouts of the most abundant messages are progressively saturating, at target masses as low as 0.5 µg. As a result, readout range is compressed as target mass increases.

ribosomal proteins. Both β-actin and GAPDH are monitored by standard probesets on all Affymetrix oligonucleotide arrays (Table 3.1). A useful application for housekeeping gene readouts is as a measure of target message degradation. Specifically, the ratio of readouts from probesets from the 5′ and 3′ ends of housekeeping genes such as GAPDH can be a good monitor of commonly occurring sample degradation or failure of processivity in the cDNA/IVT process (see Section 3.3.3.1 for details).

Another use of housekeeping genes is for normalization. This method assumes that the designated housekeeping genes were expressed at constant levels in all samples and therefore uses them as a reference point. Readouts from arrays are then universally scaled so as to satisfy the assumption. The practice of using a small number of housekeeping genes for normalization is frequently criticized[9] because the assumption that such genes are truly constant in expression is often not justified. Furthermore, typical housekeeping genes are often among the most highly expressed genes, thus normalization that utilizes their readouts as a reference can be especially subject to artifacts caused by limited dynamic range (see Section 3.2.2.2). Nevertheless, in some cases housekeeping genes can be a useful normalization control if a relatively large number of them are available and they are appropriate for the experiment in question (see Section 3.4.3).

#### 3.2.3.2 Polyadenylated Spikes

Perhaps the most direct method for detecting problems in the entire process of target preparation, array hybridization, and data reduction is the spiking of *in vitro* transcribed polyadenylated control transcripts into total or polyadenylated RNA from an experimental sample.[9] Such spikes can be prepared in pools and an aliquot of the pool added into each isolated RNA. In this way, the spikes

**TABLE 3.1**
**Housekeeping, Polyadenylated, and cRNA Spikes for Array QC[a]**

| Type of Control | Transcript | ATCC Accession | Affymetrix Gene Qualifier | QC Metrics |
|---|---|---|---|---|
| Housekeeping | beta-Actin | NA | depends on array design | 5′–3′ ratios |
| | GAPDH | NA | depends on array design | |
| Polyadenylated spike | LYS | 87482 | AFFX-LysX-5, -3, -M | 5′–3′ ratios |
| | PHE | 87483 | AFFX-PheX-5, -3, -M | Rank order |
| | THR | 87484 | AFFX-ThrX-5, -3, -M | |
| | TRP | 87485 | AFFX-TrpnX-5, -3, -M | |
| cRNA spike | DAPM | 87826 | AFFX-DapX-M | Array sensitivity |
| | DAP5 | 87827 | AFFX-DapX-5 | Standard curve slope |
| | CRE5 | 87832 | AFFX-CreX-5 | |
| | BIOB5 | 87825 | AFFX-BioB-5 | |
| | BIOD3 | 87830 | AFFX-BioDn-3 | |
| | BIOB3 | 87828 | AFFX-BioB-3 | |
| | CRE3 | 87835 | AFFX-CreX-3 | |
| | BIOC5 | 87833 | AFFX-BioC-5 | |
| | BIOC3 | 87834 | AFFX-BioC-3 | |
| | DAP3 | 87831 | AFFX-DapX-3 | |
| | BIOBM | 87829 | AFFX-BioB-M | |

[a] Where applicable, the "ATCC Accession" column gives the American Type Culture Collection accession number for related plasmid construct. The "Affymetrix Gene Qualifier" column indicates the label of the probeset on Affymetrix commercial arrays that corresponds to each transcript.

experience all stages of the sample preparation and array readout processes. Ideally, the pool of spikes will contain a variety of messages of different length and different concentrations that are not present in the samples to be tested. The readout from such spikes can be used to test for processivity and linearity of amplification and for the success of the array hybridization. We have used the *Bacillus subtilis* transcripts LYS, PHE, THR, and TRP (Table 3.1) for this purpose and have found them to be useful QC indicators in some situations.

### 3.2.3.3 Labeled cRNA Spikes

To validate the performance of an array, independent of possible sample-related QC problems, spiking of labeled control transcripts into cRNA is a very useful method. As with polyadenylated spikes, cRNA spikes can be prepared as a mixture of a number of distinct transcripts at a range of concentrations to form a standard curve. An aliquot of the pooled standard curve mixture can be added to targets prior to hybridization. The readout of cRNA spikes will indicate specific problems of the array hybridization or readout process, as opposed to failures of the sample preparation process. Nonlinearities in array response, indicative of dynamic range limits, are readily observed in the form of saturation-like curvature in the standard curve. The slope of the standard curve in units of intensity per transcript abundance is one indicator of array response. In addition, the readout of the cRNA spikes can be used to estimate array sensitivity and normalize results among arrays and array designs (see Section 3.4.3). We have used a pool of 11 bacterial transcripts for this purpose[10] and found them to be invaluable for QC purposes (Table 3.1). Specific QC metrics that can be derived from the readouts of cRNA spikes are discussed in Section 3.3.

## 3.3 AFTER THE HYBRIDIZATION: QUALITY CONTROL TESTS FOR THE ARRAY READOUT

### 3.3.1 Inspection of the Array Image

Critical review of the array readout is the focus of almost all array QC methods, simply because most problems, if present, will manifest themselves at this stage. Examples of such problems include:

- Circular-shaped artifacts resulting from poor fluid flow during hybridization or staining
- Fluorescent debris on the array surface, introduced during hybridization or staining
- Spatial gradients in the staining intensity, often occurring in the same direction in which fluid flow occurred during the staining procedures
- Manufacturing defects, such as poor spot or feature definition or scratches
- Misregistration of the gridding used by image-analysis software to segment array features

Visual inspection of the array image is the most basic, and effective, QC method. While it may seem obvious that all arrays should be examined by eye, in the context of large experiments involving hundreds of arrays visual inspection can become a chore that is easily neglected. However, even a rapid inspection of the scanned array image by a trained observer can quickly identify the majority of problems that degrade data, usually more easily and sensitively than automated QC methods.

### 3.3.2 Metrics for Measuring Overall Array Image Quality

Three core QC metrics calculated by the Affymetrix software MicroArray Suite (MAS) are strong predictors of array readout problems. While these particular numbers are specific to the Affymetrix platform, they are essentially measures of signal and noise, and alternative array software will generally compute analogous figures.[11]

#### 3.3.2.1 Average Intensity/Scaling Factor

The average intensity is simply the average hybridization intensity of all features on the array. If software applies an automatic scaling to bring the average intensity of arrays to a common value, then the scaling factor represents the multiplicative factor that was applied to a particular array (larger scaling factors indicate that an array was originally dimmer). The target value for average intensity will vary from one platform to another, but for oligonucleotide arrays average intensity of arrays within an experiment should not vary much more than threefold. Variations in average intensity beyond this range often indicate sample or array problems. Perhaps the most common problem is a "dim" array, which is often a symptom of poor target preparation. While variations in average intensity can nominally be removed by normalization (see Section 3.4.3), extreme variations will cause nonlinear, possibly gene-specific differences between samples due to background noise and the limited dynamic range of the arrays (see Section 3.2.2.2). Most normalization procedures cannot effectively remove such nonlinear effects; therefore, the problematic arrays should be excluded from further analysis.

#### 3.3.2.2 Number of Present Calls

The Affymetrix MAS software computes an *absolute decision* for each gene: absent, present, or marginal. The absolute decision provides an empirical estimate as to whether a message is detectable in the target. Using MAS v.4.0, under default settings, the false-positive rate is on the order of ~2%, while the true positive rate is typically around 20% for transcripts at concentrations as low as 0.25 pM and rises until it is >90% for transcripts above ~4 to 8 pM. The number of present calls for replicate samples within an experiment should typically be within ~10% of one another. More dramatic variation (especially reduced present call counts) for some samples often indicates that the corresponding targets or arrays are of questionable quality. Note that real biological variation among samples can also cause relatively large differences in present call counts, which should be recognized when analyzing present calls from nonreplicate samples.

#### 3.3.2.3 Background Level

In the MAS software, the *Q statistic* is a measure of the standard deviation of the 2% of features on the array with the lowest intensity. As such, it nominally represents a measure of background noise. As one would expect, Q scales with the average intensity of arrays, such that brighter arrays also tend to have higher Q values. In MAS v.4.0, there tends to be an inverse correlation between Q and the number of present calls for any readout, because the present call metric incorporates Q as a threshold value for making a present call. At a similar level of average intensity, unusually large Q values are a strong indicator of a quality problem with hybridizations. For example, we have found that poor-quality arrays or arrays with staining irregularities are sometimes distinguished by relatively high Q values.

#### 3.3.2.4 QC Measures Generated by Other Array Software

The Dchip software[12,13] generates three very useful QC measures based on the scanned array image. First, this software reports the median intensity of all features and the percentage of genes called present on each array. These are essentially the same as the first two QC figures derived from the MAS software, described previously. In addition, Dchip uses the standard errors of estimates from a linear statistical model to flag probesets that behave inconsistently (called *array outliers*) or single probes that behave abnormally (called *probe outliers*). These outlier calls are sensitive to cross-hybridizing probes and manufacturing or image defects on the arrays and serve as an efficient way to isolate poor-quality results. Dchip also offers a useful visualization of each array, in which the array and probe outliers are directly highlighted on an intensity-normalized image of each chip. Table 3.2 summarizes QC figures that are derived from the array image.

### 3.3.3 Readouts of Housekeeping Controls and Spiked Messages

#### 3.3.3.1 Housekeeping 5′–3′ Ratios

On commercial Affymetrix arrays, 5′ and 3′ regions are separately tiled for several housekeeping controls such as β-actin and GAPDH. As mentioned in Section 3.2.3.1, the ratio of signals from the 5′ and 3′ probesets within a single transcript can reveal failures of processivity during cDNA or IVT processes (i.e., the synthesized target was of insufficient length) or problems of target degradation. When considering 5′–3′ ratios, it is important to note that the absolute value of the ratio for any particular pair of 5′ and 3′ probesets will be specific to the associated transcript, due to probe-specific effects. That is, the 5′–3′ ratio may be consistently higher or lower for one housekeeping transcript compared to another, solely due to differences in the sensitivity of the associated probes. So, one should not concentrate on the absolute value of the ratios, but rather should look for inconsistencies in the ratio for any single housekeeping transcript across a set of arrays. For example, we have used threshold 5′–3′ ratios of 0.5 for the Affymetrix β-actin and GAPDH probesets as a QC warning signal; we often reserve arrays or samples with lower ratios for more detailed QC examination.

**TABLE 3.2**
**Image-Based QC Metrics[a]**

| Array QC Metric | Software Source | Description | Sensitive to | Ref. |
|---|---|---|---|---|
| Average intensity | Affymetrix MAS4/5 | Mean intensity of all probesets on array | Inadequate sample mass, poor sample preparation, poor-quality array, hybridization, staining, or scanning | b |
| Number of present calls | Affymetrix MAS4/5 | Count of messages called present on array | Inadequate sample mass, poor sample preparation, poor-quality array, hybridization, staining, or scanning | b |
| Q (noise) | Affymetrix MAS4/5 | Standard deviation of dimmest 2% of array features | Inadequate sample mass, poor sample preparation, poor-quality array, hybridization, staining, or scanning | b |
| Dchip array outliers | Dchip | Flag for probesets with inconsistent behavior | Cross hybridization to probes, image defects | c |
| Dchip probe outliers | Dchip | Flag for probes with inconsistent behavior | Cross hybridization to probes, image defects | c |

[a] Five QC metrics that we have found useful for evaluating the quality of array results are shown.
[b] Affymetrix, *Affymetrix GeneChip Analysis Suite 3.3 User Guide*, Affymetrix, Santa Clara, CA, 1999.
[c] Li, C. and Wong, W.H., *Proc. Natl. Acad. Sci. USA*, 98(1), 31–36, 2001.

#### 3.3.3.2 Polyadenylated Spikes

We have primarily used polyadenylated spikes as a general verification of the success of the target preparation and hybridization processes. Typically, we have spiked up to four distinct polyadenylated bacterial transcripts, ranging in length from 1 to 2.5 kb (Table 3.1), into starting total RNA. Specifically, we have looked for the correlation of the readout with the input spike concentrations, and the 5′–3′ ratios of the spikes as a monitor of process quality. A larger pool of distinct polyadenlyated spikes could potentially be used for estimation of the limit of detection of the array readout, estimation of process noise, linearity of response, and normalization of data from multiple samples.

#### 3.3.3.3 cRNA Spikes

Spikes added to the target cRNA have been used primarily for two purposes: estimation of array sensitivity, and normalization of array readouts (see Section 3.4.3). Because these spikes are added to the target mixture just prior to hybridization, their signals are sensitive to these processes only and are not affected by the success or failure of earlier target preparation steps. By observing the absolute decisions for each of the cRNA spikes, the sensitivity of an array can be estimated by a logistic regression.[10] For commercial oligonucleotide arrays, we typically find the sensitivity of detection to be about 1:100,000 messages; we flag arrays with sensitivity poorer than 1:50,000 as QC failures. Variation in the sensitivity estimate across the arrays in an experiment might indicate an inconsistent dynamic range.

## 3.4 REDUCING THE DATA: ENSURING THE QUALITY OF THE SUMMARIZED DATA

### 3.4.1 Alternative Data Reduction Methods

The central purpose of array experiments is to generate accurate, reproducible measurements of transcript abundances or transcript abundance differences from biological RNAs. To achieve this goal on different array platforms (spotted cDNA arrays, photolithographic oligonucleotide arrays, printed long oligonucleotide arrays, etc.), it is not surprising that different algorithms for computing transcript abundances from raw intensity data have been developed.

At the highest level, different platforms utilize different sample preparation and readout methods. For example, spotted cDNA arrays traditionally apply a mixture of two differentially labeled biological samples to an array and then use the ratio of the hybridization intensities in two channels to a single spotted feature to estimate the ratio of the transcript abundances. In contrast, Affymetrix arrays utilize a single-color staining methodology and summarize the hybridization intensity of multiple probes for each monitored transcript.

Even at the more specific, lower level, multiple data reduction methods have been proposed for each of the most commonly used array platforms. For example, both spotted and oligonucleotide arrays offer a number of different approaches to background subtraction, normalization, signal summary, etc. Some of them work better than others in particular experimental contexts, and many of them are still being developed. Each method embodies a specific set of assumptions that may or may not be appropriate for a specific experiment. The differences in performance between various methods mean that selection of a particular method becomes in some sense a QC decision. Because this area of research is so dynamic, specific algorithms for data reduction that we describe later are likely to become obsolete; however, because they are important to understand, we will survey several different types of current data reduction techniques.

Because data reduction methods are changing so rapidly, it is perhaps most valuable to understand how to evaluate different methods, as opposed to simply demonstrating that one method is currently superior in a given context. Given the central purpose of transcript quantitation as described above, it seems reasonable that the major part of the utility of a given data reduction method can be captured by standard measures of reproducibility, sensitivity, and specificity. For our purposes here, we define ***reproducibility*** as the variance among repeated measurements of the same transcript, ***sensitivity*** as the fraction of expression changes among samples that are correctly detected, and ***specificity*** as the fraction of non-changes that are correctly detected. Evaluation of reproducibility requires replication of controlled experiments for single samples, and measurement of sensitivity and specificity requires spiking or related experiments where known abundances of distinct transcripts are present in the target. In the following sections, we describe some methods that are applied during the data reduction process for oligonucleotide arrays and give examples of how these methods fare on the grounds of reproducibility, sensitivity, and specificity.

### 3.4.2 Probe Summary Methods

For oligo arrays, probe summary is the process by which the signals from a number (typically 10 to 20) of PM and MM probes are combined and reduced to a single signal readout that is proportional to transcript abundance. Currently, two closely related methods are offered by the Affymetrix MAS software (versions 4 and 5). Alternative methods, such as the PMLOG/RMA algorithm,[14] have recently been described by other groups.

The MAS software probe summary method is based on the empirical observation that the difference between the signals from perfect match and mismatch probes (PM – MM) is proportional to transcript abundance.[2] In MAS v.4.0, this (PM – MM) difference is computed for each probe pair in a probeset, outliers are discarded, and the summarized probeset readout (called the *average difference*) is calculated as the average of the (PM – MM) values. One of the problematic parts of

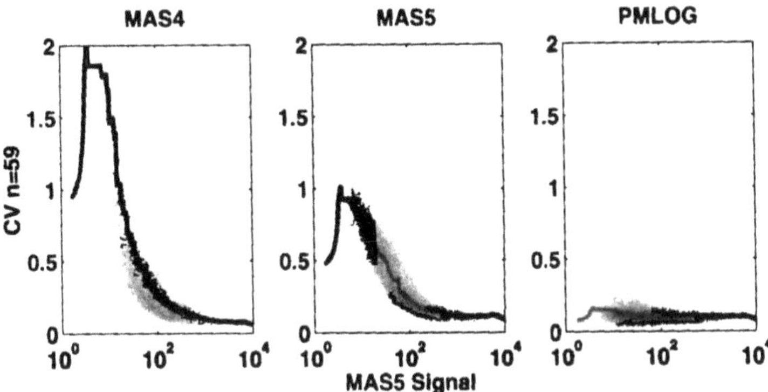


**FIGURE 3.2** Reproducibility of alternative probe summary methods. Affymetrix MAS 4, Affymetrix MAS 5, and PMLOG probe summary methods were applied to a set of 59 arrays from an Affymetrix Latin square spiking experiment (http://www.affymetrix.com/analysis/download_center2.affx). The coefficients of variation (CVs) of the readouts for all genes are plotted against the mean readout (abundance) for each of the 3 probe summary methods. The lines through the points are running medians through each set of data. In this case, use of PMLOG or MAS 5 reduced the variability of signals for low-abundance messages compared to MAS 4.

the average difference calculation is that in practice, a significant proportion of MM probes yield stronger signals than their PM partners, thus the average difference can be reported as less than zero, a result that cannot be correct and is difficult to process. In MAS v.5.0, the calculation is similar to MAS v.4.0, except that no outliers are removed, a more robust estimate of the average (PM – MM) is used, and MM values that are "suspicious" (e.g., larger than their corresponding PM partner) are imputed. The MAS v.5.0 probe-summarized value is called *signal*, and due to imputation of MM values it is never negative. Finally, in the PMLOG/RMA algorithm,[14] MM probes are not utilized except to compute a global background, and the probe-summarized value is essentially an estimate of the median PM value.

Figure 3.2 compares the MAS v.4.0 average difference, MAS 5.0 signal, and PMLOG probe summaries using data from a controlled spiking experiment carried out at Affymetrix. In this case, both MAS v.5.0 and PMLOG are more reproducible for the majority of relatively low-abundance transcripts, suggesting that the MM probe signals contribute a significant amount of noise to the average difference. The increase in reproducibility is reflected in the sensitivities of the methods for detecting differential expression. In one comparison of the sensitivity and specificity for the detection of fold changes of less than threefold in the same experiment, we found the sensitivities at 95% specificity of the methods to be 0.68 (MAS 4), 0.80 (MAS 5), and 0.91 (PMLOG/RMA).

### 3.4.3 Normalization Methods

Normalization is the process by which the readouts from multiple arrays are scaled so as to make them comparable. The Affymetrix MAS software offers two normalization methods. The first is a linear global normalization where the trimmed mean average difference for each array is scaled to a common level. This method makes the assumption that the mean expression level of samples should be the same, which may not be appropriate for highly perturbed samples, arrays monitoring a relatively small number of genes, or any set of coexpressed genes or if samples were run on different array designs. On the other hand, such a simple method may work well when monitoring many genes across mildly perturbed samples. The second method offered by MAS is a linear subset normalization, where the mean average difference of each array is scaled in order to equalize the mean average difference of a user-selectable subset of housekeeping or spiked probesets. Normalization by this method is most appropriate if one has a reasonably sized set of housekeeping or control transcripts in each target that are known to be present at the same levels in all samples.

One important application where the global and subset normalizations can differ greatly in performance is in the normalization of data from different array designs (i.e., arrays that monitor different but possibly overlapping sets of genes). In this application, global normalization can be systematically biased by differences in the mean expression levels of genes on the different designs. Specifically, if the true average expression level of genes on one design is higher than another, global normalization will inappropriately scale down data from the design with the greater mean expression level. In contrast, subset normalization can correctly normalize different designs, as the expression of a defined subset can be equal regardless of the superset of genes that is being monitored.

A number of alternative normalization procedures have also been described in the literature. Methods based on calibrating each array with a standard curve formed from the cRNA spikes described in Section 3.2.3.3 have been described.[10] To use cRNA spike signals for normalization, we form a standard curve from the signals of 11 spiked cRNAs that are typically present at levels from 0.5 to 150 pM and fit the curve by a generalized linear model to yield a calibration factor for the array (see Figure 3.3).[10] In this way, intensity signals can be converted to estimates of transcript abundance in parts per million. Such absolute abundance estimates are more informative than relative transcript measurements, such as the ratios typically generated from spotted array analyses. However, it is important to note that such absolute abundance estimates for genes with robust signals can be subject (on typical oligo arrays) to gene-specific uncertainties on the order of two- to threefold due to sequence-specific differences in the hybridization characteristics of individual 25-mer probes. This standard curve-based normalization method, like the MAS methods mentioned above, is linear, applying a single global scaling factor to each array in a dataset. Figure 3.4 compares the reproducibility of replicate hybridizations normalized by global scaling and two variants of standard curve normalization. Because of the limitations of the linear dynamic range of most array platforms (discussed in Section 3.2.2.2), other methods utilize nonlinear, or piecewise linear, scaling functions. For example, running median or spline normalizations are implemented in the Dchip software,[12] and quantile-based methods have also been described.[15] The nonlinear methods have the potential to remove common technical artifacts such as variable nonlinear saturation among a set of arrays; on the other hand, the major drawback of these methods is the danger of "over normalization." The additional degrees of freedom that are allowed in the nonlinear scaling methods can overfit the data, leading to inappropriate removal of true biological variation.

### 3.4.4 Inter-Sample Comparisons of Reduced Data

An investigator usually has a clear understanding of how the readouts from different samples in an experiment should relate to one another. For example, in a well-constructed experiment with sufficient replication, it is clear that replicate samples should have close to identical readouts. And, because most array designs monitor thousands of genes, even biologically different samples may be expected to have readouts for many genes that are fairly similar. Thus, inter-sample comparisons can be an efficient QC method to identify problematic arrays quickly. For this purpose, we find a heat-map representation of the Pearson correlation coefficients between arrays to be a fast way to zero in on questionable data in experiments involving tens to hundreds of arrays. The correlation coefficients are computed between each pair of arrays in the experiment and then presented in a color-coded matrix format, where color is proportional to the correlation coefficient. Problematic data typically appear in the form of one or more arrays that correlate poorly with all others in the set (see Figure 3.5 for an example). Other diagnostic plots can also be very helpful, such as simple scatterplots of one array against another or the rotated scatter "M-A" plots,[16] which show the log ratio of readouts in two samples plotted against a measure of average abundance.

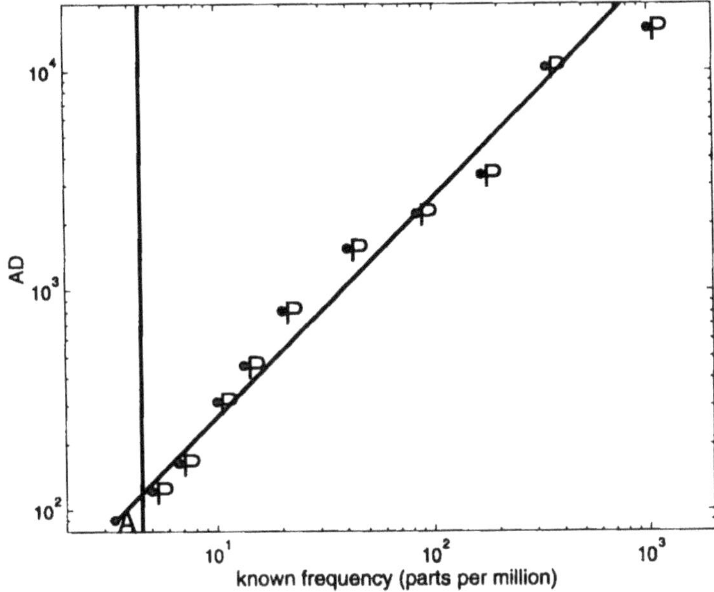


**FIGURE 3.3** A standard curve constructed from 11 *in vitro* synthesized cRNAs. cRNAs were pooled at 11 concentrations from 0.1 to 30 pM and added into complex cRNA targets. Hybridization intensity (average difference, or AD) is plotted against spike concentration, expressed in parts per million (ppm). Concentrations in ppm were calculated based on the known mass of cRNA in the target, assuming an average cRNA length of 1 kb. The response over 3 logs is close to linear, with a curvature suggesting saturation of PM signals from the more abundant transcripts. The solid line is a generalized linear model fit to the subset of readouts that were called present, indicated by a "P" on the plot. The sensitivity of the array shown was estimated by logistic regression of the present/absent calls to be 4.5 ppm (solid vertical line). (From Hill, A.A. et al., *Genome Biol.*, 2(12), research0055.1–0055.13, 2001. With permission.)

## 3.5 SUMMARY

Microarrays have established themselves as a powerful method for highly parallel gene expression monitoring. Nevertheless, the use of appropriate QC methods remains critical if arrays are to reach their potential for elucidating important biological systems. Array data quality is dependent on a number of factors, including the quality of sample preparation, overall array quality and consistency, and the limits of dynamic range for current arrays. Control reagents can be spiked into samples at appropriate stages of the sample preparation and array readout process. The readouts from those spiked reagents are sensitive indicators of anomalous samples or arrays. Visual inspection and other QC metrics derived from the array image after scanning can also reveal anomalous results efficiently. A number of methods exist for reducing raw array image data to estimates of relative or absolute message level, and many more are currently in development. Each of these normalization and/or probe summary methods makes different assumptions about the data, and careful consideration of those assumptions will allow scientists to select appropriate methods for their data. Once raw array data have been reduced, relatively simple inter-sample comparisons can often reveal common technical artifacts. Although the application of QC methods such as the ones described here does entail some effort on the part of investigators, the time and labor will be well rewarded with consistent, compelling, and valuable expression data.

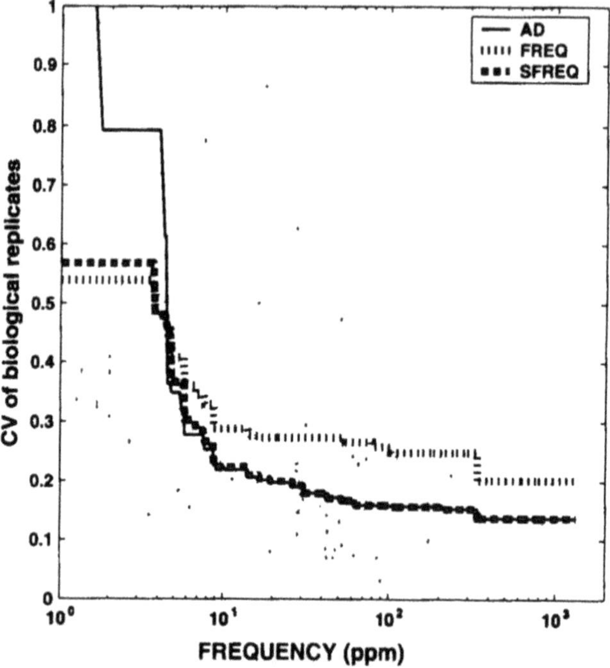


**FIGURE 3.4** Reproducibility of alternative normalization methods. MAS 4 global scaling (AD), frequency normalization (FREQ), and scaled frequency normalization (SFREQ)[10] were applied to the same set of *C. elegans* array readouts. The CV of each normalized readout for each gene is plotted as a function of the message frequency. For legibility, only 1% of the observed CVs are plotted as points. Lines are running medians through the four datasets. In this case, the FREQ method, based on calibration using spiked messages, was somewhat less reproducible than either SFREQ or AD, due to experimental variation in the effective concentration of spiked messages among hybridizations.

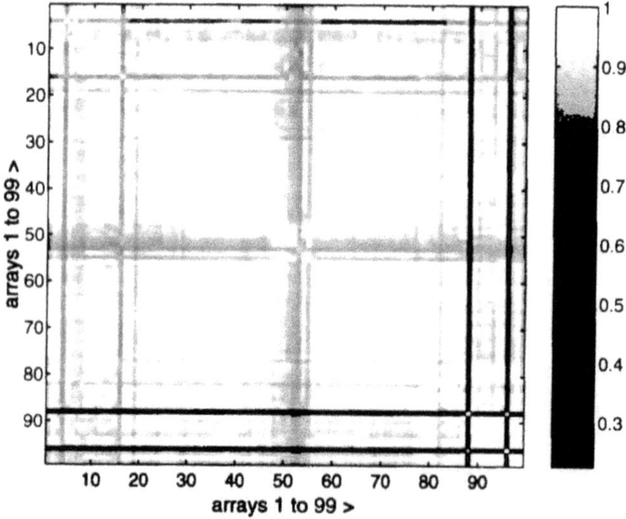


**FIGURE 3.5** Heat map of inter-sample correlations. Pearson correlation coefficients ($r$) were computed between each pair among 99 samples, and the results are displayed in a matrix format. The majority of samples are correlated with one another at the level of $r \cong 0.9$. In contrast, samples 88 and 96 are very poorly correlated ($r < 0.6$) with all other samples except one another. Review of the data revealed that these samples were abnormally dim, and they were subsequently flagged as QC outliers.

## ACKNOWLEDGMENTS

The *C. elegans* data described were generated in a collaboration with Craig Hunter (Department of Molecular and Cellular Biology, Harvard University).

## REFERENCES

1. Schena, M. et al., Quantitative monitoring of gene expression patterns with a complementary DNA microarray, *Science*, 270(5235), 467–470, 1995.
2. Lockhart, D.J. et al., Expression monitoring by hybridization to high-density oligonucleotide arrays, *Nat. Biotechnol.*, 14(13), 1675–1680, 1996.
3. Brazma, A. et al., Minimum information about a microarray experiment (MIAME)-toward standards for microarray data, *Nat. Genetics*, 29(4), 365–371, 2001.
4. Byrne, M.C., Whitley, M.Z., and Follettie, M.T., Preparation of mRNA for expression monitoring, in *Current Prototcols in Molecular Biology*, Ausubel, F.M., Brent, R., Kingston, R.E., Moore, D.D., Seidman, J.G., and Struhl, K., Eds., John Wiley & Sons, New York, pp. 22.2.1–22.2.13.
5. Baugh, L.R. et al., Quantitative analysis of mRNA amplification by *in vitro* transcription, *Nucleic Acids Res.*, 29(5), e29, 2001.
6. Chudin, E. et al., Assessment of the relationship between signal intensities and transcript concentration for Affymetrix GeneChip arrays, *Genome Biol.*, 3(1), research0005.1–0005.10, 2001.
7. Ramdas, L. et al., Sources of nonlinearity in cDNA microarray expression measurements, *Genome Biol.*, 2(11), research0047.1–0047.7, 2001.
8. Affymetrix, *Affymetrix GeneChip Expression Analysis Technical Manual*, Santa Clara, CA, 2000.
9. Eickhoff, B. et al., Normalization of array hybridization experiments in differential gene expression analysis, *Nucleic Acids Res.*, 27(22), e33, 1999.
10. Hill, A.A. et al., Evaluation of normalization procedures for oligonucleotide array data based on spiked cRNA controls, *Genome Biol.*, 2(12), research0055.1–0055.13, 2001.
11. Wang, X., Ghosh, S., and Guo, S.-W., Quantitative quality control in microarray image processing and data acquisition, *Nucleic Acids Res.*, 29(15), e75, 2001.
12. Li, C. and Wong, W.H., Model-based analysis of oligonucleotide arrays: expression index computation and outlier detection, *Proc. Natl. Acad. Sci. USA*, 98(1), 31–36, 2001.
13. Schadt, E.E. et al., Analyzing high-density oligonucleotide gene expression array data, *J. Cell. Biochem.*, 80, 192–202, 2000.
14. Irizarry, R.A. et al., Exploration, normalization, and summaries of high density oligonucleotide array probe level data, *Biostatistics*, (accepted for publication).
15. Bolstad, B., Probe level quantile normalization of high density oligonucleotide array data, http://stat-www.berkeley.edu/~bolstad/stuff/qnorm.pdf.
16. Dudoit, S. et al., Statistical methods for identifying differentially expressed genes in replicated cDNA microarray experiments, *Statistica Sinica*, 12, 111–140, 2000.

# 4 An Introduction to Cluster Analysis

*Frederick Immermann and Youping Huang*

CONTENTS



0-8493-1334-1/03/$0.00+$1.50

## 4.1 INTRODUCTION

### 4.1.1 Goals and Uses of Cluster Analysis

Cluster analysis is the study of partitioning objects into homogeneous groups. Suppose there is a collection of objects, each of which is described by a set of variables. Cluster analysis seeks a solution to the problem of classifying the objects into groups in such a way that:

- An object belongs to one group only (although this constraint is relaxed in some types of cluster analysis).
- Objects in the same group, or cluster, resemble each other.
- Objects in different clusters are dissimilar.

In some applications, the objective of a cluster analysis is to discover "natural" groupings of the objects that reflect evolutionary or functional relationships among the objects; some of the cluster analyses done in toxicogenomics research have this objective. In other applications, the objective of a cluster analysis may be merely to divide the objects into a manageable number of groups that are convenient for some purpose.

Unlike many of the statistical procedures used in pharmaceutical research, cluster analysis methods are mostly used in the exploratory phase of research, with no *a priori* hypotheses to test. Cluster analysis is intended largely for generating hypotheses rather than testing hypotheses. This explains the general absence in cluster analysis of the *p*-values that are associated with significance tests in many other areas of statistics.

### 4.1.2 Class Discovery vs. Class Prediction

Before proceeding, we want to distinguish between two different classification goals. In *class discovery*, the objects under study are not associated with any preexisting classification. There is no *a priori* knowledge about the number or composition of the classes. The data analyst seeks to divide the objects into groups and then obtain a deeper understanding about the objects on the basis of the grouping. This is the situation for which many standard clustering methods are useful. In the fields of artificial intelligence and machine learning, cluster analysis is referred to as *unsupervised learning*.

In *class prediction*, a classification is associated *a priori* with at least some of the objects on which measurements can be made. The goal in class prediction is to develop rules to assign an object to a class based on the values of its variables and then to assess the performance of the rules when they are applied to a new set of objects. Regression and discriminant analysis are statistical methods used to address this goal. In artificial intelligence, the goal is addressed by methods referred to as *supervised learning*.

### 4.1.3 General Strategy for Cluster Analysis

The following series of steps are typical for any cluster analysis.

- Identify the set of objects to be clustered; this often includes identifying some objects to omit from the cluster analysis, based on the values of their variables.

- Make adjustments to the data to accommodate differences in measurement scales or in the distributions of values.
- Select the measure to use to quantify the similarity between objects.
- Choose the clustering method(s) most appropriate for your purposes.
- Apply the cluster method(s).
- Visualize the results.

At this point, the scientist or analyst must decide if the results make sense and are potentially useful, or if another round of analysis using a different similarity measure or cluster method is needed.

Unfortunately, no widely agreed upon tools or tests are available that one can rely on to indicate that any particular clustering is the best one possible. For this reason, a good strategy is to cluster the objects using a variety of methods and/or similarity measures. If several different methods produce similar solutions, the analyst should have more confidence that the results reflect a true cluster structure in the data. Widely different solutions might be taken as evidence against any clear-cut cluster structure.

### 4.1.4 Scope of this Chapter

This chapter focuses on methods and techniques used by the gene expression profiling community. Gene expression assays yield a quantitative measure of expression for each gene. Measurements are made on several samples, with each sample (or, preferably, set of replicate samples) representing a different "treatment" (e.g., tumor type, the amount of time since exposure to some environment, vaccine type, etc.). We therefore focus on clustering methods that are appropriate when all data are numeric and all measurements are in the same units. The chapter focus is further restricted to methods concerned with class discovery, rather than those directed at class prediction. A number of books, including those by Massart and Kaufman[1] and Everitt,[2] are available to provide the interested reader more detailed information on cluster analysis.

### 4.1.5 Software

Given the volume of data involved and the complexity of the computations, cluster analysis is not something done using a hand calculator and pad of paper. Fortunately, a variety of software packages provide the ability to perform one or more types of cluster analyses. Some of the software packages, such as Eisen's Cluster and Treeview programs (http://www.microarrays.org/software),[3] Genespring, and Spotfire DecisionSite for Functional Genomics, are specifically designed to work with gene expression data. Cluster analysis capabilities are also included in many general-purpose statistical applications, including SAS, JMP, S-Plus, SPSS, and MATLAB. Our intent in this chapter is not to recommend a particular software package, as the choice of software is usually governed by considerations other than just clustering abilities. Instead, we try to provide the background necessary to permit informed use of most packages that perform cluster analysis.

## 4.2 GENERAL CONSIDERATIONS

### 4.2.1 Observations and Variables

In some cluster analyses of gene expression data, the genes are considered the objects to be clustered, and samples are the variables. However, the computational machinery underlying most clustering methods will also work if the problem is transposed — that is, if samples are the objects to be clustered and genes are the variables. The two approaches address different questions about the experimental data. If the set of samples represents replicates within treatments as well as different treatments, clustering arrays may provide insight into the repeatability of the assay. If the samples represent different tumor types, clustering samples may provide a useful classification of tumors. Some software packages, such as Eisen's Cluster program, GeneSpring, and Spotfire DecisionSite for Functional Genomics, make it easy to cluster the data in either or both directions. In other packages, such as SAS and JMP, the format of the input data file determines whether genes or samples will be regarded as the objects to be clustered.

### 4.2.2 Preliminary Treatment of the Data

What comes out of a cluster analysis depends on what goes in! Add or delete some objects, or change the scaling of the variables, and the resultant clustering will be altered, sometimes drastically. This is the reason for the first two steps in our general strategy for cluster analysis: data filtering and data adjustment.

#### 4.2.2.1 Data Filtering

The main goal in this first step of the analysis is to identify the subset of genes in which there is some variation in expression level across arrays. In most experiments, many genes are not expressed in any of the samples. Keeping those genes with little or no variation in expression can substantially affect the clustering results. It also considerably increases computation time and clutters any visual display of the clustering results. For these reasons, before doing anything else, it is important to identify and eliminate those genes that have little variation in expression levels. Standard filters used for this purpose include the following:

- Is the gene expressed (present) in at least one sample? This filter is often used for data from Affymetrix chips, because Affymetrix's GeneChip software makes an explicit "present/absent" call for each gene.
- Is the expression level for the gene greater than some minimal value in at least one sample? Often, the minimal value is chosen to reflect the lowest level of expression believed to be reliably measured by the assay.
- Is there evidence that expression levels for a gene differed among samples? This can be addressed by looking at:
  - The ratio of the maximum value to the minimum value for a gene. A rule-of-thumb may be that you are not interested in anything less than a twofold difference between treatments. If the maximum value observed for a gene is not twice as large as the minimum, then it is not worth looking further at this gene.
  - The difference between the maximum value and minimum value for a gene. This is to catch genes with low expression levels that may pass a minimum-fold-difference filter but do not have a high enough expression level in any sample to be interesting (e.g., a minimum expression level of 6 and a maximum of 15).
  - The standard deviation of the values for a gene. Using a statistical summary measure of variability rather than a rule-of-thumb criterion may be useful, but only if the variance of the expression measurement is known not to be related to the mean. This is often not the case unless the data have first been log-transformed. Using this type of filter also relies on an underlying assumption that measurement variability is the same for all genes. This assumption has not been well tested to date.

Sometimes it is preferable to filter the data prior to sending it to the software used for cluster analysis, perhaps in Excel or using a SQL or Brio query tool. In other cases, the cluster analysis software itself provides easy-to-use filtering capabilities.

#### 4.2.2.2 Data Adjustments

After identifying the subset of genes that have suitable variation in expression levels, the analyst needs to decide whether the variables (i.e., expression levels) should be transformed, rescaled, standardized, or normalized in some way. Note that definitions of these terms, especially *standardized* and *normalized*, are not universally agreed upon. The same term can be used for slightly or substantially different concepts (see below) in different software packages or by different authors. When using a new piece of software, it pays to read the fine print (if it exists).

For most gene expression data, the first thing often done at this point is to log-transform the data. This is true both for fluorescent intensity ratios from cDNA arrays and for expression measures (e.g., average difference or signal) from Affymetrix oligonucleotide arrays. The reason is the same in both situations: fold-change differences in expression are of interest, rather than simple arithmetic differences. Generally, when looking at fold-change differences, statistical analyses are best done on data expressed on a log scale.

*Example 1*

Suppose we have run a time-profile experiment with samples collected 0, 3, 6, and 9 hours after exposure to some stimulus. Samples were assayed using Affymetrix chips, which yielded the expression levels shown in Table 4.1 and Figure 4.1 for three genes, arbitrarily labeled A, B, and C. If we transform the expression levels to the $\log_{10}$ scale, we get the values shown in Table 4.2 and Figure 4.2. Note that Figures 4.1 and 4.2 provide rather different visual impressions of what is going on in these data.

**TABLE 4.1**
**Mean Expression Levels for Three Genes in Example 1**

| | Time from Exposure (hours) | | | |
|---|---|---|---|---|
| Gene | 0 | 3 | 6 | 9 |
| A | 500 | 1200 | 1500 | 1600 |
| B | 150 | 160 | 150 | 80 |
| C | 50 | 170 | 185 | 195 |

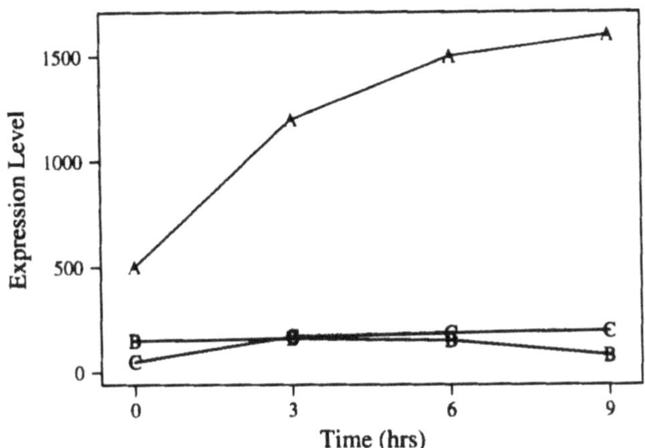


FIGURE 4.1 Expression level vs. time for genes A, B, and C in Example 1.

**TABLE 4.2**
**Mean $\log_{10}$ Expression Levels for Three Genes in Example 1**

| | Time from Exposure (hours) | | | |
|---|---|---|---|---|
| Gene | 0 | 3 | 6 | 9 |
| A | 2.70 | 3.08 | 3.18 | 3.20 |
| B | 2.18 | 2.20 | 2.18 | 1.90 |
| C | 1.70 | 2.23 | 2.27 | 2.29 |

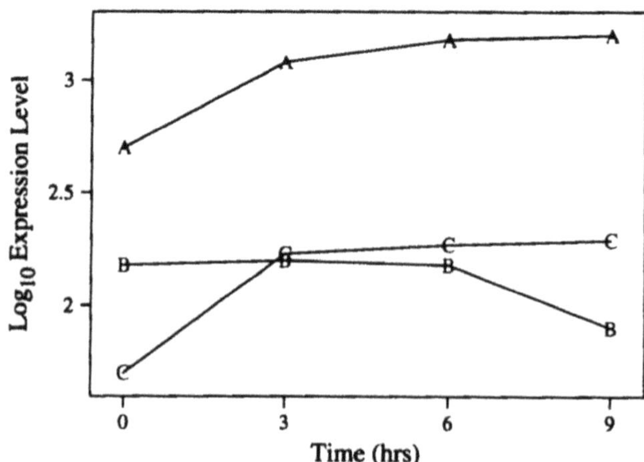


**FIGURE 4.2** $\text{Log}_{10}$-transformed expression level vs. time for genes A, B, and C in Example 1.

#### *4.2.2.2.1 Adjustments for Sample-to-Sample Variation*

Another thing the analyst might need to do is adjust for differences among samples that reflect sample-to-sample variation that is independent of the experimental treatments. This is an area in which considerable research is ongoing,[4–7] but as yet no single best approach to the problem has been widely agreed upon. A few of the simpler approaches that have been used involve adjusting for differences in:

- *Background levels of expression, as measured by negative controls.* This adjustment is made by subtracting the background level for each sample from all other measurements on that sample.
- *Levels of positive controls or housekeeping genes.* Adjustment involves expressing all measurements as a multiple of the positive control if used with raw data, or as the difference from the positive control if used with log-transformed data.
- *Overall levels due to different concentrations of bulk DNA or some other factor.* Adjustment involves expressing all measurements as a multiple of (for raw data) or difference from (for log-transformed data) the overall mean, trimmed mean, or median for the sample.

Before making any of these adjustments, it is important to understand whether they should be used with raw or log-transformed data. Which sample-to-sample adjustments must be done depends on the type of experiment run and whether any adjustments have already been done by preliminary data processing software.

**TABLE 4.3**
**Mean-Centered $\log_{10}$ Expression Levels for Three Genes in Example 1**

| Gene | Time from Exposure (hours) | | | |
|---|---|---|---|---|
| | 0 | 3 | 6 | 9 |
| A | −0.341 | 0.040 | 0.137 | 0.165 |
| B | 0.061 | 0.089 | 0.061 | −0.212 |
| C | −0.423 | 0.109 | 0.146 | 0.168 |

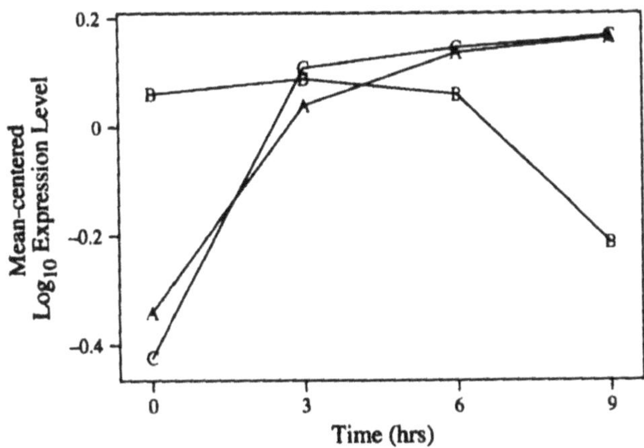


FIGURE 4.3 Mean-centered $\log_{10}$ expression level vs. time for genes A, B, and C in Example 1.

Some software packages (e.g., GeneSpring and Spotfire DecisionSite for Functional Genomics) refer to these types of adjustments as *normalizations*. Other packages (e.g., Eisen's Cluster program) refer to such adjustments as *centering* the data.

#### *4.2.2.2.2 Adjustments for Gene-to-Gene Variation*

The analyst may also want to adjust for differences among genes in their general level of expression. This approach is used when the only interest is in comparing the shapes of the gene profiles across the treatments, without taking into account differences among genes in expression levels. Some software packages (e.g., Genespring, Spotfire DecisionSite for Functional Genomics, Eisen's Cluster program) provide options for these types of adjustments.

*Example 1 (cont.)*

Suppose we just want to compare the shapes of the profiles for the three genes. For each gene, if we subtract the average $\log_{10}$ expression level from the $\log_{10}$ expression level for each time point, we get the values shown in Table 4.3 and Figure 4.3. Figure 4.3 makes it clear just how similar the changes in expression over time were in genes A and C.

#### *4.2.2.2.3 Standardizing the Variance*

One other adjustment the analyst may want to make is to standardize the variance. The reason is that variables or objects with large variances tend to have more effect on the resulting clusters than do those with small variances. The data may have to be transformed so that equal differences are of equal practical importance. This is especially true if variables are measured in different units. Fortunately, when working with gene chip data, all measurements are in the same units. For numeric variables, the most popular variance-equalizing transformation is to standardize each variable to

**TABLE 4.4**
**Z-Score $\log_{10}$ Expression Levels for Three Genes in Example 1**

| | Time from Exposure (hours) | | | |
|---|---|---|---|---|
| Gene | 0 | 3 | 6 | 9 |
| A | −1.460 | 0.170 | 0.585 | 0.705 |
| B | 0.432 | 0.630 | 0.432 | −1.493 |
| C | −1.494 | 0.385 | 0.514 | 0.595 |

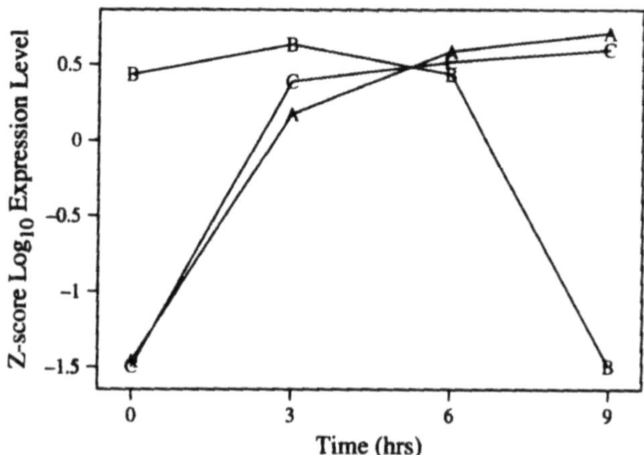


**FIGURE 4.4** Z-score for $\log_{10}$ expression level vs. time for genes A, B, and C in Example 1.

unit variance. Each variable is often standardized to a zero mean at the same time. This can be accomplished by creating a new variable $Z = (x - m)/s$ (sometimes called a Z-score) from the original variable $x$, where $m$ and $s$ are the sample mean and sample standard deviation of $x$, respectively. This type of adjustment is called *normalization* in Eisen's Cluster program and is included in Spotfire DecisionSite for Functional Genomics as a normalization option.

*Example 1 (cont.)*

Now we want to adjust the profiles of the three genes so that they have the same overall level of variability. For each gene we compute the Z-score for each $\log_{10}$ expression level and get the values shown in Table 4.4 and Figure 4.4. In Figure 4.4, the profiles for genes A and C are still very similar, but the changes in the profile for B are now more pronounced than they were in the previous plot of the mean-centered data (Figure 4.3).

Standardization of the variances may not always be desired or needed. Standardized variances may have the disadvantage of overemphasizing genes that have minor changes in expression levels, thereby diluting differences among genes for the characteristics that are the best discriminators. In subsequent steps in the analysis, the choice of similarity measure (discussed later) may make prior variance standardization unnecessary, because some measures of similarity automatically adjust for difference in variances. Any nonlinear transformations of the variables, such as the Z-score transformation, may change the clustering and should be approached with caution.

### 4.2.3 Measures of Similarity

Once the data are filtered and adjusted, the next question to be answered in a cluster analysis is what measure will be used to quantify the similarity of any two objects. If all variables are numeric, the similarity of two objects is generally quantified using either a distance or a correlation.

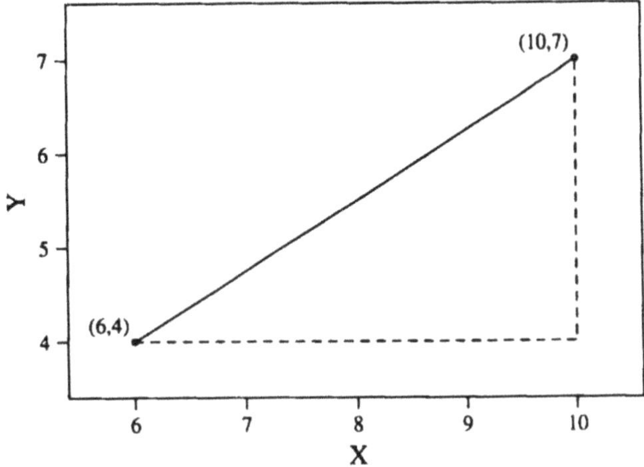


FIGURE 4.5 Distance between two points in two-dimensional space.

#### 4.2.3.1 Distances

Suppose each object to be clustered has measurements for $d$ variables. Each object can be thought of as a point (or vector) in $d$-dimensional space. The similarity of two objects is then defined by how close they are in this space. The most commonly used distance measure is the *Euclidean distance*. Many people first learned about a simple version of Euclidean distance in high-school geometry when they were taught about the Pythagorean theorem. This theorem says that the length of the hypotenuse of a right triangle is the square root of the sum of the squared lengths of the other two sides. For example, suppose there are two points in a two-dimensional space, one at (6,4) and the other at (10,7). How far apart are they (Figure 4.5.)?

In Figure 4.5, the bottom of the triangle is $10 - 6 = 4$ units long, and the side of the triangle is $7 - 4 = 3$ units long. So, the hypotenuse, which is the straight-line distance between the two points, is:

$$d = \sqrt{4^2 + 3^2} = 5$$

The Euclidean distance is just the generalization of this approach for $d$-dimensions. The Euclidean distance between points $x = (x_1, x_2, \ldots, x_d)$ and $y = (y_1, y_2, \ldots, y_d)$ is therefore expressed as:

$$\sqrt{\sum_{j=1}^{d} (x_j - y_j)^2}\,.$$

Euclidean distances are always $\geq 0$. The more similar two objects are, the smaller will be the Euclidean distance between them.

Another distance that is used occasionally is the "city block" or "Manhattan" distance. In the two-dimensional case, this is the distance you would have to travel over a rectangular grid (such as the streets and avenues of mid-town Manhattan) to get from one point to the other. In Figure 4.5, the Manhattan distance between (10,7) and (6,4) is $4 + 3 = 7$. In the $d$-dimensional case, the city block distance has the form:

$$\sum_{j=1}^{d} \left| x_j - y_j \right|.$$

The results of clustering using city block distances are usually similar to those for Euclidean distances. In this chapter, $D(x, y)$ or $\|x - y\|$ will be used to denote distance between objects $x$ and $y$.

### 4.2.3.2 Correlations

An alternative type of measure for quantifying similarity is a correlation coefficient. Several types of correlation coefficients are usually offered in the software packages developed specifically for expression profiling. The naming conventions for the different types of correlation coefficients are often not consistent across software packages. The most common types of correlation coefficients are described below.

#### *4.2.3.2.1 Pearson's Product Moment Correlation*

This is the standard correlation coefficient of statistics undergraduate textbooks. It ranges from 1, which indicates a perfect correlation between two objects, to −1 for two objects that are perfectly inversely correlated. A correlation coefficient of 0 indicates a complete lack of correlation between the objects. The formula for Pearson's correlation coefficient between two objects $X = \{x_1, x_2, \ldots, x_d\}$ and $Y = \{y_1, y_2, \ldots, y_d\}$ is:

$$r = \frac{1}{d}\sum_{j=1}^{d}\left(\frac{x_i - \overline{X}}{\sigma_x}\right)\left(\frac{Y_i - \overline{Y}}{\sigma_y}\right),$$

where $\overline{X}$ is the mean of the $X$ values, and $\sigma_x$ is the standard deviation.

Expression profiles for two genes that have identical "shapes" across samples will have a Pearson correlation of 1. This is so even if the profiles have different overall magnitudes of variation, because Pearson's correlation adjusts for differences in the variance. Therefore, when the clustering is based on Pearson's correlation, the data do not have to be adjusted to standardize the variance prior to clustering. Pearson's correlation is referred to as *centered correlation* in Eisen's Cluster program and as *correlation* in the Spotfire DecisionSite for Functional Genomics package.

#### *4.2.3.2.2 Absolute Correlation*

The absolute correlation is the absolute value of the correlation coefficient, and therefore ranges from 0 to 1. Objects with perfectly opposite profile shapes have an absolute correlation of 1, as do objects with identical profile shapes. Use absolute correlation to cluster genes with expression levels that go down when the expression levels of other genes go up.

#### *4.2.3.2.3 Uncentered Correlation*

Uncentered correlation coefficients are similar to Pearson coefficients, except that terms in the numerator and denominator are not centered on their respective means. The formula for the uncentered coefficient is

$$r = \frac{\frac{1}{d}\sum_{i=1}^{d} X_i Y_i}{\sqrt{\frac{1}{d}\sum_{i=1}^{d} X_i^2} \cdot \sqrt{\frac{1}{d}\sum_{i=1}^{d} Y_i^2}}.$$

Two objects with identically shaped profiles that are offset relative to each other will have a Pearson coefficient of 1, but an uncentered correlation of less than 1. Uncentered correlation is called *cosine correlation* in the Spotfire DecisionSite for Functional Genomics package and *standard correlation* in Genespring.

**TABLE 4.5**
**Euclidean Distance Matrix for $\log_{10}$ Expression Levels for Three Genes in Example 1**

| Gene | A | B | C |
|---|---|---|---|
| A | 0 | 1.932 | 1.839 |
| B | 1.932 | 0 | 0.622 |
| C | 1.839 | 0.622 | 0 |

**TABLE 4.6**
**Pearson's Correlation Matrix for $\log_{10}$ Expression Levels for Three Genes in Example 1**

| Gene | A | B | C |
|---|---|---|---|
| A | 0 | −0.441 | 0.989 |
| B | −0.441 | 0 | −0.357 |
| C | 0.989 | −0.357 | 0 |

*4.2.3.2.4 Spearman's Rank Correlation*

Spearman's rank correlation is a nonparametric counterpart to Pearson's correlation coefficient. The formula is similar, except that the actual data values are replaced by the ranks of each object's values. This has the effect of downweighting the influence of unusual or outlying values on the correlation.

#### 4.2.3.3 Comparison of Similarity Measures

The choice of similarity measure can have a substantial impact on the results of a cluster analysis. So which one should you use? The answer depends in part on what features of the data you consider important for assessing similarity. Examining similarity measures for the three genes in Example 1 may be enlightening.

*Example 1 (cont.)*

Let's compare distances and correlations obtained from the expression data for three genes measured at four timepoints. We start by computing distances and correlations for the $\log_{10}$-transformed expression levels shown in Figure 4.2 (Tables 4.5 and 4.6). Euclidean distances based on the $\log_{10}$-transformed expression levels indicate that genes B and C are much closer to each other than are genes A and B or genes A and C. However, if we use Pearson's correlation to measure similarity, genes A and C are very similar (the correlation coefficient almost equals 1), genes A and B are negatively correlated, and genes B and C are negatively correlated. But, what if we compute the same measures for mean-centered $\log_{10}$ frequencies instead (Figure 4.3 and Tables 4.7 and 4.8)?

The conclusions based on Euclidean distances for the mean-centered $\log_{10}$ frequencies (Table 4.7) are very different from the previous conclusions based on uncentered data (Table 4.5). Now genes A and C are closest to each other, and genes B and C are about as far apart from each other as are genes A and B. Notice that the Pearson's correlations (Table 4.8) are unchanged from those computed earlier on uncentered data (Table 4.6). For the centered data, conclusions about similarities between genes are qualitatively the same whether Euclidean distance or Pearson correlations are used.

**TABLE 4.7**
**Euclidean Distance Matrix for Mean-Centered $\log_{10}$ Expression Levels for Three Genes in Example 1**

| Gene | A | B | C |
|---|---|---|---|
| A | 0 | 0.558 | 0.108 |
| B | 0.558 | 0 | 0.621 |
| C | 0.108 | 0.621 | 0 |

**TABLE 4.8**
**Pearson's Correlation Matrix for Mean-Centered $\log_{10}$ Expression Levels for Three Genes in Example 1**

| Gene | A | B | C |
|---|---|---|---|
| A | 0 | –0.441 | 0.989 |
| B | –0.441 | 0 | –0.357 |
| C | 0.989 | –0.357 | 0 |

The bottom line is that the different measures of similarity assess different aspects of the data. If your primary interest is in patterns in the direction of changes in expression levels, regardless of the magnitude of expression, then a correlation measure is apt to be most appropriate. If the magnitude of the changes or the expression level is important, a distance measure may be more appropriate. As you choose which measure to use, remember that some measures can be greatly affected by prior adjustments to the data.

Massart and Kaufman[1] offer the generalization that, "Distances detect differences best and correlations are often better to find similarities" (p. 28).

### 4.2.4 Principal Component Analysis

Prior to applying any clustering method, some graphical representations of the data should be generated and examined. The types of graphs that are commonly used include histograms for single variables and scatterplots for pairs of variables. A useful technique for visualizing data in a cluster analysis application is principal component analysis (PCA). Classical principal component analysis is commonly employed to obtain a low-dimensional mapping of the data. Often, the data can be reduced from $d$-dimensions to a representation that in a small number of dimensions (e.g., four to six) manages to capture most of the variability in the original data. Scatterplots in these new dimensions often provide some insight into groupings inherent in the data; however, principal component analysis does not guarantee that the best variables for clustering will be identified.

Principal component analysis is a technique that finds a rotation of the original variables about the center of the data such that the variances of the new variables are maximized. The new variables are all linear combinations of the original variables, constrained to each have unit length and represent jointly perpendicular directions. That is, suppose we have $d$ variables, $v_1, v_2, \ldots, v_d$. Principal component analysis transforms these into a new set of $d$ variables, $pc_1, pc_2, \ldots, pc_d$, with the following properties:

- Each $pc_i$ is a linear combination of the original variables,

$$pc_i = a_{i1}\left(v_1 - \bar{v}_1\right) + a_{i2}\left(v_2 - \bar{v}_2\right) + \cdots + a_{id}\left(v_d - \bar{v}_d\right).$$

- The sum of squares of the $a_{ij}$ ($j$ = 1 to $d$) is 1.
- The first transformed variable, $pc_1$, has the greatest variance.
- The second transformed variable, $pc_2$, is uncorrelated with $pc_1$ and has the next greatest variance; this pattern repeats with $pc_3$, then $pc_4$, etc.

The derived variables $pc_i$ are called *principal components* or *principal component scores*. The coefficients $a_{ij}$ that define the new variables are called *loadings*. The principal components are sorted in decreasing order of the variance they account for. When the first few account for a large proportion of the variance in the original variables, they may be used as a low-dimensional summary of the original data.

Let's look at an example to see how graphics and principal component analysis can help us understand the structure of a real gene expression dataset.

*Example 2*

An experiment was run to examine expression profiles in a viral G-protein system using Affymetrix GeneChips. Duplicate chips were run on samples collected 0, 0.5, 4, and 24 hours after cells in culture were exposed to the G protein. Filtering out genes that were absent on all eight arrays or had expression levels of less than 15 ppm on all eight arrays left 3044 genes. These were further subset, using analysis of variance (ANOVA) and a 2.5 fold-difference requirement, to 392 genes felt to be worthy of further consideration. The expression levels were $\log_{10}$ transformed, and the means were computed for each gene for each time point.

Histograms and scatterplots of the mean $\log_{10}$-transformed expression levels are displayed in Figure 4.6. It is obvious from the scatterplots in Figure 4.6 that responses at 0 hour and 0.5 hour are highly correlated.

A principal component analysis was performed on the $\log_{10}$-transformed expression levels. The loadings of principal components are presented in Table 4.9.

The first two principal components can be expressed as:

$$\text{PC1} = 0.519(x_0 - \bar{x}_0) + 0.561(x_{0.5} - \bar{x}_{0.5}) + 0.511(x_4 - \bar{x}_4) + 0.394(x_{24} - \bar{x}_{24})$$

$$\text{PC2} = -0.501(x_0 - \bar{x}_0) - 0.358(x_{0.5} - \bar{x}_{0.5}) + 0.362(x_4 - \bar{x}_4) + 0.700(x_{24} - \bar{x}_{24}).$$

The four loadings for the first principal component are similar in magnitude and direction, indicating that this component essentially represents the overall or mean response. The second component loadings suggest that this component reflects the difference between the first two time points (i.e., 0 and 0.5 hours) and last two time points (i.e., 4 and 24 hours). In this example, the first two principal components account for 90% of the variance in the original data; therefore, these two new variables include much of the information contained in the dataset.

The first two principal components, PC1 and PC2, are plotted in Figure 4.7 for the 392 genes. Figure 4.7 clearly shows two groupings, one lying above and the other lying below a horizontal line through PC2 = 0.

## 4.3 HIERARCHICAL METHODS

### 4.3.1 Overview

#### 4.3.1.1 Goal of Hierarchical Methods

In a hierarchical or nested clustering, the objects are grouped using a series of successive steps. The groupings are such that one cluster can be entirely contained within another cluster, but no other kind of overlap between clusters is permitted. There are two techniques for making the

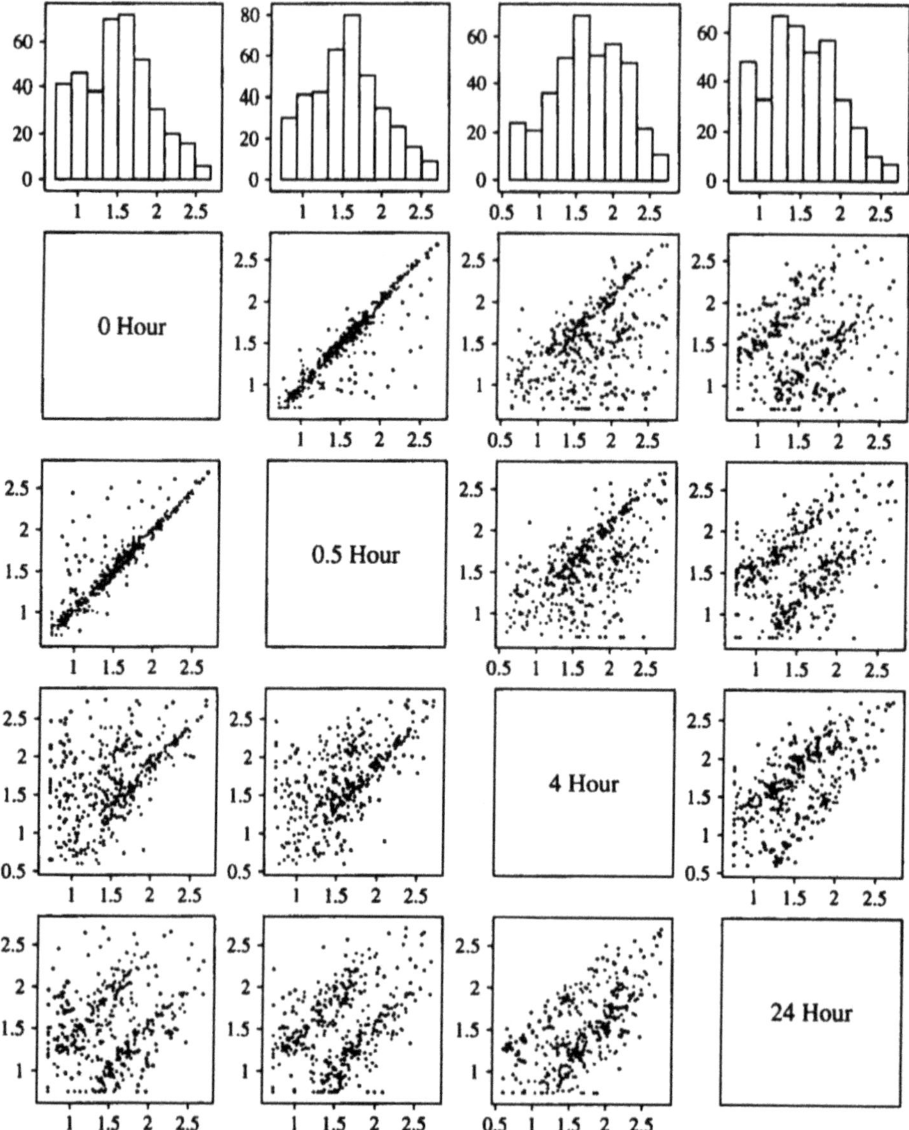


FIGURE 4.6 Scatterplot matrix and histograms of data for 392 genes in Example 2; separate histograms for each time point and scatterplots for each pairwise combination of time points.

**TABLE 4.9**
**Principal Component Loadings for Data from Example 2**

| Hour | PC1 | PC2 | PC3 | PC4 |
|---|---|---|---|---|
| 0 | 0.519 | –0.501 | 0.175 | 0.671 |
| 0.5 | 0.561 | –0.358 | 0.129 | –0.735 |
| 4 | 0.511 | 0.362 | –0.776 | 0.077 |
| 24 | 0.394 | 0.700 | 0.593 | 0.063 |

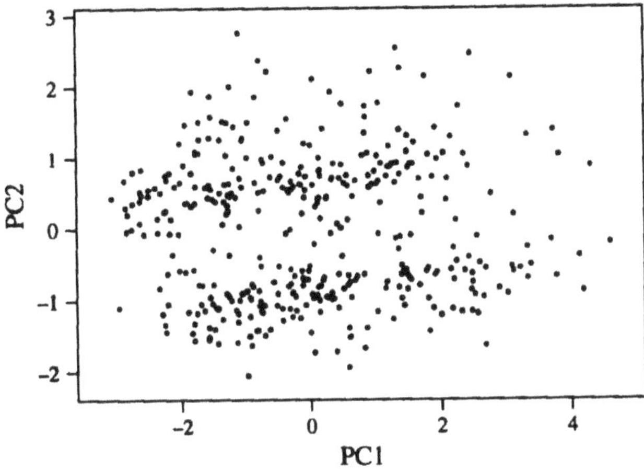


FIGURE 4.7 Scatterplot of principal component 1 vs. principal component 2 for the 392 genes in Example 2.

groupings: agglomerative methods and divisive methods. An agglomerative method proceeds by a series of successive *fusions* of objects, until at the final step all objects are in a single cluster. A divisive method operates in the opposite direction, by making a series of successive *splits* starting with the entire set of objects. Because divisive methods are computationally intensive and therefore far less popular, we will not discuss them any further.

#### 4.3.1.2 Simple Example

To get an idea of how an agglomerative clustering method works, let's revisit Example 1, with data for two additional genes to make it more interesting.

*Example 1 (cont.)*

We will work with the mean-centered, $\log_{10}$-tranformed expression levels for the three original genes in this example, plus two new genes, D and E (Table 4.10 and Figure 4.8).

For this example, we cluster using Euclidean distance as the similarity measure. The distances for the five genes are listed in Table 4.11 (only the top half of the similarity matrix is shown, as the matrix is symmetrical).

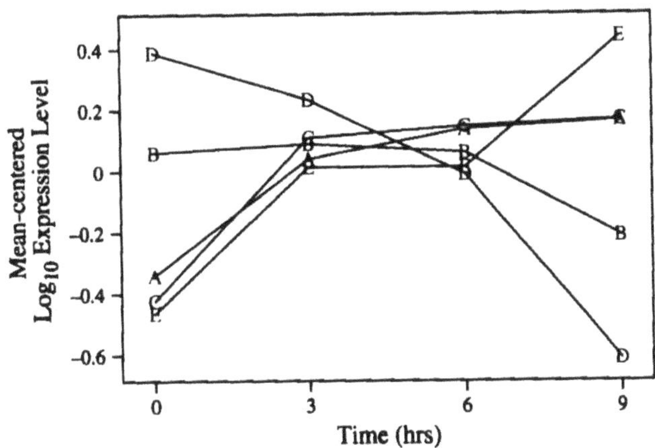


FIGURE 4.8 Mean-centered $\log_{10}$ expression level vs. time for genes A, B, C, D, and E in Example 1.

**TABLE 4.10**
**Mean-Centered $\log_{10}$ Expression Levels for Five Genes in Example 1**

| Gene | Time from Exposure (hours) | | | |
|---|---|---|---|---|
| | 0 | 3 | 6 | 9 |
| A | −0.341 | 0.040 | 0.137 | 0.165 |
| B | 0.061 | 0.089 | 0.061 | −0.212 |
| C | −0.423 | 0.109 | 0.146 | 0.168 |
| D | 0.388 | 0.233 | −0.010 | −0.612 |
| E | −0.464 | 0.013 | 0.013 | 0.439 |

**TABLE 4.11**
**Euclidean Distance Matrix for Mean-Centered $\log_{10}$ Expression Levels (Each Gene in Example 1 in a Separate Cluster)**

| Gene | A | B | C | D | E |
|---|---|---|---|---|---|
| A | — | 0.558 | 0.108 | 1.092 | 0.326 |
| B | — | — | 0.621 | 0.541 | 0.841 |
| C | — | — | — | 1.143 | 0.319 |
| D | — | — | — | — | 1.371 |
| E | — | — | — | — | — |

**TABLE 4.12**
**Euclidean Distance Matrix for Mean-Centered $\log_{10}$ Expression Levels for Four Clusters**

| Gene | A + C | B | D | E |
|---|---|---|---|---|
| A + C | — | 0.589 | 1.117 | 0.322 |
| B | — | — | 0.541 | 0.841 |
| D | — | — | — | 1.371 |
| E | — | — | — | — |

The smallest distance is between genes A and C (0.108), so they form the first cluster. We join them and compute the distance from our new A + C cluster to each of the remaining genes. We can define the distance from the new cluster to each of the remaining genes in a number of ways. For this discussion, we will use the average of the distances of A and C to each gene. For example, the distance between A and D is 1.092, and the distance between C and D is 1.143, so the distance between A + C and D is (1.092 + 1.143)/2 = 1.117. We form a new similarity matrix with one less row and one less column that contains the updated distances (Table 4.12):

We now repeat the process. The smallest distance in the new matrix is between A + C and E (0.322), so they form the new cluster at this step. The similarity matrix gets updated as before, yielding the matrix shown in Table 4.13. At the next iteration, B and D are combined and we recalculate the similarity matrix, obtaining the matrix shown in Table 4.14. We have finished our hierarchical, agglomerative clustering; now we need a way to display our results visually.

**TABLE 4.13**
**Euclidean Distance Matrix for Mean-Centered $\log_{10}$ Expression Levels for Three Clusters**

| Gene | A + C + E | B | D |
|---|---|---|---|
| A+C+E | — | 0.715 | 1.244 |
| B | — | — | 0.541 |
| D | — | — | — |

**TABLE 4.14**
**Euclidean Distance Matrix for Mean-Centered $\log_{10}$ Expression Levels for Two Clusters**

| Gene | A + C + E | B + D |
|---|---|---|
| A + C + E | — | 0.979 |
| B + D | — | — |

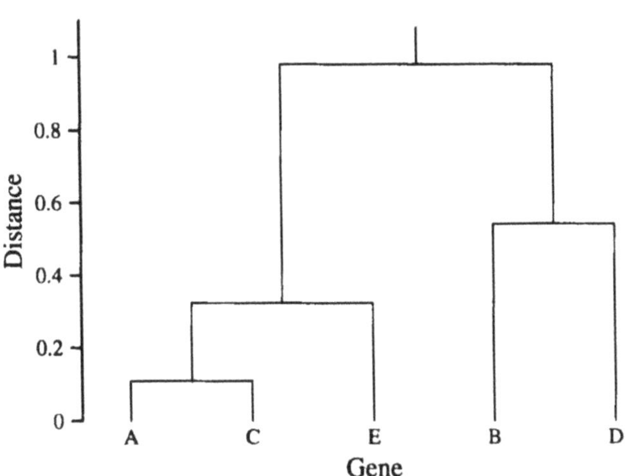


FIGURE 4.9 Dendrogram for genes A, B, C, D, and E in Example 1.

### 4.3.1.3 Dendrograms

A hierarchical clustering sequence is often represented by a two-dimensional tree-like diagram known as a *dendrogram*. All individual objects are arranged at one side of the diagram. Each step of fusion is linked until a single cluster is formed at the other side of the diagram.

*Example 1 (cont.)*

The dendrogram in Figure 4.9 is a representation of the clustering results summarized in the similarity matrices presented in Tables 4.11 through 4.14. When looking at the dendrogram, it is important to remember that the left-to-right ordering of genes within a cluster is arbitrary. That is, rather than a left-to-right ordering of A C E B D, the ordering could be C A E B D, or even D B C A E. The clustering is defined by the connections between the objects, rather than the terminal ordering of the objects.

Because we arranged the dendrogram as an upside-down tree, it is easy to see why an agglomerative hierarchical method is sometimes called a ***bottom-up*** approach. It starts with each object at the bottom and works upward by successively joining a pair of groups.

For hierarchies containing many objects, a horizontal rather than vertical arrangement of the dendrogram is often preferable. Many of the software packages that perform hierarchical clustering list the objects vertically along one side of the diagram or screen, rather than horizontally.

In the dendrogram in Figure 4.9, the location of each fusion is plotted on a scale of distance between clusters (or individuals). An alternative scale that is often used is the squared multiple correlation, $R^2$, which is the proportion of variance accounted for by the clusters. It can be calculated by the following formula:

$$R^2 = 1 - \frac{\sum_{j=1}^{g} \sum_{k \in C_j} \left\| x_k - \bar{x}_{C_j} \right\|^2}{\sum_{i=1}^{n} \left\| x_i - \bar{x} \right\|^2}.$$

In some fields, such as numerical taxonomy, the main areas of inquiry concern the historical or evolutionary relationships of the objects. In such fields, the entire dendrogram is an output of primary interest. However, in other fields, including gene expression profiling, hierarchical clustering is used primarily to partition the objects into a manageable number of groups, and the complete dendrogram itself is of lesser interest.

When the primary goal of the clustering is to partition the objects into groups, the dendrogram is used as an intermediate visual summarization. One reviews the dendrogram and works down the links, from the single cluster that contains all objects to the level at which the desired number of clusters are obtained. For the dendrogram in Figure 4.9, if you wanted to group the genes into two clusters, you would go from the top down to a distance of slightly less than 1. The two clusters would be A + C + E and B + D. If you wanted three clusters, you would go down a little further, to a distance of around 0.5, and get clusters A + C + E, B, and D.

Rather than using a predefined number of clusters, one could group using a predefined distance or $R^2$. A distance would be chosen that provides the desired degree of similarity among clustered objects. Alternatively, an $R^2$ would be selected that indicates that the clustering "explains" the desired fraction of the total variability.

The investigator wishing to have a solution with an "optimal" number of clusters will need to decide on a particular stage at which to stop. This is very similar to the case of determining the "best" number of clusters with optimization methods and is discussed in Section 4.4.1.

### 4.3.2 Agglomerative Methods

As we have seen, an agglomerative hierarchical clustering procedure produces a series of partitions of the data. The initial partition consists of $n$ single-member clusters. Then, two closest clusters are merged iteratively until a single cluster containing all $n$ objects is left. Fusions made at each step are irrevocable. When an agglomerative algorithm has joined two individuals, they cannot subsequently be separated. At the beginning of the algorithm, there are $n$ single-member clusters, and $R^2 = 1$. At the end, one cluster includes all individuals, and $R^2 = 0$. Differences between agglomerative methods arise because of the different ways of defining distance between an individual object and a group containing several objects, or between two groups of objects.

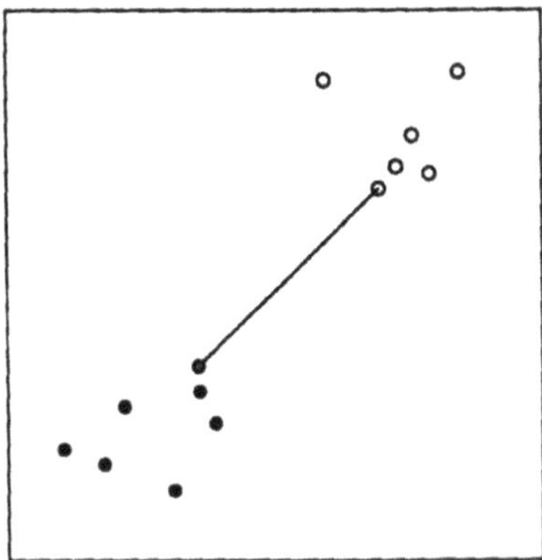

**FIGURE 4.10** Single linkage distance between two clusters.

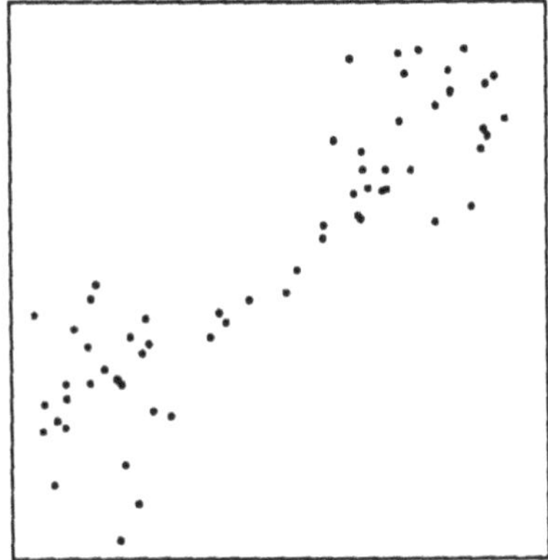

**FIGURE 4.11** Chaining of two clusters.

#### 4.3.2.1 Types of Hierarchical Methods

##### *4.3.2.1.1 Single-Linkage Clustering*

One of the simplest agglomerative hierarchical clustering methods is single linkage, also known as the nearest neighbor technique. It defines the distance between two clusters as the minimum distance between an object in one cluster and an object in the other cluster. Figure 4.10 shows an example of the single-linkage distance between two clusters. When a small number of objects lies between two relatively distinct clusters, a phenomenon known as *chaining* may cause single-linkage clustering (also median clustering) to fail to resolve the two clusters. The objects shown in Figure 4.11 would fail to resolve into two clusters using single-linkage clustering. Flynn[8] indicates that single-linkage clustering is particularly good for identifying "outliers" — those genes that have unusual profiles.

##### *4.3.2.1.2 Complete-Linkage Clustering*

The complete-linkage or furthest neighbor clustering is the opposite of single linkage. Its distance between two groups is the maximum distance between an object in one cluster and an object in the other cluster. Figure 4.12 shows an example of the complete linkage distance between two clusters. Complete linkage tends to produce clusters with roughly equal diameters and is also very sensitive to even moderate outliers.[9]

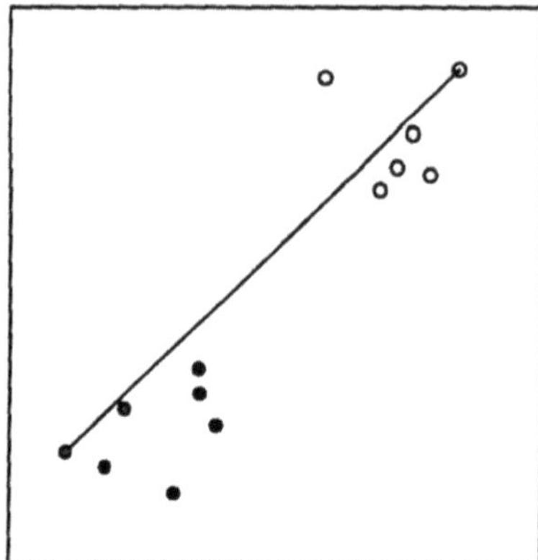

FIGURE 4.12 Complete linkage distance between two clusters.

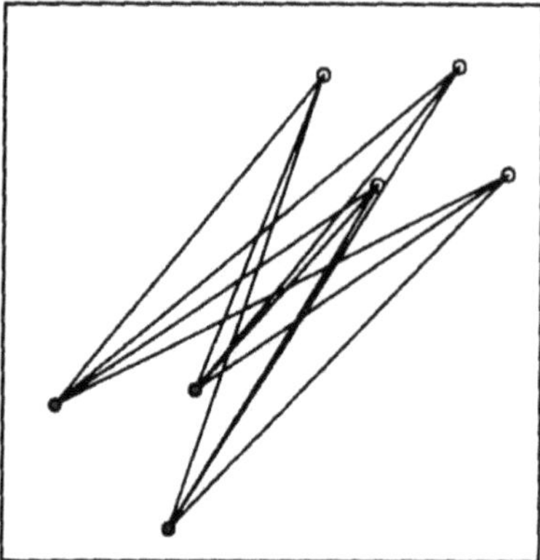

FIGURE 4.13 Components of average linkage distance between two clusters.

#### *4.3.2.1.3 Average-Linkage Clustering*

In average-linkage clustering, the distance between two clusters is defined as the average of the distances between all pairs of objects in which one of the pair is in one cluster, and the other of the pair is in the other cluster. An example showing the components of the average-linkage distance between two clusters in shown in Figure 4.13. Two types of average-linkage clustering are available: UPGMA (unweighted pair-group method with arithmetic mean) and WPGMA (weighted pair-group method with arithmetic mean). They differ in how they account for differences between clusters in the number of elements. WPGMA is the more commonly used variant. Average linkage tends to produce clusters with similar variances, and to join clusters with small variances.[9]

#### *4.3.2.1.4 Centroid Clustering*

With this method, groups once formed are represented by their mean values for each variable — that is, their centroids. The distance between two clusters is defined as the distance between their centroids. An example of the centroid distance between two clusters is shown in Figure 4.14. The centroid method is more robust to outliers than many other hierarchical methods.[9]

#### *4.3.2.1.5 Ward's Method*

The methods discussed so far are all based on joining items with the smallest distance between them. Ward's method is based on a different approach. In this method, at each step one considers all the clusters that might be formed from pairs of the existing clusters. The cluster with the lowest

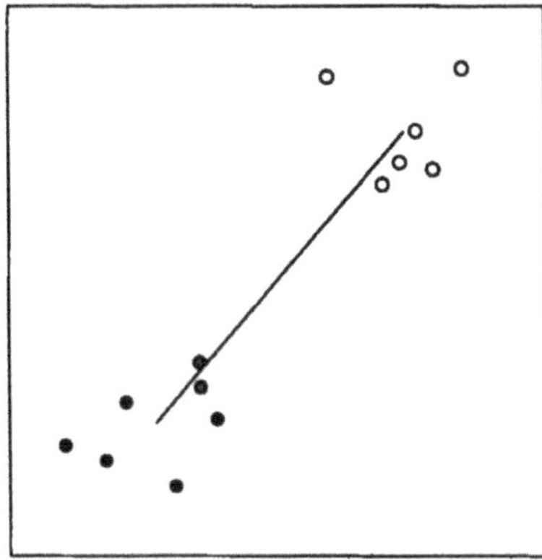

**FIGURE 4.14** Centroid linkage distance between two clusters.

increase in heterogeneity is retained for the next step. Heterogeneity is measured using the sum of the squared distances between each item in the cluster and the centroid of the cluster. Minimizing the increase in heterogeneity is equivalent to minimizing the loss of information associated with forming the new cluster. For cluster $Y_i$ we define its information loss as:

$$\sum_{x \in Y_i} \|x - y_i\|^2,$$

where $y_i$ is the mean vector of cluster $Y_i$. Ward's method seeks to merge two clusters at each step such that the minimum increase in information loss is achieved. Suppose $Y_i$ and $Y_j$ are two clusters, and $y_i$ and $y_j$ are their mean vectors, respectively. Ward's method is the same as defining distance:

$$D(Y_i, Y_j) = \frac{\|y_i - y_j\|^2}{\frac{1}{n_i} + \frac{1}{n_j}},$$

where $n_i$ and $n_j$ are the numbers of objects in $Y_i$ and $Y_j$, respectively. Ward's method tends to produce clusters with roughly the same number of objects and is very sensitive to outliers.[9]

#### *4.3.2.1.6 Two-Stage Density Linkage*

***Density linkage*** refers to a class of clustering methods based on nonparametric probability density estimates. In these methods, a dissimilarity measure, $d^*$, is computed between all pairs of objects that are considered adjacent by some criteria. The measure $d^*$ is based on a particular class of density estimates. The $d^*$s are then clustered using single linkage. ***Two-stage density linkage*** is a modification of this approach. The modification ensures that all objects are assigned to clusters with at least $m$ members, where the user specifies $m$. Density linkage and two-stage density linkage are similar to single linkage in that they can detect elongated or irregular clusters but perform better than single linkage with compact clusters.[9]

### **4.3.2.2 Comparisons of Procedures**

So, which procedure should you use? The answer depends in part on what you are trying to achieve in the clustering. As noted previously, if you are trying to identify genes that behave unusually, single-linkage clustering may work best. If you are trying to find the most "logical" clusters, Massart and Kaufman[1] report that average linkage or Ward's method typically yield the best results. However, for any particular dataset, it is worth trying several methods and then examining the results to determine which method provides the most logical, informative clusters.

*Example 2 (cont.)*

To get a sense for how the choice of methods affects the results of a cluster analysis, we clustered the 392 genes from the study described above in Example 2 based on $\log_{10}$-transformed expression levels, with the four time points as our variables. A dendrogram showing the results of clustering the 392 genes using Ward's method is shown in Figure 4.15. Note that the tree is horizontal, with the 392 genes on the right side, and that the scale for this dendrogram is in units of $R^2$, which is the proportion of the total variance.

How sensible is this clustering? Suppose we define clusters based on a cut at $R^2 = 0.385$; that is, we have two clusters that account for 38.5% of the variance. Now consider the scatterplot of the first two principal components, with the cluster assignment identified for each of the 392 genes, as shown in Figure 4.16.

Profile plots of $\log_{10}$-transformed expression levels vs. time for each of the genes in clusters 1 and 2 provide an alternative visualization of the clustering results (Figure 4.17).

Using Ward's method, cluster 1 includes genes with low overall responses (i.e., low values of PC1), and cluster 2 includes genes with high overall responses. But, is this of interest? Are these the two clusters we would define "by eye" in the original scatterplot of the first two principal components (Figure 4.7)?

In that original scatterplot, note that the shape of the two clusters is almost elliptical. Hierarchical clustering using two-stage density linkage is able to identify elongated or irregular clusters, and we get the results displayed in the Figure 4.18. Profile plots for the two clusters defined using two-stage density linkage are shown in Figure 4.19. Using two-stage density linkage, cluster 1 represents genes whose responses decrease over time (i.e., PC2 less than 0), and cluster 2 represents genes whose responses increase over time. This seems to be a more satisfying and informative clustering of the data than the clustering obtained using Ward's method.

## 4.4 OPTIMIZATION METHODS

The basic idea of optimization techniques is to define an index for partitioning a collection of objects into a fixed number of clusters. The value of the index is indicative of the quality of a particular clustering. Associating a number with each partition enables us to compare any two partitions. It quantifies the preferability of one partition over another partition.

Once a quality index has been selected, we need to find a partition that achieves an optimal value of the index. It may sound simple in theory because we have only a finite number of objects and a fixed number of clusters. But, as the number of objects grows, the number of possible partitions increases rapidly. The number of distinct partitions is more than 2 million with 15 objects and 3 clusters, more than 45 trillion with 20 objects and 4 clusters, and about $10^{68}$ with 100 objects and 5 clusters.

The very large number of possible partitions makes it prohibitive to perform a thorough search for the optimal value of the index, even with the help of rapid advancements in computer technology. Therefore, consideration needs to be given to how to choose a partition that leads to optimization.

The solution is found in the technique of local optimization. The basic idea is to define a neighborhood for each partition. Beginning with an initial partition, search its neighborhood and move from the present partition to a new partition in the neighborhood for which the quality index is optimal. This enables us to limit the scope of search. If the neighborhood is still too large to search thoroughly, pick the first partition discovered that can improve the quality index from its present value. The search is repeated until moving to a new neighboring partition can no longer

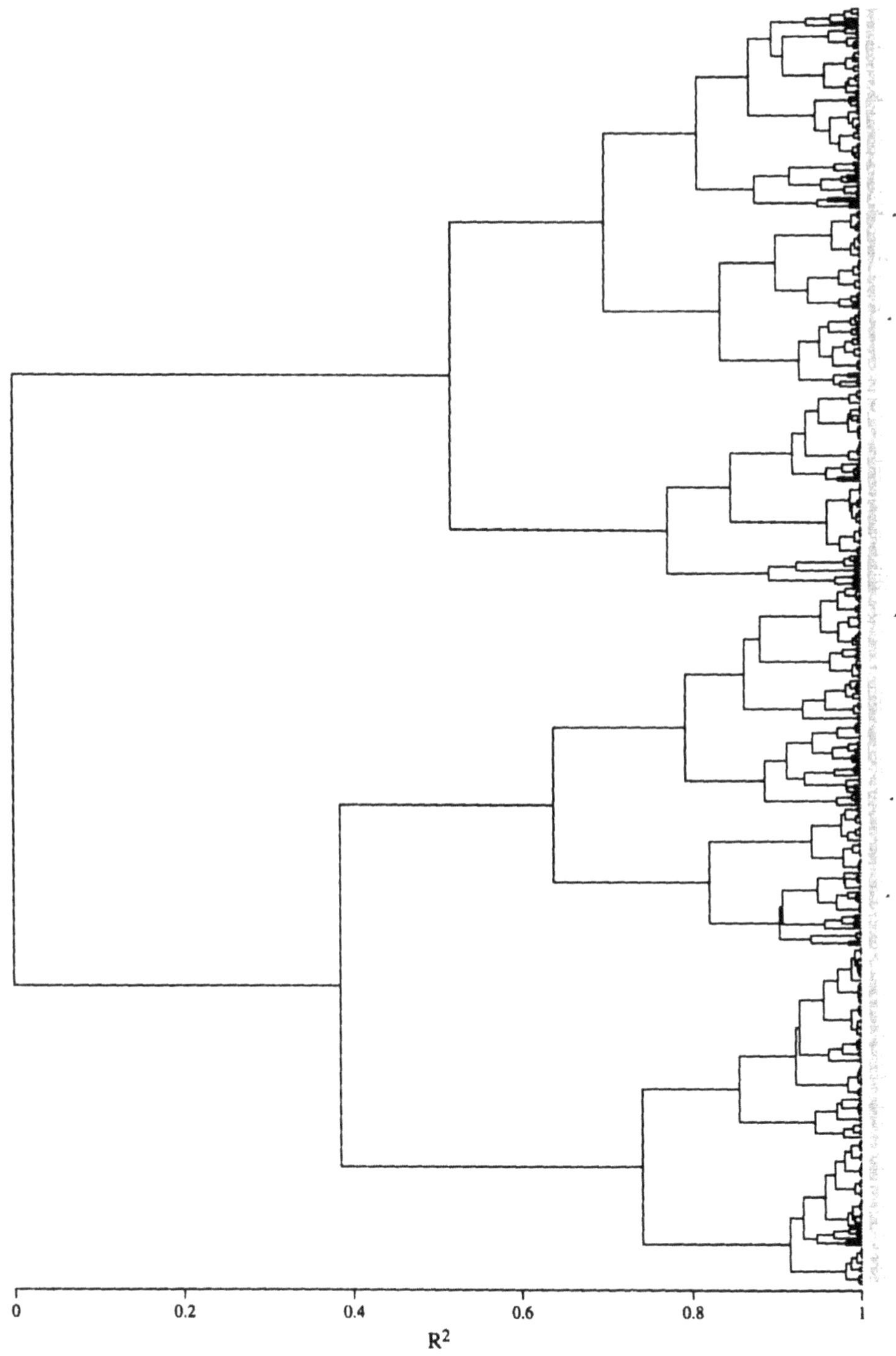


FIGURE 4.15 Dendrogram of the 392 genes in Example 2, based on Ward's method.

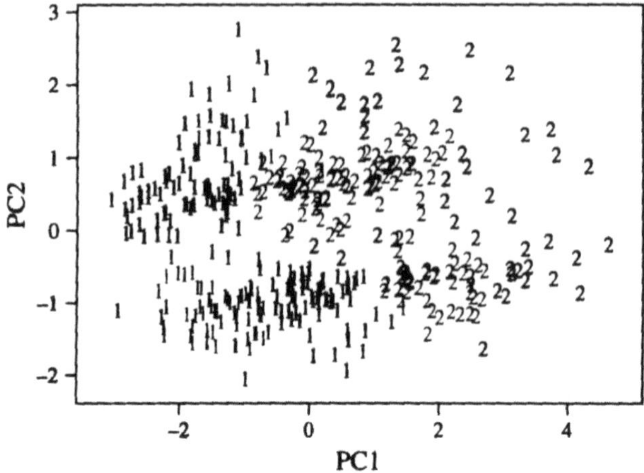


**FIGURE 4.16** Scatterplot of the first two principal components for the 392 genes in Example 2, after defining two clusters based on hierarchical clustering using Ward's method. Plotting symbols identify cluster assignments.

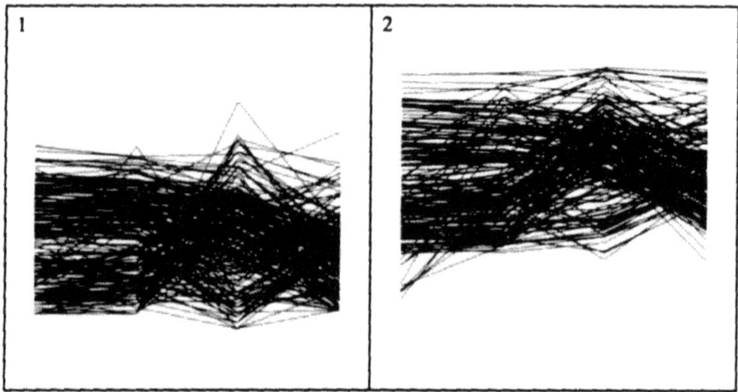


**FIGURE 4.17** Profile plots for the genes in the two clusters defined based on hierarchical clustering of the 392 genes in Example 2 using Ward's method. Each profile plot shows the $\log_{10}$ expression level vs. time for each gene in the cluster.

improve the quality index. Of the many different ways to define a meaningful index function, one that is commonly used in gene expression work is the well-known *k*-means clustering.

### 4.4.1 *k*-Means Clustering

Suppose each object in a given dataset is characterized by a *d*-dimensional vector. Let us consider the problem of partitioning *n* objects into *g* clusters, where *g* is predefined. We define an index function of any partition in terms of distances in *d*-dimensional space.

First we use Euclidean distance to describe this index function. We know that the distance between two points $x(1)$ and $x(2)$ in *d*-dimensional space is:

$$D(x(1),x(2)) = \sqrt{\sum_{j=1}^{d}\left(x_j(1)-x_j(2)\right)^2}.$$

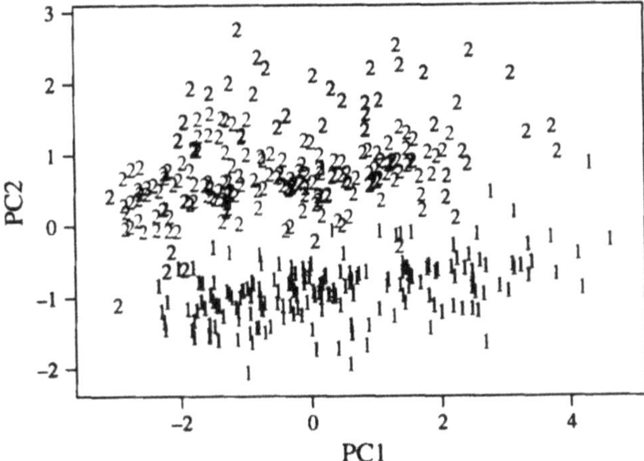


**FIGURE 4.18** Scatterplot of the first two principal components for the 392 genes in Example 2, after defining two clusters based on hierarchical clustering using two-stage density linkage. Plotting symbols identify cluster assignments.

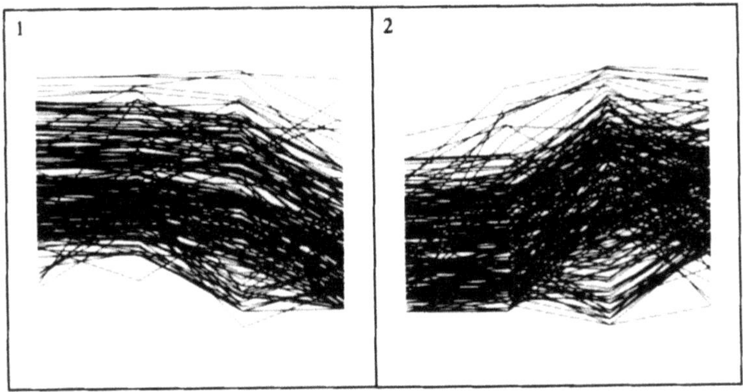


**FIGURE 4.19** Profile plots for the genes in the two clusters defined based on hierarchical clustering of the 392 genes in Example 2 using two-stage density linkage. Each profile plot shows the $\log_{10}$ expression level vs. time for each gene in the cluster.

For a cluster $Y$ in the partition, we define its mean $y$ such that its $j$th variable is the average of the $j$th variable of all objects in the cluster. The reason we have chosen the means of variables to represent a cluster is that such a point minimizes the sum of Euclidean distances to all objects in the cluster. We define the distance from object $x$ to cluster $Y$ as the distance from $x$ to the mean of $Y$:

$$D(x,Y) = D(x,y) = \sqrt{\sum_{j=1}^{d}\left(x_j - y_j\right)^2}.$$

For each object $x$, denote $y(x) = (y_1(x), \ldots, y_d(x))$ as the mean of the cluster to which $x$ belongs. Now, we are ready to define an index function — the error of a partition of $n$ objects $\mathbf{X}$ into $g$ clusters $\mathbf{Y}$:

$$\text{error}(\mathbf{X},\mathbf{Y}) = \sum_{i=1}^{n} D^2\left(x(i), y\left(x(i)\right)\right).$$

The general procedure is to search for a partition that reduces the error function by moving objects from one cluster to another. Note that the error function is defined through the sum of squared Euclidean distances from the cluster mean *y*, which minimizes the sum of squared Euclidean distances to all objects in the cluster and so minimizes the error function.

The *k*-means algorithm starts with *g* initial clusters, possibly randomly defined, and then moves objects between those clusters with the goal to (1) minimize variability within clusters, and (2) maximize variability between clusters. The algorithm has the following steps.

1. Define initial clusters $y(1), \ldots, y(g)$. Compute the cluster means and the initial error.
2. For each object *x*, compute for every cluster *y* the change in error in switching *x* from $y(x)$, the cluster to which it currently belongs, to *y*. If the error decreases in one or more clusters, transfer *x* to the cluster where the minimum error is attained.
3. Repeat step 2 until no object is transferred.

A different technique implemented by the SAS FASTCLUS procedure[10] is called *nearest centroid sorting*, and it works as follows.

1. A set of points called *cluster seeds* is selected as a first guess of the means of the clusters. The seeds may be selected from the data. Either the number of clusters or the minimum distance between clusters can be specified in the selection.
2. Temporary clusters are formed by assigning each object to the nearest seed, with an option of updating the cluster seed by the current mean of the cluster.
3. If iterations are to be used for recomputing cluster seeds, clusters are formed by assigning each object to the nearest seed, as shown in Figure 4.20. After all objects are assigned, the cluster seeds are replaced by the cluster means. This step is repeated until the changes in the cluster seeds become small or zero, or the number of iterations exceeds a prespecified limit.
4. Final clusters are formed by assigning each object to the nearest seed.

The SAS FASTCLUS procedure is intended for use with large datasets, with 100 or more objects. With small datasets, the results may be highly sensitive to the order of the objects in the dataset.

Some software applications, including Spotfire DecisionSite for Functional Genomics, offer a wide selection of initialization options, including random assignment, user-defined centroids, and data-based methods. Other software packages, including Eisen's Cluster program, do not offer the user any choice of initialization options. Documentation for some software does not clearly indicate how the initial clusters are defined.

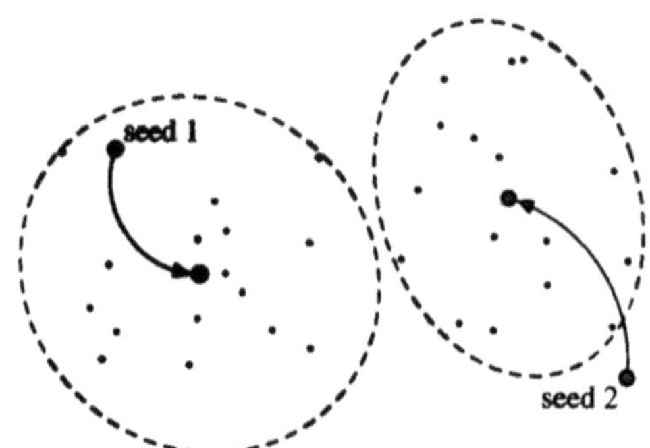


**FIGURE 4.20** Updating cluster seeds with cluster means in *k*-means clustering.

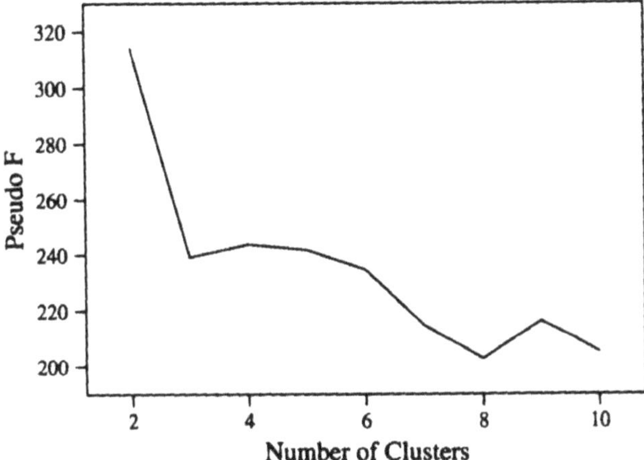


**FIGURE 4.21** Pseudo $F$ vs. number of clusters, defined using $k$-means clustering for Example 2.

Cluster analysis using any optimization method requires the number of clusters in a dataset to be preselected, but the analyst generally does not know *a priori* what the best number of clusters is. A variety of methods have been suggested to guess or estimate the best number of clusters in a dataset. Some of the solutions are relatively informal and involve defining a clustering criterion and inspecting its values computed over a range of numbers of clusters. Local peaks are usually taken as suggestive of a particular number of groups. Such criteria generated by SAS include the pseudo $F$ and the cubic clustering criterion (CCC).[10] CCC values are not valid with correlated variables. Other criteria described in the gene expression literature include the Gap statistic,[11] the figure of merit (FOM),[12] and the Bayes information criterion (BIC) for model-based clustering.[13] Most of the software packages that focus on analysis of gene expression data currently do not compute any clustering criteria to permit quantitative comparison of clustering results.

As yet, no solution to the problem of defining the optimum number of clusters has emerged that is completely satisfactory for gene expression data. However, for illustrative purposes we will look at the behavior of the pseudo $F$ criterion.

*Example 2 (cont.)*

Let's look at using $k$-means to cluster the 392 genes in our G-protein example. Because responses at 0 hour and 0.5 hour are correlated, we cannot trust the CCC values we get from PROC FASTCLUS in SAS. We ran the $k$-means analysis 9 times, with the number of clusters ranging from 2 to 10. The values of pseudo $F$ are plotted in Figure 4.21. Local peaks of pseudo $F$ occur at 2, 4, and 9. They are plausible stopping points for the number of clusters.

When the number of clusters is two, the two clusters generally represent genes with high and low overall responses, which can be seen from the scatterplot of PC1 and PC2 in Figure 4.22 and the profile plots in Figure 4.23.

The scatterplot and profile plots for four clusters are shown in Figures 4.24 and 4.25, respectively. With $k = 4$, $k$-means clustering is able to begin to distinguish between different profile shapes over time. An appreciable amount of heterogeneity can still be seen in the profiles in some clusters. Table 4.15 provides a rough description of the clusters.

For $k = 9$ clusters, Figure 4.26 shows the scatterplot of PC1 and PC2, and corresponding profile plots are shown in Figure 4.27. These clusters provide clear distinctions between shapes of expression profiles over time, as well as the general level of expression.

We can see from the scatterplots of principal components for this example that the clusters identified by $k$-means clustering are roughly spherical. Based on simulation studies, $k$-means

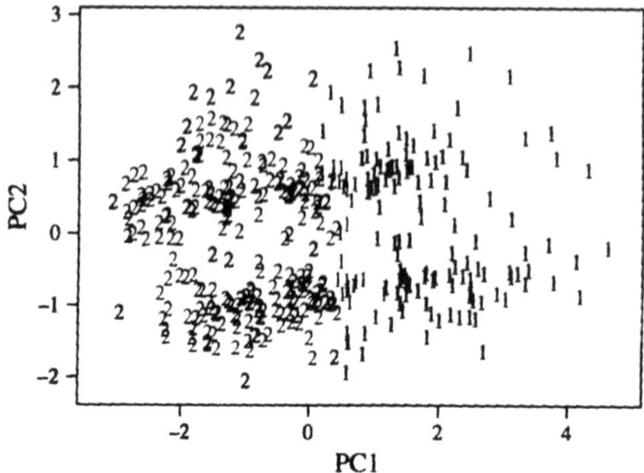


FIGURE 4.22 Scatterplot of the first two principal components for the 392 genes in Example 2, after defining two clusters based on *k*-means clustering. Plotting symbols identify cluster assignments.

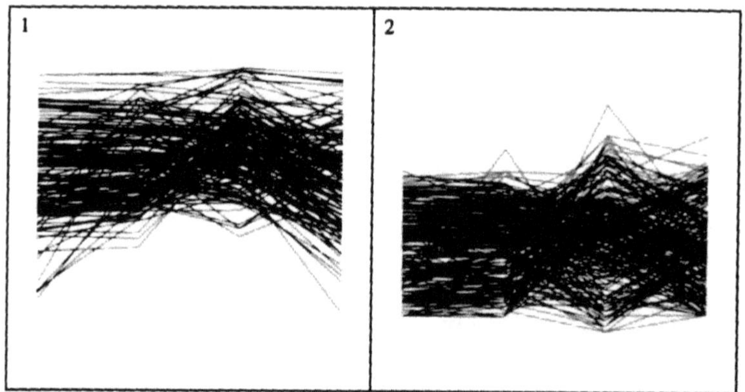


FIGURE 4.23 Profile plots for the 392 genes in Example 2 after defining two clusters using *k*-means clustering. Each profile plot shows the $\log_{10}$ expression level vs. time for each gene in the cluster.

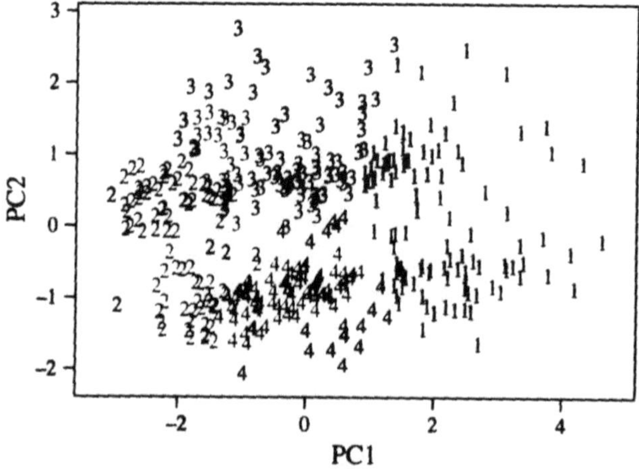


FIGURE 4.24 Scatterplot of the first two principal components for the 392 genes in Example 2, after defining four clusters based on *k*-means clustering. Plotting symbols identify cluster assignments.

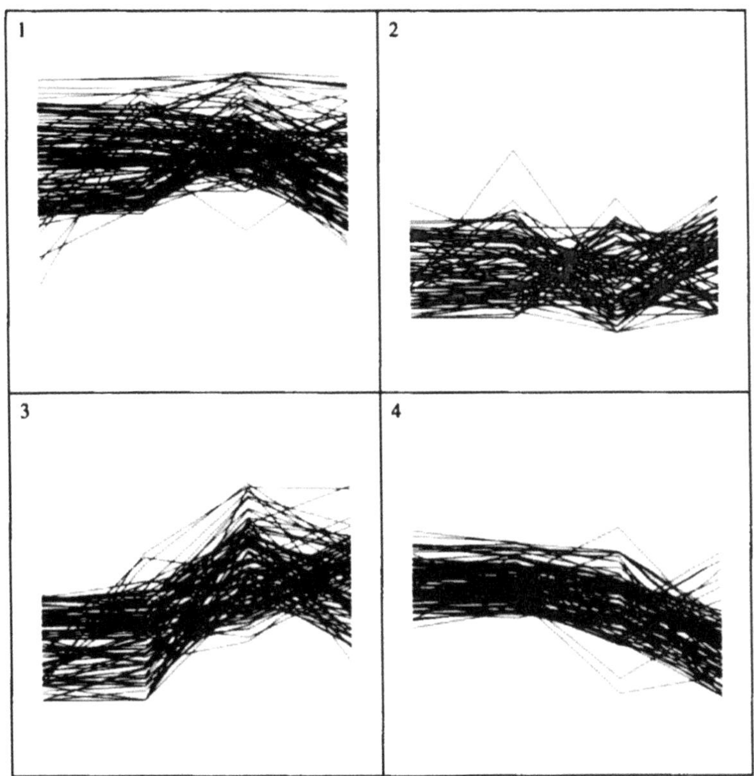


FIGURE 4.25 Profile plots for the 392 genes in Example 2 after defining four clusters using *k*-means clustering. Each profile plot shows the $\log_{10}$ expression level vs. time for each gene in the cluster.

**TABLE 4.15**
**Descriptions of Four Clusters Generated Using *k*-Means Clustering on Data from Example 2**

| Cluster | Overall Expression Level | Change over Time |
|---|---|---|
| 1 | High | No change |
| 2 | Low | No change |
| 3 | Low to average | Increase |
| 4 | Low to average | Decrease |

clustering usually works correctly when clusters are approximately spherical but fails when clusters are highly elongated or irregular.

## 4.5 SELF-ORGANIZING MAPS

The use of self-organizing map (SOM) methodology as a clustering device for gene expression data was first described by Tamayo et al.[14] An SOM is a tool to visualize large datasets in a high-dimensional space $\mathbf{R}^d$ by a grid in a two-dimensional space (i.e., a plane). The type of grid can be either rectangular (Figure 4.28) or hexagonal (Figure 4.29).

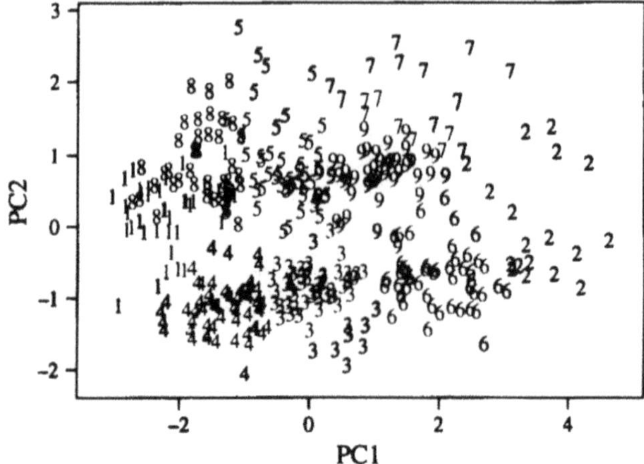


**FIGURE 4.26** Scatterplot of the first two principal components for the 392 genes in Example 2, after defining nine clusters based on *k*-means clustering. Plotting symbols identify cluster assignments.

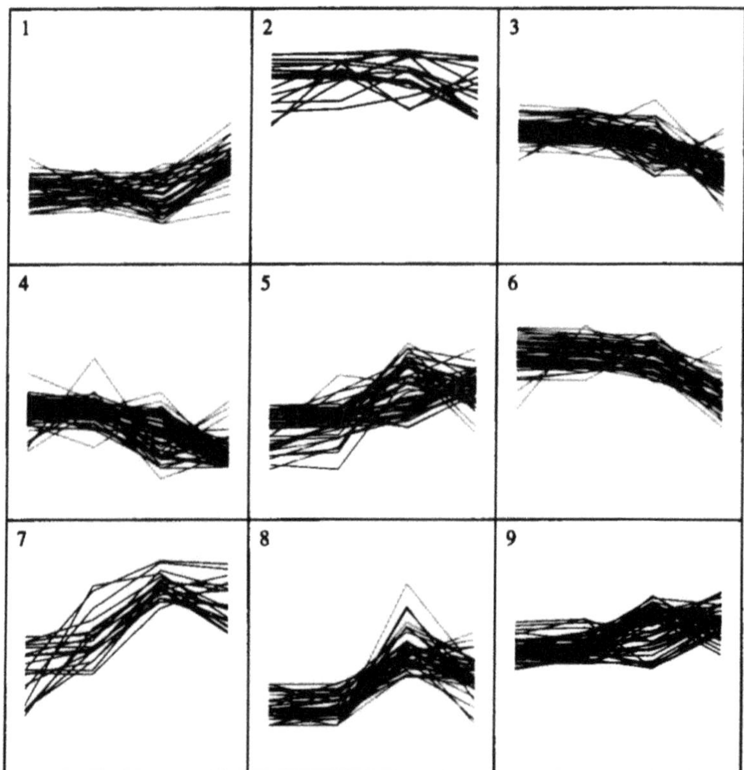


**FIGURE 4.27** Profile plots for the 392 genes in Example 2 after defining nine clusters using *k*-means clustering. Each profile plot shows the $\log_{10}$ expression level versus time for each gene in the cluster.

The points in the grid are called nodes. Each node $r_i$ is associated with a reference point $m_i$ in $\mathbf{R}^d$, which in turn represents some cluster in the data. The SOM approach attempts to use the topographical layout of the grid to organize the clusters in the data (Figure 4.30). It is hoped that the proximity in nodes reflects the proximity of reference points, hence the proximity of clusters.

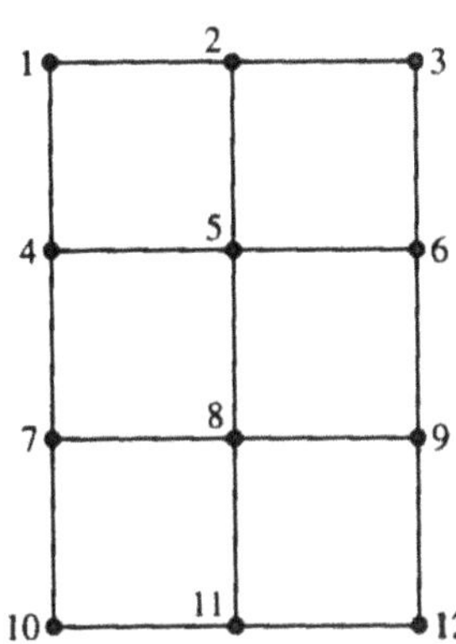


FIGURE 4.28 A 3-by-4 rectangular grid for an SOM.

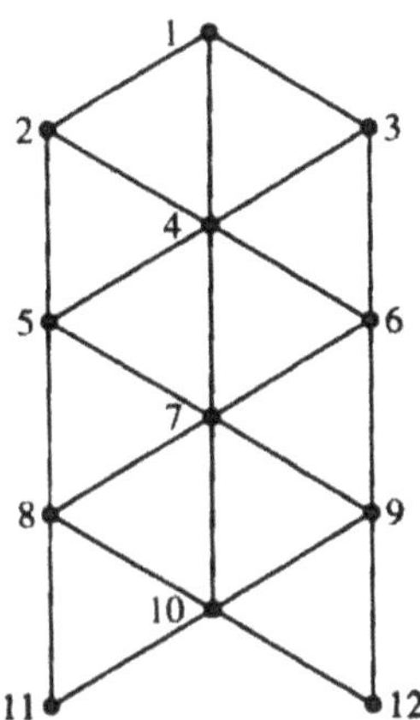


FIGURE 4.29 A 3-by-4 hexagonal grid for an SOM.

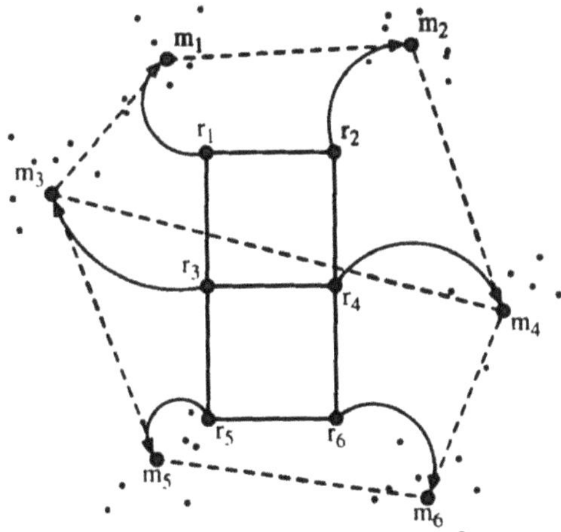


FIGURE 4.30 Assignments of reference points $m_i$ to grid nodes $r_i$ in a 2-by-3 SOM. (Adapted from Tamayo, P. et al., *Proc. Natl. Acad. Sci. USA*, 96, 2907, 1999. With permission.)

To begin generating an SOM, a configuration for the grid is selected. An initial reference point $m_i(0)$ is assigned at random to each node $r_i$. During the self-organizing process, the reference points are updated iteratively.

At iteration $t$, a point $x(t)$ in the data is selected. The best matching reference point $m_c(t)$ is found such that the distance from observation $x(t)$ to the reference point $m_c(t)$ minimizes the distance from the observation $x(t)$ to all reference points. Then, reference points are updated from $m_i(t)$ to $m_i(t + 1)$ by:

$$m_i(t+1) = m_i(t) + h\left(d\left(r_i, r_c\right), t\right) \cdot \left(x(t) - m_i(t)\right),$$

where $h(d(r_i, r_c), t)$ is the so-called neighborhood kernel. The right-hand side of the formula moves a reference point $m_i(t)$ toward an observation $x(t)$ by some ratio $h$. Some special cases are (1) if the ratio equals 0, $m_i(t)$ does not move; (2) if the ratio equals 1, $m_i(t)$ moves to $x(t)$; and (3) if the ratio equals 0.5, $m_i(t)$ moves to the midpoint between $m_i(t)$ and $x(t)$ (Figure 4.31).

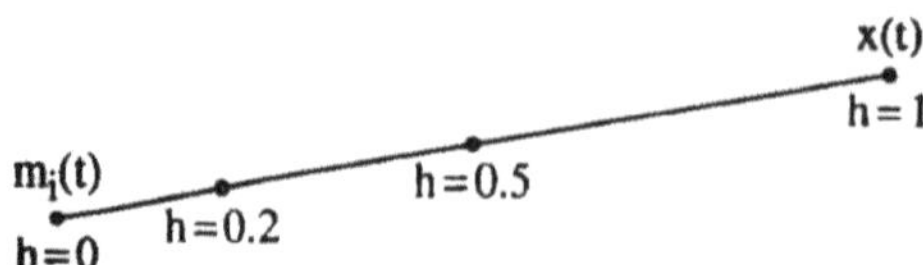


**FIGURE 4.31** Updated positions for reference point $m_i(t)$ based on different neighborhood kernel values.

Different choices exist for the neighborhood kernel. The kernel usually decreases and tends to 0 when the distance between the two nodes $r_c$ and $r_i$ increases, or when the iteration index $t$ increases. Therefore, the ratio by which a reference point $m_i(t)$ moves toward the observation $x(t)$ selected at each iteration depends on the distance between the node $r_i$ associated with $m_i(t)$ and the node $r_c$ that best represents $x(t)$; the larger the distance, the smaller the ratio.

The updating process is performed for each observation in a randomly ordered sequence. If the number of observations in the data is not large enough, the sequence will be repeated. An alternative to recycling of data is to pick an observation at random from the dataset at each iteration.

Using the SOM approach requires choosing both the number of nodes and their geometric configuration. At this time, no quantitative criteria or rules are in use to help guide this process in gene expression applications. In practice, analysts that routinely use SOM initially generate an SOM for a configuration that has worked well in the past. They then modify either the number of nodes or the shape until they obtain an SOM in which most, if not all, clusters contain multiple objects, and the objects within each cluster appear relatively homogeneous in a visual display of the SOM.

*Example 2 (cont.)*

We now apply the SOM methodology to the $\log_{10}$-transformed frequencies of Example 2, the 392-gene, G-protein dataset, with a configuration of four rows and three columns. The response profiles for the genes in each cluster are displayed in Figure 4.32.

We can see that genes with high overall response levels are at the top of the grid, those with low overall response levels at the bottom, those with increasing response at the left, and those with decreasing response at the right.

## 4.6 SUMMARY

For toxicogenomic applications, cluster analysis is an exploratory tool, useful for discovering groupings of genes or groupings of samples that provide insights into underlying biological processes. Cluster analysis methods, also known as unsupervised learning methods, create groupings based only on the data. They are distinct from class prediction, or supervised learning, methods, which seek to create prediction rules using classification information available *a priori* for a set of samples.

The data analyst performing a cluster analysis is required to make a series of decisions about:

- Which method to use (e.g., hierarchical, *k*-means, SOM)
- The type of similarity measure to use (e.g., distance or correlation)
- Particular technical details, such as which linkage algorithm to use
- The number of clusters to generate

In this chapter, we have provided some insights on characteristics of the different methods, similarity measures, and technical nuances. Unfortunately, no generally accepted rules or quantitative measures exist that will always guide the analyst to select the optimal choices for the series of decisions that must be made. Therefore, the analyst usually needs to iterate through various combinations of methods and settings, examining the output for each iteration and looking for consistent, sensible

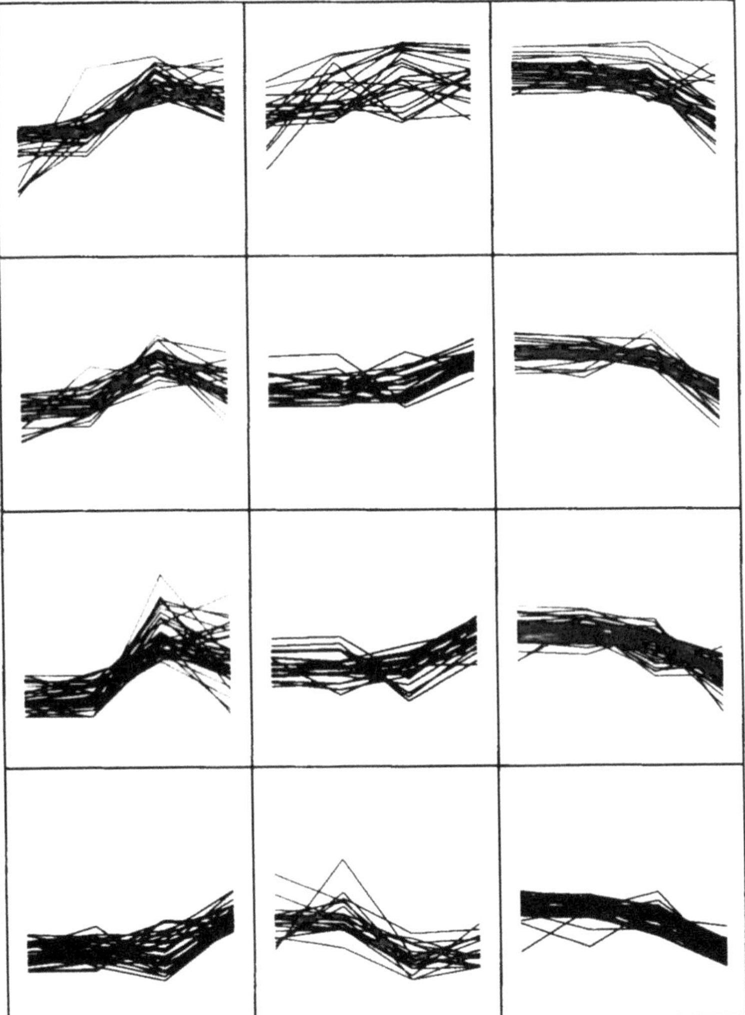

FIGURE 4.32 Cluster assignments from a 3-by-4 SOM run on $\log_{10}$ expression levels for the 392 genes in Example 2.

results. Over time, previous experiences in clustering gene expression data will enable the analyst to more efficiently obtain good clustering results. The "best" clustering results are those that provide logical insights into the questions of interest and suggest fruitful avenues for further research.

## REFERENCES


1. Massart, D.L. and Kaufman, L., *The Interpretation of Analytical Chemical Data by the Use of Cluster Analysis*, John Wiley & Sons, New York, 1983.
2. Everitt, B.S., *Cluster Analysis*, Third ed., Edward Arnold, London, 1993.
3. Eisen, M.B. et al., Cluster analysis and display of genome-wide expression patterns, *Proc. Natl. Acad. Sci. USA*, 95, 14863, 1998.
4. Hill, A.A. et al., Evaluation of normalization procedures for oligonucleotide array data based on spiked cDNA controls, *Genome Biol.*, 2(12), research0055.1, 2001.
5. Irizzary, R.A. et al., Exploration, normalization, and summaries of high density oligonucleotide array probe level data, *Biostatics* (in press).

6. Amaratunga, D. and Cabrera, J., Analysis of data from DNA microchips, *J. Am. Stat. Assoc.*, 96, 1161, 2001.
7. Yang, Y.H. et al., Normalization for cDNA microarray data: a robust, composite method addressing single and multiple slide systematic variation, *Nucleic Acids Res.*, 30, e15, 2002.
8. Flynn, J., Hierarchical clustering and heat maps, *SpotFire Array Explorer Webcast Series*, http://www.spotfire.com/news_events/webcast_page.asp?id =21, 2001.
9. SAS Institute, Inc., The CLUSTER procedure, in *SAS OnlineDoc® Version 8, SAS/STAT User's Guide*, SAS Institute, Inc., Cary, NC, 1999.
10. SAS Institute Inc., The FASTCLUS procedure, in *SAS OnlineDoc® Version 8, SAS/STAT User's Guide*, SAS Institute, Inc., Cary, NC, 1999.
11. Tibshirani, R., Walther, G., and Hastie, T., Estimating the number of clusters in a dataset via the gap statistic, in *Technical Report 208*, Department of Statistics, Stanford University, 2000.
12. Yeung, K.Y., Haynor, D.R., and Ruzzo, W.L., Validating clustering for gene expression data, *Bioinformatics*, 17, 309, 2001.
13. Yeung, K.Y. et al., Model-based clustering and data transformations for gene expression data, *Bioinformatics*, 17, 977, 2001.
14. Tamayo, P. et al., Interpreting patterns of gene expression with self-organizing maps: methods and application to hematopoietic differentiation, *Proc. Natl. Acad. Sci. USA*, 96, 2907, 1999.

# *Section 2*

## *Expression Profiling and Toxicology: The Advent of Toxicogenomics*

# 5 Transcriptional Profiling in Toxicology

*Julia Scheel, Marie-Charlotte von Brevern, and Thorsten Storck*

CONTENTS



## 5.1 INTRODUCTION

Toxicogenomics can be described as a marriage of classical toxicology assays and the new discipline of expression profiling – a union between the treatment of laboratory animals and the attempt to extract meaning from thousands of colored spots on a DNA chip. While not founded on a necessarily apparent relationship, one powerful element drives these two disciplines together: the necessity to speed up risk assessment for new chemical entities (NCEs). Classical toxicity testing is a major bottleneck in the drug development process, and this situation has been exacerbated with the advent of combinatorial libraries, high-throughput screening robots, and the exponential increase in potential drug targets triggered by genomics testing at the stages of early drug development. The net effect of these advances has been the identification of an unprecedented number of potential drug candidates. Because later stages of development can cope with only a limited number of compounds, increased input (from the early discovery end) does not, by default, result in a higher number of successful candidates, thus it is crucial to identify the most promising ones early on.[1]

0-8493-1334-1/03/$0.00+$1.50

Although many companies have implemented high-throughput screening procedures for selecting NCEs on the basis of their pharmacological efficacy and ADME characteristics (i.e., absorption, distribution, metabolism, and excretion properties of a compound), tools for the rapid analysis or prediction of adverse effects of NCEs are lagging behind. Undesired toxicity continues to account for about one third of compound failures during the developmental process.[2] Because classical toxicological testing[3] is too time consuming to meet the requirements of early stages of drug development, alternative strategies are needed.

Several mechanism-based, short-term *in vitro* assays for genotoxicity, metabolism, and drug–drug interactions are available. In addition, computational or *in silico* toxicology methods are also used to predict (quantitative) structure–activity relationships (QSARs). Expert systems such as DEREK operate on the basis of rules established by experts in the field, whereas statistical approaches such as TOPKAT require the input of experimental data obtained with training sets of compounds. MULTICASE, another commercially available software that is widely used, combines a QSAR portion with an artificial expert system. These approaches have been shown to be highly predictive for toxicological endpoints with well-defined mechanisms of toxicity (e.g., genotoxicity), but because the availability of large training sets is crucial for their performance, their range of applications has been limited.[2,4,5] Further progress in this area is still hampered by the lack of high-quality toxicological data.[6]

Because of the propagation of transcription profiling techniques over the last few years, toxicogenomics is attracting increasing attention as a possible alternative for early screening.[2,7–32] Toxicogenomics is based upon the fact that most, if not all, toxicologically relevant outcomes are preceded by changes in gene expression (see Figure 5.1). The concept of applying expression profiling to the field of toxicology has several important implications.

First of all, the amount of time and resources required for the toxicological analysis of a compound can in theory be drastically reduced. As a particularly striking example, a standard rodent cancer bioassay includes the analysis of over 1000 animals, costs millions of dollars, and takes almost 4 years.[28] This long time is necessary because tumorigenesis is a slow process, and it may take years before the classical toxicological endpoint (development of a detectable tumor) is reached. However, tumorigenic compounds may exert their effects much earlier upon entering their target cells. If their early effects are preceded or accompanied by characteristic and reproducible alterations in gene expression profiles over this much shorter period, it is theoretically possible that characteristic changes in RNA transcript levels could be observed within a few hours, days, or weeks of exposure to a tumorigenic compound.

Another important benefit arises from the fact that toxicogenomic studies are performed at the molecular level. Assessing the transcript levels of several hundreds to thousands of genes, these studies analyze how compounds interfere with or affect the molecular machinery of their target

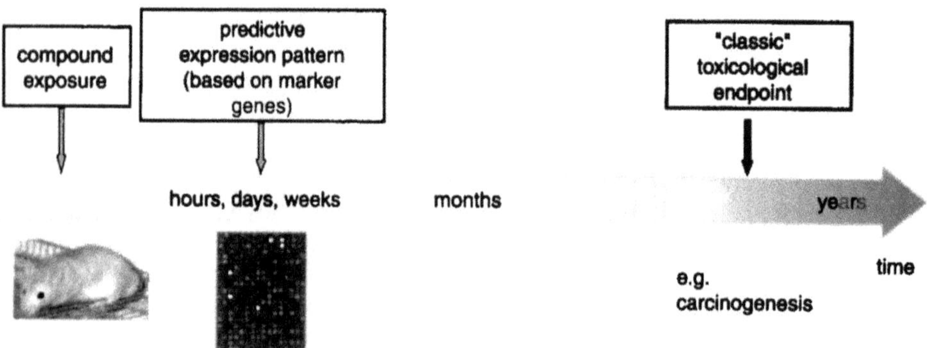


FIGURE 5.1 Toxicologically relevant outcomes are preceded by changes in gene expression.

cells; therefore, toxicogenomic data may hold additional insights into cellular mechanisms of toxicity beyond those derived from classical toxicological analyses based on whole organ pathology or survival endpoints. Molecular profiling of toxicants may enhance our basic understanding of some of the major issues in risk assessment (e.g., species differences; see Chapter 15) and may eventually reveal new ways to overcome these hurdles.

The generation and collection of data are only part of the exercise, but it is becoming more apparent that understanding the toxicological meaning of particular changes in a gene expression profile is an even greater challenge. In order to learn the language of the transcriptome, researchers are currently trying to unravel the relevant correlations between changes in the transcription patterns of target organs and the respective toxicological endpoints that are preceded (and therefore, may be predicted) by these changes. In order to find such correlations, suitable model systems (e.g., rat liver, hepatocytes) have been challenged with a number of reference compounds eliciting well-defined toxicological endpoints. In the few pioneering studies of this kind published to date, gene sets were identified that displayed characteristic changes in correlation with the endpoints examined, providing an initial proof of principle regarding the concept of toxicogenomics.[33-37]

The goal of this chapter is to provide the reader a cursory introduction to several fundamental aspects of toxicogenomic analysis; many of the topics highlighted in this chapter are covered in greater detail in subsequent sections. After reviewing important aspects concerning the various transcription profiling methods available and the model systems used, different methods of data analysis are highlighted. Finally, the distinguishing features of predictive vs. mechanistic approaches are described. In the last part of this chapter, we provide our own company's approach to high-throughput toxicogenomic analysis. This chapter is therefore intended to serve as a bridge between general aspects of expression profiling analysis and the recent applications of expression profiling to problems in toxicology. More in-depth discussions of mechanistic and predictive toxicogenomic studies and issues are covered in detail in the remaining sections of this book.

## 5.2 TRANSCRIPTION PROFILING METHODS IN TOXICOLOGY

Transcript levels can be analyzed by a broad range of methods, including common techniques such as Northern hybridization and quantitative reverse transcriptase–polymerase chain reaction (RT-PCR). Both methods, however, are applicable for the analysis of a small number of candidate genes only. An additional drawback with Northern hybridizations is their relative insensitivity, which limits their use in expression analysis as the majority of mRNAs in a cell are of low abundance.[38] In the case of rare transcripts or if the amount of starting RNA is limited, it is more suitable to apply quantitative real-time PCR (QRT-PCR) for differential gene expression analysis. This technology is especially useful when a small number of genes are monitored for a large number of samples. A better fit for the specific needs in the exploratory field of toxicogenomics, however, can be found in the methods for transcriptional profiling which allow for the simultaneous analysis of thousands of genes, thus providing the researcher with a much broader view of the transcriptome. In general, transcription profiling technologies can be classified into closed and open systems (see Figure 5.2).

### 5.2.1 CLOSED SYSTEM APPROACHES

The most important closed system used today are DNA microarrays (or DNA chips), which permit the analysis of transcript levels for the set of genes displayed on their surface.[25,39,40] (A number of reviews focusing on different aspects of microarray technology are compiled at http://www.nature.com/ng/chips_interstitial.html.) Different formats of microarrays are in use, either carrying cDNA fragments or oligodeoxynucleotides (see Chapters 1 and 2). Broadly built toxicogenomic studies usually include a number of test compounds, multiple dosages, and treatment periods, as well as replicate experiments. This makes it necessary to use transcription profiling methods with suitable throughput in order to cope with these larger sample numbers. DNA microarrays, being

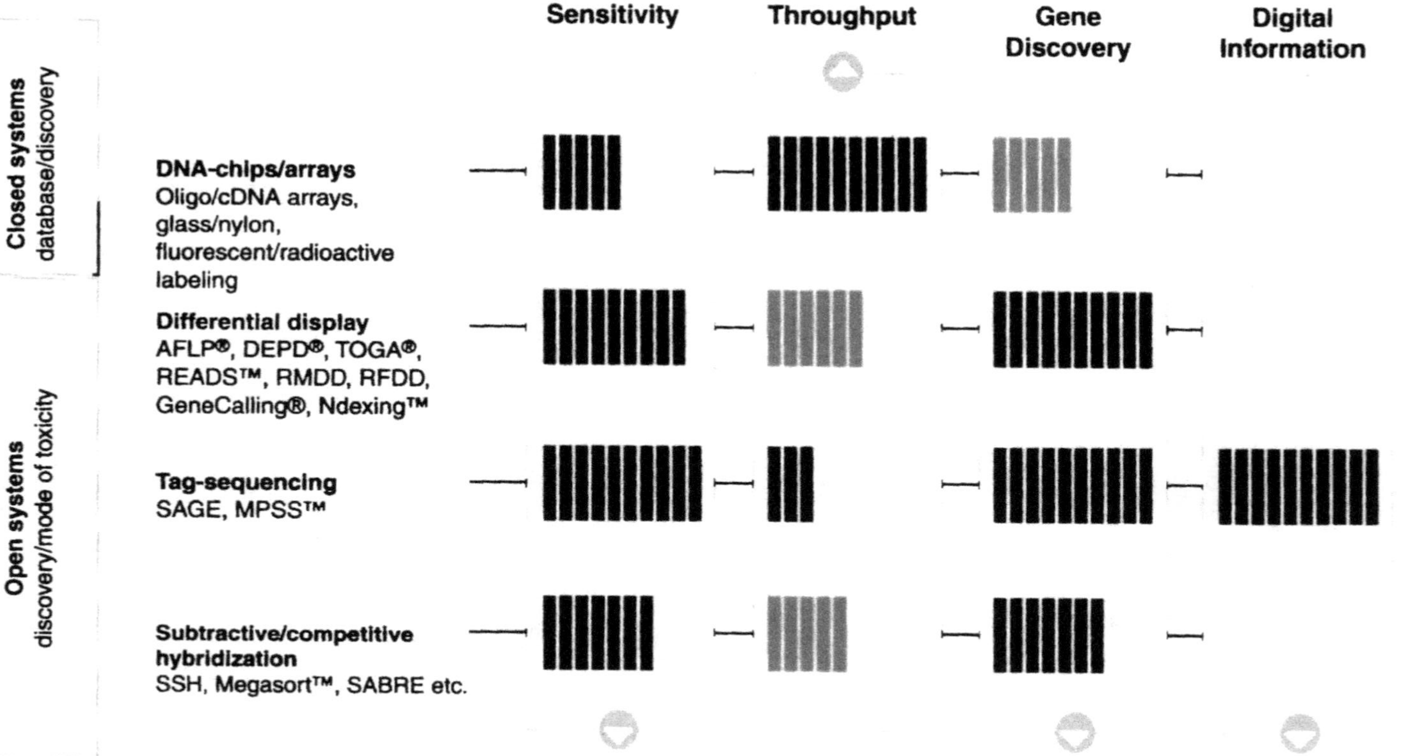


**FIGURE 5.2** Overview of transcription profiling technologies. Advantages and disadvantages of different types of transcription profiling methods are depicted. The darker the column displayed by a certain technology type in a certain field (sensitivity, throughput, gene discovery, or digital information), the better its performance with respect to the specific gene expression feature.

superior to all open systems with respect to throughput, have therefore become the first choice for this kind of study (for references, see Chapter 10). Frequently, microarray data are confirmed using quantitative or semiquantitative RT-PCR and/or Northern blotting for a limited number of genes.

Because the amount of information one can derive from a microarray experiment depends on the level of genome characterization and annotation of the organism investigated, at present not all toxicologically relevant organisms can be studied equally well with microarrays. Databases for toxicologically important species such as dogs and monkeys are sparse to nonexistent,[41] thus limiting the use of microarray experiments for their analysis. Therefore, the main focus of microarray-based experiments in the field of toxicogenomics has inevitably been on human, mouse, and rat to date.

Microarrays used in toxicogenomic studies vary widely in composition and the number of genes they display, ranging from 43 to 44,000 genes (see Chapter 10, Tables 10.1 to 10.3). While *more* does not necessarily imply *better*, the relevance of the data for a particular study depends on the composition of the selected gene set. There has been a trend toward using toxicology-focused arrays that display genes selected for their known or suspected involvement in cellular responses to toxic insults.[28,33–36] However, at the dawning of this novel discipline, our knowledge about toxicologically relevant gene regulation is still very limited. Thus, it has become a fruitful strategy to include open methods in these studies in order to discover novel toxicologically relevant genes that may represent important pieces in the complex puzzle that toxicogenomicists are just starting to assemble.

### 5.2.2 Open System Approaches

Open systems such as differential display technologies,[42] subtractive hybridization-based methods,[43] and the tag-sequencing methods MPSS™ (massively parallel signature sequencing)[44] and SAGE (serial analysis of gene expression)[45] are designed to circumvent the limitations of a preselected gene set and allow, in principle, for the analysis of the entire transcriptome.

The various versions of the differential display method (Figure 5.2) have a common principle, which is the conversion of sample mRNA populations into representative cDNA fragments. Subpopulations containing a manageable number of fragments are size separated by electrophoresis and visualized by different DNA staining or labeling methods. Genes that are differentially expressed between treated and control samples can be identified through comparison of the signal intensities of the corresponding cDNA fragments. In most versions of differential display, cDNAs are subsequently identified by excision, cloning, and sequencing of the respective bands. Although differential display is probably the most popular of all open approaches used today in the field of toxicogenomics (see Chapter 10, Table 10.4), several versions of this method have the disadvantage of yielding high rates of false positives, sometimes as high as 93%.[46,47] Similar problems arise with subtractive hybridization technologies, which involve hybridization of mRNA/cDNA from one population (tester) to excess mRNA/cDNA from another (driver). The unhybridized tester fraction containing the differentially expressed genes is then separated from the hybridized common sequences either physically or chemically or by selective PCR techniques. Typically, these procedures provide high numbers of potentially regulated clones. However, in order to identify the true positives it is often necessary to include secondary screening steps,[29] which render these technologies labor intensive and time consuming. Therefore, microarrays have also been propagated as a screening tool for identifiying truly regulated clones derived from such open transcription profiling studies.[48] Finally, enhanced derivatives of differential display have been developed (e.g., Axaron's Restriction-Mediated Differential Display [RMDD]) that have reduced the number of false positives to a negligible level.

The major distinction of tag-sequencing methods from all other transcription profiling techniques is their ability to measure absolute transcript levels. In analogy to most other transcription profiling methods, both SAGE and MPSS™ start with the conversion of mRNA into cDNA. In order to identify and count the transcripts present in an mRNA sample, a small sequence stretch of each cDNA molecule (referred to as "SAGE tag" or "MPSS signature") is determined. For SAGE-analyses, this is done by conventional sequencing of polymerized tags in a serial process. With MPSS™, up to 1 million signatures are obtained in parallel by cloning and sequencing of the cDNA molecules fixed to microbeads, providing an in-depth view of the transcriptome.[44]

Our currently limited knowledge about toxicologically relevant genes may, to some extent, be compensated by open methods.[29] Because they are not restricted to any preselected gene set, open transcription profiling technologies can generate very valuable data with respect to mechanistic studies on toxicity. They can serve as starting points for the building and verification of novel hypotheses on the mode of toxicity of a certain compound or compound class. However, because all of these technologies are substantially more laborious than the microarray system, open methods are not well suited to the population of large toxicogenomic reference databases,[29,30,49–52] and their application has been restricted to analyses of a single or very few compounds.

Most toxicogenomic studies involving open approaches have been performed using differential display or subtractive hybridization technologies, while only a few tag-sequencing expression analyses have been published (see Chapter 10). Novel genes identified with these methods are potentially relevant marker genes and may be transferred onto a microarray in order to create a tool for higher throughput analysis. This strategy has successfully been employed by our group at Axaron to identify predictive expression patterns for specific classes of rodent tumor promoters. Potential marker genes identified by applying MPSS™ and RMDD to selected samples are subsequently included into the gene set of our toxicology-focused TOXaminer™ microarray, which is routinely used for the analysis of all samples in our studies.

## 5.3 MODEL SYSTEMS FOR TOXICOGENOMIC ANALYSES

Basically, the same model systems that are used in classical toxicological studies can be used for toxicogenomic analyses. While the ultimate goal of toxicogenomics (and for toxicology in general) is to increase product safety for humans, the majority of studies are currently performed in rodents, despite the fact that the human predictability of standard rodent tests shows an overall concordance of only 45%.[2] The two main model systems in toxicogenomic studies are rat liver changes assessed *in vivo* and primary rat hepatocytes cultured *in vitro*. Both of these model systems, along with others, are briefly summarized later. Detailed characterization of the utility of these model systems in expression profiling studies is provided in Chapters 6 and 7.

### 5.3.1 *In Vivo* Systems

Most toxicogenomic analyses to date have been focused on rodent liver, the main target organ for toxic insults of vertebrates. The fact that effects of toxicants on rodent liver physiology have been studied extensively and an enormous amount of relevant toxicological data has been published is a great advantage for the evaluation of toxicogenomic studies. In order to elicit a specific endpoint and to identify related surrogate markers, appropriate reference compounds, application routes, and dosing regimens have to be chosen. To a large extent, data concerning the effects of toxicants on rodent liver is available in the literature and can be used for study design. *In vivo* exposures are broadly applied to test for liver toxicity and are widely used in toxicogenomic studies (see Chapter 10, Tables 10.1 to 10.4).

For the analysis of most tissue samples, it should be kept in mind that the cellular complexity of tissues has an impact on the result of the transcription analysis, because expression changes in target cells may be diluted or even masked by the presence of unresponsive surrounding cells. A drastic example in this respect is the effect of Alloxan on the pancreas. Alloxan specifically targets the β-cells, which make up less than 2% of pancreatic cells. None of the existing transcription profiling technologies is capable of monitoring changes in the transcriptome of target cells at such a dilution,[17] unless pure target cell populations are procured prior to the analysis. A number of microdissection technologies, including, for example, laser-capture microdissection (LCM),[53,54] have been developed to selectively isolate defined cells or cell populations.[55] Tissue microdissection can be applied to tissue sections of frozen tissues as well as cytological preparations. RNA from microdissected tissue samples has been radioactively labeled and hybridized to cDNA arrays, but at least 5000 to 50,000 cells are required for this type of analysis, unless amplification procedures are applied.[56,57]

The ability to explore histopathology archives with transcription profiling technologies would be of great value for toxicogenomic studies. For this purpose, efforts have been made to extract RNA samples from archival specimens of paraffin-embedded tissues, which were then subjected to gene expression analyses.[58] These studies demonstrated that the original fixation protocol used on the sample influenced both RNA integrity and yield.[58,59] Total RNA isolated from formalin-fixed specimens was used in RT-PCR and microarray analyses, although it was significantly degraded. Gene regulation could be determined qualitatively for some genes in a few samples, but reproducible transcription profiles could not be obtained.

### 5.3.2 *In Vitro* Systems

With respect to higher throughput and minimization of costs and compound requirements, cellular systems are attractive alternative models for toxicogenomic studies. A general advantage of cellular systems is that they allow for a standardized dose-finding strategy in which compounds are first analyzed for their cytotoxicity to determine an effective dose equivalent (e.g., $ED_{30}$ or $ED_{50}$).[27,33] However, the value of this strategy has been disputed, as the focus on cytotoxic effects leads to the application of relatively high exposure levels. In consequence, a compound may not be assigned to the correct, more relevant mode-of-action class, which would be observed at lower doses.[60] Established cell lines such as the human hepatoma HepG2 are, in terms of handling and availability, the most convenient system for toxicogenomic studies.[33,61,62] However, because their metabolic properties differ from typical hepatocytes in some respects, their value as a model for the physiology of hepatocytes is currently unclear.[24]

Primary hepatocytes are well suited for toxicogenomic studies[37] as they retain a certain level of metabolic activity; however, due to inter-animal variations and a sensitive isolation procedure, batch to batch variation is high, which would likely cause significant alterations in the expression profiles induced by a test compound. Refined protocols for the cryoconservation of hepatocytes allow for long-term storage and better availability of hepatocytes and may also play an important role in future toxicogenomic analyses.

In general, cellular systems do not permit the study of toxic effects that are based on the interaction of different tissues — for instance, a compound whose toxicity is exerted via interference with hormonal regulation circuits. Additionally, many toxicological endpoints involve more than a single cell type, which makes it impossible to detect certain effects when using only a single type of cultured cell. Researchers have tried to overcome these problems by establishing co-cultures of different cell types. For example, co-cultures of hepatocytes with nonparenchymal cells have been developed[63] and tested for their use as bioartificial livers.[64,65] Those systems, frequently referred to as *bioreactors*, might find use in toxicogenomic studies as *in vitro* systems with increased complexity and improved metabolic competence compared to primary hepatocytes.

Precision-cut liver slices maintain the *in situ* architecture of the liver, allow a relatively high sample throughput, and combine several of the desirable characteristics of the liver and hepatocytes system.[66,67] Although toxicogenomic analyses on slice models have been announced at several meetings in recent years, to our knowledge no peer-reviewed report on this topic has been published to date.

## 5.4 IMPORTANT ASPECTS OF DATA ANALYSIS

The equipment and software tools necessary for microarray analyses are in continual refinement. While a number of advanced clustering algorithms and statistical approaches are available as mentioned in Chapter 4, the analysis of data compiled during toxicogenomic studies still represents a major challenge, and no standard procedures have yet been established.[68] Despite the uncertainties regarding the vagaries of increasingly complex data analysis approaches, two basic questions of interest arise when analyzing lists of regulated genes discovered in toxicogenomic studies.

The first question probes the physiological meaning of observed changes in transcript levels following toxicant exposure. In order to facilitate this kind of analysis, we have assigned functional categories to all genes on our TOXaminer™ microarray. Using our proprietary query tools, which organize all categories in a hierarchical tree-like fashion (Figure 5.3), the researcher can zoom back and forth from a global perspective on one side to a detailed view on the level of the individual gene on the other. Splitting up the genes according to functional categories has become a common approach in toxicogenomic studies to visualize the overall effect of compounds.[33–35,69] Additional database modules that map gene expression data onto biochemical pathways, as well as the integration of supplementary toxicological and clinical chemistry data, further help to connect gene expression to cell physiology.

The second question is concerned with the problem of identifying compounds that elicit similar changes in transcription profiles. In order to detect overall resemblance in expression changes and to identify groups of compounds that affect cell physiology in a related fashion, correlation methods or cluster analyses have been used in the majority of studies.[33,34,36,37,70] Hierarchical clustering algorithms, which produce nested arrays of clusters by repeated aggregation of smaller clusters, are widely used.[68] Encouragingly, groups of compounds that have been clustered in this way often coincide with their assignment to known mode-of-action classes.[33,34,36,37,70] These results are the basis for the development of larger toxicogenomic reference databases for the prediction of mode-of-action categories and, thus, for defining early toxicological clues for NCEs. However, cluster analysis must be dealt with cautiously, as it has been shown to produce significantly different results for the same dataset depending on the selection of the gene subset included in the calculation[33] and on the clustering algorithm used. As mentioned in Chapter 4 and in various other publications,[34,36] it has been proposed to apply different clustering methods (e.g., principal component analysis,[71] multidimensional scaling,[72] global geometric framework,[73] local linear embedding[74]) to a gene list in order to identify significant clusters.

## 5.5 ESSENTIAL ISSUES IN STUDY DESIGN

Although a number of issues are important for study design, we would like to emphasize two main aspects in this introductory chapter: first, the robustness of the transcriptional changes observed in model systems and, second, the comparability of data derived from different sources.

### 5.5.1 Robustness/Reproducibility of Model Systems

Testing for reproducibility is crucial with respect to reliability and statistical robustness of the results. In order to assess the importance of replicate experiments, Burczynski et al.[33] treated HepG2 cells with cisplatin in 13 separate experiments, each time according to the same protocol. Comparison of the corresponding transcriptional profiles showed that, among the 250 genes examined, only a small percentage of genes were regulated consistently throughout all 13 experiments, whereas regulation was more variable for a larger set of genes (genes displaying up- and downregulation in response to equal doses of cisplatin). The differences in consistency did not depend on the magnitude of the induction event and consequently may reflect an intrinsic baseline variability of the HepG2 cells. The authors propose that model systems used for toxicogenomic analyses should

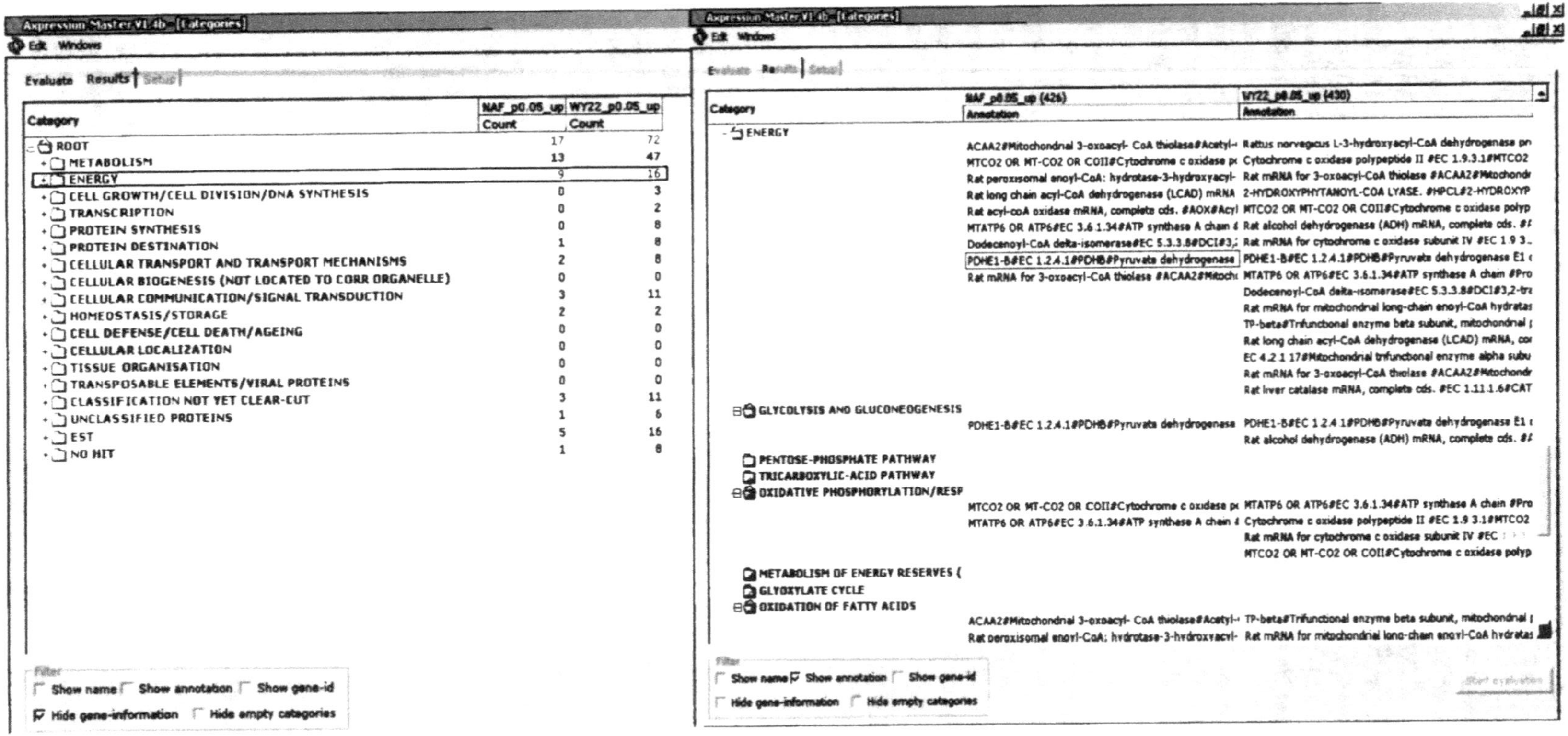
Expression Master V1.4b - [Categories]
Edit Windows
Evaluate Results Setup
Category
NAF_p0.05_up Count
WY22_p0.05_up Count
ROOT 17 72
METABOLISM 13 47
ENERGY 9 16
CELL GROWTH/CELL DIVISION/DNA SYNTHESIS 0 3
TRANSCRIPTION 0 2
PROTEIN SYNTHESIS 0 8
PROTEIN DESTINATION 1 8
CELLULAR TRANSPORT AND TRANSPORT MECHANISMS 2 8
CELLULAR BIOGENESIS (NOT LOCATED TO CORR ORGANELLE) 0 0
CELLULAR COMMUNICATION/SIGNAL TRANSDUCTION 3 11
HOMEOSTASIS/STORAGE 2 2
CELL DEFENSE/CELL DEATH/AGEING 0 0
CELLULAR LOCALIZATION 0 0
TISSUE ORGANISATION 0 0
TRANSPOSABLE ELEMENTS/VIRAL PROTEINS 0 0
CLASSIFICATION NOT YET CLEAR-CUT 3 11
UNCLASSIFIED PROTEINS 1 6
EST 5 16
NO HIT 1 8
Filter
Show name
Show annotation
Show gene-id
Hide gene-information
Hide empty categories
NAF_p0.05_up (426) Annotation
WY22_p0.05_up (430) Annotation
ENERGY
GLYCOLYSIS AND GLUCONEOGENESIS
PENTOSE-PHOSPHATE PATHWAY
TRICARBOXYLIC-ACID PATHWAY
OXIDATIVE PHOSPHORYLATION/RESP
METABOLISM OF ENERGY RESERVES (
GLYOXYLATE CYCLE
OXIDATION OF FATTY ACIDS

**FIGURE 5.3** Functional categorization of rat liver expression changes after treatment with the two peroxisome proliferators Wy-14,643 (WY) and Nafenopin (NAF). The left window shows a view of the main branches of the category tree, with the figures representing the number of regulated genes in each of those branches. "Root" displays the total number of genes in the list of choice. Categories headed by "+" can be further fanned out to subcategories, as shown in the right window. This window additionally displays the gene annotations (including all available synonyms) of the differentially expressed genes, aiding the comparison of similarities and differences between compounds on a functional level.

first be monitored for the robustness and reproducibility of their transcription profiles. Similar fluctuations in the expression level of certain genes noted in yeast necessitated the application of an error model. This statistical tool facilitates correction of the significance of the regulation events with respect to the fluctuations of the respective genes observed in the control samples.[75]

In most studies, the strategy of pooling tissue samples of different animals in a treatment group prior to hybridization is pursued in order to counterbalance the inter-individual variability — and for cost-saving reasons as well. However, pooling may cause misinterpretation if one of the animals shows an extremely different response compared to the others. Recently, the value of biological replicates (i.e., the analysis of single animals) to account for inter-individual differences in susceptibility toward certain compounds has become apparent. Typically, transcription profiles of different animals treated with the same compound are highly similar as revealed by hierarchical cluster analysis.[76] Others have noted that single-animal experiments and pooled experiments have clustered together.[36] However, depending on the compounds to which the animals are exposed, inter-individual differences can be much more pronounced. Exposing rat livers to different doses of a particular compound (e.g., morphine), we found that, while most of the animals of a treatment group are represented in the same cluster, transcription profiles of some animals from the low-dose group cluster with the high-dose group, indicating a significantly higher susceptibility of certain animals.

To assess natural variations in unstimulated murine gene expression, Pritchard et al.[77] compared the expression profiles of kidneys, livers, and testes of six untreated C57BL6 mice. The percentage of differentially expressed genes on a cDNA microarray comprising 5406 clones was determined to be 0.8% in liver, 1.9% in testis, and 3.3% in kidney. Animal-to-animal variability is generally greater than variability associated with spotting, hybridization, or data acquisition[78] (Axaron, unpublished results). In general, a minimum of three replicates is recommended for statistical and, in the case of biological replicates, biological reasons.

### 5.5.2 Comparability of Expression Profiling Data Across Different Sources

A second topic is the problem of comparability of data from different sources. Researchers and reviewers often face the problem of not being able to compare data from different study reports, even if the experimental settings are very similar. This is mainly due to the fact that the information given about the experimental design and the data analysis chain is highly heterogeneous and often fragmentary. Similar problems arise if gene expression data from different technological platforms are to be integrated. To overcome the problem of nonexistent standards for presenting microarray data, a proposal for the minimum information required about a microarray experiment (MIAME) has been made.[79] The authors suggest that researchers include in every report the description of experimental design, array design, samples, hybridization conditions, measurements, and normalization controls. This seems to be a reasonable first step toward a better comparability and integration of different studies and platforms. Another problem encountered when comparing large sets of regulated genes derived from different studies is the use of various synonyms for the same gene, often making it difficult to find a specific gene of interest within a gene list. At Axaron, we therefore include all synonyms available from the public databases in our gene annotation.

## 5.6 MECHANISTIC VS. PREDICTIVE TOXICOGENOMICS

### 5.6.1 Mechanistic Toxicogenomic Approaches

Two general approaches toward toxicogenomics can be distinguished. On the one hand, some studies are designed to investigate a specific mechanistic question and analyze the effects of only one or a small number of compounds. The results of these studies provide valid answers for specific questions, yet cross-comparisons between different studies related to the same compound or compound class are often difficult due to differences in study design. Chapters 8 and 9 discuss expression profiling of specific pathways of intense interest to toxicologists (transcription pathways regulated by the xenobiotic response element and the antioxidant response element). However, many other studies have used expression profiling to probe mechanisms of other toxicologically relevant compounds. We provide an overview of those mechanistic toxicogenomic studies in Chapter 10.

### 5.6.2 Predictive Toxicogenomic Approaches

The second type of approach involves the analysis of a large number of compounds in order to be able to assess the toxic potential of, eventually, any unknown compound of interest. It is referred to as *predictive toxicogenomics* in which analyses generate initial reference datasets suited for the categorization of future unknown compounds according to their functional class. In addition to our own company's approach to predictive toxicogenomics summarized at the end of this chapter, a selection of pioneering studies and approaches in the field of predictive toxicogenomics are covered in the chapters in Section 5.

Ideally, toxicogenomics databases (see Chapter 11) should be filled with gene expression profiles corresponding to well-characterized toxicological endpoints (e.g., tumor promotion, genotoxicity, certain types of liver necrosis). For this reason, it is necessary to select compounds that elicit very specific phenotypes, the expression profiles of which can serve as references for compounds with previously unknown or incompletely elucidated mechanisms of toxicity. The main goal is to generate a tool for the rapid prediction of the mode-of-action class of those compounds. Obviously, this does not exclude the option to exploit these datasets for mechanistic questions as well. To date, despite the enormous interest in predictive toxicogenomics, the number of published studies of a large number of compounds is still limited due to high costs and intellectual property issues.

The essential starting point of every predictive dataset is a careful design that takes into account which level of prediction is to be achieved. In Figure 5.4, the crucial building blocks of such a toxicogenomics dataset are displayed. Dose-range and application route must be chosen considering the half-lives and effect levels of the respective test compounds. Screening and comparison of a series of compounds requires standardized test conditions. However, the principal dilemma in these types of studies is balancing a desire to test a reasonably low number of variables (for reasons of throughput and comparability between compounds) with the need to analyze as many time points and dosages as possible in order to identify expression patterns with relevance to the specific endpoint of concern.

### 5.6.3 Example of Predictive Approaches at Axaron

At Axaron, we have chosen the rat as the premier species to use to build up a predictive database. Initially, we have focused on analyzing the rat liver as the major organ for metabolism and therefore as the main target for toxic insults. Because no reliable short-term test system is available for nongenotoxic carcinogens (NGCs) — in contrast to genotoxic compounds — we decided to characterize this particular endpoint in the hopes of being able to predict the tumor-promoting potential of NCEs by solely toxicogenomic means.

One example of a hierarchical clustering derived from this study is depicted in Figure 5.5. Female Fischer 344 rats were treated with a number of known tumor promoters, including enzyme inducers and peroxisome proliferators (PPs) as well as cyto- and genotoxic compounds. For this clustering, a 72-hr treatment period was chosen with dosings below the MTD. RNA pools of five animals in each treatment group were hybridized on our TOXaminer™ microarray comprised of approximately 1500 genes (see also Section 5.2.2). Differentially regulated genes were identified using a modified $t$-test ($p \leq 0.05$[81]). The clustering is based on a list of 211 genes that are significantly

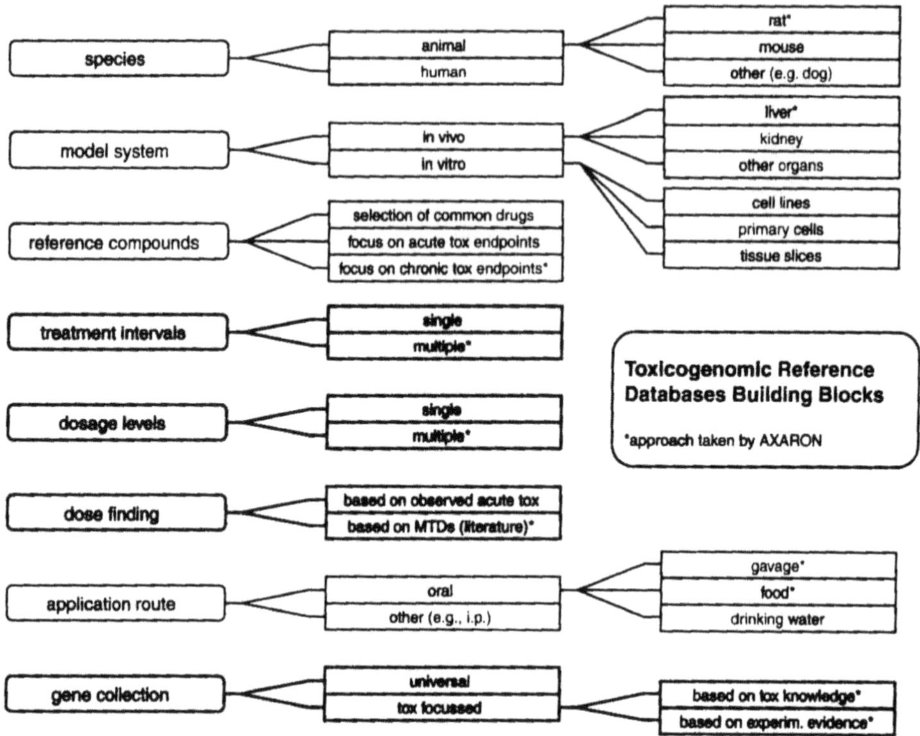


**FIGURE 5.4** Toxicogenomic reference databases building blocks.

regulated in either one of the treatments and that additionally show at least twofold regulation. The hierarchical cluster analysis shows that in most cases compounds with similar modes of action cluster together. The three enzyme inducers phenobarbital (PB), alpha-hexachlorocyclohexane (alpha-HCH), and cyproterone acetate (CPA), the latter also acting as a hormone, form one cluster; the two peroxisome proliferators Wy-14,643 (WY) and nafenopin (NAF) form another. Two other clusters are built by the three remaining genotoxic toluenes 2,4-dinitroluene (2,4-DNT), 2,4,-diaminotoluene (2,4-DAT), and 2,6-diaminotoulene (2,6-DAT) on the one side and by the cytotoxic compounds furan (FU) and carbon tetrachloride (CT) on the other. Expression profiles of the hormone ethinyl estradiol and of 2,6-dinitrotoluene (2,6-DNT) do not show significant similarity to any of the other compounds in this kind of analysis. Additionally, we adapted the approach of Thomas et al.[80] to determine genes that were able to discriminate between a group of tumor promoters (two peroxisome proliferators, three enzyme inducers, two cytotoxic compounds, one hormone, and the two toluene derivatives 2,4-DAT and 2,6-DNT) and another group of non-tumor promoters (morphine, gabapentin, and the other two toluene derivatives 2,4-DNT and 2,6-DAT). Hierarchical clustering of the test compounds over the 21 discriminator genes obtained is shown in Figure 5.6.

At Axaron, toxicogenomic studies routinely include multiple dosage levels and treatment periods. The additional effort is significant, but it pays off in a solid basis for the comparison of compounds with different ADME characteristics or of different model systems. Comparing gene expression profiles at different doses and time points aids in identifying treatments that are most suited for the specific discrimination between functional classes.

We and others[33,36,37] have shown that it is possible to distinguish between compounds belonging to different functional classes, given a careful study design and suitable data analysis methods. It has also been shown that, in principle, blinded samples can be classified correctly according to their mode of action. The more compounds with similar endpoints that are included in a database,

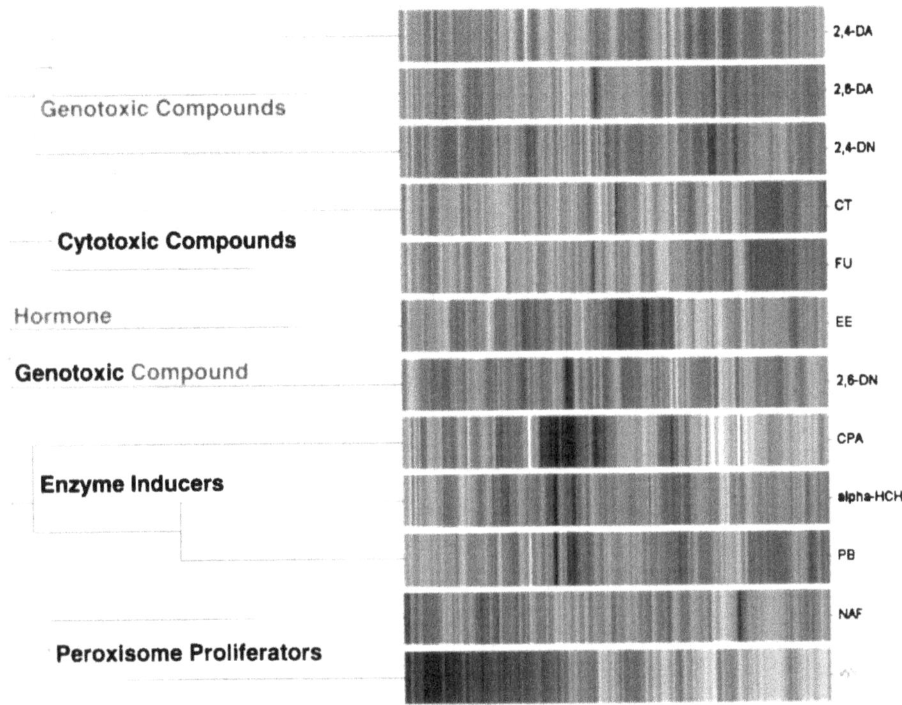


**FIGURE 5.5** Hierarchical clustering of 12 tumor promoters and cyto- and genotoxic compounds. Black bars represent upregulated genes and white bars represent downregulated genes compared to the average expression of all experiments. The Pearson correlation coefficient was set to 0.75. Experimental settings and abbreviations are explained in the text. (See color insert following page 112.)

the more accurate predictive analyses will become. To us, it seems reasonable to focus on those endpoints that are of highest concern in later stages of drug development — for example, tumor promotion and teratogenicity.

## 5.7 SUMMARY

With toxicogenomics becoming increasingly attractive to both academia and industry, many kinds of open and closed transcription profiling methods have been applied to toxicological questions over the last few years. Encouragingly, part of the data generated has been consistent with previous knowledge and can thus be seen as a proof of concept for the new technologies. In addition, a wealth of unexpected gene regulation has been detected, demonstrating the exciting complexity of toxicological responses to be explored in the future. To date, it is clear that the potential use of gene transcripts as biomarkers for toxic or pharmacological effects remains to be established.

In order to become broadly applicable as a predictive tool, toxicogenomics databases will have to include transcription profiles for a large number of compounds. The results from the first few published pilot studies have demonstrated that, under appropriate experimental settings, compound-induced changes in transcription profiles correlate well with the known mode of action of the compound. Numerous future studies will be required to validate and extend these findings and determine whether predictive toxicogenomics will become a viable discipline in the future.

By now, transcription profiling can provide evidence in favor of or in contradiction to certain mechanistic hypotheses and can provide new hints on the mechanism of toxicity of a compound. Currently, however, toxicogenomic data alone are insufficient to deduce complete sequences of toxic events. Supplementary information derived from methods such as histopathology,

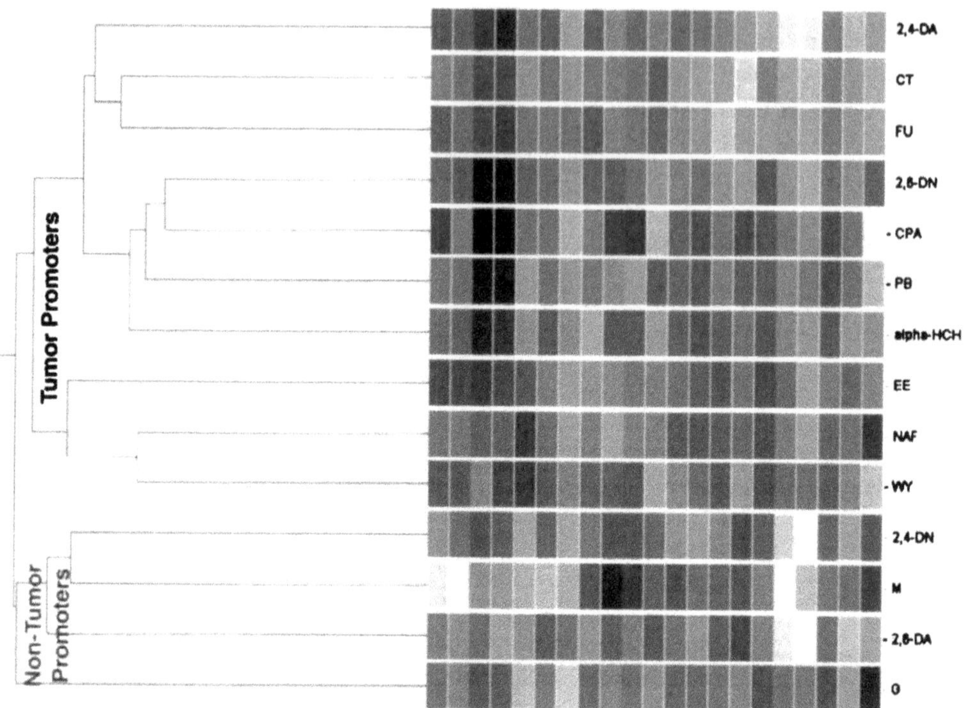


**FIGURE 5.6** Hierarchical clustering based on genes discriminating between tumor promoting and non-tumor-promoting agents. Black bars represent upregulated genes and white bars represent downregulated genes compared to the average expression of all experiments included. The Pearson correlation coefficient was set to 0.95. Experimental settings and abbreviations are explained in the text.

immunohistochemistry, proteomics, or metabolic studies are likely to further improve the interpretation of gene expression data. Taken together, transcription profiling has already become an important contribution to toxicological research. The analyses performed so far have revealed several crucial parameters of study design that may affect the relevance of the data generated and will aid us in improving future experiments. However, joint efforts by the scientific community to develop common standards for gene expression data will be necessary to extract as much physiological meaning as possible out of the toxicogenomics data generated. Following this and other investments, such as enlargement of the databases and refinement of bioinformatic tools, the rewards gained from toxicogenomics will quickly dwarf the effort spent.

## ACKNOWLEDGMENTS

We would like to thank Elard Jacob for the successful cooperation during our joint tumor promoter study.

## REFERENCES

1. Drews, J., Drug discovery: a historical perspective, *Science*, 287(5460), 1960, 2000.
2. Johnson, D.E. and Wolfgang, G.H., Predicting human safety: screening and computational approaches, *Drug Discov. Today*, 5(10), 445, 2000.
3. Gelbke, H.P., Hildebrand, B., and Kerfoot, E.J., Toxicological evaluation methods for the registration of chemicals, in *Toxicology*, Marquardt, H., Schäfer, S., McClellan, R., and Welsch, F., Eds., Academic Press, New York, 1999, p. 1115.

4. Durham, S.K. and Pearl, G.M., Computational methods to predict drug safety and liabilities, *Curr. Opin. Drug Disc. Dev.*, 4(1), 110, 2001.
5. Cronin, M.T., Prediction of drug toxicity, *Farmaco*, 56(1–2), 149, 2001.
6. Barratt, M.D. and Rodford, R.A., The computational prediction of toxicity, *Curr. Opin. Chem. Biol.*, 5(4), 383, 2001.
7. Hamadeh, H.K., Amin, R.P., Paules, R.S., and Afshari, C.A., An overview of toxicogenomics, *Curr. Issues Mol. Biol.*, 4(2), 45, 2002.
8. Aardema, M.J. and MacGregor, J.T., Toxicology and genetic toxicology in the new era of "toxicogenomics": impact of "-omics" technologies, *Mutat. Res.*, 499(1), 13, 2002.
9. Furness, L.M., Analysis of gene and protein expression for drug mode of toxicity, *Curr. Opin. Drug Discov. Dev.*, 5(1), 98, 2002.
10. Storck, T., von Brevern, M.C., Behrens, C.K., Scheel, J., and Bach, A., Transcriptomics in predictive toxicology, *Curr. Opin. Drug Discov. Dev.*, 5(1), 90, 2002.
11. Ulrich, R. and Friend, S.H., Toxicogenomics and drug discovery: will new technologies help us produce better drugs?, *Nat. Rev. Drug Discov.*, 1, 84, 2002.
12. Boguslavsky, J., Genes hold the key to understanding toxicity, *Drug Discov. Dev.*, 2001.
13. Burchiel, S.W., Knall, C.M., Davis, J.W., 2nd, Paules, R.S., Boggs, S.E., and Afshari, C.A., Analysis of genetic and epigenetic mechanisms of toxicity: potential roles of toxicogenomics and proteomics in toxicology, *Toxicol. Sci.*, 59(2), 193, 2001.
14. Clarke, P.A., te Poele, R., Wooster, R., and Workman, P., Gene expression microarray analysis in cancer biology, pharmacology, and drug development: progress and potential, *Biochem. Pharmacol.*, 62(10), 1311, 2001.
15. Hamadeh, H.K., Bushel, P., Paules, R., and Afshari, C.A., Discovery in toxicology: mediation by gene expression array technology, *J. Biochem. Mol. Toxicol.*, 15(5), 231, 2001.
16. Heck, D.E., Roy, A., and Laskin, J.D., Nucleic acid microarray technology fortoxicology: promise and practicalities, *Adv. Exp. Med. Biol.*, 500, 709, 2001.
17. Fielden, M.R. and Zacharewski, T.R., Challenges and limitations of gene expression profiling in mechanistic and predictive toxicology, *Toxicol. Sci.*, 60(1), 6, 2001.
18. Iannaccone, P.M., Toxicogenomics: "the call of the wild chip," *Environ. Health Perspect.*, 109(1), A8, 2001.
19. Kaminski, N., Allard, J., and Heller, R.A., Use of oligonucleotide arrays to analyze drug toxicity, *Ann. N.Y. Acad. Sci.*, 919, 1, 2000.
20. Lovett, R.A., Toxicogenomics. Toxicologists brace for genomics revolution, *Science*, 289(5479), 536, 2000.
21. Pennie, W.D., Use of cDNA microarrays to probe and understand the toxicological consequences of altered gene expression, *Toxicol. Lett.*, 112–113, 473, 2000.
22. Pennie, W.D., Tugwood, J.D., Oliver, G.J., and Kimber, I., The principles and practice of toxigenomics: applications and opportunities, *Toxicol. Sci.*, 54(2), 277, 2000.
23. Steiner, S. and Anderson, N.L., Expression profiling in toxicology: potentials and limitations, *Toxicol. Lett.*, 112–113, 467, 2000.
24. Robertson, D.G. and Bulera, S.J., High-throughput toxicology: practical considerations, *Curr. Opin. Drug Discov. Dev.*, 3(1), 42, 2000.
25. Rockett, J.C. and Dix, D.J., DNA arrays: technology, options and toxicological applications, *Xenobiotica*, 30(2), 155, 2000.
26. Afshari, C.A., Nuwaysir, E.F., and Barrett, J.C., Application of complementary DNA microarray technology to carcinogen identification, toxicology, and drug safety evaluation, *Cancer Res.*, 59(19), 4759, 1999.
27. Farr, S. and Dunn, R.T., 2nd, Concise review: gene expression applied to toxicology, *Toxicol. Sci.*, 50(1), 1, 1999.
28. Nuwaysir, E.F., Bittner, M., Trent, J., Barrett, J.C., and Afshari, C.A., Microarrays and toxicology: the advent of toxicogenomics, *Mol. Carcinog.*, 24(3), 153, 1999.
29. Rockett, J.C., Esdaile, D.J., and Gibson, G.G., Differential gene expression in drug metabolism and toxicology: practicalities, problems and potential, *Xenobiotica*, 29(7), 655, 1999.
30. Rodi, C.P., Bunch, R.T., Curtiss, S.W., Kier, L.D., Cabonce, M.A., Davila, J.C., Mitchell, M.D., Alden, C.L., and Morris, D.L., Revolution through genomics in investigative and discovery toxicology, *Toxicol. Pathol.*, 27(1), 107, 1999.

31. Todd, M.D. and Ulrich, R.G., Emerging technologies for accelerated toxicity evaluation of potential drug candidates, *Curr. Opin. Drug Discov. Dev.*, 2(1), 58, 1999.
32. Bailey, D.S., Bondar, A., and Furness, L.M., Pharmacogenomics: it's not just pharmacogenetics, *Curr. Opin. Biotechnol.*, 9(6), 595, 1998.
33. Burczynski, M.E., McMillian, M., Ciervo, J., Li, L., Parker, J.B., Dunn, R.T., 2nd, Hicken, S., Farr, S., and Johnson, M.D., Toxicogenomics-based discrimination of toxic mechanism in HepG2 human hepatoma cells, *Toxicol. Sci.*, 58(2), 399, 2000.
34. Bulera, S.J., Eddy, S.M., Ferguson, E., Jatkoe, T.A., Reindel, J.F., Bleavins, M.R., and De La Iglesia, F.A., RNA expression in the early characterization of hepatotoxicants in Wistar rats by high-density DNA microarrays, *Hepatology*, 33(5), 1239, 2001.
35. Gerhold, D., Lu, M., Xu, J., Austin, C., Caskey, C.T., and Rushmore, T., Monitoring expression of genes involved in drug metabolism and toxicology using DNA microarrays, *Physiol. Genomics*, 5(4), 161, 2001.
36. Waring, J.F., Jolly, R.A., Ciurlionis, R., Lum, P.Y., Praestgaard, J.T., Morfitt, D.C., Buratto, B., Roberts, C., Schadt, E., and Ulrich, R.G., Clustering of hepatotoxins based on mechanism of toxicity using gene expression profiles, *Toxicol. Appl. Pharmacol.*, 175(1), 28, 2001.
37. Waring, J.F., Ciurlionis, R., Jolly, R.A., Heindel, M., and Ulrich, R.G., Microarray analysis of hepatotoxins *in vitro* reveals a correlation between gene expression profiles and mechanisms of toxicity, *Toxicol. Lett.*, 120(1–3), 359, 2001.
38. Zhang, L., Zhou, W., Velculescu, V.E., Kern, S.E., Hruban, R.H., Hamilton, S.R., Vogelstein, B., and Kinzler, K.W., Gene expression profiles in normal and cancer cells, *Science*, 276(5316), 1268, 1997.
39. Gerhold, D., Rushmore, T., and Caskey, C.T., DNA chips: promising toys have become powerful tools, *Trends Biochem. Sci.*, 24(5), 168, 1999.
40. Pilarsky, C.P., Schmitt, A.O., Dahl, E., and Rosenthal, A., Microarrays-changes and challenges, *Curr. Opin. Mol. Therap.*, 1(6), 727, 1999.
41. Pennisi, E., Genomics. Rat genome off to an early start, *Science*, 289(5483), 1267, 2000.
42. Martin, K.J. and Pardee, A.B., Principles of differential display, *Meth. Enzymol.*, 303, 234, 1999.
43. Diatchenko, L., Lau, Y.F., Campbell, A.P., Chenchik, A., Moqadam, F., Huang, B., Lukyanov, S., Lukyanov, K., Gurskaya, N., Sverdlov, E.D., and Siebert, P.D., Suppression subtractive hybridization: a method for generating differentially regulated or tissue-specific cDNA probes and libraries, *Proc. Natl. Acad. Sci. USA*, 93(12), 6025, 1996.
44. Brenner, S., Johnson, M., Bridgham, J., Golda, G., Lloyd, D.H., Johnson, D., Luo, S., McCurdy, S., Foy, M., Ewan, M., Roth, R., George, D., Eletr, S., Albrecht, G., Vermaas, E., Williams, S.R., Moon, K., Burcham, T., Pallas, M., DuBridge, R.B., Kirchner, J., Fearon, K., Mao, J., and Corcoran, K., Gene expression analysis by massively parallel signature sequencing (MPSS) on microbead arrays, *Nat. Biotechnol.*, 18(6), 630, 2000.
45. Velculescu, V.E., Zhang, L., Vogelstein, B., and Kinzler, K.W., Serial analysis of gene expression, *Science*, 270(5235), 484, 1995.
46. Thai, S.F., Allen, J.W., DeAngelo, A.B., George, M.H., and Fuscoe, J.C., Detection of early gene expression changes by differential display in the livers of mice exposed to dichloroacetic acid, *Carcinogenesis*, 22(8), 1317, 2001.
47. Harris, A.J., Shaddock, J.G., Manjanatha, M.G., Lisenbey, J.A., and Casciano, D.A., Identification of differentially expressed genes in aflatoxin B1-treated cultured primary rat hepatocytes and Fischer 344 rats, *Carcinogenesis*, 19(8), 1451, 1998.
48. Satoh, T., Toyoda, M., Hoshino, H., Monden, T., Yamada, M., Shimizu, H., Miyamoto, K., and Mori, M., Activation of peroxisome proliferator-activated receptor-gamma stimulates the growth arrest and DNA-damage inducible 153 gene in non-small cell lung carcinoma cells, *Oncogene*, 21(14), 2171, 2002.
49. Sun, Y., Identification and characterization of genes responsive to apoptosis: application of DNA chip technology and mRNA differential display, *Histol. Histopathol.*, 15(4), 1271, 2000.
50. Bhattacharjee, A., Lappi, V.R., Rutherford, M.S., and Schook, L.B., Molecular dissection of dimethylnitrosamine (DMN)-induced hepatotoxicity by mRNA differential display, *Toxicol. Appl. Pharmacol.*, 150(1), 186, 1998.

51. Kegelmeyer, A.E., Sprankle, C.S., Horesovsky, G.J., and Butterworth, B.E., Differential display identified changes in mRNA levels in regenerating livers from chloroform-treated mice, *Mol. Carcinog.*, 20(3), 288, 1997.
52. Rockett, J.C., Esdaile, D.J., and Gibson, G.G., Molecular profiling of non-genotoxic hepatocarcinogenesis using differential display reverse transcription-polymerase chain reaction (ddRT-PCR), *Eur. J. Drug Metab. Pharmacokinet.*, 22(4), 329, 1997.
53. Luo, L., Salunga, R.C., Guo, H., Bittner, A., Joy, K.C., Galindo, J.E., Xiao, H., Rogers, K.E., Wan, J.S., Jackson, M.R., and Erlander, M.G., Gene expression profiles of laser-captured adjacent neuronal subtypes, *Nat. Med.*, 5(1), 117, 1999.
54. Ohyama, H., Zhang, X., Kohno, Y., Alevizos, I., Posner, M., Wong, D.T., and Todd, R., Laser capture microdissection-generated target sample for high-density oligonucleotide array hybridization, *Biotechniques*, 29(3), 530, 2000.
55. Walch, A., Specht, K., Smida, J., Aubele, M., Zitzelsberger, H., Hofler, H., and Werner, M., Tissue microdissection techniques in quantitative genome and gene expression analyses, *Histochem. Cell. Biol.*, 115(4), 269, 2001.
56. Leethanakul, C., Patel, V., Gillespie, J., Pallente, M., Ensley, J.F., Koontongkaew, S., Liotta, L.A., Emmert-Buck, M., and Gutkind, J.S., Distinct pattern of expression of differentiation and growth-related genes in squamous cell carcinomas of the head and neck revealed by the use of laser capture microdissection and cDNA arrays, *Oncogene*, 19(28), 3220, 2000.
57. Sgroi, D.C., Teng, S., Robinson, G., LeVangie, R., Hudson, J.R., Jr., and Elkahloun, A.G., *In vivo* gene expression profile analysis of human breast cancer progression, *Cancer Res.*, 59(22), 5656, 1999.
58. Lewis, F., Maughan, N.J., Smith, V., Hillan, K., and Quirke, P., Unlocking the archive: gene expression in paraffin-embedded tissue, *J. Pathol.*, 195(1), 66, 2001.
59. Shibutani, M., Uneyama, C., Miyazaki, K., Toyoda, K., and Hirose, M., Methacarn fixation: a novel tool for analysis of gene expressions in paraffin-embedded tissue specimens, *Lab. Invest.*, 80(2), 199, 2000.
60. Corton, J.C. and Stauber, A.J., Toward construction of a transcript profile database predictive of chemical toxicity, *Toxicol. Sci.*, 58(2), 217, 2000.
61. Frueh, F.W., Hayashibara, K.C., Brown, P.O., and Whitlock, J.P., Use of cDNA microarrays to analyze dioxin-induced changes in human liver gene expression, *Toxicol. Lett.*, 122(3), 189, 2001.
62. Harries, H.M., Fletcher, S.T., Duggan, C.M., and Baker, V.A., The use of genomics technology to investigate gene expression changes in cultured human liver cells, *Toxicol. in Vitro*, 15(4–5), 399, 2001.
63. Bhatia, S.N., Balis, U.J., Yarmush, M.L., and Toner, M., Effect of cell–cell interactions in preservation of cellular phenotype: cocultivation of hepatocytes and nonparenchymal cells, *FASEB J.*, 13(14), 1883, 1999.
64. Sauer, I.M., Obermeyer, N., Kardassis, D., Theruvath, T., and Gerlach, J.C., Development of a hybrid liver support system, *Ann. N.Y. Acad. Sci.*, 944, 308, 2001.
65. Tsiaoussis, J., Newsome, P.N., Nelson, L.J., Hayes, P.C., and Plevris, J.N., Which hepatocyte will it be? Hepatocyte choice for bioartificial liver support systems, *Liver Transpl.*, 7(1), 2, 2001.
66. Lupp, A., Danz, M., and Muller, D., Morphology and cytochrome P450 isoforms expression in precision-cut rat liver slices, *Toxicology*, 161(1–2), 53, 2001.
67. Muller, D., Steinmetzer, P., Pissowotzki, K., and Glockner, R., Induction of cytochrome P450 2B1-mRNA and pentoxyresorufin O-depentylation after exposure of precision-cut rat liver slices to phenobarbital, *Toxicology*, 144(1–3), 93, 2000.
68. Hess, K.R., Zhang, W., Baggerly, K.A., Stivers, D.N., and Coombes, K.R., Microarrays: handling the deluge of data and extracting reliable information, *Trends Biotechnol.*, 19(11), 463, 2001.
69. Cadet, J.L., Jayanthi, S., McCoy, M.T., Vawter, M., and Ladenheim, B., Temporal profiling of methamphetamine-induced changes in gene expression in the mouse brain: evidence from cDNA array, *Synapse*, 41(1), 40, 2001.
70. Scherf, U., Ross, D.T., Waltham, M., Smith, L.H., Lee, J.K., Tanabe, L., Kohn, K.W., Reinhold, W.C., Myers, T.G., Andrews, D.T., Scudiero, D.A., Eisen, M.B., Sausville, E.A., Pommier, Y., Botstein, D., Brown, P.O., and Weinstein, J.N., A gene expression database for the molecular pharmacology of cancer, *Nat. Genet.*, 24(3), 236, 2000.
71. Joliffe, I.T., *Principal Component Analysis*, Springer-Verlag, New York, 1986.
72. Cox, T. and Cox, M., *Multidimensional Scaling*, Chapman & Hall, London, 1994.

73. Tenenbaum, J.B., de Silva, V., and Langford, J.C., A global geometric framework for nonlinear dimensionality reduction, *Science*, 290(5500), 2319, 2000.
74. Roweis, S.T. and Saul, L.K., Nonlinear dimensionality reduction by locally linear embedding, *Science*, 290(5500), 2323, 2000.
75. Hughes, T.R., Marton, M.J., Jones, A.R., Roberts, C.J., Stoughton, R., Armour, C.D., Bennett, H.A., Coffey, E., Dai, H., He, Y.D., Kidd, M.J., King, A.M., Meyer, M.R., Slade, D., Lum, P.Y., Stepaniants, S.B., Shoemaker, D.D., Gachotte, D., Chakraburtty, K., Simon, J., Bard, M., and Friend, S.H., Functional discovery via a compendium of expression profiles, *Cell*, 102(1), 109, 2000.
76. Hamadeh, H.K., Bushel, P.R., Jayadev, S., Martin, K., DiSorbo, O., Sieber, S., Bennett, L., Tennant, R., Stoll, R., Barrett, J.C., Blanchard, K., Paules, R.S., and Afshari, C.A., Gene expression analysis reveals chemical-specific profiles, *Toxicol. Sci.*, 67(2), 219, 2002.
77. Pritchard, C.C., Hsu, L., Delrow, J., and Nelson, P.S., Project normal: defining normal variance in mouse gene expression, *Proc. Natl. Acad. Sci. USA*, 98(23), 13266, 2001.
78. Bartosiewicz, M., Trounstine, M., Barker, D., Johnston, R., and Buckpitt, A., Development of a toxicological gene array and quantitative assessment of this technology, *Arch. Biochem. Biophys.*, 376(1), 66, 2000.
79. Brazma, A., Hingamp, P., Quackenbush, J., Sherlock, G., Spellman, P., Stoeckert, C., Aach, J., Ansorge, W., Ball, C.A., Causton, H.C., Gaasterland, T., Glenisson, P., Holstege, F.C., Kim, I.F., Markowitz, V., Matese, J.C., Parkinson, H., Robinson, A., Sarkans, U., Schulze-Kremer, S., Stewart, J., Taylor, R., Vilo, J., and Vingron, M., Minimum information about a microarray experiment (MIAME): toward standards for microarray data, *Nat. Genet.*, 29(4), 365, 2001.
80. Thomas, R.S., Rank, D.R., Penn, S.G., Zastrow, G.M., Hayes, K.R., Pande, K., Glover, E., Silander, T., Craven, M.W., Reddy, J.K., Jovanovich, S.B., and Bradfield, C.A., Identification of toxicologically predictive gene sets using cDNA microarrays, *Mol. Pharmacol.*, 60(6), 1189, 2001.
81. Baldi, P. and Long, A.D., A Bayesian framework for the analysis of microarray expression data: regularized *t*-test and statistical inferences of gene changes, *Bioinformatics*, 17(6), 509, 2001.

# Section 3

## Model Systems in Toxicogenomic Studies

# 6 Microarrays in Drug Metabolism and Toxicology: Rat Liver Responses to Prototypical CYP Inducers

*David Gerhold, Jian Xu, and Thomas Rushmore*

CONTENTS



0-8493-1334-1/03/$0.00+$1.50

## 6.1 INTRODUCTION

Oligonucleotide DNA microarrays were evaluated for measuring classical and nonclassical genomic responses to prototypical cytochrome P450 (CYP) inducers by monitoring global expression profiles of drug metabolism genes in rat liver. This study[1] was performed to investigate the feasibility of using microarray technology to minimize the long and expensive process of testing drug candidates for safety in animals. Gene expression "profiles" were generated from livers of rats treated with 3-methylcholanthrene (3MC), phenobarbital (PB), dexamethasone (DEX), or clofibrate (CLOF) relative to vehicle-treated controls. The mRNA responses were measured using a Merck Drug Safety Chip employing the 25-mer oligodeoxynucleotide microarray technology from Affymetrix, Inc. (Santa Clara, CA). This microarray design included drug metabolism genes such as the cytochromes P450 (*CYP*), epoxide hydrolases (*EH*), UDP-glucuronosyltransferases (*UGT*), glutathione sulfotransferases (*GST*), sulfotransferases (*ST*), and drug transporter genes. Additional genes included on the array addressed broader safety issues including peroxisomal genes, stress responses, and carbohydrate and lipid metabolism. Transcriptional responses measured with this array were in excellent agreement with published observations. Additional gene regulatory responses were noted that characterized drug metabolism effects, metabolic effects, or stress responses to these compounds. In order to critically evaluate this microarray technology for further drug safety investigations, we assayed the gene specificity of the microarray for genes in four *CYP* subfamilies and checked microarray observations for six genes using TaqMan® quantitative real-time polymerase chain reaction (QRT-PCR). The gene chip technology from Affymetrix was capable of distinguishing human *CYP* genes up to a threshold of approximately 90% DNA identity. QRT-PCR technology for six genes demonstrated that microarray data were qualitatively reliable but underestimated -fold changes when an mRNA species was below a threshold of detection in one of the samples. These data provide a foundation for use of microarray technology in assaying gene expression responses relevant to drug metabolism and toxicology.

## 6.2 BACKGROUND

### 6.2.1 The Liver

Mammals metabolize xenobiotics and hormones primarily in the liver. The liver also coordinates energy metabolism, integrating levels of circulating glucose and lipids with glycogen stores in muscle and liver, triglyceride stores in adipose, and cholesterol stores throughout the body. In addition to direct enzymatic transformations of sugars and lipids, the liver also homeostatically regulates energy metabolism by breaking down and producing steroid, eicosanoid, and peptide hormones.[2] This integrative homeostatic function of the liver means that gene regulatory events in liver reflect the detailed state of the organism; thus, alterations in liver gene expression in response to a given treatment, termed a gene expression *profile*, can give us detailed insights into the effects of drugs on the organism.

### 6.2.2 Drug Metabolism Phases I to III

Xenobiotics and steroid hormones are metabolized in three phases.[3] In phase I, drug-metabolizing enzymes modify substrates directly via oxidation, hydroxylation, or dealkylation. Phase I drug metabolism, usually by one or two cytochrome P450 enzymes, is rate limiting for most drugs. In phase II, drugs are covalently linked to sulfate, glutathione, or carbohydrates by conjugating enzymes. Finally, in phase III, drugs are directionally transported from plasma to liver, and drug conjugates from liver into the bile or urine by transporters.

### 6.2.3 Transcriptional Regulation of Drug Metabolism Genes

Genes encoding drug-metabolizing enzymes are transcriptionally regulated by a network of soluble receptor-transcription factors that recognize both xenobiotics and endogenous compounds (Figure 6.1).[2] These receptor–transcription factors include the aryl hydrocarbon receptor (AHR), a basic helix–loop–helix transcription factor that heterodimerizes with cotranscription factor aryl hydrocarbon receptor nuclear translocator (ARNT). They also include nuclear receptors that regulate transcription as heterodimers with the retinoid X receptor (RXR), in concert with tissue-specific transcription factors, and transcription factors that respond to various stresses and cytokines.[2] Nuclear receptors that regulate drug-metabolizing genes include the pregnane X receptor (PXR), the constitutive androstane receptor (CAR), the glucocorticoid receptor, the peroxisome proliferator activated receptor alpha (PPARα), the liver X receptor (LXR), and the Farnesoid X-activated receptor (FXR). Such a dynamic transcriptional regulatory system provides an opportunity for study by microarray analysis.

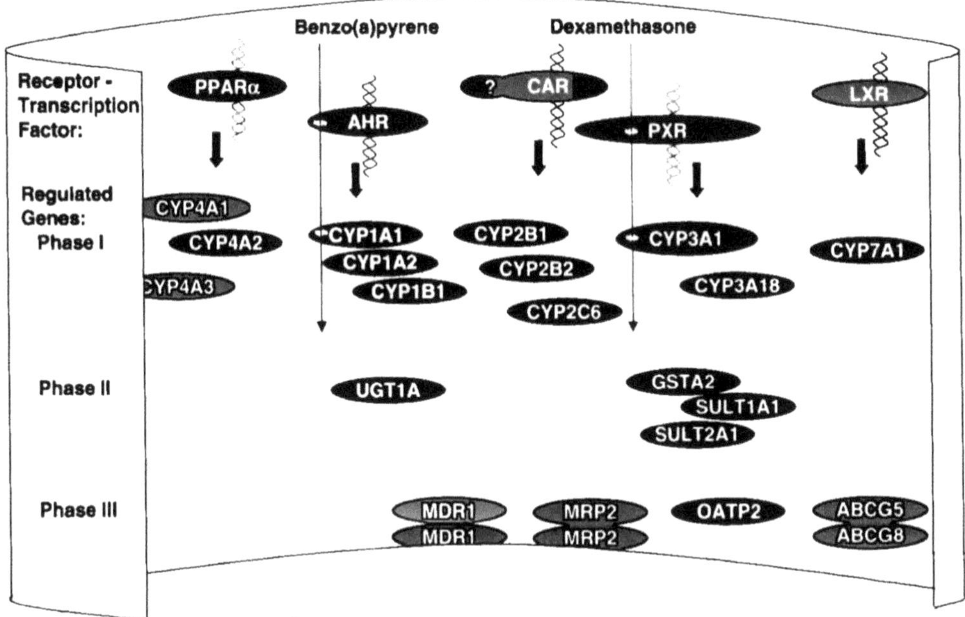


**FIGURE 6.1** Drug metabolizing genes are regulated by a network of receptor–transcription factors. These gene regulatory relationships were derived from the literature for the rat. Physical space in the schematic is occupied by compounds according to structures, by receptors according to ligand-binding specificities, and by enzymes according to substrate specificities. Xenobiotic compounds bind to receptors (intersected by solid arrows) to activate transcription of genes with receptor-response elements in their promoters. Colored arrows indicate the transcriptional regulation of drug metabolizing genes (below arrow) by the receptor (above arrow). Note that this schematic is a gross oversimplification. Some compounds activate several receptors, several receptors can regulate the same genes, and only a few induced genes are shown for simplicity. Some commercial drug compounds are not significantly metabolized in the liver, and some compounds may elude all receptors. The substrates for drug conjugating and drug transporter proteins are largely unknown. Phenobarbital triggers induction of CYP genes by CAR indirectly, as it does not bind to CAR. References are cited in the text except for MDR1,[46] OATP2,[47] CYP7A1,[48] ABCG5, and ABCG8[49] and for the metabolism of benzo(*a*) pyrene[50] and dexamethasone.[10] *Abbreviations:* PPAR, peroxisome-proliferator-activated receptor; AHR, aryl hydrocarbon receptor; PXR, pregnane X receptor; CAR, constitutive androstane receptor; LXR, liver X receptor; CYP, cytochrome P450; GST, glutathione-*S*-transferase; UGT, UDP-glucuronosyltransferase; ABC, ATP-dependent transporter; SULT, sulfotransferase; MDR, multidrug resistance protein; OATP, organic anion transporter protein; and MRP, multidrug-resistance-associated protein. Note that MRP2 and cMOAT are synonymous. (See color insert following page 112.)

Although many examples of drug metabolism genes are known to be regulated directly by a nuclear receptor, microarray analyses are beginning to reveal a global picture of how the liver responds to xenobiotics. Figure 6.1 schematically illustrates some general relationships between metabolism of xenobiotics and the transcriptional regulatory network that governs drug metabolism gene expression in the rat liver. These relationships were compiled from published literature for genes that appear to be directly regulated by each receptor. The top plane of the drawing represents two-dimensional chemical space such that each compound occupies space according to its structure, each receptor occupies space according to its ligand-binding specificity, and each enzyme occupies space according to its substrate specificity. While this schematic is a gross oversimplification and our knowledge of the networks is incomplete, we will point out some useful principles and limitations of the schema. A compound often, but not always, transcriptionally induces genes that metabolize that compound. A compound may activate several receptors, one receptor, or potentially no receptors. Similarly, compounds may be metabolized competitively by several (phase I) enzymes, by one enzyme, or by no enzymes. Several receptors can regulate the same genes (see Figure 6.1, MDR1 and MRP2), and only a few induced genes are shown for simplicity. The specific substrates for drug conjugating and drug transporter proteins are unknown for most compounds; hence, phase II and III metabolism of the compounds is not shown. It is important to note that drug metabolism genes are also regulated posttranscriptionally. Xenobiotics are known to regulate drug metabolism genes posttranscriptionally, directly or indirectly, by the following mechanisms: altering message stability, inhibiting enzyme activity, or altering protein stability.[4]

### 6.2.4 Drug Interactions

Because genetic variations (polymorphisms) are common in *CYP* genes, conjugating genes, and transporter genes in the human population, it would be useful to be able to monitor regulation of these enzymes in clinical subjects. If a drug is dependent on a single drug metabolizing enzyme for its turnover and an individual has inactive alleles for that gene, the drug may persist in that individual so as to cause toxicity or other adverse reactions. For example, inactive alleles of *CYP2C9* can lead to persistence and life-threatening toxicity of warfarin levels in serum.[5] Thus, it is necessary to identify the enzymes that metabolize a drug candidate prior to exposure to a human clinical trials.

A drug frequently induces expression at the transcriptional level of the proteins responsible for its metabolism.[4] Thus, the gene expression profile of rat liver or human hepatocytes treated with a drug may yield clues to how the drug is metabolized. While it may be postulated that genes induced by a drug metabolize that drug, many counterexamples are known, such that these guilt-by-association hypotheses must be evaluated by conventional enzyme and mass spectrometric assays. Because existing enzymatic and mass spectrometric analyses often fail to distinguish gene subfamily members involved in metabolism of a particular drug, DNA microarray assays can thus provide complementary information.

### 6.2.5 Reference Databases for Transcriptional Profiling

DNA microarray assays can also be used to compare transcriptional changes caused by drugs or other xenobiotics to elucidate mechanisms of adverse events.[6] In support of this concept, several investigators have demonstrated that liver toxins acting by similar mechanisms give rise to similar gene expression profiles.[7–9] This observation suggests the following comparative strategy for determining how an adverse event arises. Provided several compounds are known that give rise to that adverse event by known mechanisms, one could generate a database of gene expression profiles for such compounds. The profile of a novel compound could then be compared to the database to associate the novel profile with a known profile and its associated mechanism. Because the relationships between physiological events and transcriptional changes are still poorly understood, once again such a guilt-by-association hypothesis will demand further evaluation by other techniques.

The liver responds to xenobiotics with a dynamic transcriptional regulatory system; such a system is readily amenable to study by microarray analysis. By profiling livers of drug-treated rats or cultured primary human hepatocytes, one can identify transcriptional induction of drug metabolism genes. The dynamic transcriptional regulation of liver genes we observed in preliminary experiments led us to design the Merck Drug Safety Chip to address drug metabolism and toxicology issues using DNA microarray technology.[1] Because cytochrome P450s (CYPs) are dynamically regulated, rate limiting for clearance of most drugs, and relevant to toxicology, we focused initially on compounds that are known to induce *CYP1A*, *CYP2C*, *CYP3A*, and *CYP4A* gene family members.[10] The techniques and experiments described in this chapter were performed to demonstrate the feasibility and validity of this approach.

### 6.2.6 Gene Specificity Evaluation

In order to assess gene regulation of *CYP* and other genes that are closely related by DNA sequence, we evaluated Affymetrix GeneChip™ technology in these studies.[11] This microarray technology employs 25-mer oligodeoxynucleotides covalently anchored to the array surface. Such short synthetic oligonucleotides allow rational avoidance of DNA sequences that are shared by several genes or repeated in the genome and allow discrimination between gene family members that may differ in only one or several nucleotides per 25-mer sequence.

## 6.3 METHODS/TECHNOLOGIES

### 6.3.1 DNA Microarray Design and Synthesis (Affymetrix Format)

Genes were selected in aggregate from rat, human, and mouse for their roles in drug metabolism, toxicology, and energy metabolism. 1443 genes were identified and included in the chip design of the Merck Drug Safety Chip. Among the 300 rat genes included in the chip design were 130 genes classified as drug-metabolizing genes, including 51 *CYP* genes and 79 phase II and phase III drug-metabolizing genes and gene splice-forms. Each gene on the chip was represented by 20 pairs of 25-mer oligodeoxyribonucleotide probes. Each probe pair consisted of a perfect match (PM) oligonucleotide that precisely matched the cognate gene sequence and a mismatch (MM) oligonucleotide that differed only in a single nucleotide mismatch in the center position of the oligonucleotide. Selection of probes and manufacture of the gene chips were performed by Affymetrix.

### 6.3.2 Rat Treatment Protocols

Sprague–Dawley rats were purchased from Charles River Laboratories (Wilmington, MA). The rats were watered and fed chow *ad libitum*. Animal treatments were reviewed and approved by the Institutional Animal Care and Use Committee (IACUC). Rats were dosed orally (p.o.) once per day for three days with vehicle (0.1% methylcellulose) or 30 mg/kg/day of 3-methyl cholanthrene (3MC), dexamethasone (DEX), phenobarbital (PB), or clofibrate (CLOF). On day four, the rats were sacrificed and the livers were harvested and frozen immediately in liquid nitrogen. Each treatment group was represented by two replicate samples, using three pooled rat livers per samples.

### 6.3.3 RNA Preparation

The liver tissue was homogenized in Trizol™ (Gibco-BRL; Gaithersburg, MD) using a polytron (Omni International; Warrenton, VT), and total RNA was isolated from each sample according to the manufacturers' instructions. Total RNA was reprecipitated using RNAmate™ (Biochain; San Leandro, CA), and poly-A mRNA was isolated using oligo-dT-decorated latex beads (Qiagen; Hilden, Germany) according to manufacturers' instructions.

### 6.3.4 Hybridization and Staining

Hybridization samples were prepared according to Affymetrix instructions as described.[12] Briefly, a primer encoding the T7 RNA polymerase promoter linked to oligo-dT$_{17}$ was used to prime double-stranded cDNA synthesis from each mRNA sample using Superscript II RNAseH$^-$ reverse transcriptase (Life Technologies; Rockville, MD). Each double-stranded cDNA sample was purified by adsorption to silica (Qiaquick™ kit, Qiagen) according to the manufacturer's instructions, then *in vitro* transcribed using T7 RNA polymerase (MEGAscript T7 kit, Ambion, Inc.; Austin, TX), incorporating biotin-UTP and biotin-CTP (Enzo Biochemicals, Inc.; New York, NY) into the resulting copy-RNA (cRNA). These cRNA transcripts were purified using RNEasy™ (Qiagen) and quantitated by measuring absorption at 260 nm/280 nm. The 5-μg mRNA samples typically yielded between 30 and 150 μg purified cRNA. cRNA samples were fragmented at 95° for 35 min in 10-mM $MgCl_2$ to a mean size of ~50 to 150 nucleotides; they were then added to hybridization buffer and hybridized to the Merck Drug Safety Chip for 16 hours at 45°. Gene chips were washed, stained with streptavidin-R-phycoerythrin, and scanned with a dedicated instrument to capture a fluorescence image (Molecular Dynamics; Sunnyvale, CA).

### 6.3.5 Gene Spike Experiments

Biotinylated RNA samples were prepared from individual *CYP* genes (Genbank entry) for addition at known concentrations into hybridization samples as follows: CYP3A4, T60335; CYP3A5, AA740526; CYP3A7, AA455159; and CYP2A6, M33318. The cDNA clones were obtained from Frank Gonzalez and Harry Gelboin, National Institutes of Health; *CYP2C19* (Genbank entry M61854) was obtained from Joyce Goldstein, National Institute of Environmental Health Sciences; and *CYP2D6* (Genbank entry number M20403) was cloned in-house. The cDNA inserts representing the various genes were amplified by PCR using primers that match DNA sequences flanking the cDNA insertion sequences in each vector: T7 primer (5′TAATACGACTCACTATAGGG), T3-pCDNA3.1 primer (for pCDNA3.1, 5′AGATGCAATTAACCCTCACTAAAGGGAGAGAAG-GCACAGTCGAGGCTGA), or a T3 primer (for EST clones, 5′ATTAACCCTCACTAAAGGGA) that encodes a T3-RNA polymerase promoter sequence. PCR products were purified by adsorption to silica (Qiaquick™ kit, Qiagen) according to manufacturer's instructions. Each PCR product was then transcribed *in vitro* to generate the antisense strand of each gene using T3 RNA polymerase for pCDNA3.1 and pT7T3D-Pac, or T7 RNA polymerase for pBluescript SK, incorporating biotin-UTP and biotin-CTP into the resulting cRNA. These cRNA transcripts were purified using RNEasy™ (Qiagen), quantitated by measuring absorption at 260 nm/280 nm, and spiked into hybridizations at known concentrations.

### 6.3.6 Microarray Data Analysis

Data from each gene chip were normalized to data from a single vehicle gene chip using global scaling based on the overall hybridization intensities. Normalization, assessments of replicates, and calculation of gene expression levels as average difference values were performed using GeneChip® v.3.1 software (Affymetrix). Each treatment group was represented by two replicate samples, using three pooled rat livers per sample and two gene chips.

## 6.4 RESULTS

### 6.4.1 Data Analysis Methods

Each gene is represented on the Merck Drug Safety Chip by 20 pairs of 25-mer oligodeoxynucleotides, or probes. One 25-mer of each pair is a perfect match to the gene sequence, and the other member of the pair is a mismatch that differs from the first only at the central (13th) nucleotide.

The expression level of a gene was represented empirically as the average of the PM minus MM values for all 25-mer pairs that lie within 3 standard deviations of the mean of the probe set for that gene.[13] In order to minimize the occurrence of false-positive observations, we used the following empirical cutoff values. The absolute expression levels (average difference values) between two hybridizations differed by ≥20 units, and by a ratio ≥2-fold. A detailed discussion of the algorithm used is described in Gerhold et al.,[1] and we have since developed a statistical error model for Affymetrix data, called SAFER, that assigns confidence or *p*-values to data points and further reduces false-positive observations.[14]

### 6.4.2 Regulation of Drug Metabolism Genes

To validate application of DNA microarray technology to drug metabolism and drug safety, four prototypical inducers of *CYP* subgene families were used to treat male Sprague–Dawley rats prior to analysis of liver samples using the Merck Drug Safety Chip. These four compounds, 3-methyl cholanthrene (3MC), phenobarbital (PB), dexamethasone (DEX), and clofibrate (CLOF), are extensively studied inducers of the rat *CYP* subfamilies *CYP1A*, *CYP2B*, *CYP3A*, and *CYP4A*, respectively.[10] The regimen of oral administration once daily for three days and harvest on day four was selected, based on prior experience, to represent a steady-state induction response in the liver after exposure to the four test compounds. Figure 6.2 shows the results from microarray analysis of mRNA from livers of rats that had been treated with one of the four probe drugs or vehicle. Data are shown for all of the 300 rat genes represented on the array that showed a reproducibly altered expression in any of the four treatments. The results are presented as the mean expression levels in average difference units,[12] a linear indicator of mRNA expression level that is arbitrary in scale. Colors were assigned to datapoints to indicate induction (red) or suppression (green) relative to vehicle control.

The four *CYP*-inducer compounds activate transcription through four known receptor–transcription factors, as shown in Figure 6.1 and described below. 3MC activates transcription of the *CYP1* family through the aromatic hydrocarbon receptor (AHR);[4] phenobarbital activates transcription of the *CYP2B* and *CYP3A* gene families through the constitutive androstane receptor (CAR);[15] dexamethasone activates transcription of the *CYP3A* and possibly *CYP2C* family through the pregnane X receptor (PXR);[16] and clofibrate activates transcription of the CYP4A family through the peroxisome proliferator activated receptor (PPARα).[17]

#### 6.4.2.1 Regulation of Drug Metabolism Genes by 3MC

3MC is an aromatic hydrocarbon that induces the expression of *CYP1A1*, *CYP1A2*, and *CYP1B1* by activating the AH receptor. The AH receptor is a helix–loop–helix protein that belongs to the period-aryl hydrocarbon receptor-single minded (PAS) family of transcription factors and regulates transcription of the CYP1 family genes. The AH receptor becomes activated by binding to an aromatic hydrocarbon in the cytosol. The complex translocates to the nucleus where it complexes with the nuclear factor Arnt. The activated ligand–receptor transcription factor complex then binds to the xenobiotic responsive element (XRE) in the *CYP1* gene promoter. Binding of the activated receptor–ligand complex activates transcription of the genes.[18] *CYP1A1* is undetectable in the liver of control (vehicle-treated) rats but was found to be highly expressed (induced) in the liver of 3MC-treated rats (Figure 6.2).[18] Induction of *CYP1A2* and *CYP1B1* was also observed in the livers from rats treated with 3MC. This is in agreement with published reports that clearly show that both *CYP1A2* and *CYP1B1* induction can occur by an AH-receptor-dependent mechanism.[18] In addition, several reports have indicated that both *CYP1A2* and *CYP1B1* induction can occur by an AH-receptor-independent mechanism, possibly occurring through the AP1 transcription complex.[18,19] *CYP2D4* was moderately induced by 3MC and by dexamethasone (Figure 6.2). Regulation of *CYP2D4* has not been previously explored.[20] Induction of several phase II enzymes — *UGT1A6*,

*GSTA1*, *GSTA2*, and *GSTM1* — was also observed in the liver recovered from rats treated with 3MC (Figure 6.2). 3MC is known to induce expression of the UDP-glucuronosyl transferase gene *UGT1A6* and glutathione-*S*-transferase gene *GSTA1*, via an Ah-receptor-dependent mechanism.[21,22] The *GSTA2* and *GSTM1* genes are both induced by the epoxide and hydroxylated metabolites of 3MC generated by CYP1A1. Both genes contain an antioxidant responsive element (ARE) in their promoter region (see Chapter 9). This *cis*-acting element has been shown to be responsive to the diol-metabolites of 3MC that can redox cycle and produce a pro-oxidative environment.[22] The glutathione-*S*-transferases *GSTA2* and *GSTM1* were previously observed to be 3MC-inducible in cultured rat hepatocytes.[23]

| Gene | Acc. # | Control | Clofib | Dex | Phenob | 3MC |
|---|---|---|---|---|---|---|
| **Phase I** | | | | | | |
| *CYP1A1* | X00469 | 1 | 1 | 2 | 2 | 93 |
| *CYP1B1* | U09540 | 1 | 1 | 0 | 1 | 32 |
| *CYP1A2* | K02422 | 110 | 40 | 51 | 43 | 530 |
| *CYP2A1* | J02669 | 27 | 64 | 15 | 15 | 10 |
| *CYP2A2* | J04187 | 140 | 190 | 43 | 59 | 74 |
| *CYP2B1* | M37134 | 2 | 47 | 14 | 270 | 0 |
| *CYP2B2* | K00996 | 7 | 40 | 33 | 130 | 16 |
| *CYP2B3* | M20406 | 74 | 35 | 28 | 25 | 25 |
| *CYP2D4* | AB008425 | 4 | 2 | 11 | 7 | 15 |
| *CYP3A1* | M10161 | 11 | 24 | 76 | 60 | 8 |
| *CYP3A18* | X79991 | 15 | 17 | 47 | 12 | 10 |
| *CYP4A1** | M14972 | 18 | 156 | 11 | 11 | 22 |
| *CYP4A2* | M57719 | 8 | 100 | 6 | 6 | 6 |
| *CYP4A3* | M33936 | 43 | 150 | 29 | 21 | 27 |
| *CYP4F1* | M94548 | 55 | 29 | 38 | 66 | 46 |
| *CYP4F4* | U39206 | 11 | 8 | 18 | 21 | 13 |
| *CYP4F6* | U39208 | 17 | 6 | 6 | 21 | 5 |
| *EHc* | X65083 | 5 | 29 | 2 | 0 | 1 |
| *EHm* | M26125 | 57 | 36 | 59 | 220 | 87 |
| **Phase II** | | | | | | |
| *UGT1A6* | J02612 | 25 | 57 | 58 | 41 | 72 |
| *UGT2B1* | M13506 | 4 | 3 | 30 | 47 | 2 |
| *GlPx1* | M21210 | 130 | 100 | 50 | 150 | 100 |
| *GSTA1* | K01931 | 150 | 66 | 250 | 270 | 270 |
| *GSTA2* | M25891 | 14 | 7 | 35 | 44 | 38 |
| *GSTA3* | S72505 | 30 | 39 | 27 | 57 | 34 |
| *GSTM1* | M11719 | 47 | 7 | 140 | 200 | 140 |
| *GSTM2* | J02592 | 54 | 22 | 79 | 76 | 59 |
| *EST* | M86758 | 56 | 86 | 27 | 48 | 44 |
| *PST* | X52883 | 170 | 250 | 290 | 101 | 90 |
| *HSST1* | M31363 | 23 | 15 | 45 | 9 | 27 |
| *HSST2* | M33329 | 8 | 7 | 60 | 6 | 7 |
| **Phase III** | | | | | | |
| *NTCP* | M77479 | | | 25 | | |
| *CMOAT* | D86086 | | 3 | | 34 | |

A

FIGURE 6.2

### 6.4.2.2 Regulation of Drug Metabolism Genes by Phenobarbital

The barbiturate phenobarbital is also capable of inducing genes via the activation of transcription. The rat genes *CYP2B1*, *CYP2B2*, and *CYP3A1/23* are all examples of genes induced after PB exposure.[10] In mouse, phenobarbital induction of *Cyp2b10* requires the phenobarbital response element (PBRE), an enhancer upstream of the *Cyp2b10* gene. CAR is a nuclear receptor that interacts with RXR to form CAR-RXR heterodimers, which bind to the PBRE in response to phenobarbital treatment, suggesting that CAR-RXR may play a role in phenobarbital induction of mouse *Cyp2b10*.[15] Binding of phenobarbital to the Ah receptor has not been established, suggesting the possible existence of an additional ligand-binding component of the Ah receptor–Arnt complex. The microarray data in Figure 6.2 shows induction of *CYP2B1*, *CYP2B2*, and *CYP3A1/23* after

| Other Genes | | | | | | |
|---|---|---|---|---|---|---|
| *HSLIP* | X51415 | 14 | 29 | 12 | 6 | 8 |
| *FACO* | J02752 | 36 | 430 | 22 | 17 | 22 |
| *BE* | K03249 | 2 | 410 | 0 | 4 | 1 |
| *CPT2* | U88295 | 12 | 41 | 13 | 9 | 10 |
| *MCACD* | J02791 | 39 | 80 | 26 | 28 | 27 |
| *KCAT* | J02749 | 42 | 680 | 19 | 60 | 60 |
| *LCAT* | U62803 | 46 | 22 | 72 | 41 | 26 |
| *APOA1* | M00001 | 240 | 140 | 560 | 340 | 260 |
| *APOA4* | X13629 | 46 | 0 | 190 | 84 | 22 |
| *APOC1* | X15512 | 600 | 320 | 440 | 740 | 300 |
| *PEPCK* | K03248 | 32 | 9 | 50 | 23 | 4 |
| *DALS* | J04044 | 1 | 27 | 1 | 5 | 6 |
| *PRCR* | D28966 | 7 | 44 | 0 | 4 | 0 |
| *HSP70* | L16764 | 14 | 64 | 22 | 21 | 15 |
| *SODxc* | X68041 | 10 | 30 | 13 | 16 | 13 |

B

**FIGURE 6.2** (Continued) Microarray gene regulation profiles of 3MC, phenobarbital, dexamethasone, and clofibrate in male rat livers. Expression data are shown in units of relative fluorescence for those genes that respond to one or more drugs within gene families that encode drug metabolism enzymes. Upregulation events are shaded black with white numerals and downregulation events are shaded gray. Events that do not meet explicit criteria (differ by ≥20 units and by a ratio ≥2-fold) but are consistent in replicate datasets are shaded dark gray with white numerals if upregulated or gray with black numerals if downregulated. Numbers represent mean expression level measurements in two separate gene chips and are given in arbitrary units. Genbank accession numbers representing each gene are given under the column heading "Acc. #". Gene abbreviations: *CYP*, cytochrome P450; *EHc*, cytosolic epoxide hydrolase; *EHm*, mitochondrial epoxide hydrolase; *UGT*, UDP-glucuronosyltransferase; *GlPx*, glutathione peroxidase; *GST*, glutathione sulfotransferase; *EST*, estradiol sulfotransferase; *PST*, phenol/aryl sulfotransferase; *HSST*, hydroxysteroid sulfotransferase; *NTCP*, Na/taurocholate cotransporting polypeptide; *CMOAT*, canalicular multispecific organic anion transporter; *FACO*, fatty acid/acyl CoA oxidase; *BE*, peroxisomal enoyl-coenzyme A hydratase/3-hydroxyacyl coenzyme A dehydrogenase "bifunctional enzyme"; *HSLIP*, hormone-sensitive lipase; *FACO*, acyl-CoA oxidase; *BE*, bifunctional enzyme; *CPT2*, carnitine palmitoyl transferase II; *MCACD*, medium-chain acyl-CoA dehydrogenase; *KCAT*, 3-ketoacyl-CoA thiolase; *LCAT*, lecithin–cholesterol acyltransferase; *APO*, apolipoproteins; *PEPCK*, phosphoenolpyruvate carboxykinase C; *DALS*, delta-aminolevulinate synthase; *PRCR*, prostacyclin- or prostaglandin I2 receptor; *HSP70*, heat shock protein 70; *SODxc*, extracellular superoxide dismutase. 8 of 20 probe pairs for *CYP4A1* were masked due to gene-nonspecific hybridization. This appears to result from a 185-bp perfect identity between the 3′ ends of the *CYP4A1* and *Apolipoprotein B* gene sequences (Genbank accession M14972 and M27440, respectively; unpublished comparison). (From Gerhold, D. et al., *Physiol. Genomics*, 5(4), 161–170, 2001. With permission.)

exposure to PB. We observed a moderate, but consistent suppression of *CYP2B3* by all four compounds. Little is known about *CYP2B3* function, but Yamada et al.[24] have reported expression and drug-mediated induction of *CYP2B3* in rat liver. *CYP2C7* is also reportedly induced by phenobarbital, but primarily in female rather than in male rats.[10] No CAR-RXR response element has been identified in the *CYP3A1* promoter region to date. In the human, both PXR and CAR receptor complexes have recently been shown to contribute to the transcriptional regulation of the *CYP3A4* gene.[25]

Induction of several phase II enzymes was also observed after treatment with PB. Significant increases in the specific mRNAs for microsomal epoxide hydrolase (EHm), *UGT2B1*, *GSTA1*, *GSTA2*, *GSTA3*, and *GSTM1* were observed in the livers recovered from rats treated with PB (Figure 6.2). PB is known to induce expression of the *CYP2B* genes by the aforementioned CAR-dependent mechanism. No PBRE sequence has been identified in the promoters for any of the phase II enzymes to date. In addition to the phase I and II enzyme induction, a significant increase in the cation transporter, *cMOAT (MRP2)*, was also observed. Several reports have been published after this work describing the regulation of *CYP3A* and *cMOAT/MRP2* genes by phenobarbital[26,27] which are in agreement with our results.

#### 6.4.2.3 Regulation of Drug Metabolism Genes by Dexamethasone

Dexamethasone is a synthetic glucocorticoid mimetic that is known to induce expression of several *CYP3A* subfamily genes. Recent reports have described the identification and cloning of a nuclear receptor *PXR* that can bind a variety of chemical structures leading to the induction of *CYP3A1/23* in the liver. This nuclear receptor has been identified in mouse (mPXR), rat (rPXR), rabbit (rbPXR), dog (dPXR), rhesus (rhPXR), and man (PAR, or SXR).[16,28] In a manner similar to that of the CAR, the PXR can also form heterodimeric complexes with the RXR and bind to a *cis*-acting sequence in the promoter region of the target genes. PXR-responsive *cis*-acting regulatory elements have been identified in the promoters of *CYP3A* genes in many of these species. In agreement with these findings, we observed a marked induction of *CYP3A1/23* and *CYP3A18* genes in the liver from rats treated with DEX (Figure 6.2). A moderate induction of *CYP2B1* and *CYP2B2* is also evident, as reported by Ronis.[29]

Induction of several of the phase II enzymes was also observed after treatment with dexamethasone. Significant increases in the specific mRNAs for *UGT1A6*, *UGT2B1*, *GSTA1*, *GSTA2*, *GSTM1*, and sulfotransferases *PST*, *HSST1*, and *HSST2* were detected in the mRNA recovered from the livers of rats treated with DEX (Figure 6.2). No *cis*-acting sequence for the PXR has been identified in the promoters of any of the phase II enzymes to date. A functional *cis*-acting sequence for the glucocorticoid receptor, or GRE, has been identified in the *GSTA1* and *GSTA2* genes.[22] In addition to the phase I and phase II enzyme induction, a significant decrease in the levels of mRNA for sulfotransferase *EST* and phase III transporter *NTCP* (Na/taurocholate cotransporting polypeptide) were also observed. Little is known about regulation of sulfotransferases (SULTs), but regulation of *NTCP*[30] has been reported in rat liver. DEX has recently been shown to be a potent agonist for the rat PXR, and the *MRP1* and *MRP2* genes have been found to be responsive to PXR agonists in liver and intestine.[31]

#### 6.4.2.4 Regulation of Drug Metabolism Genes by Clofibrate

Clofibrate is a lipid-lowering agent that triggers peroxisome proliferation in rodents. The increase in enzyme expression that accompanies the proliferation of peroxisomes is mediated through a fourth nuclear receptor, the peroxisome proliferator-activated receptor (alpha) (PPARα). Peroxisome proliferators (PPs) are considered to be agonists of PPARα receptor. As described previously for the CAR and PXR receptors, PPARα also forms heterodimeric complexes with RXR. The complex can bind agonist and interact with the *cis*-acting sequence (PPRE) in the promoters of responsive genes. PPARα is also activated by cellular long-chain fatty acid derivatives, stimulating β-oxidation of long-chain fatty acids in peroxisomes. A recent report suggests the possibility of an interaction between the activation of PPARα by an agonist and the p38 MAP kinase pathway.[32]

Clofibrate and other PPARα agonists are known to induce members of the *CYP4A* subfamily. Figure 6.2 shows transcriptional induction by CLOF of *CYP4A1*, *CYP4A2*, and *CYP4A3*. Additionally, a small but significant increase in the expression of *CYP2B1*, *CYP2B2*, and *CYP3A1/23* was also observed. No PPRE has been identified in either of the *CYP2B* genes or in the promoter of the *CYP3A1/23* gene, but the possibility of cross-talk among the PXR, CAR, and PPARα receptors has been described.[32]

Induction of several of the phase II enzymes was also observed after treatment with CLOF. A significant increase in the specific mRNA for *UGT1A6* and sulfotransferase *PST* was observed in the liver of rats treated with clofibrate (Figure 6.2). No *cis*-acting PPRE sequence has been identified in the promoters of any of the phase II enzymes to date. We also observed a significant decrease in the mRNA for several of the phase II and III enzymes. Significant decreases were seen for *GSTA1*, *GSTM1*, *GSTM2*, and the transporter *cMOAT* (*MRP2*). Little is known about their regulation in rat liver by peroxisome proliferators.

Each of the four compounds examined in this report directly regulates several drug metabolism genes through binding to a nuclear receptor or the AH receptor, thus activating the promoters of the genes. The data in Figure 6.2 indicate that each compound also regulates (up and down) a number of other drug metabolism genes by other mechanisms. Many of these are likely to be indirect mechanisms, including competition of the agonized-nuclear receptor for other transcription factors, such as RXR, CBP/p300, histone acetylases, and others. Such indirect mechanisms may also include primary regulation of a gene product or endogenous metabolic intermediate that alters metabolism of the natural ligands for nuclear receptors, altering expression of secondary genes.

### 6.4.3 Metabolic Energy and Stress-Related Gene Regulation

In addition to drug metabolizing genes, Figure 6.2 shows the induction or suppression of genes that relate to toxicological events and genes that regulate carbohydrate and lipid metabolism. 3MC and CLOF treatments, for example, downregulate lecithin–cholesterol acyltransferase (*LCAT*), apolipoprotein CI (*APOC1*), apolipoprotein AIV (*APOA4*), and phosphoenolpyruvate carboxykinase C (*PEPCK*). These regulatory events suggest an inertia toward decreased lipid turnover and decreased gluconeogenesis mediated by *PEPCK* in the liver. Consistent with the data for 3MC, AH receptor agonists have been found to cause lipid accumulation within hepatocytes *in vivo*[33] and to regulate lipid metabolism via the Ah receptor in cultured fibroblasts,[34] although the mechanism remains unclear. Induction of apolipoprotein genes *APOA1* and *APOA4* by phenobarbital and dexamethasone was also observed (Figure 6.2), consistent with increased high-density lipoprotein levels and increased apolipoprotein AI expression in response to these agents in previous reports.[35,36]

Clofibrate is known to increase lipoprotein uptake globally and increase fatty acid β-oxidation in both peroxisomes and mitochondria in liver.[37] The increased lipoprotein uptake is reflected in Figure 6.2 by induction of hormone-sensitive lipase (*HSLIP*). Similarly, decreased lipid levels in serum are suggested by downregulation of apolipoprotein genes *APOA1*, *APOA4*, and *APOC1*. Clofibrate suppression of *APOA1* expression is opposite that in human but is consistent with previous studies in male rodents.[38] Increased lipid β-oxidation in response to CLOF is indicated by induction of peroxisomal genes acyl-CoA oxidase (*FACO*), bifunctional enzyme (*BE*), and 3-ketoacyl-CoA thiolase (*KCAT*) and mitochondrial genes carnitine palmitoyl transferase II (*CPT2*) and medium-chain acyl-CoA dehydrogenase (*MCACD*).

Induction of mitochondrial delta-aminolevulinate synthase (*DALS*) indicates increased heme synthesis. The increase in heme synthesis likely supports increased CYP4A-enzyme synthesis, mitochondrial electron transport protein synthesis, or both. Fatty acid β-oxidation by peroxisomes and mitochondria are thought to cause oxidative stress by producing reactive oxygen intermediates.

Evidence of oxidative stress includes induction of the extracellular Cu–Zn superoxide dismutase (*SODxc*), and heat shock protein 70 (*HSP70*), as observed in hepatic oxidative stress caused by $CCl_4$.[39] Induction of the *SODxc* gene by peroxisome proliferators has not been previously reported, although Yoo[40] reports induction of the rat intracellular Cu–Zn superoxide dismutase (*SOD1*) gene by arachidonic acid.

Clofibrate induction of the prostacyclin- or prostaglandin $I_2$ receptor (*PRCR*) is also shown in Figure 6.2. Prostaglandin $I_2$ mediates blood vessel dilation and inhibits platelet activation.[41] Induction of *PRCR* may thus be a marker for the antithrombotic activity of clofibrate.[41] Induction of *PRCR* appears to be part of a larger shift in the balance of prostaglandin species in clofibrate treated liver. *CYP4A1*, *CYP4A2*, and *CYP4A3* genes are induced by clofibrate (Figure 6.2) and show varying degrees of omega-hydroxylase activity on prostaglandins[42] as well as fatty acids. The latter fatty acid omega-hydroxylase activity commits fatty acids to peroxisomal degradation. *CYP4F* also can form or degrade various prostaglandins,[43,44] although their activities are not well characterized. Figure 6.2 shows a trend toward downregulation of the *CYP4F1*, *CYP4F4*, and *CYP4F6* genes by CLOF. Thus, although the physiological implications are not clear, it appears likely that regulation of the *CYP4A-*, *CYP4F-*, and *PRCR* genes by clofibrate may decrease thrombotic activity by altering the prostaglandin balance in rats.

### 6.4.4 Gene Specificity Evaluation

Given the importance of identifying identities of induced genes, we evaluated the ability of the Affymetrix technology to discriminate between closely related genes using several *CYP* gene families.[1] Synthetic biotinyl-RNAs were prepared by *in vitro* transcription of cDNA clones for the following human genes: *CYP2A6*, *CYP2C19*, *CYP2D6*, *CYP3A4*, *CYP3A5*, and *CYP3A7*. These genes were used in hybridizations in the absence of any complex mammalian RNAs, using one gene from each subfamily per hybridization. Each gene was used in hybridizations at 30 pM, corresponding to roughly 1 part in 5000 in an mRNA population or about 60 transcripts per mammalian cell. DNA homology relationships between the 3′ 1000 bp of these human cytochrome P450s include DNA:DNA identities of 88 to 92% between *CYP2A6* and *CYP2A13* genes, between *CYP2C9* and *CYP2C19* genes, and between *CYP3A4*, *CYP3A5*, and *CYP3A7* genes.

Probes representing the *CYP2A6*, *CYP2C19*, *CYP2D6*, *CYP3A4*, and *CYP3A7* genes hybridized faithfully and selectively to their cognate genes. The *CYP3A5* probe set, however, showed significant heterologous hybridization to *CYP3A4* and *CYP3A7*. On average, the *CYP3A5* probes showed a 2.5-fold reduced hybridization to *CYP3A7* and a 10-fold reduced hybridization to *CYP3A4*. A detailed examination of the data identified a subset of 12 probe pairs that hybridized to the *CYP3A5* gene robustly and with 10-fold selectivity relative to *CYP3A4* and *CYP3A7*.[1]

### 6.4.5 Quantitation of mRNA Levels by QRT-PCR

An independent method, TaqMan® QRT-PCR,[45] was used to assay mRNA levels for six selected genes from this study. The mRNA levels were determined in each sample for *CYP1A1*, *CYP3A18*, *CYP4A1*, *GSTM1*, *BE*, and *APOA1* relative to an internal control, 18s rRNA.[1] The TaqMan® technique was selected as the gold standard for mRNA measurements due to its high gene specificity and sensitivity, requiring only a few mRNA molecules for detection. The data for these six genes were qualitatively consistent for the two techniques. While data for relatively abundant mRNAs/genes were also in quantitative agreement, mRNA species at very low abundance in vehicle- or drug-treated samples yielded artifactually compressed -fold change ratios in microarray data relative to the QRT-PCR data. The extreme example was the induction of *CYP1A1* mRNA by 3MC, detected at ≥9.3-fold by microarray but revealed to actually be ~5000-fold by QRT-PCR. Thus, microarrays can be used to search many genes for those that respond to a stimulus, and QRT-PCR can be used in a complementary way to confirm and extend observations on a few selected genes.

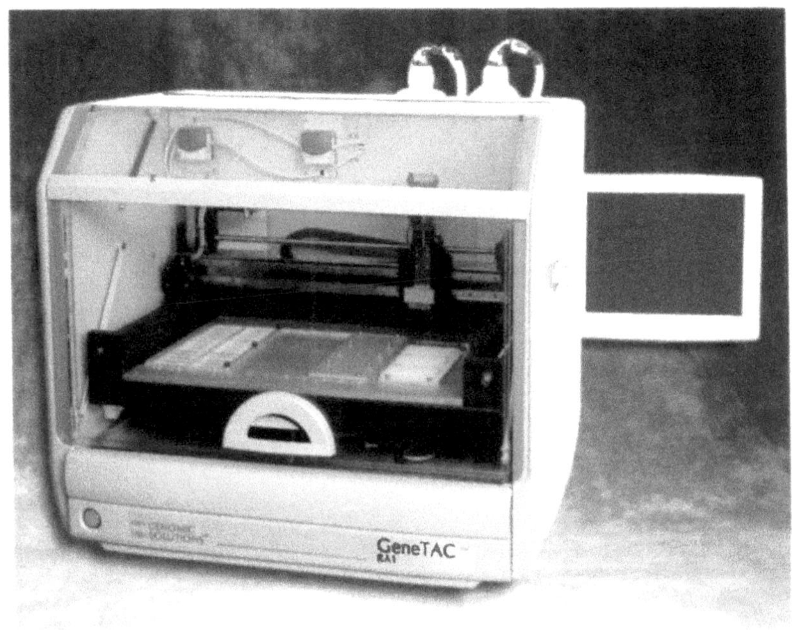


COLOR FIGURE 1.2 The GeneTac RA1 spotter. (Photograph courtesy of Genomics Solutions; Ann Arbor, MI.)

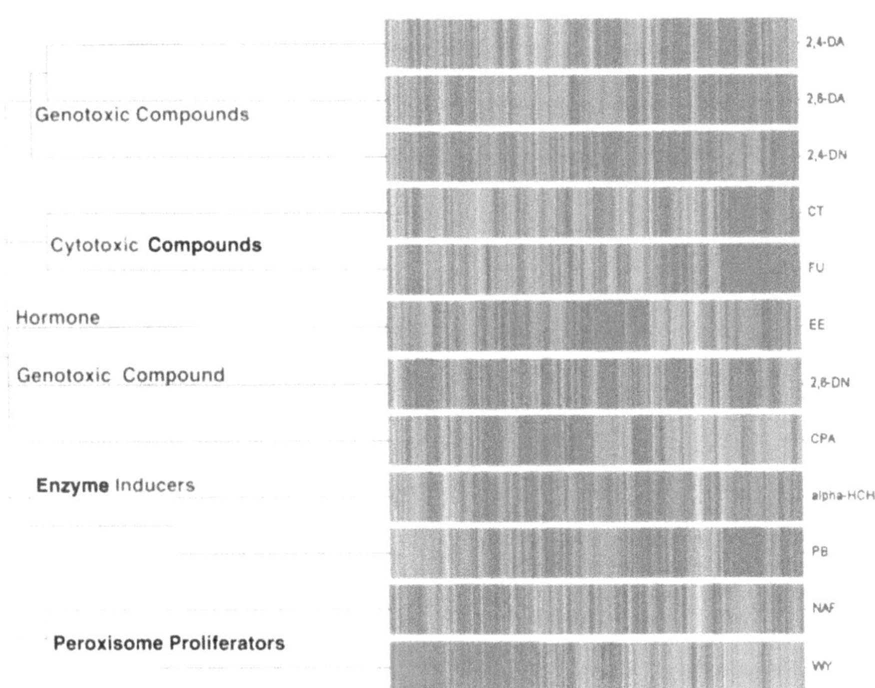


COLOR FIGURE 5.5 Hierarchical clustering of 12 tumor promoters and cyto- and genotoxic compounds. Red bars represent upregulated genes and green bars represent downregulated genes compared to the average expression of all experiments. The Pearson correlation coefficient was set to 0.75. Experimental settings and abbreviations are explained in the text.

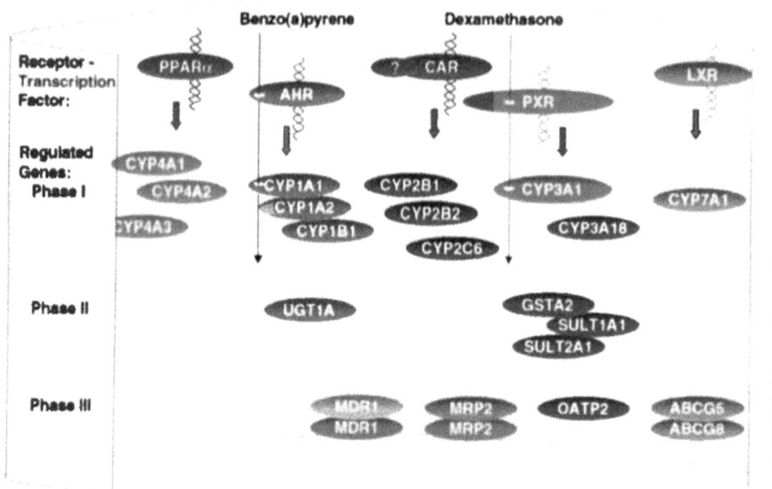


**COLOR FIGURE 6.1** Drug metabolizing genes are regulated by a network of receptor–transcription factors. These gene regulatory relationships were derived from the literature for the rat. Physical space in the schematic is occupied by compounds according to structures, by receptors according to ligand-binding specificities, and by enzymes according to substrate specificities. Xenobiotic compounds bind to receptors (intersected by solid arrows) to activate transcription of genes with receptor-response elements in their promoters. Colored arrows indicate the transcriptional regulation of drug metabolizing genes (below arrow) by the receptor (above arrow). Note that this schematic is a gross oversimplification. Some compounds activate several receptors, several receptors can regulate the same genes, and only a few induced genes are shown for simplicity. Some commercial drug compounds are not significantly metabolized in the liver, and some compounds may elude all receptors. The substrates for drug conjugating and drug transporter proteins are largely unknown. Phenobarbital triggers induction of CYP genes by CAR indirectly, as it does not bind to CAR. References are cited in the text except for MDR1, OATP2, CYP7A1, ABCG5, and ABCG8 and for the metabolism of benzo(*a*) pyrene and dexamethasone. *Abbreviations:* PPAR, peroxisome-proliferator-activated receptor; AHR, aryl hydrocarbon receptor; PXR, pregnane X receptor; CAR, constitutive androstane receptor; LXR, liver X receptor; CYP, cytochrome P450; GST, glutathione-*S*-transferase; UGT, UDP-glucuronosyltransferase; ABC, ATP-dependent transporter; SULT, sulfotransferase; MDR, multidrug resistance protein; OATP, organic anion transporter protein; and MRP, multidrug-resistance-associated protein. Note that MRP2 and cMOAT are synonymous.

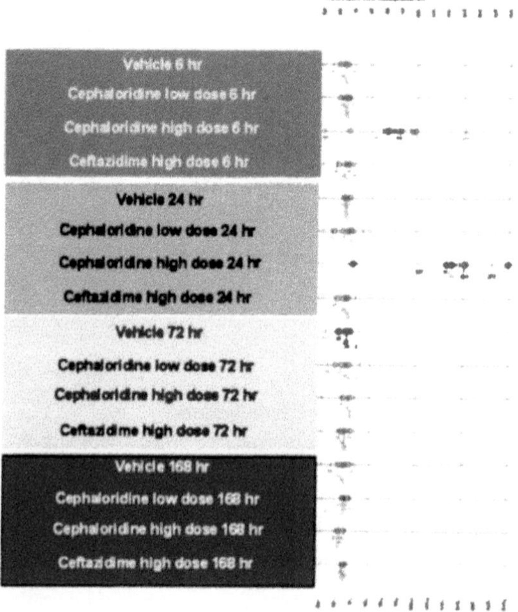


**COLOR FIGURE 12.6** SAR distinctions. A box plot generated with the GeneExpress® software system contrast analysis tool illustrates the average difference values for each treatment condition (circles). Each treatment condition (far left) contains multiple samples corresponding to individual animals within that group. A 95% CI and median are indicated for each set of samples. For example, treatment-condition cephaloridine, high dose 6 hr, has a CI between 64 and 139. The median is 116 for this same group and is indicated within the cluster of samples.

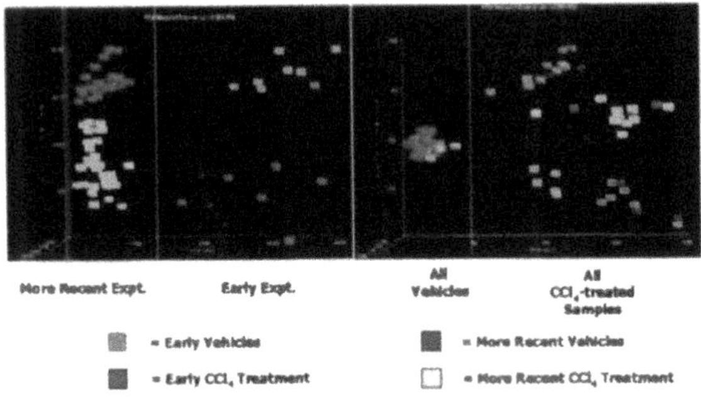


**COLOR FIGURE 12.18** Two PCA visualizations show that gene selection is necessary for removing confounding factors of variability unrelated to toxicity. The left plot shows the principal separation of samples due to the time the experiment was conducted when using all genes. The right plot shows the principal separation of samples due to toxicity. In fact, there appear to be no observable differences between studies in the first three components. Within each plot, the highest contributing factor to the overall variability is shown on the $x$-axis and is known as the *first component*. The $y$-axis shows the second-highest component, and the $z$-axis shows the third-highest variability component.

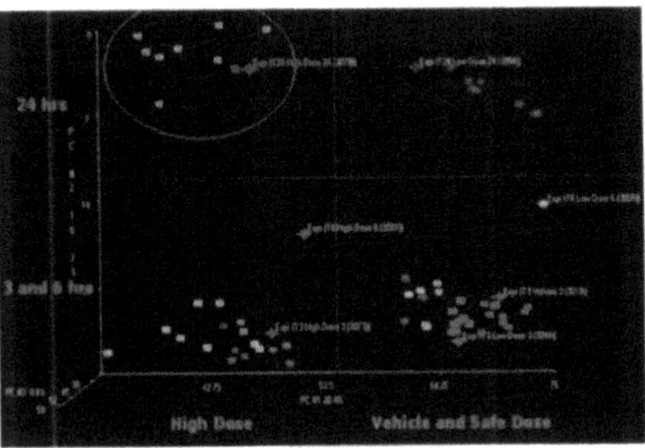


**COLOR FIGURE 12.22** Principal components analysis (PCA) of APAP data is visualized for all animals in the study using those genes that change due to APAP administration at the highest dose. The homogeneity of the gene expression response is characterized by both a toxic response and a toxic response related to sacrifice time post-dose.

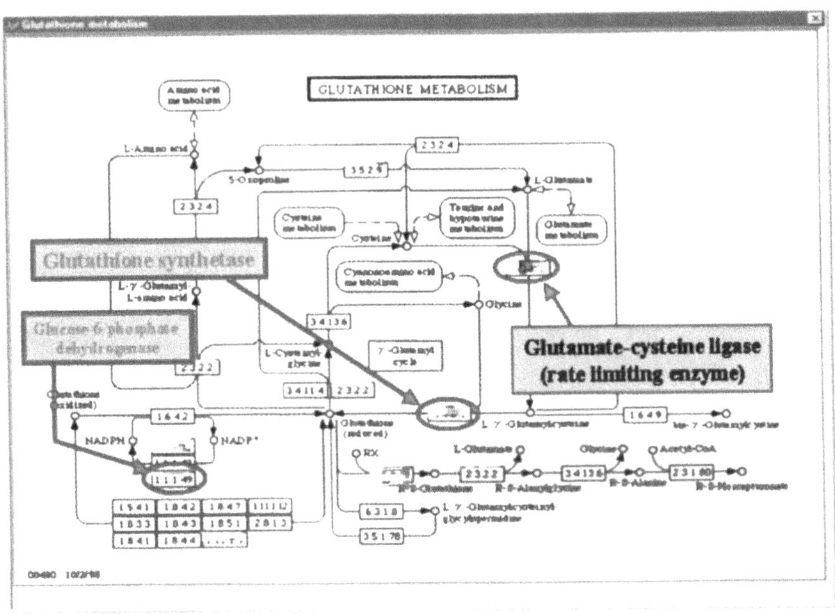


**COLOR FIGURE 12.24** Glutathione pathway. The fold-change values for genes that encode proteins in the glutathione pathway are illustrated in the context of the KEGG glutathione metabolism pathway. Overexpressed genes are brown, and the underexpressed genes are green.

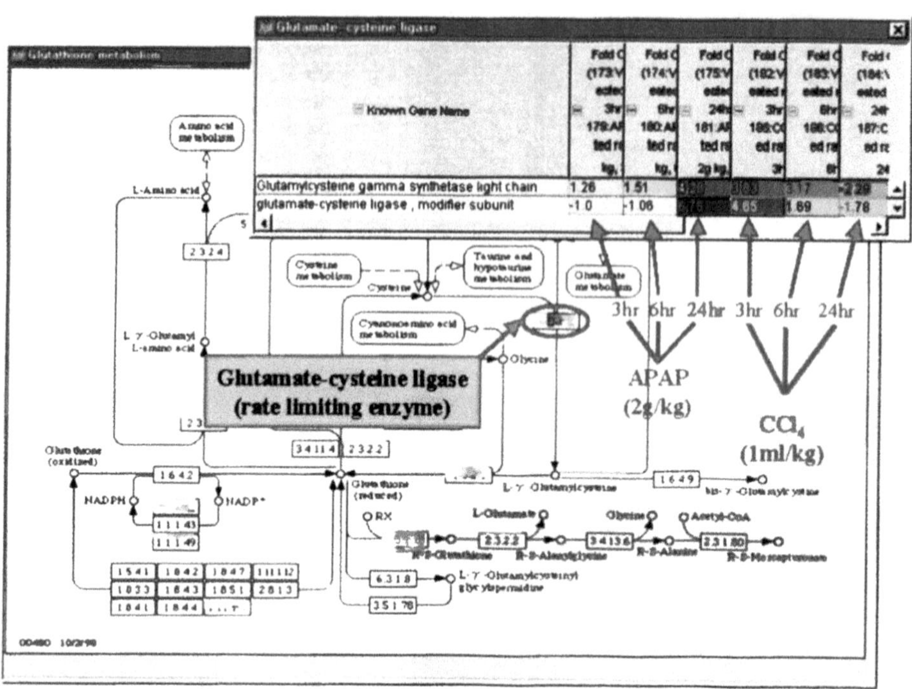


COLOR FIGURE 12.25 The rate-limiting enzyme for the glutathione pathway, glutamine-cysteine ligase, is upregulated upon APAP treatment at 24 hr and by $CCl_4$ treatment at 3 hr, as shown in the upper right-hand box.

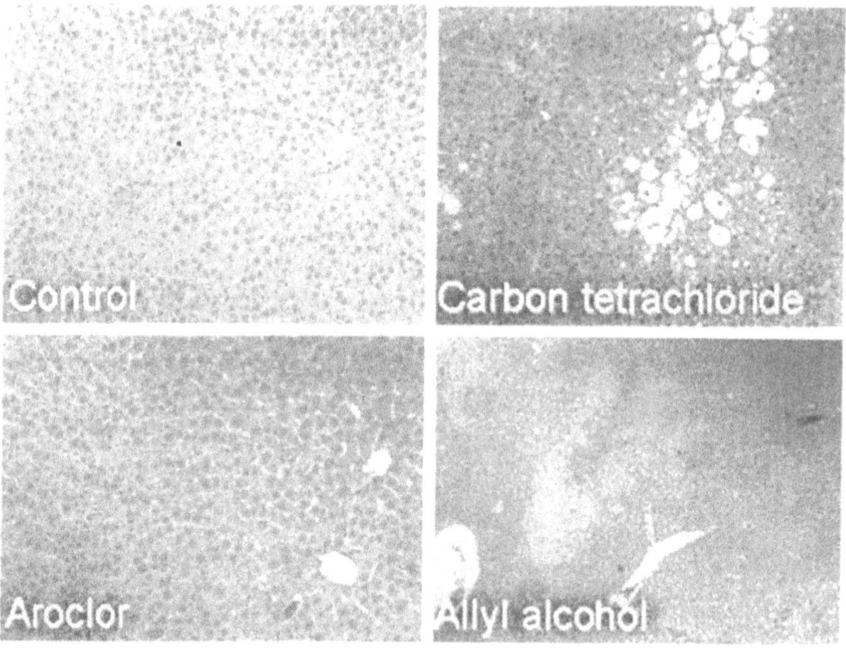


COLOR FIGURE 13.1 Slide showing liver sections from control or rats treated with carbon tetrachloride, Aroclor 1254, or allyl alcohol. All sections are shown at 100× magnification except for allyl alcohol, which is shown at 50×. The central vein is located to the right in each slide.

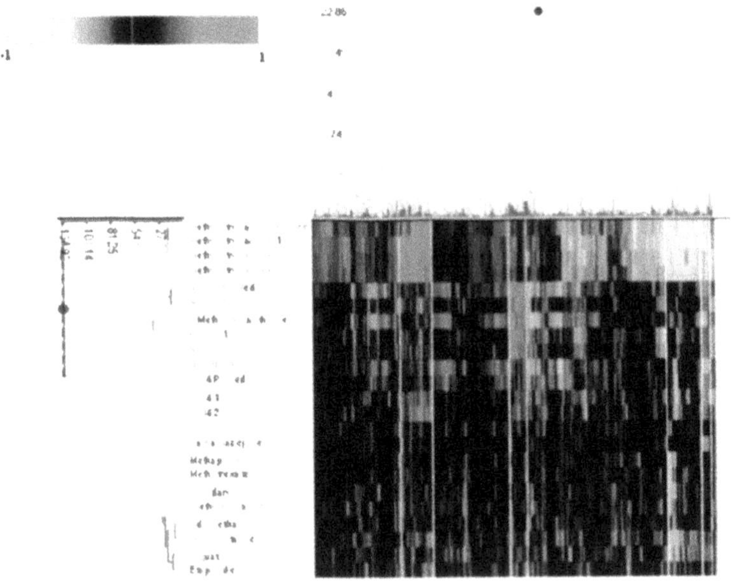


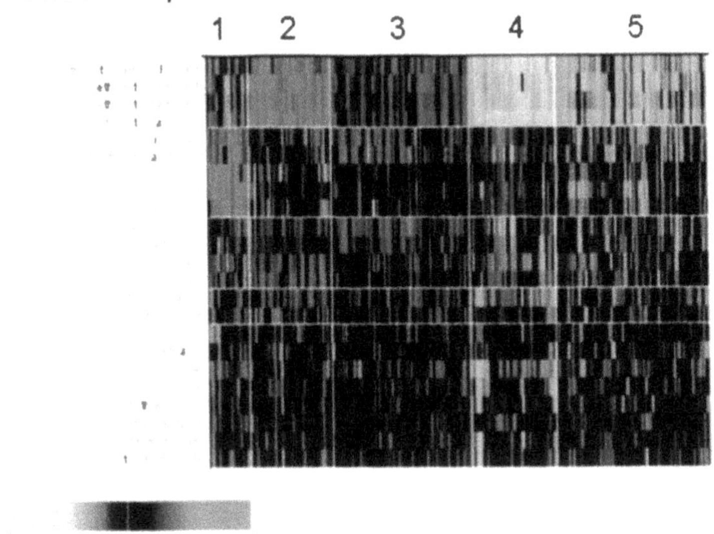


**COLOR FIGURE 13.5** (**A**) Two-dimensional graph showing the expression profiles of pooled samples from 15 hepatotoxins and expression profiles from individual animals treated with Aroclor 1254, diethylnitrosamine, and carbon tetrachloride grouped by divisive clustering. (**B**) Gene expression profile clusterings from pooled and individual samples. The profiles were grouped using *k*-means clustering.

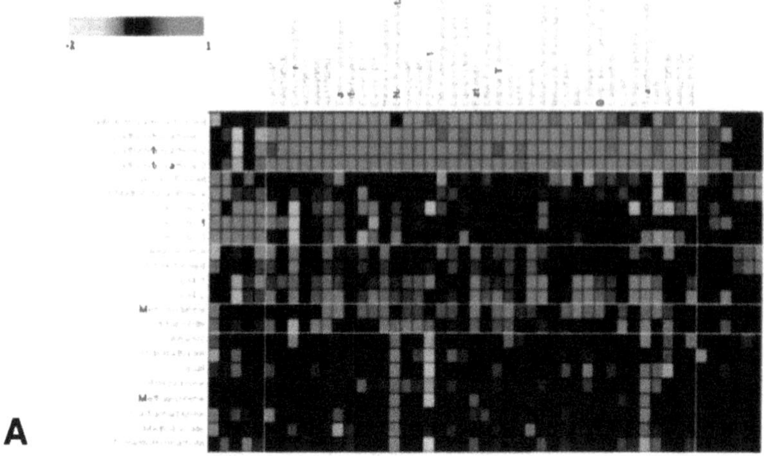


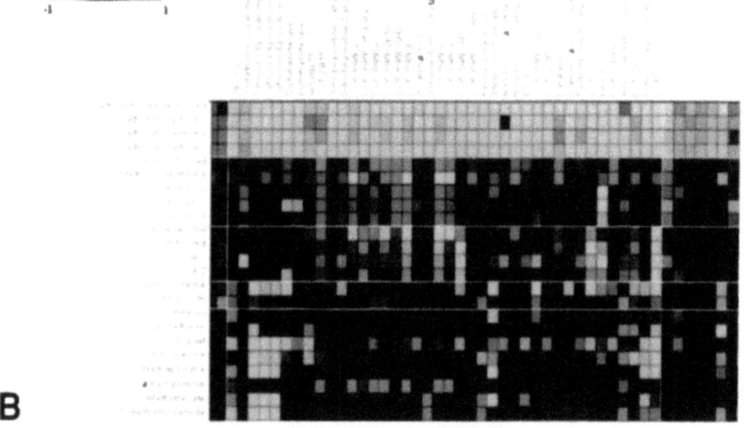


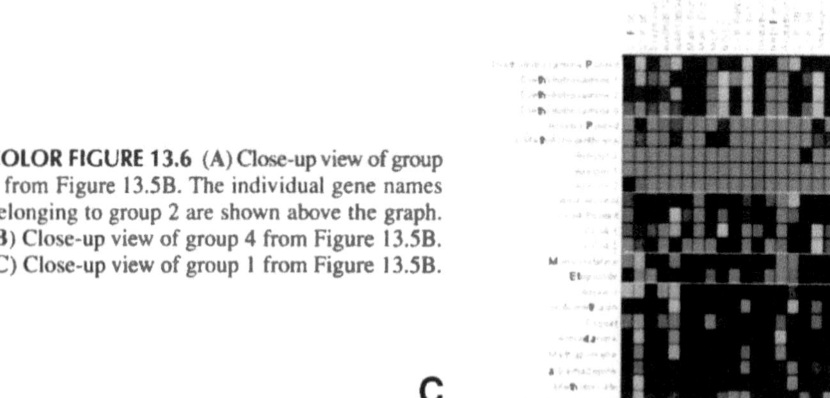


**COLOR FIGURE 13.6** (**A**) Close-up view of group 2 from Figure 13.5B. The individual gene names belonging to group 2 are shown above the graph. (**B**) Close-up view of group 4 from Figure 13.5B. (**C**) Close-up view of group 1 from Figure 13.5B.

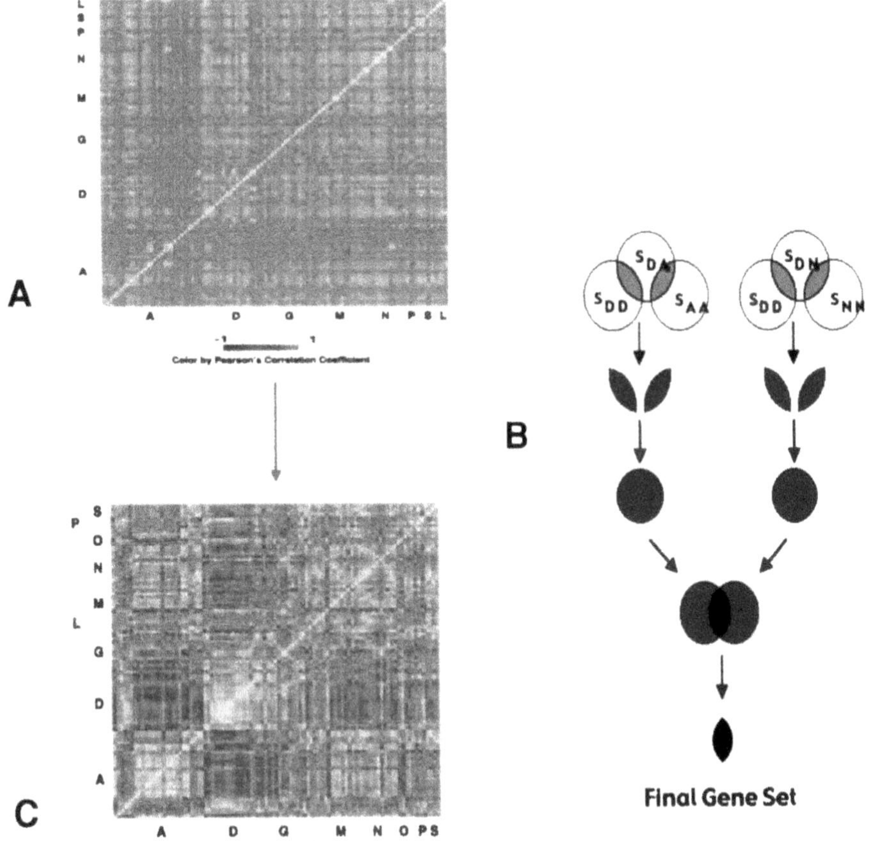


COLOR FIGURE 14.1 An early supervised approach to toxicogenomic analysis. (**A**) HepG2 cells were treated with various cytotoxic compounds and relative changes in gene expression were ascertained using microarray analyses. Comparisons of the gene expression patterns between each pair of two treatments were made by Pearson's correlation coefficient and depicted graphically by ordering in a supervised fashion the toxicant expression profiles on the *x* and *y* axes by the assigned mechanisms of action. The color in each cell of the plot reflects the similarity between the two experiments, resulting in a similarity matrix that is symmetric about the main diagonal (identical experiment compared to itself). The letters designate the following toxicant classes: A, antiinflammatory; D, DNA-damagers; G, gene synthesis inhibitors; L, low dose DNA damaging agents (nontoxic), M, metabolic poisons; N, nongenotoxic controls; O, other; P, peroxisomal proliferators; S, steroids. The initial supervised analysis was performed across the entire database of 100 toxic compounds using all genes on the microarray (~250). (**B**) The optimization algorithm depicted is a "brute force" type of supervised-learning algorithm that relies on the processing power of the computer to perform calculations on all pairs of experiments in all pairs of two groups. The groups of treatments from which data were used for algorithm optimization were DNA damaging agents, antiinflammatories, and nongenotoxic controls. By focusing on one gene at a time and determining the effect of each gene on the average correlation coefficient, the algorithm selected genes that maximized the average intra-group correlation coefficient and minimized the average inter-group correlation coefficient for all pairs of groups in our analysis. Application of the algorithm resulted in six group/group comparisons and their associated optimized gene sets: Gene Set$_{\text{DNA damagers, DNA damagers}}$, $S_{DD}$; Gene Set$_{\text{DNA damagers, Antiinflammatories}}$, $S_{DA}$; Gene Set$_{\text{DNA damagers, Notoxic controls}}$, $S_{DN}$; Gene Set$_{\text{Antiinflammatories, Antiinflammatories}}$, $S_{AA}$; Gene Set$_{\text{Antiinflammatories, Notoxic controls}}$, $S_{AN}$; Gene Set$_{\text{Notoxic controls, Notoxic controls}}$, $S_{NN}$. The genes in the resulting six gene sets were reduced further by taking the intersection of the union of the intersection of the initial genes sets as displayed graphically in the Venn diagrams. (**C**) Supervised analysis of all compounds in the database using genes identified by the algorithm-based optimization gene set for distinguishing cisplatin and diflunisal/flufenamic acid. The distinction between toxic doses of DNA damaging agents and NSAIDs is much clearer, providing evidence that genes identified in this supervised analysis may serve as predictors of toxicity in subsequent comparisons of other members of these toxicant classes. (Adapted from Burczynski, M.E. et al., *Toxicol. Sci.*, 58, 399–415, 2000. With permission.)

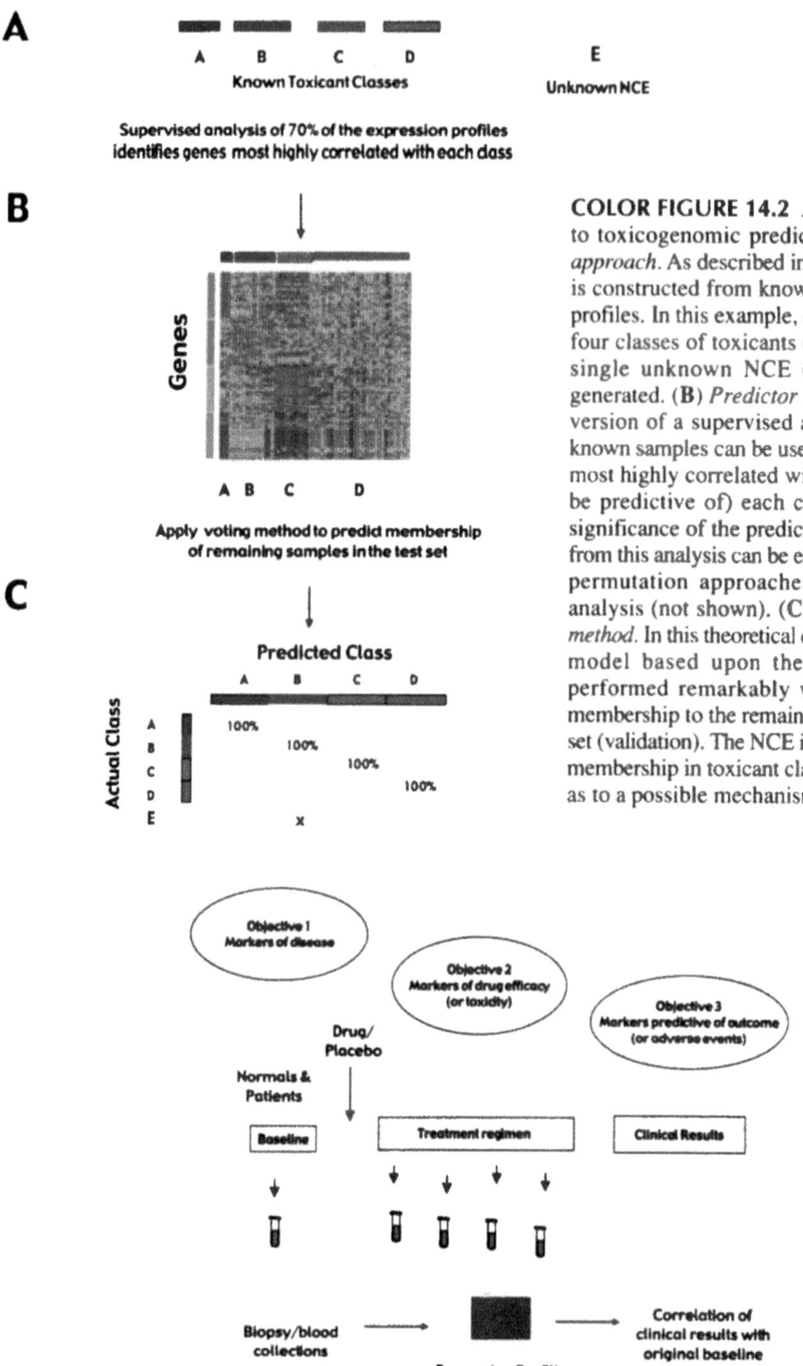


**COLOR FIGURE 14.2** A multiclass approach to toxicogenomic prediction. (**A**) *Supervised approach*. As described in the text, a training set is constructed from known toxicant expression profiles. In this example, expression profiles for four classes of toxicants (classes A to D) and a single unknown NCE (class E) have been generated. (**B**) *Predictor gene selection*. In one version of a supervised approach, 70% of the known samples can be used to identify the genes most highly correlated with (and most likely to be predictive of) each class of toxicant. The significance of the predictor gene sets resulting from this analysis can be estimated using random permutation approaches and neighborhood analysis (not shown). (**C**) *Results for a voting method*. In this theoretical example, the predictive model based upon the training set above performed remarkably well when assigning membership to the remaining samples in the test set (validation). The NCE in class E was assigned membership in toxicant class B, providing a lead as to a possible mechanism of toxicity.

**COLOR FIGURE 14.3** Objectives of clinical pharmacogenomic analyses. This schematic depicts the objectives of clinical pharmacogenomic studies. The first objective, identification of markers of disease, is achieved by comparing baseline expression profiles from normal individuals and patients enrolled in the clinical response. The second objective, identification of markers of drug efficacy, is achieved by comparing expression profiles over time in drug and placebo-administered patients and correlating them with measures of clinical response. The third objective, identification of markers predictive of outcome, is obtained by comparing baseline and/or longitudinal expression profiles of patients that responded to therapy with those who failed to respond to therapy. As noted in the text, in the event of toxicity during the clinical trial, similar supervised methods can be applied to identify biomarkers of drug toxicity or markers predictive of adverse events.

## 6.5 SUMMARY

We evaluated an oligonucleotide microarray technology for determining how the rat liver responds transcriptionally to drug treatments. The responses observed using the microarray confirmed many previous observations using these much-studied compounds. The global nature of the microarray assays also enabled additional observations to be made and in some cases confirmed using QRT-PCR. Novel gene regulation responses observed included induction of markers that characterize toxicological responses such as *HSP70* and *SODxc*. This study validated microarray technology for assessing induction of drug metabolism genes by xenobiotics and suggests utility in investigating toxicological responses to xenobiotics.

Microarrays provide detailed indicators reflecting the state of the tissue being studied; hence, microarrays are an excellent resource for generating hypotheses regarding the etiology of adverse events. The ability to explain adverse events or predict phenotypic outcomes from gene expression profiles is currently in its infancy, however. Interpreting those indicators may require a reference database of expression profiles for each adverse event, measurement of physiological parameters, and software tools to correlate these parameters with gene expression profiles. Because microarrays do not resolve direct from indirect (e.g., homeostatic) effects of xenobiotic treatments, detailed biochemical studies are likely to be required to evaluate hypotheses. Integration of microarray technology with classical pathology and biochemistry techniques shows promise for accelerating development of safe and effective medicines.

## REFERENCES


1. Gerhold, D., Lu, M., Xu, J., Austin, C., Caskey, C.T., and Rushmore, T., Monitoring expression of genes involved in drug metabolism and toxicology using DNA microarrays, *Physiol. Genomics*, 5(4), 161–170, 2001.
2. Karpen, S.J., Nuclear receptor regulation of hepatic function, *J. Hepatol.*, 36(6), 832–850, 2002.
3. Gonzalez, F.J., Overview of experimental approaches for study of drug metabolism and drug–drug interactions, *Adv. Pharmacol.*, 43, 255–277, 1997.
4. Whitlock, J.P., Induction of cytochrome P450 enzymes that metabolize xenobiotics, in *Cytochrome P450s, Structure, Mechanism, and Biochemistry*, 2nd ed., Ortiz de Montellano, P.R., Ed., Plenum Press, New York, 1995, p. 367.
5. Haining, R.L., Hunter, A.P., Veronese, M.E., Trager, W.F., and Rettie, A.E., Allelic variants of human cytochrome P450 2C9: baculovirus-mediated expression, purification, structural characterization, substrate stereoselectivity, and prochiral selectivity of the wild-type and I359L mutant forms, *Arch. Biochem. Biophys.*, 333(2), 447–458, 1996.
6. Marton, M.J., DeRisi, J.L., Bennett, H.A., Iyer, V.R., Meyer, M.R., Roberts, C.J., Stoughton, R., Burchard, J., Slade, D., Dai, H., Bassett, D.E., Jr., Hartwell, L.H., Brown, P.O., and Friend, S.H., Drug target validation and identification of secondary drug target effects using DNA microarrays [see comments], *Nat. Med.*, 4(11), 1293–1301, 1998.
7. Burczynski, M.E., McMillian, M., Ciervo, J., Li, L., Parker, J. B., Dunn, R.T., 2nd, Hicken, S., Farr, S., and Johnson, M.D., Toxicogenomics-based discrimination of toxic mechanism in HepG2 human hepatoma cells, *Toxicol. Sci.*, 58(2), 399–415, 2000.
8. Pennie, W.D., Woodyatt, N.J., Aldridge, T.C., and Orphanides, G., Application of genomics to the definition of the molecular basis for toxicity, *Toxicol. Lett.*, 120(1–3), 353–358, 2001.
9. Waring, J.F., Jolly, R.A., Ciurlionis, R., Lum, P.Y., Praestgaard, J.T., Morfitt, D.C., Buratto, B., Roberts, C., Schadt, E., and Ulrich, R.G., Clustering of hepatotoxins based on mechanism of toxicity using gene expression profiles, *Toxicol. Appl. Pharmacol.*, 175(1), 28–42, 2001.
10. Correia, M.A., Rat and human liver cytochromes P450, in *Cytochrome P450s, Structure, Mechanism, and Biochemistry*, 2nd ed., Ortiz de Montellano, P.R., Ed., Plenum Press, New York, 1995, pp. 610–611.
11. Fodor, S.P., Rava, R.P., Huang, X.C., Pease, A.C., Holmes, C.P., and Adams, C.L., Multiplexed biochemical assays with biological chips, *Nature*, 364(6437), 555–556, 1993.

12. Lockhart, D.J., Dong, H., Byrne, M.C., Follettie, M.T., Gallo, M.V., Chee, M.S., Mittmann, M., Wang, C., Kobayashi, M., Horton, H., and Brown, E.L., Expression monitoring by hybridization to high-density oligonucleotide arrays [see comments], *Nat. Biotechnol.*, 14(13), 1675–1680, 1996.
13. Der, S.D., Zhou, A., Williams, B.R., and Silverman, R.H., Identification of genes differentially regulated by interferon alpha, beta, or gamma using oligonucleotide arrays, *Proc. Natl. Acad. Sci. USA*, 95(26), 15623–8, 1998.
14. Holder, D., Pikounis, V., Raubertas, R., Svetnik, V., and Soper, K., Statistical analysis of high density oligonucleotide arrays: a SAFER approach, *Proc. Am. Stat. Assoc*, 2002.
15. Honkakoski, P., Zelko, I., Sueyoshi, T., and Negishi, M., The nuclear orphan receptor CAR-retinoid X receptor heterodimer activates the phenobarbital-responsive enhancer module of the CYP2B gene, *Mol. Cell. Biol.*, 18(10), 5652–5658, 1998.
16. Bertilsson, G., Heidrich, J., Svensson, K., Asman, M., Jendeberg, L., Sydow-Backman, M., Ohlsson, R., Postlind, H., Blomquist, P., and Berkenstam, A., Identification of a human nuclear receptor defines a new signaling pathway for CYP3A induction, *Proc. Natl. Acad. Sci. USA*, 95(21), 12208–12213, 1998.
17. Simpson, A.E., The cytochrome P450 4 (CYP4) family, *Gen. Pharmacol.*, 28(3), 351–359, 1997.
18. Quattrochi, L.C., Shih, H., and Pickwell, G.V., Induction of the human CYP1A2 enhancer by phorbol ester, *Arch. Biochem. Biophys.*, 350(1), 41–48, 1998.
19. Sakuma, T., Ohtake, M., Katsurayama, Y., Jarukamjorn, K., and Nemoto, N., Induction of CYP1A2 by phenobarbital in the livers of aryl hydrocarbon-responsive and -nonresponsive mice, *Drug Metab. Dispos.*, 27(3), 379–384, 1999.
20. Hiroi, T., Imaoka, S., Chow, T., and Funae, Y., Tissue distributions of CYP2D1, 2D2, 2D3 and 2D4 mRNA in rats detected by RT-PCR, *Biochim. Biophys. Acta*, 1380(3), 305–312, 1998.
21. Bock, K.W., Gschaidmeier, H., Heel, H., Lehmkoster, T., Munzel, P.A., Raschko, F., and Bock-Hennig, B., AH receptor-controlled transcriptional regulation and function of rat and human UDP-glucuronosyltransferase isoforms, *Adv. Enzyme Regul.*, 38, 207–222, 1998.
22. Rushmore, T.H., King, R.G., Paulson, K.E., and Pickett, C.B., Regulation of glutathione S-transferase Ya subunit gene expression: identification of a unique xenobiotic-responsive element controlling inducible expression by planar aromatic compounds, *Proc. Natl. Acad. Sci. USA*, 87(10), 3826–3830, 1990.
23. Maheo, K., Antras-Ferry, J., Morel, F., Langouet, S., and Guillouzo, A., Modulation of glutathione S-transferase subunits A2, M1, and P1 expression by interleukin-1β in rat hepatocytes in primary culture, *J. Biol. Chem.*, 272(26), 16125–16132, 1997.
24. Yamada, H., Minematsu, Y., Nakamura, T., Mise, M., Fujisaki, H., and Oguri, K., Brucine as a potent inducer of CYP2B3, the third member of the CYP2B subfamily P450 in rats, *Biol. Pharm. Bull.*, 19(2), 291–293, 1996.
25. Goodwin, B., Hodgson, E., D'Costa, D.J., Robertson, G.R., and Liddle, C., Transcriptional regulation of the human CYP3A4 gene by the constitutive androstane receptor, *Mol. Pharmacol.*, 62(2), 359–365, 2002.
26. Shibutani, M., Takahashi, N., Kobayashi, T., Uneyama, C., Masutomi, N., Nishikawa, A., and Hirose, M., Molecular profiling of genes up-regulated during promotion by phenobarbital treatment in a medium-term rat liver bioassay, *Carcinogenesis*, 23(6), 1047–1055, 2002.
27. Ueda, A., Hamadeh, H.K., Webb, H.K., Yamamoto, Y., Sueyoshi, T., Afshari, C.A., Lehmann, J.M., and Negishi, M., Diverse roles of the nuclear orphan receptor CAR in regulating hepatic genes in response to phenobarbital, *Mol. Pharmacol.*, 61(1), 1–6, 2002.
28. Kliewer, S.A., Moore, J.T., Wade, L., Staudinger, J.L., Watson, M.A., Jones, S.A., McKee, D.D., Oliver, B.B., Willson, T.M., Zetterstrom, R.H., Perlmann, T., and Lehmann, J.M., An orphan nuclear receptor activated by pregnanes defines a novel steroid signaling pathway, *Cell*, 92(1), 73–82, 1998.
29. Ronis, M.J., Rowlands, J.C., Hakkak, R., and Badger, T.M., Altered expression and glucocorticoid-inducibility of hepatic CYP3A and CYP2B enzymes in male rats fed diets containing soy protein isolate, *J. Nutr.*, 129(11), 1958–1965, 1999.
30. Kullak-Ublick, G.A., Regulation of organic anion and drug transporters of the sinusoidal membrane, *J. Hepatol.*, 31(3), 563–573, 1999.

31. Brady, J.M., Cherrington, N.J., Hartley, D.P., Buist, S.C., Li, N., and Klaassen, C.D., Tissue distribution and chemical induction of multiple drug resistance genes in rats, *Drug Metab. Dispos.*, 30(7), 838–844, 2002.
32. Roberts, R.A., Evidence for cross talk between PPARα and p38 MAP kinase, *Toxicol. Sci.*, 68(2), 270–274, 2002.
33. Barak, Y., Nelson, M.C., Ong, E.S., Jones, Y.Z., Ruiz-Lozano, P., Chien, K.R., Koder, A., and Evans, R.M., PPAR gamma is required for placental, cardiac, and adipose tissue development, *Mol. Cell*, 4(4), 585–595, 1999.
34. Alexander, D.L., Ganem, L.G., Fernandez-Salguero, P., Gonzalez, F., and Jefcoate, C.R., Aryl-hydrocarbon receptor is an inhibitory regulator of lipid synthesis and of commitment to adipogenesis, *J. Cell Sci.*, 111(pt. 22), 3311–3322, 1998.
35. Elshourbagy, N.A., Boguski, M.S., Liao, W.S., Jefferson, L.S., Gordon, J.I., and Taylor, J.M., Expression of rat apolipoprotein A-IV and A-I genes: mRNA induction during development and in response to glucocorticoids and insulin, *Proc. Natl. Acad. Sci. USA*, 82(23), 8242–8246, 1985.
36. Reddy, M.N., Effect of anticonvulsant drugs on plasma total cholesterol, high-density lipoprotein cholesterol, and apolipoproteins A and B in children with epilepsy, *Proc. Soc. Exp. Biol. Med.*, 180(2), 359–363, 1985.
37. Staels, B., Dallongeville, J., Auwerx, J., Schoonjans, K., Leitersdorf, E., and Fruchart, J.C., Mechanism of action of fibrates on lipid and lipoprotein metabolism, *Circulation*, 98(19), 2088–2093, 1998.
38. Vu-Dac, N., Chopin-Delannoy, S., Gervois, P., Bonnelye, E., Martin, G., Fruchart, J.C., Laudet, V., and Staels, B., The nuclear receptors peroxisome proliferator-activated receptor alpha and Rev-erbalpha mediate the species-specific regulation of apolipoprotein A-I expression by fibrates, *J. Biol. Chem.*, 273(40), 25713–25720, 1998.
39. Schiaffonati, L. and Tiberio, L., Gene expression in liver after toxic injury: analysis of heat shock response and oxidative stress-inducible genes, *Liver*, 17(4), 183–191, 1997.
40. Yoo, H.Y., Chang, M.S., and Rho, H.M., Induction of the rat Cu/Zn superoxide dismutase gene through the peroxisome proliferator-responsive element by arachidonic acid, *Gene*, 234(1), 87–91, 1999.
41. Metz, G., Sim, A.K., McCraw, A.P., and Cleland, M.E., Effect of etofylline clofibrate on experimental thrombus formation and prostacyclin activation, *Arzneimittelforschung*, 36(9), 1363–1365, 1986.
42. Aoyama, T., Hardwick, J. P., Imaoka, S., Funae, Y., Gelboin, H.V., and Gonzalez, F.J., Clofibrate-inducible rat hepatic P450s IVA1 and IVA3 catalyze the omega- and(omega-1)-hydroxylation of fatty acids and the omega-hydroxylation of prostaglandins E1 and F2 alpha, *J. Lipid Res.*, 31(8), 1477–1482, 1990.
43. Bylund, J., Finnstrom, N., and Oliw, E.H., Gene expression of a novel cytochrome P450 of the CYP4F subfamily in human seminal vesicles, *Biochem. Biophys. Res. Commun.*, 261(1), 169–174, 1999.
44. Jin, R., Koop, D.R., Raucy, J.L., and Lasker, J.M., Role of human CYP4F2 in hepatic catabolism of the proinflammatory agent leukotriene B4, *Arch. Biochem. Biophys.*, 359(1), 89–98, 1998.
45. Wang, T. and Brown, M.J., mRNA quantification by real time TaqMan polymerase chain reaction: validation and comparison with RNase protection, *Anal. Biochem.*, 269(1), 198–201, 1999.
46. Johnson, D.R. and Klaassen, C.D., Regulation of rat multidrug resistance protein 2 by classes of prototypical microsomal enzyme inducers that activate distinct transcription pathways, *Toxicol. Sci.*, 67(2), 182–189, 2002.
47. Guo, G.L., Staudinger, J., Ogura, K., and Klaassen, C.D., Induction of rat organic anion transporting polypeptide 2 by pregnenolone-16alpha-carbonitrile is via interaction with pregnane X receptor, *Mol. Pharmacol.*, 61(4), 832–839, 2002.
48. Gupta, S., Pandak, W.M., and Hylemon, P.B., LXR alpha is the dominant regulator of CYP7A1 transcription, *Biochem. Biophys. Res. Commun.*, 293(1), 338–343, 2002.
49. Remaley, A.T., Bark, S., Walts, A.D., Freeman, L., Shulenin, S., Annilo, T., Elgin, E., Rhodes, H.E., Joyce, C., Dean, M., Santamarina-Fojo, S., and Brewer, H.B., Jr., Comparative genome analysis of potential regulatory elements in the ABCG5-ABCG8 gene cluster, *Biochem. Biophys. Res. Commun.*, 295(2), 276–282, 2002.
50. Kim, J.H., Stansbury, K.H., Walker, N.J., Trush, M.A., Strickland, P.T., and Sutter, T.R., Metabolism of benzo[*a*]pyrene and benzo[*a*]pyrene-7,8-diol by human cytochrome P450 1B1, *Carcinogenesis*, 19(10), 1847–1853, 1998.

# 7 Characterization of Hepatocytes and Their Use as a Model System in Toxicogenomics*

*Thomas K. Baker, Marnie A. Higgins, Mark A. Carfagna, and Timothy P. Ryan*

**CONTENTS**

## 7.1 INTRODUCTION

Hepatotoxicity is a major hurdle that must be cleared in bringing candidate molecules through the drug development process and is the most common reason for withdrawal of compounds from the market. Better models of hepatotoxicity could streamline drug development, resulting in safer compounds with less likelihood of producing undesired effects in humans. The use of cultured primary hepatocytes to model hepatotoxicity has proven to be a valuable tool; however, questions remain with regard to functional differences observed in primary hepatocytes relative to the intact liver. Parenchymal hepatocytes are complex, highly differentiated cells that comprise over 80% of the normal liver mass and 60% of the total liver cell population.[1,2] The parenchymal hepatocyte performs most specialized functions of the liver, including lipid metabolism, detoxification of exogenous xenobiotics, regulation of urea, and production of plasma proteins. The majority of proteins involved in liver homeostasis are produced in parenchymal hepatocytes, and their abundance is primarily regulated at the transcriptional level.[3] In the normal liver, most differentiated hepatocytes are quiescent.

With the development of suitable enzymatic methods for the isolation of parenchymal hepatocytes,[4–6] researchers in a variety of scientific disciplines have demonstrated the utility of *in vitro* hepatocyte models. Isolated hepatocytes have proven useful in the evaluation of drugs with regard to protein binding,[7] metabolism,[8–10] mechanisms of xenobiotic action,[11,12] oxidative stress,[13–16] apoptosis,[17–19] and neoplasia.[20,21] The application of hepatocyte models results in reduced animal utilization, provides a bridge for extrapolation to human effects, and allows for higher experimental throughput. Although hepatocyte models are useful mechanistic tools, researchers have been challenged by the fact that there is a loss of specialized liver function in cultured hepatocytes, a process often referred to as dedifferentiation. Cultured hepatocytes have typically been described as dedifferentiated based upon the investigation of a few key cellular processes or hepatocellular markers. For example, much effort has been expended to better understand the effects of cell culture on cytochrome P450 (CYP) expression and the metabolism of xenobiotics. Studies show that a progressive loss of CYP mRNA and protein content with time in culture occurs in isolated hepatocytes.[22,23] During the isolation process and thereafter, hepatocytes must respond to new environmental stresses through physiological and morphological adaptations, in which an altered steady biological state is achieved to preserve cellular viability. Further stresses that invoke responses in the hepatocyte result from insult with foreign agents, and these responses can be used to model the response in an intact liver. Comprehensive evaluations of the underlying processes driving physiological and morphological hepatocyte adaptations have not been conducted. Adaptation includes the induction and repression of genes that are required for cellular survival in their new environment, and transcriptional profiling is a tool that can be exploited to understand these changes *en masse*. Recent advances in gene expression technology provide a tool to perform comprehensive, quantitative comparisons at the transcriptional level of thousands of genes simultaneously. Currently, gene microarrays can be constructed with cloned cDNA,[24] or with synthetic oligonucleotide probes.[25,26] By labeling cellular RNA and monitoring hybridization to arrayed probes, transcriptional profiling provides the opportunity to evaluate thoroughly the processes of dedifferentiation, adaptation, and responses to compound administration in cultured hepatocytes.

In this chapter, we demonstrate that valuable insight can be gained regarding cellular dedifferentiation and effects of compounds on hepatocytes via gene expression analysis. Parallel expression monitoring was used to characterize mRNA changes in hepatocyte cultures over time and following treatment with nuclear hormone receptor (NHR) ligands. Cluster analysis of time course data revealed a classic hepatocyte dedifferentiation response. Hepatocyte dedifferentiation was confirmed by changes observed with phase I metabolizing enzymes, while genes involved in maintaining cellular processes such as glutathione status, sugar metabolism, and production of extracellular matrix proteins demonstrated changes in expression consistent with adaptation to the culture environment. The status of genes involved in proliferation was consistent with the maintenance of cultured hepatocytes in a quiescent, nonproliferative state. When ligands of the NHR family of transcription factors were added to these quiescent hepatocytes, responses characteristic of the *in vivo* situation were reproduced. For example, hepatocytes treated with the peroxisome proliferator activated receptor α (PPARα) agonist fenofibrate produced gene expression changes representative of previously described *in vivo* effects of this compound on lipid metabolism. Transcript changes were robust and ligand specific, such that hierarchical clustering of gene expression profiles grouped NHR ligands (PPARα, PPARγ, and retinoid X receptor [RXR]) according to ligand specificity. These findings illustrate how hepatocytes can be used as powerful predictive tools to model *in vitro* toxicology and pharmacology.

## 7.2 EXPERIMENTAL PROCEDURES

### 7.2.1 Chemicals

Gentamicin, LD-L lactate dehydrogenase reagent, Leibovitz's L-15, Hank's balanced salt solution magnesium- and calcium-free sodium bicarbonate, glucose, N-[2-hydroxyethyl]piperazine-N′-[2-ethanesulfonic acid] (HEPES), ethylene glycol-bis[β-aminoethyl ether]-N,N,N′N′-tetraacetic acid (EGTA), and L-glutamine were purchased from Sigma Chemical Co. (St. Louis, MO). Collagenase type D and insulin-transferrin-sodium selenite supplement were purchased from Boehringer Mannheim (Indianapolis, IN). William's media E (WE) and certified fetal bovine serum (FBS) were purchased from Gibco BRL (Gaithersburg, MD). Hepatocyte maintenance media (HMM) was purchased from Clonetics (San Diego, CA). Wy14,643 ([4-chloro-6-(2,3-xylidino)-2-pyrimidinyl-thiol]-acetate) was purchased from Sigma. Fenofibrate, troglitazone, and rosiglitazone were synthesized at Eli Lilly and Company (Indianapolis, IN) and retinoid X receptor (RXR) agonists (LG 100268 and LG100324) were synthesized at Ligand Pharmaceuticals, Inc. (San Diego, CA).[27,28] All other chemicals not specifically mentioned were purchased at the highest available quality through Sigma.

### 7.2.2 Hepatocyte Isolation and Culture

Hepatocytes were isolated from 8- to 10-week-old male $F_{344}$ rats (Harlan Sprague–Dawley; Indianapolis, IN) using a modified two-stage portal vein perfusion technique.[4–6,29] Hepatocyte viability was assessed to be greater than 90% using Trypan blue dye exclusion. Cells were plated at a density of $2.75 \times 10^6$ cells per Falcon Primaria™ 100-mm culture dish (Lincoln Park, NJ) in WE containing 2% FBS, 5.0 μg/mL insulin, 5.0 μg/mL transferrin, 5.0 ng/mL sodium selenite, 10 ng/mL dexamethasone, 2 mM L-glutamine, and 50 μg/mL gentamicin (WE complete) and incubated in a 37°C, 100%-humidified environment (5% $CO_2$, 95% air). Following an attachment period of 4 hours, the culture media was replaced with serum-free HMM containing 0.5 μg/mL insulin, 39 ng/mL dexamethasone, 50 μg/mL gentamicin, and 50 ng/mL amphotericin B.

### 7.2.3 Cellular Viability Assessment

Samples (50 μL) of cell culture media were collected and examined for lactate dehydrogenase (LDH) leakage using a Roche, Cobas Mira™ clinical chemistry analyzer (Montclair, NJ) with Sigma LD-L lactate dehydrogenase reagent. None of the compounds evaluated produced significant cytolethality at the concentrations used in these studies (data not shown).

### 7.2.4 Sample Collection

Whole liver samples were collected prior to cannulation by ligating and removing the caudate lobe. Isolated hepatocytes were collected for analysis at various time points throughout the isolation process. The first samples were collected from freshly isolated cell suspensions following a 50×-*g* centrifugation. Subsequently, samples were collected following a 4-hour attachment period in WE complete. The remaining samples were collected at 12, 24, 48, and 72 hours after plating. Hepatocyte viability analysis showed that all cultures maintained greater than 95% viability over the 72-hour culture duration (data not shown). Primary hepatocytes were treated with NHR ligands: fenofibrate (100 μM), Wy14,643 (100 μM), troglitazone (50 μM), rosiglitazone (100 μM), or RXR agonists (30 μM) for 48 hours. These doses produced significant peroxisomal β-oxidation when tested in this model system (data not shown). Control cells were cultured in the presence of vehicle alone (1% DMSO). To evaluate the ability of *in vitro* models to predict fenofibrate-induced changes in rodents, $F_{344}$ rats (80–100 g) were treated with 300 mg/kg fenofibrate daily for 4 consecutive days. Liver samples were snap-frozen in liquid nitrogen and stored at –70°C. Peroxisomal β-oxidation was assessed using the method of Lazarow,[30] and gene expression analysis was performed using approximately 100 mg of snap-frozen liver as described below.

### 7.2.5 Sample Analyses

Total RNA was isolated from whole livers by direct homogenization of 100 mg tissue in 1.0 mL RNA STAT-60 (Tel-Test; Friendswood, TX) or from cell pellets or hepatocyte cultures by first removing media, then adding 1.0 mL RNA STAT-60 to each centrifuge tube or 100-mm culture dish. Cells were collected by scraping, and further steps in the isolation process were performed according to the manufacturer's recommended protocol. Subsequent purification and size selection of total RNA employed RNeasy columns (Qiagen; Valencia, CA) as suggested by the manufacturer. Reverse transcription–polymerase chain reaction (RT-PCR) was performed for transcript confirmation using a hot-start technique (Superscript II RT + Platinum Taq polymerase; Gibco), with 1 μg total RNA as starting material and an annealing temperature of 55°C for 25 to 35 cycles, depending on the individual transcript being amplified. Labeling of total RNA for microarray analysis was performed as described below. Double-stranded cDNA was synthesized from 22 μg of total RNA using Superscript II and an oligo T-7-(dT)24 primer. cRNA was synthesized using a primer that contained a T-7 RNA polymerase site that was labeled with biotin-11-CTP and biotin-16-UTP using a BioArray T-7 polymerase labeling kit (Enzo; Farmingdale, NY). Hybridizing, washing, antibody amplification, and staining of probe arrays were performed as per the Affymetrix technical manual (rev. 4). Experiments were performed using either rat genomic microarrays (RGU34A) or rat toxicology subarrays (RTU34) and visualized using GeneChip v.3.3 software (Affymetrix; Santa Clara, CA). Chip fluorescence was normalized by scaling total chip fluorescence intensities to a common value of 1500 prior to comparison. Because Affymetrix oligonucleotide microarrays utilize multiple perfect-match and mismatch oligonucleotides to determine expression levels, an algorithm within GeneChip software was used to determine the presence and abundance of each individual transcript. This absolute analysis of each individual microarray was performed to determine expression levels of individual genes. Absolute analyses of individual microarrays were followed by comparison analyses between individual microarrays within the GeneChip software to determine expression differences between samples. Data filtering was performed using Data Mining Tool (DMT) software (Affymetrix; Santa Clara, CA). Once comparisons between individual microarrays were performed, the time course dataset was organized using self-organizing maps (SOMs),[29,30] a nonlinear generalization of principal components analysis[30] and hierarchical clustering developed in-house and written in C and Perl.

When utilizing SOMs as a clustering tool, it was important to prevent extremely large fold-change values from skewing the results, so all fold changes that were greater than 20 and less than –20 were set to 20 and –20, respectively. Because DMT software calculates both positive and negative fold changes, a fold change of –1 is equal to a fold change of +1, making two probe sets with potentially the same fold change appear very different. In order to make such probe sets equivalent for both clustering and display purposes, the value 1 was subtracted from each positive fold change and 1 was added to each negative fold change. The data were normalized separately for each experiment with the maximum value set to +1 and the minimum value set to –1. This effectively filtered the data so probe sets that had very little change were grouped together in neighboring clusters. SOMs were run for 100 epochs using a time-varying learning rate[31,34,35] and a time-varying Gaussian neighborhood function.[31,36,37] Due to the nature of the SOMs algorithm, the number of clusters for the SOMs was specified *a priori*. For these data, an 8×8 matrix producing 64 clusters was specified. This choice was a compromise between too many clusters, which would negate the effect of clustering, and too few, which would allow too much variance within each cluster. The Gaussian neighborhood function imparted a two-dimensional ordering on clusters, so nearby clusters contained transcripts that were more similar than clusters that were farther apart on the map.

Expression data used to group NHR agonists (fold change and average difference) were analyzed using a proprietary hierarchical clustering algorithm. Cluster metrics included: (1) Pearson's correlation coefficient with cluster subtype average linkage, and (2) Euclidean distance with Ward's minimum variance method. Transcripts determined to be "present" by the Affymetrix algorithm with a minimal two-fold change or average difference intensity ≥2500 in at least one treatment group were included in the clustering dataset.

## 7.3 RESULTS

### 7.3.1 Array Data Confirmation and Validation

Three approaches were used to verify the array data collected in the hepatocyte differentiation study. First, RT-PCR of select transcripts with various expression patterns was performed with the same RNA used with rat genomic microarrays. As can be seen in Figure 7.1, the qualitative expression patterns of five genes chosen for RT-PCR analysis matched the expression patterns from the microarray analyses. This includes mRNAs that were temporally upregulated (annexin, glutathione S-transferase $\pi$, $\beta$-actin), and mRNAs whose expression was temporally downregulated (CYP 2C23, ST1C1). RT-PCR of 18s ribosomal RNA served as a control gene whose transcript levels did not vary at any times investigated in this study.

A second method used to support data from the primary microarray experiment utilized a separate RNA preparation (different animal/cultures) hybridized to rat toxicology subarrays. These Affymetrix-designed subarrays (RTU34) contained a 900-probe subset of the larger rat genomic arrays (RGU34A) and were utilized as a means to confirm expression on a scale that could not be achieved with transcript-by-transcript methods, such as RT-PCR. The overall expression levels (percent of total detectable transcripts) observed in both experiments were in agreement with other experiments analyzing hepatocyte gene expression with Affymetrix arrays in our laboratory. On average, 28% of the transcripts probed were detectable in experiments employing genomic arrays, and 34% of the transcripts probed were detectable in experiments employing toxicology subarrays (Table 7.1). The design of this subarray focused upon genes for which the expression would be of interest to toxicologists, resulting in a probe bias for highly regulatable and liver-specific genes. When employing these subarrays in hepatic studies, our laboratory consistently detects a greater percentage of transcripts than with arrays designed for genomic coverage. As an additional measure of reproducibility, we found that 79% of the transcripts were detectable in both primary and repeat experiments, demonstrating reasonable experimental reproducibility between studies (data not shown). A summary of the number of transcripts for which the expression levels changed at critical points in both the primary (Column 2) and repeat experiments (Column 3) can be found in Table 7.1. This table was constructed by first removing transcripts from the datasets for which expression was undetectable in both compared samples, followed by setting a fold-change threshold of two as the criterion for inclusion. At the earliest time point, the number of transcripts for which the levels were elevated exceeded the number that were repressed on both the rat genomic microarray and rat toxicology subarray. At later time points, induced and repressed transcripts were somewhat

A.

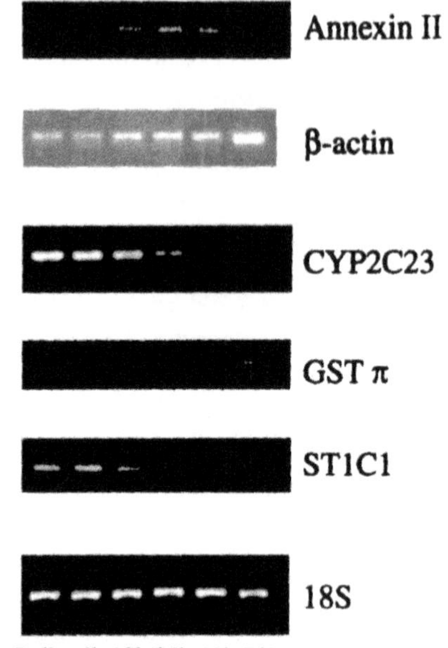


B.

| | Pellet | 4H | 12H | 24H | 48H | 72H |
|---|---|---|---|---|---|---|
| Annexin II | 1 | 1.2 | 8.5 | 20 | 20 | 20 |
| Beta-actin | 1 | 1.4 | 2.0 | 2.8 | 2.2 | 2.3 |
| CYP2C23 | 1 | -1.2 | -1.1 | -2.8 | -20 | -20 |
| GST π | 1 | -1.6 | -1.8 | -1.3 | -2 | 7 |
| STIC1 | 1 | -1.6 | -2.8 | -20 | -20 | -20 |

**FIGURE 7.1** (A) PCR primers were designed from sequences defined by Affymetrix for each transcript and used in RT-PCR confirmation studies. Reactions contained 1.0 μg total RNA and were conducted with an annealing temperature of 55°C for 25 (β-actin), 30 (STIC1, Annexin II, CYP2C23), or 35 (GST-π) cycles. 18s ribosomal RNA was used as a control gene for visual normalization. (B) Comparative analyses of the same eight transcript levels as determined by oligonucleotide arrays. All values are expressed as fold changes relative to the cell pellet using Affymetrix Gene Chip software, as described in the Experimental Procedures section.

equalized on the rat genomic microarray. On the rat toxicology subarray, however, this trend actually reversed at all time points beyond 12 hours, reflecting the bias on the subarray for probes associated with drug metabolism which were generally repressed in this study. Because every gene on the toxicology subarray was present on the larger rat genomic microarray, a direct comparison of transcript changes in repeat experiments could be made. As can be seen in Table 7.1, changes in gene expression between experiments at each time agreed quite well, with nearly 80% of repressed transcripts reproducibly changing using the relatively stringent criteria in these separate studies.

A third method of investigating experimental variability was performed by rehybridizing labeled RNA from initial experiments to rat toxicology subarrays. As can be seen in Table 7.2, transcript levels were similar between separate hybridizations of the same RNA sample. Transcripts that were highly up- or downregulated (fold-change threshold ≥5) behaved similarly in rehybridization experiments (98% decrease, 95% increase) and in repeat experiments (85% decrease, 89% increase) (Table 7.2). Because agreement was higher when the same samples were hybridized to different arrays than when the experiment was repeated in entirety, it can be inferred that some of the variability observed between experiments is biological in nature. This conclusion is supported by

**TABLE 7.1**
**Gene Expression Changes at Critical Times in the Culture Process**

| | Genomic Array (RGU34A) | Toxicology Subarray (RTU34) | Agreement between Common Genes |
|---|---|---|---|
| Total probes | ~8,700 | ~900 | ~900 in common |
| Detectable transcripts | 27.7 ± 2.3% | 33.9 ± 5.5% | 28.3 ± 5.7% |
| Hepatocyte-enriched cell pellet: | | | |
| 4 hours after plating | D = 92 | D = 18 | D: 28% |
| | I = 170 | I = 40 | I: 60% |
| 12 hours after plating | D = 393 | D = 90 | D: 69% |
| | I = 253 | I = 35 | I: 66% |
| 24 hours after plating | D = 405 | D = 116 | D: 74% |
| | I = 406 | I = 46 | I: 61% |
| 48 hours after plating | D = 361 | D = 105 | D: 80% |
| | I = 464 | I = 85 | I: 45% |
| 72 hours after plating | D = 360 | D = 100 | D: 79% |
| | I = 468 | I = 74 | I: 49% |

*Note:* The absolute number of transcripts determined to be increased (I) and decreased (D) at each time point were determined for both the primary experiment performed on the rat genomic arrays and the secondary experiments performed on the rat toxicology subarrays. Agreement was determined by comparing transcripts (I or D) from both toxicology and genomic arrays with the number of transcript changes (I or D) from toxicology subarrays using the criteria described in the Experimental Procedures section (>2×).

**TABLE 7.2**
**Gene Expression Changes Common in Repeat and Replicate Experiments**

| Fold-Change Threshold | Rehybrization Experiment (% Agreement) | Repeat Experiment (% Agreement) |
|---|---|---|
| 0-fold | D = 82 | D = 74 |
| | I = 34 | I = 27 |
| 2-fold | D = 75 | D = 63 |
| | I = 43 | I = 58 |
| 5-fold | D = 98 | D = 85 |
| | I = 95 | I = 89 |

*Note:* Transcripts were determined to be differentially expressed using Affymetrix GeneChip software, as described in the Experimental Procedures section, and sorted by the fold-change score for increased (I) and decreased (D) transcripts. Differentially expressed transcripts were then compared to corresponding changes in the primary experiment employing rat genomic arrays using the same criteria and expressed as % agreement.

the findings of Cohen et al.,[38] who employed this technology to monitor cellular responses to a pathogenic microorganism.

### 7.3.2 Clustering of Temporal Hepatocyte Data

The data obtained from each comparison outlined in Table 7.1 were grouped using SOMs, and individual transcripts were placed into one of 64 clusters based upon their temporal expression pattern as described in the Experimental Procedures section. All 64 clusters and associated gene changes are contained in a supplemental data section that can be found at http://pubs.acs.org/subscribe/journals/crtoec/supmat/index.html. Given that thousands of transcripts changed in these studies, only general trends are discussed herein. Figure 7.2 contains two representative clusters from the entire 64-cluster diagram and the transcript expression patterns at each time point investigated. These clusters were chosen because they contain many transcripts associated with either the extracellular matrix (Figure 7.2A) or the CYP family (Figure 7.2B), two protein groups discussed extensively later. In order to address the dedifferentiation and adaptation processes, we compared expression profiles between the enriched hepatocyte pellet (baseline sample) and various times in culture (variable sample) to determine fold-change values. Comparison of the enriched hepatocyte pellet to whole-liver tissue revealed gene expression changes associated with removal of cell types during the isolation process.

Trends in expression patterns were observed in many functional classes of genes, and those observed with phase I enzymes were exemplified by the CYP superfamily presented in Figure 7.3. As can be seen in Figure 7.3, most CYP genes were downregulated over time. CYP1 gene expression (open bars) was substantially downregulated between 24 and 48 hours, while CYP2 and CYP4 families (gray and black bars, respectively) were downregulated earlier, with the most dramatic

A.

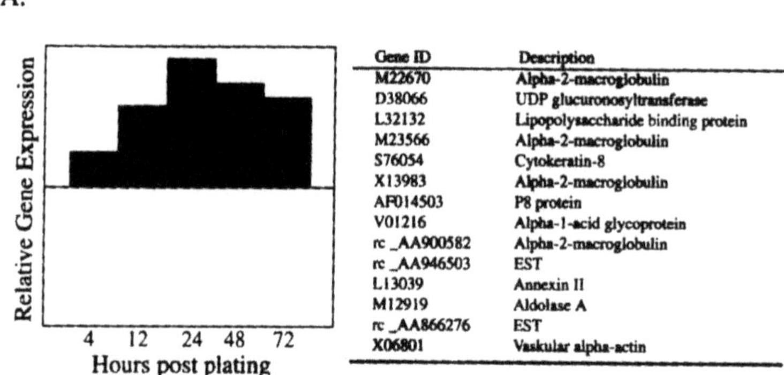


| Gene ID | Description |
|---|---|
| M22670 | Alpha-2-macroglobulin |
| D38066 | UDP glucuronosyltransferase |
| L32132 | Lipopolysaccharide binding protein |
| M23566 | Alpha-2-macroglobulin |
| S76054 | Cytokeratin-8 |
| X13983 | Alpha-2-macroglobulin |
| AF014503 | P8 protein |
| V01216 | Alpha-1-acid glycoprotein |
| rc _AA900582 | Alpha-2-macroglobulin |
| rc _AA946503 | EST |
| L13039 | Annexin II |
| M12919 | Aldolase A |
| rc _AA866276 | EST |
| X06801 | Vaskular alpha-actin |

B.

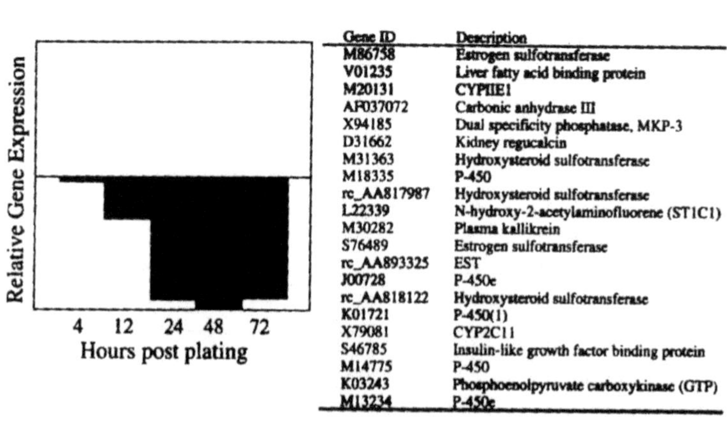


| Gene ID | Description |
|---|---|
| M86758 | Estrogen sulfotransferase |
| V01235 | Liver fatty acid binding protein |
| M20131 | CYPIIE1 |
| AF037072 | Carbonic anhydrase III |
| X94185 | Dual specificity phosphatase, MKP-3 |
| D31662 | Kidney regucalcin |
| M31363 | Hydroxysteroid sulfotransferase |
| M18335 | P-450 |
| rc_AA817987 | Hydroxysteroid sulfotransferase |
| L22339 | N-hydroxy-2-acetylaminofluorene (ST1C1) |
| M30282 | Plasma kallikrein |
| S76489 | Estrogen sulfotransferase |
| rc_AA893325 | EST |
| J00728 | P-450e |
| rc_AA818122 | Hydroxysteroid sulfotransferase |
| K01721 | P-450(1) |
| X79081 | CYP2C11 |
| S46785 | Insulin-like growth factor binding protein |
| M14775 | P-450 |
| K03243 | Phosphoenolpyruvate carboxykinase (GTP) |
| M13234 | P-450e |

**FIGURE 7.2** Cluster analysis of transcripts over time using SOMs. Two representative clusters show genes for which the transcript levels increase (A) and decrease (B) with time. Filtering and clustering analyses were performed as described in the Experimental Procedures section. All transcripts represented in each cluster are listed with the GeneBank accession number. In total, 64 clusters with different transcriptional patterns were assembled, all of which can be viewed (including the entire population of transcripts with expression change values at each time point) in the supplemental data section (http://pubs.acs.org/subscribe/journals/crtoec/supmat/index.html).

changes being observed between the 12- and 24-hour time points. The expression of CYP3 family members (cross-hatch) remained constant throughout the 72-hour culture period.

Transcript changes for phase II biotransformation enzymes were grouped by family and further organized by subfamily where appropriate (Table 7.3). Unlike the CYPs, which were predominately

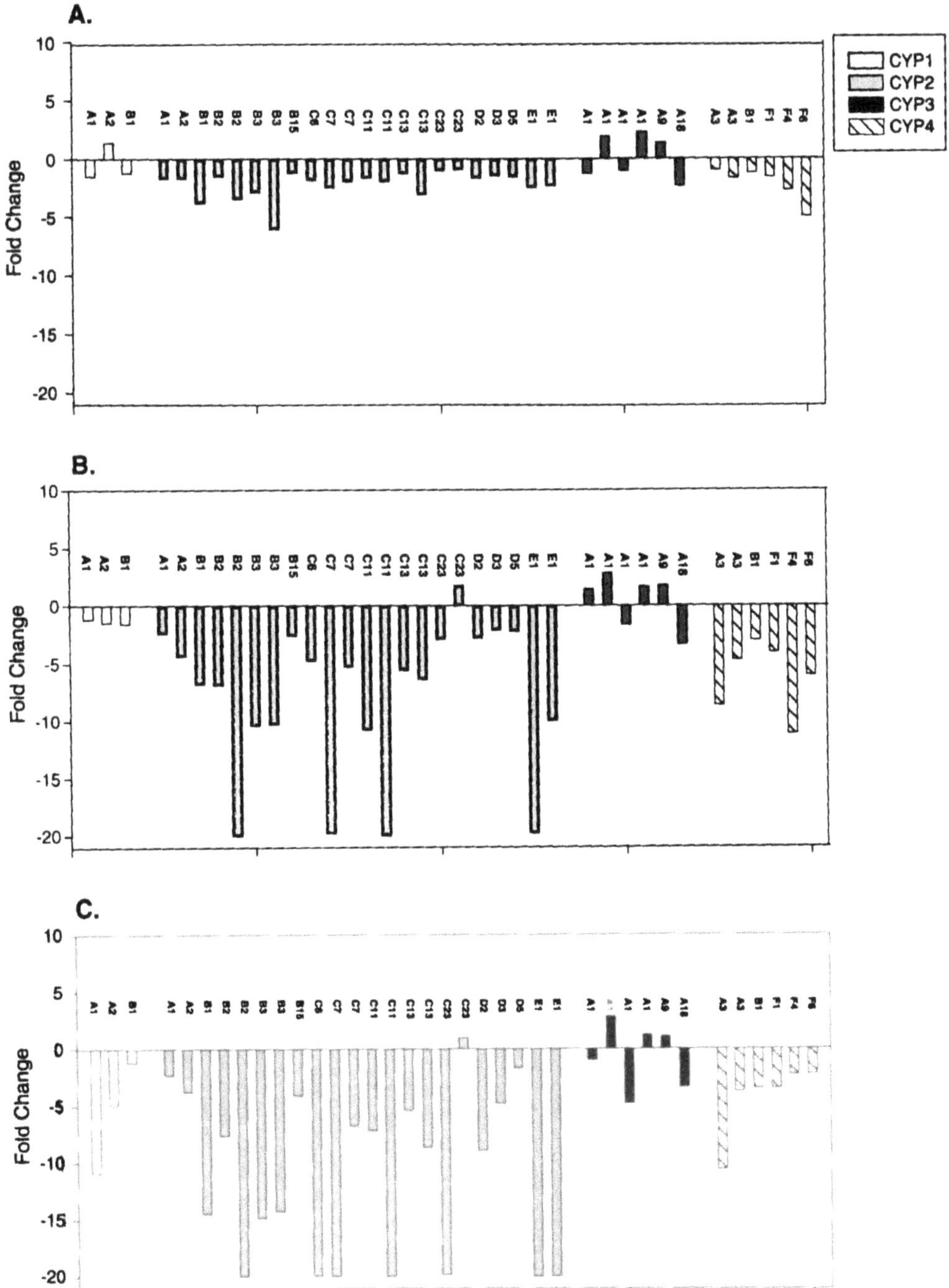


FIGURE 7.3 Evaluation of cyctochrome P450 gene expression at (A) 12, (B) 24, and (C) 48 hours after isolation. Data are presented as the fold change from the enriched hepatocyte pellet. Open bars represent CYP1, gray bars represent CYP2, black bars represent CYP3, and cross-hatch bars represent CYP4 transcripts.

**TABLE 7.3**
**Transcript Profiling of Phase II Metabolizing Enzymes by Family**

| Probe Set | Gene Identification | Cluster[a] | 4 Hr | 12 Hr | 24 Hr | 48 Hr | 72 Hr |
|---|---|---|---|---|---|---|---|
| | | **A. Glucuronidation** | | | | | |
| D38066 | UDP-GT 1B2F rat DNA for UDP-GT | 1 | 4.8 | 9.4 | 20 | 16.6 | 17.8 |
| D38065 | UDP-GT 1B1E rat DNA for UDP-GT | 9 | 2.5 | 2.7 | 4.6 | 10.8 | 12.8 |
| D83796 | Rat U-GT1 mRNA for UDP-GT | 18 | 2 | 3.5 | 7.1 | 4.6 | 5.5 |
| S56937 | Rat 3MC-inducible UDP-GT | 18 | 1.8 | 4.4 | 3.6 | 4.6 | 5.6 |
| J02612 | Rat mRNA for UDP-GT | 19 | 1.6 | 3 | 4.7 | 3.2 | 3.6 |
| J05132 | Rat 3MC-inducible UDP-GT mRNA | 19 | 1.8 | 3.4 | 3.3 | 3.5 | 4 |
| D38061 | UDP-GT 1A1A rat DNA for UDP-GT | 21 | –1.1 | 2.1 | 2.2 | 1.9 | 1.1 |
| D38062 | UDP-GT 1A2B rat DNA for UDP-GT | 28 | 1.7 | 1.2 | 2.2 | 2.7 | 1.9 |
| M13506 | Rat phenobarbital-inducible UDP-GT mRNA | 37 | –1.2 | –1.5 | –2.3 | –2 | 1.8 |
| M33747 | UDP-GT21 | 38 | –1.3 | –1.3 | –2.5 | –1.8 | –3.6 |
| M31109 | Rat UDP-GT mRNA complete cds | 38 | –1.5 | –1.8 | –2.1 | –2.4 | –2.1 |
| Y00156 | Rat mRNA for hepatic microsomal UDP-GT | 38 | –1.2 | –1.8 | –1.8 | –2.2 | –2 |
| X03478 | Rat liver mRNA for androsterone UDP-GT | 38 | –1.3 | –1.3 | –1.7 | –2 | –1.8 |
| S53936 | Rat bilirubin-specific UDP-GT | 42 | 1 | –1.3 | –1.1 | 2.2 | –1.7 |
| D38069 | UDP-GT 1B5I rat DNA for UDP-GT | 44 | 1.4 | –1.6 | –1.3 | –1.3 | –1.8 |
| M33746 | Rat UDP-GTr-5 mRNA | 44 | –1.3 | –1.3 | –1.7 | –1.8 | –1.7 |
| | | **B. Sulfation** | | | | | |
| D89375 | Rat mRNA for ST1B1 | 34 | –1.6 | –1.2 | 2.2 | 3.5 | 3.4 |
| L19998 | Rat minoxidil ST mRNA | 42 | –1.4 | –1.6 | 1.3 | 2.1 | 1.6 |
| L19998 | Rat minoxidil ST mRNA | 50 | –1.6 | –1.7 | –2 | 1.7 | 1.3 |
| AF022729 | HNK-1 sulfotransferase mRNA | 52 | –1.2 | –2.5 | –2.6 | –3 | –1.4 |
| X63410 | Rat mRNA hydroxysteroid ST | 53 | –1.8 | –7.2 | –7.2 | –3.9 | –4.8 |
| M33329 | Rat hydroxysteroid Sta mRNA | 53 | –1.5 | –5.3 | –3.3 | –2.5 | –3 |
| D50564 | Mercaptopyruvate ST | 62 | –6.9 | –13.6 | –3.2 | –11.5 | –8.3 |
| D14987 | HSS1 rat mRNA for hydroxysteroid ST | 63 | –1.9 | –13.8 | –20 | –12.3 | –14.6 |
| D14988 | HSS2 rat mRNA for hydroxysteroid ST | 63 | –1.5 | –7.3 | –13.2 | –6.3 | –8.4 |
| D14989 | HSS3 Rat mRNA for hydroxysteroid ST | 63 | –1.8 | –11.5 | –15.7 | –11.4 | –11 |
| AI169695 | Hydroxysteroid ST | 63 | –1.6 | –8.1 | –12.7 | –6.5 | –9.6 |
| M31363 | HSST rat hydroxysteroid ST mRNA | 64 | –1.8 | –15 | –20 | –20 | –20 |

| | | | | | | | |
|---|---|---|---|---|---|---|---|
| | | **C. GSH Conjugation** | | | | | |
| X02904 | mRNA for GST-P subunit (pi) | 25 | −1.7 | −1.8 | −1.3 | −1.9 | 7 |
| X04229 | mRNA for GST-Yb subunit (mu) | 28 | 1.1 | 1.7 | 1.4 | 2.3 | 2.6 |
| J02592 | GST-Yb subunit mRNA (mu) | 35 | 1.2 | 1.2 | 1.3 | 1.4 | 1.5 |
| J03914 | GST-Yb subunit gene (mu) | 35 | 1.1 | 1.6 | 1.6 | 1.5 | 1.6 |
| K01932 | Liver GST-Yc subunit mRNA (alpha) | 38 | −1.3 | −1.3 | −2.7 | −2.4 | −2.8 |
| S72505 | Fetal liver GST-Yc1 subunit (alpha) | 38 | −1.1 | −1.6 | −2.4 | −2.7 | −2.4 |
| X78848 | GST-Yc1 mRNA (alpha) | 38 | −1.1 | −1.1 | −2.8 | −2.5 | −2.6 |
| J03752 | GST mRNA | 44 | −1.1 | −1.1 | −1.8 | −1.7 | −2 |
| S82820 | Hepatoma cell GST-Yc2 subunit mRNA (alpha) | 45 | −2.2 | −3.2 | −5 | −1.1 | −1.3 |
| X62660 | mRNA for GT subunit 8 | 46 | −1.3 | −2.6 | −5.5 | −4.2 | −2.8 |
| K00136 | GST-Ya subunit (alpha) | 48 | 1.2 | −1.1 | −3 | −5 | −11 |
| X62660 | mRNA for GT subunit 8 | 52 | −1.1 | −3.7 | −3 | −2.9 | −2.1 |
| D10026 | mRNA for GST-Yrs-Yrs | 61 | −1.6 | −9.2 | −2.2 | −2.1 | −1.8 |
| | | **D. Methylation** | | | | | |
| U60882 | Rat protein arginine N-MT (PRMT-1) mRNA | 14 | 1.8 | 1.6 | 1.3 | 1.6 | 1.9 |
| M93257 | Rat catechol-O-MT mRNA | 24 | −1.6 | −1.5 | −1.4 | −2.6 | 1.1 |
| M76704 | Rat O-6-methylguanine-DNA MT (RMGMT) mRNA | 25 | 1.1 | 1.5 | 4 | 3.6 | 6 |
| X06150 | Rat mRNA for glycine MT | 32 | −1.3 | −1.9 | −2.8 | −4.9 | −3.2 |
| M60753 | Rat catechol-O-MT mRNA | 43 | −1.2 | −1.1 | −1.3 | 1.2 | 1.7 |
| X08056 | Rat gene for guanidinoacetate MT | 45 | −1.4 | −2.7 | −3.7 | −1.4 | −1.3 |
| J03588 | Rat guanidinoacetate MT mRNA | 46 | −1.3 | −3.2 | −6.3 | −1.5 | −1.3 |
| L14441 | Rat phosphotidylethanolamine N-MT mRNA | 47 | −1.3 | −2.2 | −5.1 | −4.8 | −5.4 |
| L20427 | Rat dihydroxypolyprenylbenzoate MT mRNA | 60 | −1.6 | −4.4 | −1.2 | −2 | −2.2 |
| AF038870 | Rat mRNA for cytosine 5 MT | 63 | 1.1 | −2.8 | −15 | −16.4 | −10.1 |
| X06150 | Rat mRNA for glycine MT | 40 | −1.3 | −1.9 | −3.9 | −8.6 | −5.1 |

*Note:* Data are presented as fold changes from hepatocyte pellet as determined using Affymetrix's Data Mining Tool.

[a] Represents associated cluster from 8×8 SOMs. **Supplemental Data Section (http://pubs.acs.org/subscribe/journals/crtoec/supmat/index.html)**

**TABLE 7.4**
**Transcript Profiling of Genes Involved in Glutathione Maintenance and Utilization**

| Probe Set | Gene Identification | Cluster[a] | 4 Hr | 12 Hr | 24 Hr | 48 Hr | 72 Hr |
|---|---|---|---|---|---|---|---|
| J05181 | γ-GCS | 8 | 4.3 | –2.2 | –1.1 | –1.7 | –2.2 |
| L38615 | Glutathione synthetase | 9 | 1.8 | 1.2 | 10.1 | 11.7 | 9.5 |
| Y00497 | Mn-SOD | 11 | 1.9 | 6.6 | 3.1 | 1.9 | 1.6 |
| U73174 | Glutathione reductase | 19 | 2.6 | 2.8 | 3.7 | 3.6 | 3.1 |
| U73174 | Glutathione reductase | 26 | 1.3 | 2.5 | 4 | 3.8 | 2.5 |
| M25157 | Cu, Zn-SOD | 40 | –1.5 | –2.3 | –4.1 | –6 | –7.1 |
| X12367 | Glutathione peroxidase | 47 | –1.5 | –7.2 | –5.4 | –5.4 | –7.4 |
| M11670 | Catalase | 63 | –1.6 | –3.3 | –18.4 | –6.1 | –6.6 |

*Note:* Data are presented as fold changes from hepatocyte pellet as determined using Affymetrix's Data Mining Tool.

[a] Represents associated cluster from 8×8 SOMs.

downregulated over time, phase II enzymes demonstrated variable expression patterns. In general, expression of glucuronyl transferases either increased or was relatively unchanged over the 72-hour culture period (Table 7.3, part A) while sulfyltransferases and methyltransferases were predominately downregulated (Table 7.3, parts B and D, respectively) when viewed as a group. Furthermore, individual genes were regulated differently within family classifications, as exemplified by transcripts encoding proteins involved in glutathione conjugation (Table 7.3, part C). Here, glutathione S-transferase (GST)-α transcripts were downregulated with increasing time in culture, while GST-μ transcripts remained unchanged. Further evaluation of the dataset revealed a sevenfold upregulation of the fetal expressed GST-π isoenzyme transcript following 72 hours in culture.

### 7.3.3 Pathway Analysis and Functional Grouping of Temporal Hepatocyte Data

An important advantage of global expression profiling compared to individual gene regulation studies is the ability to monitor changes in entire metabolic pathways simultaneously. Three such pathways of particular interest were identified within this dataset. The first pathway of note included genes involved in the production and utilization of glutathione. Gamma-glutamylcysteine synthetase (γ-GCS) was upregulated four-fold at 4 hours, and glutathione synthetase was upregulated ten-fold at 24, 48, and 72 hours (Table 7.4). Genes responsible for utilizing glutathione (GSH) in the cellular redox cycle (GSH peroxidase, GSSG reductase, GST) were variably expressed (Tables 7.3 and 7.4). Figures 7.4 and 7.5 depict the 24-hour gene expression changes in gluconeogenesis and glycolysis, respectively. In general, the genes in the gluconeogenic pathway were downregulated while the genes involved in glycolysis were upregulated.

A distinct observation within this dataset was the consistent and progressive upregulation of genes encoding proteins associated with the cytoskeleton and extracellular matrix (EM). In all, more than 40 different genes were induced, while few were repressed. Included in the list in Table 7.5 are mRNAs coding for proteins involved in the synthesis and maintenance of connective tissue and the basement membrane, proteins involved in actin synthesis and function, and others involved in matrix signaling, maintenance of shape, and cell motility. We found it difficult to draw lines of distinction in genes known to have multiple and sometimes diverse function when constructing this particular table. For example, several genes in Table 7.5 are involved in cell cycle progression through mitosis.

Genes involved in hepatocellular proliferation and differentiation were also noted based on their chronological involvement in the cell cycle as depicted in Schematic 7.1 and Table 7.6.

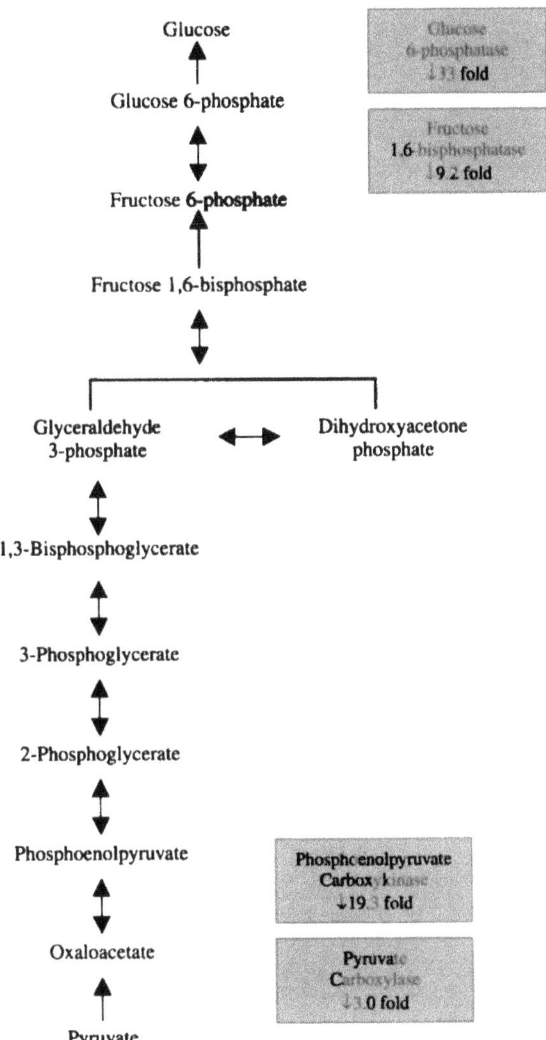


**FIGURE 7.4** Gluconeogenic pathway depicting important catalytic enzymes responsible for the formation of glucose from pyruvate. The fold change in gene expression level is given for each enzyme at the 24-hour time point. The fold change in gene expression levels can be found for remaining time points in the supplemental data section (http://pubs.acs.org/subscribe/journals/crtoec/supmat/index.html).

Hepatocellular genes involved in cellular priming were mostly unchanged at the mRNA level throughout the culture process. Exceptions were found that included two insulin-responsive genes, insulin growth factor binding protein I (upregulated 10-fold maximally), insulin-like growth factor I (downregulated 20-fold maximally), and interleukin-6 (IL-6) receptor ligand-binding protein (upregulation maximal-early) (Table 7.6, part A). Transcription factors involved in cell proliferation, such as nuclear factor kappa B (NF-κB), CCAAT/enhancer-binding protein (C/EBP), and hepatocyte nuclear factor (HNF), were found to be variably modulated at the early time points (Table 7.6, part B). Immediate-early genes were primarily unaffected or upregulated throughout the culture process. Exceptions included junB (downregulated 6-fold at 48 and 72 hours) and Egr-1 (Krox 24) (downregulated 4- to 6-fold at 12, 24, 48, and 72 hours after plating) (Table 7.6, part C). A few delayed-early genes, such as clone-6 (CL-6), hepatic Arg-Ser protein (HRS), and B-cell leukemia/lymphoma x gene (Bcl-x) were variably modulated with early upregulation of

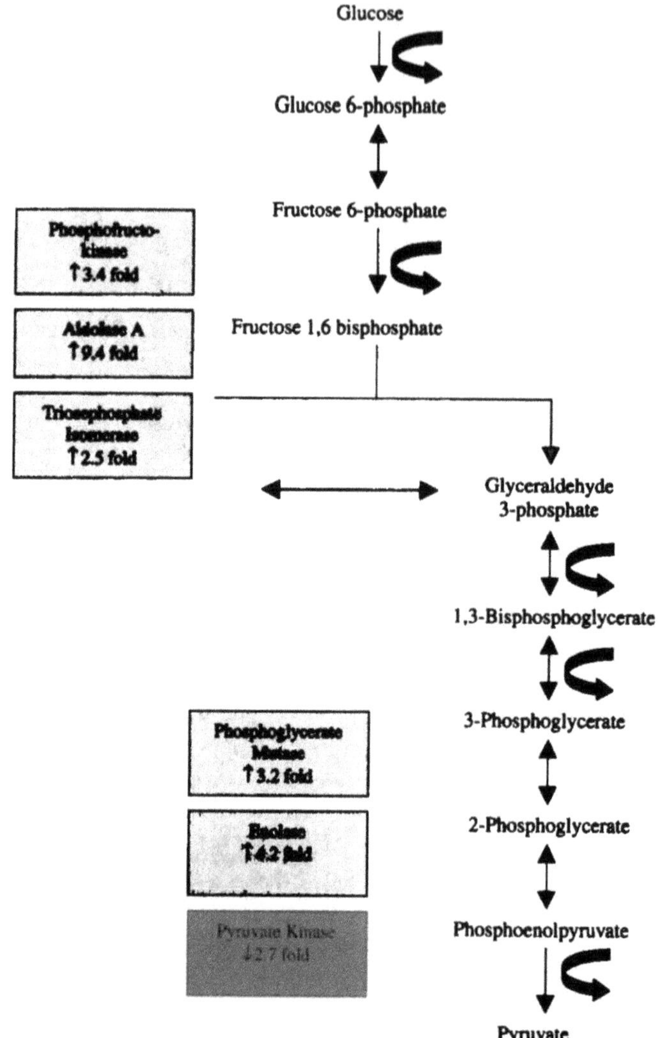


**FIGURE 7.5** Glycolytic pathway and important catalytic enzymes responsible for the breakdown of glucose. The fold change in gene expression level is given for each enzyme at the 24-hour time point. The fold change in gene expression levels can be found for remaining time points in the supplemental data section (http://pubs.acs.org/subscribe/journals/crtoec/supmat/index.html).

CL-6, early downregulation of HRS, and overall upregulation of Bcl-x (Table 7.6, part D). Cell-cycle progression requires a balance between positive and negative regulatory pathways. Evaluation of genes involved in cell-cycle regulatory pathways showed that, during culture, hepatocytes altered the expression of genes associated with the negative regulatory pathway of the cell cycle. Examples include $G_0/G_1$ switch gene (Gos-2) (downregulated 18-fold at 12 hours) and p53/p21 (upregulated throughout the culture process) (Table 7.6, parts C and E).

### 7.3.4 *In Vitro/In Vivo* Comparisons

Treatment with the peroxisome proliferator fenofibrate produced robust changes in transcripts associated with lipid metabolism in both cultured hepatocytes and rat liver. As shown in Table 7.7, the majority of genes associated with fatty acid metabolism were induced in both models, and the magnitude of change was usually similar between *in vitro* and *in vivo* experiments. For example,

**TABLE 7.5**
**Cellular Cytoskeleton and Extracellular Matrix**

| Probe Set | Gene Identification | Cluster[a] | 4 Hr | 12 Hr | 24 Hr | 48 Hr | 72 Hr |
|---|---|---|---|---|---|---|---|
| L13039 | Annexin II | 1 | 1.2 | 8.5 | 20 | 20 | 20 |
| X06801 | Vascular a-actin | 1 | 4.2 | 6 | 13.4 | 12.8 | 12.7 |
| S76054 | Cytokeratin 8 | 1 | 4.6 | 20 | 17.5 | 10.5 | 11.1 |
| AA946503 | α2-Microglobulin | 1 | 6.1 | 7.6 | 2 | 7.1 | 6.1 |
| S61868 | Ryudocan | 4 | 4.5 | 1.4 | 2.2 | 1.9 | 2.1 |
| D38056 | Protein b61 | 4 | 6.3 | 2.3 | 2 | 1.9 | 1.6 |
| U27201 | TIMP3: tissue inhibitor of metalloproteinase | 6 | 2.1 | 1.5 | 1.5 | 2.5 | 2.5 |
| X62952 | Vimentin | 9 | −2.8 | 6 | 9.2 | 11 | 18.3 |
| X07648 | Amyloidogenic glycoprotein | 9 | −1.1 | 2.6 | 6 | 9.2 | 14.8 |
| X54617 | Myosin regulatory light chain | 9 | 1.6 | 5.9 | 8.4 | 11.1 | 14.3 |
| M28259 | Fibronectin | 10 | 2.5 | 3.8 | 6.7 | 6.2 | 7 |
| X80130 | α-Actin | 10 | 3.9 | 6.2 | 6.1 | 5 | 7 |
| AA894200 | Myosin | 12 | 1.9 | 3.5 | 2.4 | 2.4 | 2 |
| D42137 | Annexin V: (lipocortin V) | 17 | −1.3 | −1.3 | 1.8 | 5.2 | 10.3 |
| AF084186 | α-Fodrin | 17 | 1.4 | −1 | 2.7 | 5.6 | 7.7 |
| X05566 | Myosin regulatory light chain | 17 | 1.5 | 3 | 4 | 4.8 | 7.4 |
| M60666 | α-Tropomycin 2 | 17 | 1.6 | 1.7 | 3.3 | 5.7 | 7.4 |
| M55534 | α-Crystallin | 17 | 2 | 1.9 | 3.3 | 5.5 | 9.1 |
| U82612 | Fibronectin 1 | 18 | 1.5 | 2.9 | 4.3 | 4.3 | 4 |
| AF097593 | N-cadherin | 18 | 2.4 | 1.1 | 5.6 | 4.5 | 4.4 |
| AF054826 | VAMP5 | 18 | 2.5 | 3.1 | 3.8 | 3.6 | 4.1 |
| U23769 | CLP36: subunit of Annexin II | 18 | 3 | 3.9 | 5 | 3 | 5.7 |
| X52815 | γ-Actin | 19 | 2.8 | 4.6 | 4.3 | 3 | 3.4 |
| AA817887 | Prolifin | 20 | 1.5 | 2.5 | 2.6 | 3 | 3.7 |
| AF051895 | Lipocortin V; Annexin V | 24 | −1.8 | −1.9 | 3 | 3.7 | 7.5 |
| U76714 | CARI | 24 | −1.3 | 1.2 | 2.8 | 2.3 | 4.6 |
| AF054618 | Cortactin isoform C | 24 | −1 | 1.7 | 3.3 | 3.2 | 5 |
| M15474 | α-Tropomoycin | 24 | 1.5 | 1.2 | 3.4 | 4.5 | 6.8 |
| AA860030 | β-Tubulin | 26 | −1.1 | 1.9 | 4 | 3.1 | 3.4 |
| L00191 | Fibronectin | 26 | 1.4 | 2.1 | 4.3 | 3.4 | 3.1 |
| AF083269 | P41-arc | 26 | 1.5 | 1.7 | 2.4 | 2.7 | 4.1 |
| J00797 | α-Tubulin | 27 | 1.2 | 2.1 | 3.2 | 1.9 | 1.9 |
| AA892333 | α-Tubulin | 27 | 1.4 | 1.8 | 3.2 | 1.3 | 1.1 |
| V01217 | β-Actin | 27 | 1.4 | 2 | 2.8 | 2.2 | 2.3 |
| D84477 | RhoA | 28 | 1.5 | 1.1 | 1.6 | 2 | 2.7 |
| S66184 | Lysyl oxidase | 33 | 1.3 | 1 | 1.5 | 2.8 | 4.4 |
| S61865 | Syndecan | 33 | 1.4 | −1.4 | 2.9 | 1.3 | 4 |
| AF080507 | Mannose-binding protein | 54 | −1.6 | −2.3 | −7.8 | −2.4 | −2.4 |
| X05023 | Mannose-binding protein | 55 | −2.1 | −5.6 | −7 | −8.2 | −6.1 |
| X70369 | Pro a 1 collagen | 55 | −1.8 | −2.7 | −7.6 | −8.7 | −4.2 |

*Note:* Data are presented as fold changes from hepatocyte pellet as determined using Affymetrix's Data Mining Tool.

[a] Represents associated cluster from 8×8 SOMs.

acyl-CoA hydrolase was increased greater than 100-fold in both experiments, whereas fatty-acid-binding protein (FABP) and delta3, delta2-enoyl-CoA isomerase transcripts were increased 14.9- and 5.9-fold *in vitro* and 6.7- and 4.8-fold *in vivo*, respectively. From a biological standpoint, it can be seen that transcript levels for the rate-limiting enzymes in the peroxisomal and mitochondrial β-oxidation pathways were markedly upregulated; acyl-CoA oxidase (ACO) levels were increased 12.6- and 3.0-fold in cultured hepatocytes and liver, respectively (Table 7.7). Similarly, carnitine palmitoyltransferase I (CPT I) mRNA was increased 3.1-fold in isolated hepatocytes and 2.0-fold in the liver. Further inspection of this dataset revealed that the majority of transcript levels common to both systems were increased, whereas only five transcripts were repressed in both systems (Table 7.8A). Comparison of the *in vitro* and *in vivo* data revealed that only a small subset of transcripts was differentially regulated (i.e., displayed opposite regulation in the two systems), with only three transcripts meeting the predetermined criteria for differential expression (Table 7.8B).

### 7.3.5 Predictive Value of Hepatocyte Culture Systems

As shown in Figure 7.6, hierarchical clustering of gene expression data from primary rat hepatocytes treated with NHR agonists revealed three nodes, each corresponding to a unique ligand specificity: PPARα (fenofibrate and Wy14,643), PPARγ (troglitazone and rosiglitazone), or RXR agonists. PPARα and RXR agonist-treated hepatocytes gave overall expression patterns that were more similar to each other than to PPARγ-treated hepatocytes, as illustrated by their proximity in the dendogram.

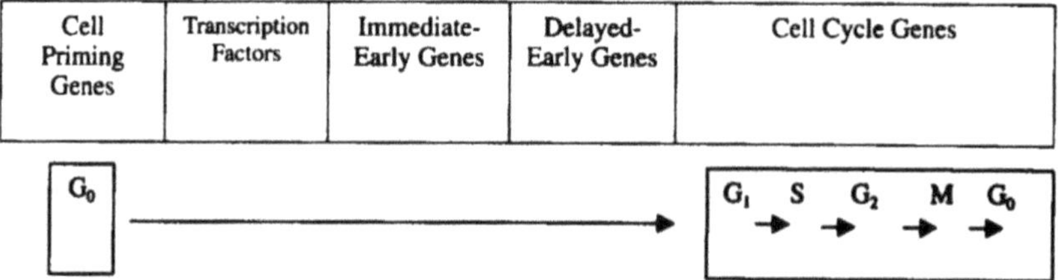


**SCHEMATIC 7.1** Representation of sequential gene induction categories in hepatocellular proliferation and the cell-cycle stage associated with gene induction.

**TABLE 7.6**
**Hepatocellular Proliferation and Differentiation**

| Probe Set | Gene Identification | Cluster[a] | 4 Hr | 12 Hr | 24 Hr | 48 Hr | 72 Hr |
|---|---|---|---|---|---|---|---|
| | | **A. Cell-Priming Genes** | | | | | |
| M58634 | IGF binding protein I | 3 | 10.1 | 5.1 | 5.4 | 3.9 | 5.4 |
| M58587 | IL-6 receptor ligand-binding protein | 4 | 5.9 | 3.4 | 3 | 3.1 | 1.9 |
| U65007 | HGF receptor | 6 | 2.1 | 1.1 | 1.4 | 2.3 | 1.6 |
| M92340 | IL-6 signal transducer | 13 | 2.3 | 3.4 | 1.7 | 1.5 | 1.3 |
| M17960 | Insulin-like growth factor II | 15 | 1.4 | 1.3 | –1.2 | –1.1 | –1.2 |
| M26744 | IL-6 mRNA | 21 | 1.1 | 2.3 | 1.6 | 2 | 1.3 |
| J04807 | Insulin II gene | 21 | 1.2 | 2 | –1 | 1.6 | 1.4 |
| M98820 | IL-1β mRNA | 23 | –1.3 | 1.2 | 1.4 | 1.4 | –1.4 |
| M15481 | Insulin-like growth factor I | 48 | –1.3 | –3.4 | –3.5 | –5.4 | –20 |
| | | **B. Transcription Factors** | | | | | |
| AB017044 | HNF-3γ | 4 | 5.2 | 5.4 | 2.6 | 2.2 | 2.5 |
| L26267 | NF-κB (p105) | 5 | 4.8 | 1.5 | 4.4 | 2.2 | 2.7 |
| S77528 | C/EBP-related transcription factor | 8 | 2.8 | –1.1 | –1.2 | –1.2 | –1.8 |
| X91810 | STAT-3 | 15 | 1.4 | 1.2 | 1.2 | –1 | 1.1 |
| X56546 | HNF-1 | 50 | –2.2 | –1.7 | 1.1 | 1.1 | 1.2 |
| L09647 | HNF-3β | 57 | –3.8 | –2.5 | –1.5 | –1.1 | –1.4 |

*(continued)*

**TABLE 7.6 (CONTINUED)**
**Hepatocellular Proliferation and Differentiation**

| Probe Set | Gene Identification | Cluster[a] | 4 Hr | 12 Hr | 24 Hr | 48 Hr | 72 Hr |
|---|---|---|---|---|---|---|---|
| | | **C. Immediate-Early Genes** | | | | | |
| L22190 | SAP A | 2 | 11.8 | 20 | 20 | 3.3 | 1.9 |
| L77843 | PRL-1 | 4 | 8.5 | 3.4 | 2.1 | 2.6 | 2.5 |
| Y00396 | c-*myc* oncogene | 5 | 2.9 | 1.3 | 4 | 2.3 | 3.4 |
| Y00396 | c-*myc* oncogene | 5 | 2.7 | 1.8 | 3 | 2.8 | 2.1 |
| X02601 | EGF/H-ras-induced 53-kDa polypeptide | 13 | 1.5 | 2.5 | 1.6 | 1.1 | 1.1 |
| D37951 | MIB1 (c-*myc* intron binding protein 1) | 15 | 1.3 | 1.5 | −1.5 | −1.2 | 1.1 |
| X06769 | Rat c-*fos* mRNA | 22 | 1.1 | 1.8 | 1.2 | −1.5 | 1.3 |
| X06769 | Rat c-*fos* mRNA | 23 | −1.2 | 1.1 | −1.3 | −1.6 | −1.9 |
| M64300 | ERK-1 | 26 | 1.5 | 1.3 | 3.1 | 3.7 | 2.5 |
| M61177 | ERK-1 | 28 | −1 | 1.4 | 1.7 | 2 | 2 |
| X54686 | PJunB | 39 | −1.2 | −1.2 | −1.8 | −6.2 | −6.4 |
| U75397 | Egr-1 (Krox 24) | 47 | 1.8 | −4 | −5.4 | −6 | −4.4 |
| AA893235 | $G_0/G_1$ switch gene 2 (Gos-2) | 62 | −12.9 | −17.5 | −2.2 | −2.1 | −2.3 |
| | | **D. Delayed-Early Genes** | | | | | |
| U05784 | MAP 1A/1B | 2 | 8.6 | 14.9 | 8.1 | 6.1 | 4.7 |
| U48596 | MAPKKK (MEKK-1) | 3 | 7.5 | 6.4 | 4.5 | 3.5 | 4.1 |
| L13619 | Insulin-induced growth response protein CL-6 | 4 | 5.5 | 1.7 | −1.1 | −2.3 | −1.8 |
| U73142 | P38 MAPK | 4 | 5 | 4.7 | 2.2 | 1.8 | 1.9 |
| U72350 | Bcl-x alpha | 5 | 2.4 | 2.7 | 2.8 | 2.5 | 3.5 |
| L13619 | Insulin-induced growth response protein CL-6 | 8 | 3.9 | 1.5 | −1.3 | −1.8 | −1.9 |
| U48596 | MAPKKK (MEKK-1) | 14 | 2 | 2 | −1.3 | 1.5 | 1.7 |
| D14591 | MAPKK | 15 | −1 | 1.5 | −1.1 | 1.2 | 1.4 |
| L04485 | MAPKK | 26 | 1.5 | 2.6 | 1.8 | 4.2 | 3.9 |
| L13635 | Insulin-induced growth response protein HRS | 60 | −2 | −3.8 | −1.6 | −1.4 | −1.1 |
| | | **E. Cell-Cycle Genes** | | | | | |
| X70871 | Cyclin G | 2 | 12.6 | 13.6 | 6 | 7.6 | 9.5 |
| U75925 | Rb susceptibility gene mRNA | 8 | 2.6 | 1.3 | 3.6 | −2.2 | −1.2 |
| X13058 | P53 nuclear oncoprotein mRNA | 10 | 6.7 | 3.5 | 9.6 | 1.4 | 7.6 |
| L41275 | P21 (WAF-1) | 11 | 4.7 | 5.4 | 6.2 | 3.8 | 4.3 |
| D14015 | Cyclin E | 13 | 1.8 | 2.3 | 1.9 | 1 | −1.2 |
| U31668 | E2F | 18 | −1.8 | −1.2 | −1.7 | 1.6 | −1.2 |
| D14015 | Cyclin E | 22 | 1.4 | 1.6 | 1.1 | −1.7 | 1 |
| D16309 | Cyclin D3 | 25 | 1.7 | 1.1 | 3.3 | 1.3 | 5.5 |
| D38560 | Cyclin-G-associated kinase | 26 | 1.7 | 2.3 | 4.4 | 2.5 | 4.5 |
| M24604 | PCNA/cyclin | 34 | 1.1 | 1.3 | −1 | 4.3 | 2.4 |
| D25233 | Rb mRNA | 35 | 1.1 | 1.2 | 2 | 1.2 | 1.2 |
| AF090306 | Rb-binding protein | 35 | −1 | 1.5 | 1.4 | 1.8 | 1.6 |
| L11007 | Cdk4 | 36 | −1.6 | 1.3 | 1.3 | 2.2 | 2.6 |
| D16309 | Cyclin D3 | 43 | −1.2 | −1.3 | −1.5 | 1.4 | 1.1 |
| D14014 | Cyclin D1 | 46 | −1.5 | −1.5 | −3.6 | −3.6 | −1.3 |
| L25785 | TGF-β1-induced transcript i4 | 62 | −5.6 | −11 | −6.5 | −2.8 | −2.9 |

**TABLE 7.7**
**Transcripts Induced Following Treatment with Fenofibrate**

| Probe Set | Gene Identification | *In Vitro* | *In Vivo* |
|---|---|---|---|
| AB010428 | Acyl-CoA hydrolase | 170.5 | 310.1 |
| AA924267 | CYP452 | 117.2 | 7.5 |
| AF034577 | Pyruvate dehydrogenase kinase isoenzyme 4 (PDK4) | 31.1 | 3.8 |
| K03249 | Peroxisomal enoyl-CoA- hydrotase-3-hydroxyacyl-CoA bifunctional enzyme | 30.5 | 4.9 |
| X07259 | CYP452 | 30.5 | 3.8 |
| J02749 | Peroxisomal 3-ketoacyl-CoA thiolase | 28.1 | 7.1 |
| M57718 | CYP4A1 | 24.3 | 2.5 |
| X60328 | Cytosolic epoxide hydrolase | 17.4 | 81 |
| J02773 | Low-molecular-weight, fatty-acid-binding protein | 14.9 | 6.7 |
| U08976 | Peroxisomal enoyl hydase-like protein (PXEL) | 13.4 | 27.4 |
| J02752 | Acyl-coA oxidase | 12.6 | 3 |
| AA893239 | 2-Hydroxyphytanoyl-CoA lyase | 11.9 | 2.3 |
| AB005743 | Fatty acid transporter | 11 | 5 |
| AB010429 | Acyl-CoA hydrolase-like protein | 10.4 | 5.4 |
| AI104882 | Cytosolic epoxide hydrolase (Ephx2) | 10.2 | 6.4 |
| Y09333 | Mitochondrial very-long-chain acyl-CoA thioesterase | 10 | 12.8 |
| AF044574 | Putative peroxisomal 2,4-dienoyl-CoA reductase (DCR-AKL) | 9 | 8.5 |
| D43623 | Carnitine palmitoyltransferase I (CPTI)-like protein | 7.4 | 4.8 |
| J02791 | Acyl-CoA dehydrogenase medium chain | 6.8 | 2 |
| U26033 | Carnitine octanoyltransferase | 6.7 | 4.3 |
| D17695 | Water channel aquaporin 3 (AQP3) | 6.2 | 14.2 |
| D00729 | Delta3, delta2-enoyl-CoA isomerase | 5.9 | 4.8 |
| D00512 | Mitochondrial acetoacetyl-CoA thiolase precursor | 5.1 | 4 |
| AI170568 | Mitochondrial 3-2trans-enoyl-CoA isomerase | 5.1 | 2.9 |
| D13921 | Mitochondrial acetoacetyl-CoA thiolase | 4.8 | 4.3 |
| AA800120 | Carnitine/acylcarnitine carrier protein | 4.3 | 4 |
| AA892864 | Monoglyceride lipase | 4.2 | 3.4 |
| AA800243 | Cell-death-inducing, DFFA-like effector a (CIDEA) | 4.1 | 5 |
| AA891916 | Membrane interacting protein of RGS16 (Mir16) | 4.1 | 5.4 |
| D00569 | 2,4-Dienoyl-CoA reductase | 4 | 4.5 |
| X65083 | Cytosolic epoxide hydrolase | 3.4 | 6 |
| X98225 | Gastrin-binding protein | 3.3 | 2.7 |
| M77184 | Parathyroid hormone receptor | 3.2 | 2.8 |
| AI008020 | Cytosolic malic enzyme | 3.2 | 5.1 |
| AJ224120 | Peroxisomal membrane protein Pmp26p (peroxin-11) | 3.1 | 4.1 |
| D16479 | Mitochondrial long-chain 3-ketoacyl-CoA thiolase beta-subunit of mitochondrial trifunctional protein | 3.1 | 2.5 |
| L07736 | Carnitine palmitoyltransferase I | 3.1 | 2 |
| J05470 | Carnitine palmitoyltransferase II (CPT II) | 2.8 | 3.2 |
| Y09332 | Cytosolic-peroxisome-proliferator-induced acyl-CoA thioesterase | 2.8 | 5.6 |
| M26594 | Malic enzyme gene | 2.8 | 4.7 |
| AA893000 | KIAA0564 protein | 2.7 | 2.7 |
| AI013834 | Peroxisomal multifunctional enzyme type II | 2.7 | 2.2 |
| AB010635 | Carboxylesterase precursor | 2.3 | 2.2 |
| AF063302 | CPT Iβ 1, 2, 3 genes, alternatively spliced products | 2.3 | 2.1 |
| AI171090 | 3-Hydroxy-3-methylglutaryl-CoA lyase (Hmgcl) | 2.3 | 2.1 |
| D16478 | Mitochondrial long-chain enoyl-CoA hydase/3-hydroxyacyl-CoA dehydrogenase alpha-subunit of trifunctional protein | 2.2 | 2.6 |

*(continued)*

**TABLE 7.7 (CONTINUED)**
**Transcripts Induced Following Treatment with Fenofibrate**

| Probe Set | Gene Identification | *In Vitro* | *In Vivo* |
|---|---|---|---|
| D30647 | Very-long-chain acyl-CoA dehydrogenase | 2.2 | 2.3 |
| M15114 | DNA polymerase alpha | 2.2 | 25.2 |
| AI171506 | Malic enzyme gene | 2.2 | 10.3 |
| AA800315 | Peroxisomal farnesylated protein (Pxf) | 2.1 | 3.4 |
| AI176621 | Iron-responsive element-binding protein | 2.1 | 2.2 |
| AA875269 | Stearoyl-CoA desaturase | 2 | 21.6 |

*Note:* Data are presented as fold changes from control, as determined using Affymetrix's Data Mining Tool.

**TABLE 7.8A**
**Transcripts Repressed Following Treatment with Fenofibrate**

| Probe Set | Gene Identification | *In Vitro* | *In Vivo* |
|---|---|---|---|
| L28135 | Glucose transporter type 2 gene | –2 | –2.1 |
| S81478 | PTPase = oxidative stress-inducible protein tyrosine phosphatase | –2 | –2.1 |
| AF026476 | Transcription factor USF-1 | –4.4 | –3.1 |
| AF080468 | Putative glycogen storage disease type 1b protein | –2.2 | –2 |
| AI235890 | MHC class I RT1.C/E | –4.6 | –2.2 |

*Note:* Data are presented as fold changes from control, as determined using Affymetrix's Data Mining Tool.

**TABLE 7.8B**
**Transcripts Differentially Expressed Following Treatment with Fenofibrate**

| Probe Set | Gene Identification | *In Vitro* | *In Vivo* |
|---|---|---|---|
| AF014503 | p8 | –2.2 | 2.1 |
| AI102795 | Pleiotrophin (heparin-binding factor) | 2.5 | –2.5 |
| AA893235 | $G_0/G_1$ switch gene 2 (Gos-2) | 3.8 | –2.6 |

*Note:* Data are presented as fold changes from control, as determined using Affymetrix's Data Mining Tool.

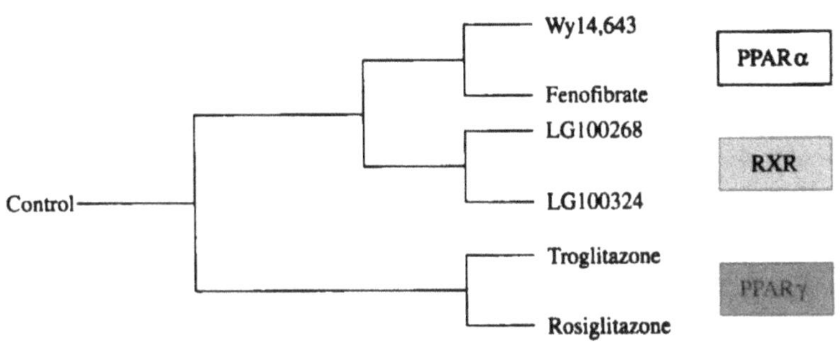


**FIGURE 7.6** Hierarchical clustering of gene expression profiles from primary rat hepatocytes treated with nuclear hormone receptor ligands (PPARα, PPARγ, or RXR) for 48 hours in culture.

## 7.4 DISCUSSION

Primary isolated hepatocytes have become a standard *in vitro* model used to advance our understanding of liver biology and xenobiotic metabolism. Ideally, *in vitro* hepatocyte models should provide an environment that allows for the maintenance of differentiated hepatocytes. The loss of specialized cellular functions that results from the culturing process is referred to as *cellular dedifferentiation*. Classic evaluations have utilized select markers (α-fetoprotein, γ-glutamyl transpeptidase (GGT), GST-π, tyrosine aminotransferase (TAT), connexin (Cx) 26, Cx 43, sorbitol dehydrogenase) or protein alterations involved in specific hepatocellular processes as general indicators of hepatocyte dedifferentiation.[38–40] Through the use of these select markers and focused studies of hepatocellular processes, a description of hepatocyte dedifferentiation as we currently view it has evolved. Using microarray technology, our objective was to demonstrate the utility of global gene expression analysis to further elucidate changes in cultured rat hepatocytes over time, and to use this new information to clarify the general understanding of hepatocyte dedifferentiation.

This study focused on transcripts that demonstrated changes at multiple time points and those that were part of functionally related sets of genes showing similar expression patterns. For direct confirmation of these temporal transcript changes, five transcripts were chosen for RT-PCR analysis, and all five reproduced the expression trends observed via microarray analysis. We chose genes with varying expression patterns for confirmation analysis to ensure that differences in both induced and repressed transcripts could be reliably reproduced. This confirmation process included multiple RT-PCR analyses of β-actin, for which the subtle, yet reproducible, increase in expression over time in culture precluded our initial attempt to use it as a housekeeping gene for normalization of RT-PCR data.

By inspecting the number of transcripts up- or downregulated at different times throughout this study, we observed patterns in the overall transcriptional behavior of hepatocytes in culture (Table 7.1). Four hours after the cells were plated, more transcripts were induced than repressed when compared to the hepatocyte-enriched cell pellet. Upon removal from their *in situ* environment and placement into culture, hepatocytes demonstrated increased gene expression of many immediate-early genes associated with the stress response including metallothionine, heme oxygenase, superoxide dismutase, and heat shock proteins (see supplemental data section at http://pubs.acs.org/subscribe/journals/crtoec/supmat/index.html). As hepatocytes aged in culture beyond 4 hours, the trend reversed, and many downregulated transcripts appeared. Downregulation was more pronounced with genes previously determined to be of toxicological significance, as the proportion of downregulated genes was greater on the toxicology subarray than that on genomic arrays at all of the later time points (Tables 7.1 and 7.2). Additionally, the downregulation process appeared to be more reproducible than the upregulation response, as illustrated by the greater percentage of transcripts in agreement between repeat and replicate experiments in both Tables 7.1 and 7.2. Collectively, by considering the excellent agreement between transcripts at individual time points and the fact that our analyses focus on transcriptional changes over time, the trends in gene expression observed in this study can be deemed reliable and reproducible.

One caution when using subarrays to study gene expression becomes evident by viewing the data presented in Table 7.1. As can be seen in this table, data derived from the toxicology subarrays is heavily biased toward genes downregulated over time. This is explained by inspecting the probe content of the rat toxicology subarray. This array is biased with many probes for transcripts encoding proteins studied by toxicologists which are preferentially downregulated under *in vitro* conditions utilized in the present study. Using a microarray containing thousands of genes, such as the rat genomic microarray employed in these studies, increases the aperture of expression analysis. Examples of gene expression changes that would have been missed if these experiments were to be conducted using the toxicology subarray include important transcripts associated with the cytoskeleton, extracellular matrix, and cell cycle (Tables 7.5 and 7.6).

In this study, transcripts were first grouped based upon their expression patterns (Figure 7.2), followed by functional classification to illustrate the underlying biological changes associated with monolayer culturing of hepatocytes. Maximally, 10% of the genes probed were affected at any singular time point, leaving 90% of the genes unaffected by the culture process. It is not the intent of this discussion to describe all transcriptional changes observed in this series of experiments; for a comprehensive view of all changes observed in this study, the reader is referred to the supplemental data contained on the ACS website. This supplement contains all transcripts that changed in at least one time point in a comprehensive cluster diagram. Several biological processes were transcriptionally affected in the study, and a few of these processes are discussed below.

### 7.4.1 Phase I P450 Metabolism

Previous studies have established that many phase I biotransformation enzymes, particularly members of the CYP family, are downregulated with time in culture.[22,23] An example of the downregulation of CYP genes can be seen in Figure 7.2B, where approximately 30% of the progressively downregulated members in this cluster are transcripts of the P450 family. Although this study confirms that most genes associated with phase I P450 metabolism are downregulated upon culturing, to generalize that all P450s are rapidly downregulated is misleading. Our studies reveal that P450 expression levels are heterogeneous in both timing and magnitude over the 72-hour culture period (Figure 7.3 and supplemental data). For example, expression of the CYP1 family (open bars) is downregulated between 24 and 48 hours, while the CYP2 and CYP4 families (gray and black bars) are downregulated between the 12- and 24-hour time points. The relatively constant expression of CYP3 (cross-hatch bars) may be due to inclusion of dexamethasone in the culture media, as dexamethasone is a CYP3 substrate. This interpretation is supported by the dramatic induction of the steroid-metabolizing cytochrome P450, CYP17α hydroxylase, and other transcripts induced in this study (also see supplemental data).[23] It should be pointed out that some P450s remain inducible under monolayer culture conditions, despite the baseline gene expression decrease. We, along with other investigators, have previously demonstrated this with members of the CYP4A subfamily following treatment with peroxisome proliferators.[42] This demonstrates that cultured hepatocytes do not lose their ability to express CYP4A genes; rather, these genes are selectively repressed under these *in vitro* conditions. In the current studies, many probes for liver-specific markers, in addition to the P450s discussed previously, were present on the microarrays. We evaluated several of these genes to determine if cultured hepatocytes downregulate the majority of liver-specific genes or if only a few of the better described families, such as the CYPs, were affected. Based upon the genes that we evaluated (i.e., α fetoprotein, vitamin-D-binding protein, α albumin), not all were differentially expressed, making a general classification of liver-specific markers difficult with the current gene expression dataset.

### 7.4.2 Phase II Metabolism

Transcriptional changes representing elements of both adaptation and dedifferentiation, as defined earlier, were observed upon monitoring the expression of genes associated with phase II biotransformation. Uridine-5′diphospho-glucuronosyltransferases (UDP-GTs) represent a family of enzymes responsible for the conjugation of many endogenous and exogenous xenobiotics. As can be seen in Table 7.3, several UDP-GT genes are upregulated during culture. Dexamethasone has previously been shown to induce UDP-GT activity and may be responsible for the changes in UDP-GT mRNA observed here.[43] Sulfotransferases (STs) catalyze the conjugation of hydroxyl groups through transfer of inorganic sulfates. ST gene expression was dramatically downregulated following 12 hours in culture. Previous work in monolayer cultured rat hepatocytes studying sulfation in the presence of acetaminophen revealed that sulfyltransferases remain active in culture given appropriate sulfate, hormone, and substrate availability.[44] The downregulation of ST gene expression in this study may be related to the absence of these factors and illustrates the importance of understanding how varying culture environments affect gene expression. GSTs catalyze the initial step of GSH conjugation, where GSH itself (discussed below) serves as a nucleophile for metabolic conversion of xenobiotics. Previous studies using primary hepatocyte cultures have shown that activity levels of GST-α and GST-μ fall, while GST-π increases with time.[45] GST-α is downregulated and GST-π is upregulated in this study, supporting these observations. In contrast to GST-α and GST-π, GST-μ mRNA levels were unchanged in the current study. This may represent a situation where mRNA expression levels are not indicative of protein content and therefore enzyme activity. The upregulation of GST-π is a classic marker of hepatocellular dedifferentiation. This gene is not normally expressed in adult liver, and its induction indicates a recapitulation of the fetal state.[41] Methylation is a phase II conjugation reaction catalyzed by methyltransferases, and in the current study methyltransferase gene expression was primarily repressed. In total, expression of phase II metabolism gene expression is illustrative of both dedifferentiation and adaptation to the culturing process employed in these studies.

### 7.4.3 GSH Production and Utilization

Hepatocellular GSH participates in critical cell survival defense mechanisms by detoxifying electrophiles and scavenging free radicals. GSH synthesis requires the catalytic activity of γ-GCS and GSH synthetase and is feedback regulated by competitive inhibition of γ-GCS by GSH. An early increase in γ-GCS mRNA most likely occurred due to lack of available cysteine.[46] Cysteine is not stable in the culture media used in these studies, so *in vitro* hepatocytes must generate cysteine intracellularly either through methionine or cystine conversion to cysteine. This lack of cysteine would explain our data showing an early induction of γ-GCS followed by a return to baseline (Table 7.4). Glutathione synthetase, which is normally in excess, was further elevated at the mRNA level in these studies. Recent studies have demonstrated that over-expression of GSH synthetase does not lead to elevated GSH levels and is not damaging to the cell, while an under-expression of GSH synthetase will lead to γ-GCS accumulation and metabolic acidosis.[47] The antioxidant function of GSH is critical for cell survival due to oxidative stresses associated with aerobic metabolism. In this study, GSH peroxidase mRNA was decreased along with other antioxidant enzymes such as catalase and Cu,Zn-superoxide dismutase, while GSSG reductase mRNA was increased (Table 7.4). The reduction of these genes in culture may render hepatocytes more susceptible to oxidative damage relative to the intact liver.

### 7.4.4 Glucose Metabolism

Lactic acid, the end product of anaerobic glucose breakdown by glycolysis in the muscle, is converted to pyruvate, and then pyruvate is converted back to glucose in the liver through gluconeogenesis (Figure 7.4). In this study, the expression of genes encoding the enzymes of gluconeogenesis was significantly downregulated. This decrease in gluconeogenic gene expression is most likely an adaptive response to the culture system and can be attributed to the high levels of glucose in the culture media (2000 mg/L) and the lack of lactic acid production. Glycolysis converts glucose into pyruvate-yielding adenosine triphosphate (ATP) (Figure 7.5). In contrast to the enzymes of the gluconeogenic pathway, cultured hepatocytes had elevated expression levels of key glycolytic regulatory enzymes. These data are consistent with a shift from gluconeogenesis to glycolysis and provide evidence of the concomitant regulation of gluconeogenesis and glycolysis as an adaptive response to the culture environment.

### 7.4.5 Cytoskeleton and Extracellular Matrix

Upon culturing, hepatocytes change shape from their *in situ* cuboidal differentiated form to a more flattened dedifferentiated structure that adheres to the bottom of the plastic culture plate.[48] This physical change in cellular shape upon adherence must be accompanied by architectural changes in structural components of the hepatocyte and be controlled by signal transducing systems in a coordinated manner. Furthermore, the culturing of cells requires severing and reestablishing cellular contacts as cells are isolated from the intact liver and then plated. Many of these contacts are mediated through proteins that comprise the extracellular matrix. Cells often undergo apoptosis when deprived of an extracellular matrix;[49] therefore, for cells to survive the process of culturing, a response to the mechanical disruption of their environment is needed. The extracellular matrix is a dynamic protein and polysaccharide network that provides cells with position and environmental information, while serving as a tissue-specific structure that controls cell function. Similarly, the actin cytoskeleton is a structure that underlies the cell membrane and contributes directly to the maintenance of cell shape.[49]

Because of the intricate interplay of this system, Table 7.5 was assembled to reveal more than 40 transcripts encoding for proteins associated with the cytoskeleton and extracellular matrix. As a general trend in these studies, these transcripts were upregulated dramatically, reflecting the need for cells to reestablish their connectivity and adapt to their new environment. It is interesting to note that some genes are transcriptionally upregulated very early, and these appear to be involved with the early process of hepatocellular reorganization. Genes induced at later time points appear to be involved with the later phases of cellular reorganization and connection. For example, the protein b61 is an immediate-early gene that binds to tyrosine kinases and has been postulated to be involved in local depolymerization of the actin cytoskeleton.[50] This protein and ryudocan, a proteoglycan involved in assembly of focal adhesions, are induced maximally at the earliest time point investigated, presumably so that cells can reassemble their cytoskeletal machinery and make way for integrin signaling and rebuilding of the actin cytoskeleton. Subsequently, synthesis of structural proteins, such as the intermediate filament proteins cytokeratin and vimentin, are heavily upregulated at and beyond 12 hours of culturing. Finally, gene products that associate with the actin cytoskeleton, such as p41-arc and α-crystallin, as well as genes involved with cross-linking the newly synthesized actin cytoskeleton with phospholipids, are induced. Molecules involved in cellular adhesion and signaling through newly synthesized or rearranged focal adhesions are also upregulated at the latest time points investigated and include transcripts for cadherins, CARI, and the signal-transducing and actin-associated protein Rho A. Collectively, these changes in gene expression clearly demonstrate a reorganization of the extracellular matrix and cellular cytoskeleton that is necessary for cells to adapt and survive in their new culture environment. In addition, it is firmly established that the extracellular matrix profoundly influences the major programs of growth and differentiation.

### 7.4.6 Cell Cycle and Differentiation

Parenchymal hepatocytes in adult liver are quiescent (held in the $G_0$ phase of the cell cycle) and replicate infrequently. In response to insult, as indicated by a change in liver mass/body mass ratios, parenchymal hepatocytes retain the capacity to proliferate. A change in this ratio can occur through chemical, infectious, ischemic, or physical injury and results in hyperplastic hepatic regeneration. The advancement of hepatocytes through the cell cycle is transcriptionally coordinated, and transcripts responsible for this regulation can be broken into categories of cell priming (hormones, cytokines), transcription (signal transducer and activator of transcription 3 [STAT-3], C/EBP), immediate-early responses (c-*fos*, c-*myc*), delayed-early responses (CL-6, HRS), and cell-cycle alterations (cyclins, retinoblastoma gene [Rb]) (Schematic 7.1). Our evaluation of changes in these transcripts revealed an overall pattern indicative of maintaining hepatocytes in a nonproliferative state. These changes include the early downregulation of the immediate-early gene Gos-2, followed by the upregulation of the cell-cycle inhibitory genes p53 and p21. Data collected from cultured human blood mononuclear cells suggest that Gos-2 is required for the transition of cells into the $G_1$ phase of the cell cycle.[51] Evaluations of liver have revealed that the cyclin-dependent kinase

(CDK) inhibitor p21 can block the hepatocyte cell cycle in $G_1$ and possibly $G_2$.[52] Expression analysis of p21 in cultured cells and mice has demonstrated a correlation between elevated p21 expression and withdrawal of cells from their proliferative cycle during cellular differentiation.[53] Such correlative data put forth the possibility that p21 could play a role in regulating cellular differentiation through negative regulation of cell-cycle genes. Therefore, an upregulation in the expression level of p21 in this study may indicate that the hepatocytes are attempting to conserve a differentiated, $G_0$ cell-cycle state.

### 7.4.7 Application of Hepatocyte Models in Toxicogenomics

Treatment of primary rat hepatocytes with the PPARα agonist fenofibrate resulted in an alteration of gene expression characteristic of the *in vivo* response in rat liver. Comparison of hepatic gene expression profiles from fenofibrate-treated rats and primary hepatocytes revealed remarkable similarities in both the affected biological pathways and the rank-order magnitude of the response. To normalize the biological response between the isolated hepatocyte model and *in vivo* systems, an *in vitro* dose of 100 μM was selected, as it best approximated the peroxisomal β-oxidation data from a live phase study in which rats were treated with 300 mg/kg fenofibrate for 4 days (38.5 vs. 46.9 nmoles NAD reduced per minute, respectively). A number of changes in genes involved with fatty acid metabolism, including ACO and CPT I, were detected both *in vivo* and in cultured hepatocytes (Table 7.7). The majority of transcripts were induced in both model systems, whereas only five genes were downregulated more than 2-fold both *in vitro* and *in vivo*. These findings are consistent with mRNA induction of enzymes involved with fatty acid metabolism. Further inspection of the data revealed a subset of transcripts differentially regulated by fenofibrate in the two models. However, only a very small number of differentially regulated transcripts changed more than two-fold in both systems, and no biological significance appeared to be associated with these changes. For example, Gos-2 was increased 3.8-fold in cultured hepatocytes; however, levels of this transcript were decreased 2.6-fold *in vivo*. The ability of this *in vitro* system to parallel the *in vivo* response to compounds suggests that cultured hepatocytes could be used as a mechanistic model in preclinical drug development.

Hierarchical clustering analysis was able to separate the NHR ligands PPARα, PPARγ, and RXR into distinct categories based upon their gene expression profiles from cultured rat hepatocytes (Figure 7.6). The choice of clustering metrics (Euclidean distance vs. Pearson's correlation) or variable type (fold change vs. average difference) did not alter the clustering outcome, suggesting that the gene expression changes that drove the clustering are sufficiently robust to differentiate mechanistic classes of compounds. A large number of genes exhibited expression patterns common to all or several of the treatment groups, presumably reflecting the promiscuity of receptor-driven transcriptional effects between these classes. Despite the considerable overlap of genes induced by PPARα, PPARγ, and RXR ligands, differences in expression patterns were sufficient to drive the classes into separate groups, demonstrating the sensitivity of this approach for screening applications. In addition, PPARα and RXR ligands clustered together in a subgroup distinct from PPARγ ligands. This subgroup clustering is likely due to the abundance of PPARα and RXR and relative lack of PPARγ receptors in hepatocytes. These data demonstrate that clustering of expression data from *in vitro* models can classify compounds according to ligand specificity. This approach is further supported by studies showing the utility of large gene expression databases containing "signature" profiles of known toxicants in the prediction of compound-mediated toxicity of novel pharmacological agents.[54,55] In total, cluster-based analysis of array data illustrates how primary hepatocytes can be used as a powerful predictive tool for toxicology and pharmacology screening.

## 7.5 SUMMARY

In summary, the data within this chapter provide a comprehensive, temporal gene-expression-based analysis of a traditional monolayer cultured hepatocyte system. From these studies, it is clear that transcriptional fingerprints at select points in time can be used to adequately describe biological responses associated with the transition of hepatocytes from the intact liver to an artificial culture environment. This study confirmed previously reported changes in gene expression of cultured hepatocytes that can appropriately be described as dedifferentiation. At the same time we were able to show other, adaptive biological responses associated with cell morphology, signaling, and metabolism. Although hepatocyte dedifferentiation was apparent, the expression levels of approximately 90% of the transcripts monitored in this study were unaffected by the culturing process, demonstrating that characteristics of the liver *in situ* are maintained in isolated hepatocytes. Collectively, these observations begin to define the cultured rat hepatocyte and will allow more informed interpretation of data derived from this *in vitro* model system.

## ACKNOWLEDGMENTS

The authors would like to thank Amy Ryan for administrative assistance provided during the preparation of this chapter.

## REFERENCES

1. Miyai, K., Structural organization of the liver, in *Hepatotoxicology*, Meeks, R., Harrison, S., and Bull, R., Eds., CRC Press, Boca Raton, FL, 1991, pp. 1–65.
2. Fawcett, D.W., The liver and gallbladder, in *Bloom and Fawcett: A Text Book of Histology*, Fawcett, D., Ed., Chapman & Hall, New York, 1994, p. 652.
3. Derman, E. et al., Transcriptional control in the production of liver-specific mRNAs, *Cell*, 23, 731, 1981.
4. Berry, M.N. and Friend, D.S., High-yield preparation of isolated rat liver parenchymal cells, *J. Cell Biol.*, 43, 506, 1969.
5. Seglen, P.O., Preparation of rat liver cells. I. Effects of $Ca^{2+}$ on enzymatic dispersion of isolated, perfused liver, *Exp. Cell Res.*, 74, 450, 1972.
6. Seglen, P.O., Preparation of isolated rat liver cells, *Meth. Cell Biol.*, 13, 29, 1976.
7. Griffin, J.M., Lipscomb, J.C., and Pumford, N.R., Covalent binding of trichloroethylene to proteins in human and rat hepatocytes, *Toxicol. Lett.*, 95, 173, 1998.
8. Billings, R.E. et al., The metabolism of drugs in isolated rat hepatocytes: a comparison with *in vivo* drug metabolism and drug metabolism in subcellular liver fractions, *Drug Metab. Dispos.*, 5, 518, 1977.
9. Shull, L.R. et al., Application of isolated hepatocytes to studies of drug metabolism in large food animals, *Xenobiotica*, 17, 345, 1987.
10. Guillouzo, A. et al., Use of human hepatocyte cultures for drug metabolism studies, *Toxicology*, 82, 209, 1993.
11. Foxworthy, P.S. and Eacho, P.I., Cultured hepatocytes for studies of peroxisome proliferation: methods and applications, *J. Pharmacol. Toxicol. Meth.*, 31, 21, 1994.
12. Anderson, K. et al., Immortalized hepatocytes as *in vitro* model systems for toxicity testing: The comparative toxicity of menadione in immortalized cells, primary cultures of hepatocytes and HTC hepatoma cells, *Toxicol. In Vitro*, 10, 721, 1996.
13. Onderwater, R.C.A. et al., Cytotoxicity of a series of mono- and di-substituted thiourea in freshly isolated rat hepatocytes: a preliminary structure–toxicity relationship study, *Toxicology*, 125, 117, 1998.

14. Westmoreland, C. et al., Ethionine toxicity *in vitro*: the correlation of data from rat hepatocyte suspensions and monolayers with *in vivo* observations, *Arch. Toxicol.*, 72, 588, 1998.
15. DeLeve, L.D. and Wang, X., Role of oxidative stress and glutathione in busulfan toxicity in cultured murine hepatocytes, *Pharmacology*, 60, 143, 2000.
16. Smith, M.T. et al., The measurement of lipid peroxidation in isolated hepatocytes, *Biochem. Pharmacol.*, 31, 19, 1982.
17. Rolfe, M., James, N.H., and Roberts, R.A., Tumor necrosis factor α (TNFα) suppresses apoptosis and induces DNA synthesis in rodent hepatocytes: a mediator of the hepatocarcinogenicity of peroxisome proliferators, *Carcinogenesis*, 18, 2277, 1997.
18. Chen, M.K. et al., Fas-mediated induction of hepatocyte apoptosis in a neuroblastoma and hepatocyte coculture model, *J. Surg. Res.*, 84, 82, 1999.
19. Li, J. et al., Nitric oxide suppresses apoptosis via interrupting caspase activation and mitochondrial dysfunction in cultured hepatocytes, *J. Biol. Chem.*, 274, 17325, 1999.
20. Baker, T.K. et al., Inhibition of gap junctional intercellular communication by 2,3,7,8-tetrachlorodibenzo-*p*-dioxin (TCDD) in rat hepatocytes, *Carcinogenesis*, 16, 2321, 1995.
21. Goll, V. et al., Comparison of the effects of various peroxisome proliferators on peroxisomal enzyme activities, DNA synthesis, and apoptosis in rat and human hepatocyte cultures, *Toxicol. Appl. Pharmacol.*, 160, 21, 1999.
22. Price, R.J. et al., Retention of xenobiotic-inducible cytochrome P450 gene expression in hepatocytes, *In Vitro Toxicol.*, 10, 365, 1997.
23. Sidhu, J.S. and Omiecinski, C.J., Modulation of xenobiotic-inducible cytochrome P450 gene expression by dexamethasone in primary rat hepatocytes, *Pharmacogenetics*, 5, 24, 1995.
24. Brown, P.O. and Botstein, D., Exploring the new world of the genome with DNA microarrays, *Nat. Genet.*, 21, 33, 1999.
25. Shoemaker, D.D. et al., Experimental annotation of the human genome using microarray technology, *Nature*, 409, 922, 2001.
26. Lipshutz, R.J. et al., High density synthetic oligonucleotide arrays, *Nat. Genet.*, 21, 20, 1999.
27. Boehm, M.F. et al., Synthesis and structure-activity relationships of novel retinoid x receptor-selective retinoids, *J. Med. Chem.*, 37, 2930, 1994.
28. Boehm, M.F. et al., Design and synthesis of potent retinoid x receptor selective ligands that induce apoptosis in leukemia cells, *J. Med. Chem.*, 38, 3146, 1995.
29. Klaunig, J.E. et al., Mouse liver cell culture I. Hepatocyte isolation, *In Vitro*, 17, 913, 1981.
30. Lazarow, P.B., Assay of peroxisomal β-oxidation of fatty acids, in *Methods in Enzymology*, Lowenstein, J.M., Ed., Academic Press, New York, pp. 315–319.
31. Kohonen, T., Self-organized formation of topologically correct feature maps, *Biol. Cybernetics*, 43, 59, 1982.
32. Kohonen, T., The self-organizing map, *Proc. IEEE*, 78, 1464, 1990.
33. Ritter, H., Self-organizing feature maps: Kohonen maps, in *The Handbook of Brain Theory and Neural Networks*, Arbib, M.A., Ed., MIT Press, Cambridge, MA, 1995, pp. 846–851.
34. Ritter, H., Martinez T., and Schulten, K., *Neural Computation and Self-Organizing Maps: An Introduction*, Addison-Wesley, Reading, MA, 1992.
35. Kohonen, T., Exploration of very large databases by self-organizing maps, in *Proc. 1997 Int. Conf. on Neural Networks*, Vol. 1, Houston, TX, 1997, pp. PL1–PL6.
36. Lo, Z.P., Fujita, M., and Bavarian, B., Analysis of neighborhood interaction in Kohonen neural networks, *Sixth Int. Parallel Processing Symp. Proc.*, Los Alamitos, CA, 1991, pp. 247–249.
37. Lo, Z.P., Yu, Y., and Bavarian, B., Analysis of the convergence properties of topology preserving neural networks, *IEEE Trans. Neural Networks*, 4, 207, 1993.
38. Cohen, P. et al., Monitoring cellular responses to *Listeria monocytogenes* with oligonucleotide arrays, *J. Biol. Chem.*, 275, 11181, 2000.
39. Klaunig, J.E. et al., Morphological and functional studies of mouse hepatocytes in primary culture, *Anatomical Rec.*, 204, 231, 1982.
40. Guguen-Guillouzo, C. and Guillouzo, A., Modulation of functional activities in cultured hepatocytes, *Mol. Cell. Biochem.*, 53/54, 35, 1983.
41. Sirica, A.E. et al., Fetal phenotypic expression by adult rat hepatocytes on collagen gel/nylon meshes, *Proc. Natl. Acad. Sci. USA*, 76, 283, 1979.

42. Sabzevari, O. et al., Comparative induction of cytochrome P4504A in rat hepatocyte culture by the peroxisome proliferators, bifonazole and clofibrate, *Xenobiotica*, 25, 395, 1995.
43. Jemnitz, K. et al., Glucuronidation of thyroxine in primary monolayer cultures of rat hepatocytes: *in vitro* induction of UDP-glucuronosyltransferases by methylcholanthrene, clofibrate, and dexamethasone alone and in combination, *Drug Metab. Dispos.*, 28, 34, 2000.
44. Seo, K.W. et al., Alteration of acetaminophen metabolism by sulfate and steroids in primary monolayer hepatocyte cultures of rats and mice, *Biol. Pharmacol. Bull.*, 22, 261, 1999.
45. Lee, S.J. and Boyer, T.D., The effects of hepatic regeneration on the expression of the glutathione S-transferases, *Biochem. J.*, 293, 137, 1993.
46. Wang, S.T. et al., Methionine and cysteine affect glutathione level, glutathione-related enzyme activities and the expression of glutathione S-transferase isozymes in rat hepatocytes, *J. Nutr.*, 127, 2135, 1997.
47. Lu, S.C., Regulation of glutathione synthesis, *Curr. Top. Cell. Reg.*, 36, 95, 2000.
48. DiPersio, C.M., Jackson, D.A., and Zaret, K.S. (1991) The extracellular matrix coordinately modulates liver transcription factors and hepatocyte morphology, *Mol. Cell Biol.*, 11, 4405, 1991.
49. Ruoslahti, E., Stretching is good for a cell, *Science*, 276, 1345, 2000.
50. Mellitzer, G., Xu, Q., and Wilkinson, D.G., Control of cell behavior by signaling through Eph receptors and ephrins, *Curr. Opin. Neurobiol.*, 10, 400, 2000.
51. Cristillo, A.D., Cyclosporin A inhibits early mRNA expression of $G_0/G_1$ *switch gene 2* (GOS2) in cultured human blood mononuclear cell, *DNA Cell Biol.*, 16, 1449, 1997.
52. Wu, H. et al., Targeted *in vivo* expression of the cyclin-dependent kinase inhibitor p21 halts hepatocyte cell-cycle progression, postnatal liver development, and regeneration, *Genes Dev.*, 10, 245, 1996.
53. Albrecht, J.H., Meyer, A.H., and Hu, M.Y., Regulation of cyclin-dependent kinase inhibitor $p21^{WAF1/Cip1/Sdi1}$ gene expression in hepatic regeneration, *Hepatology*, 25, 557, 1997.
54. Waring, J.F, et al., Clustering of hepatotoxins based on mechanism of toxicity using gene expression profiles, *Toxicol. Appl. Pharmacol.*, 175, 28, 2001.
55. Hamadeh, H.K. et al., Prediction of compound signature using high density gene expression profiling, *Toxicol. Sci.*, 76, 232, 2002.

# Section 4

## Mechanistic Toxicogenomics

# 8 Molecular Signatures of Dioxin Toxicity

*J. Kevin Kerzee, Craig R. Tomlinson, Jennifer L. Marlowe, and Alvaro Puga*

## CONTENTS



## 8.1 INTRODUCTION

Gene expression profiling using microarray technology is now commonplace in toxicological research. Decreased costs and increased efficiency and reliability, along with an extensive number of genes represented in multispecies arrays, make high-throughput genomics an extremely attractive option. Microarrays are most commonly used in toxicology to examine global gene expression patterns resulting from an acute toxic exposure. Gene array profiling need not be limited to classical studies of dose vs. response but should be expanded to include studies examining tissue, sex, age, and species differences within chemical treatment profiles. As datasets are accumulated, methods for the analysis of coordinately regulated genes in the study of toxic outcomes will be expanded to incorporate the complex regulatory networks of chemically induced responses and chemically mediated disease processes. By deciphering the links between toxic exposures and transcriptional responses, toxicologists will be able to use high-throughput methods to distinguish gene expression profiles, use those profiles to delineate the role of toxic agents in the disease process, and ultimately devise methods to prevent their adverse effects.

Tetrachlorodibenzo-*p*-dioxin (TCDD), the prototypical dioxin compound and ubiquitous environmental contaminant, originates as a byproduct of incineration, industrial waste, and incomplete combustion of fossil fuels. Dioxin accumulates in organisms primarily through low levels of consumption through the food chain.[1,2] The prevalence of this chemical in the environment has resulted in it being one of the most studied environmental contaminants. The numerous toxic effects associated with dioxin and dioxin-like compounds are suspected to result from alterations of highly regulated genes, leading to critical changes in the normal physiology of the organism. Despite detailed observations of the biological effects of dioxin exposure, the exact mechanisms leading to these effects are widely speculative and relatively unknown. Toxicologists now have the ability

0-8493-1334-1/03/$0.00+$1.50

to examine altered gene expression profiles on multiple levels, including cell-, tissue-, and organism-specific effects, through the use of macro- and micro-gene arrays.

## 8.2 DIOXIN AND THE ARYL HYDROCARBON RECEPTOR

It is generally accepted that the vast majority of the effects of dioxin arise from its ability to bind the aryl hydrocarbon receptor (AHR), a member of a family of transcription factors containing basic helix–loop–helix and Per–ARNT–Sim homology domains (bHLH-PAS). Founding members of the family include the mammalian AHR nuclear translocator (ARNT) protein (HIF-1β), the *Drosophila* neurogenic protein Sim ("*single-minded*"), and the *Drosophila* circadian rhythm protein Per ("*period*").[3–6] In response to agonist binding in the PAS domain, the cytosolic AHR undergoes a conformational change, translocates to the nucleus, dissociates from two 90-kDa heat shock proteins (HSP90) and XAP2,[7–10] and dimerizes with ARNT.[3,11–13] Binding of this heterodimeric complex to AHR responsive elements (AHREs; also known as dioxin responsive elements [DREs] or xenobiotic-responsive elements [XREs]) in the promoter region of dioxin-responsive genes results in an alteration in their rate of transcription (Figure 8.1). Many genes respond to dioxin-mediated activation of the AHR in this manner, including several phase I and phase II detoxification genes, such as CYP1A1, CYP1A2, CYP1B1, GST-Ya subunit, NAD(P)H menadione oxidoreductase-1, UDP-glucuronosyltransferase-1A6, and aldehyde-3-dehydrogenase-3.[14–18]

The classic toxicological example of dioxin-mediated gene regulation is the induction of *CYP1A1*. Much of what is known of AHR biology results from studies designed to analyze the mechanism of *CYP1A1* transactivation by the dioxin-activated AHR. The molecular and biological aspects of the AHR and its role in P450 induction have been extensively studied.[19,20] The ligand-bound AHR transactivates *CYP1A1*, as well as *CYP1A2* and *CYP1B1*, leading to metabolic activation of those ligands — namely, the polycyclic aromatic hydrocarbons (PAHs).

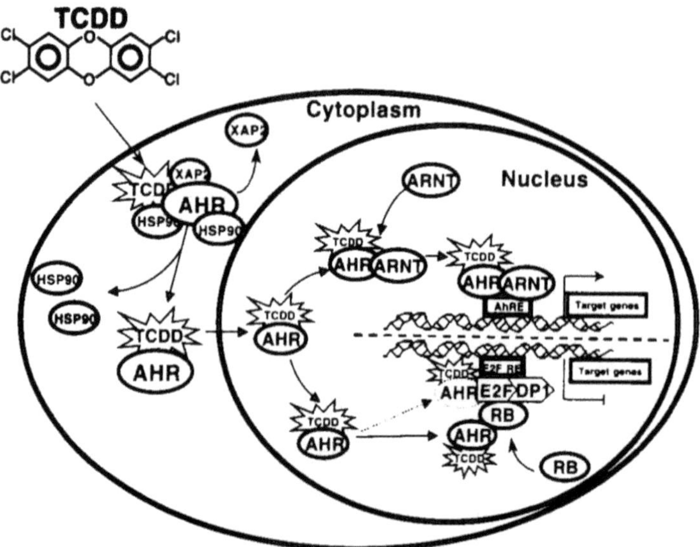


**FIGURE 8.1** Mechanisms of dioxin-induced gene alterations. In response to dioxin binding to the cytosolic AHR, a conformational change results in dissociation from two 90-kDa heat shock proteins (HSP90) and XAP2, translocation to the nucleus, and dimerization with ARNT. Binding of this heterodimeric complex to AHREs (see top of nucleus) in the promoter region of dioxin-responsive genes results in an alteration in their rate of transcription. In the lower nucleus, the ligand-activated AHR can also interact with RB protein (solid line) and/or E2F (dotted line). This interaction is critical for the transcriptional regulation/repression of E2F-dependent and cell-cycle-specific genes.

The generation of transgenic and gene knockout mice has allowed the analysis of the effects of TCDD on specific genotypes. Of particular interest has been the development of *Ahr*-knockout mice by three separate laboratories[21–23] and the development of *CYP1A1*-, *CYP1A2*-, and *CYP1B1*-null mice.[24–26] Several laboratories are attempting to generate mice lacking multiple genes by cross-breeding these different knockout mice, an effort hindered by the close linkage between the *CYP1A1* and *CYP1A2* genes.[27,28] *CYP1A1* and *CYP1A2* are separated by 23 kb and oriented in opposite directions, indicating that the 5′ flanking region is shared between these two genes. The presence of multiple AHREs upstream of both genes suggests that the regulatory elements that control dioxin-mediated *CYP1A1* expression may also control the regulation of *CYP1A2* by AHR agonists.[28] TCDD-mediated toxic and pathological effects are not observed in $Ahr^{-/-}$ mice at doses 10 times higher than those that result in severe toxicity in $Ahr^{+/+}$ mice.[29] Similarly, benzo[*a*]pyrene (BaP), another prototypical AHR agonist and potent tumor initiator and promoter, does not induce lesions in the livers or skin of mice lacking the AHR.[30] AHR knockout mice have allowed researchers to study the role of the AHR on the expression of genes important for metabolism of such xenobiotics and, considering the role of the Ah receptor in other cellular processes, of genes involved in cell-cycle regulation, signal transduction, and apoptosis.

Many of the effects of dioxin, or the lack thereof, may best be explained by differences in AHR binding affinities within and among different species. In mice, the $Ahr^{b1}$, $Ahr^{b2}$, and $Ahr^{b3}$ alleles code for a variant of the AHR with higher affinity for ligand than the AHR encoded by the $Ahr^{d}$ allele. The $Ahr^{b1}$ allele occurs in C57, C58, and MA/My inbred strains, whereas the $Ahr^{b2}$ allele is carried by the C3H, BalB/cBy, and A inbred strains. The $Ahr^{b3}$ allele is present in *Mus caroli*, *Mus spretus*, and MOLF/Ei, and the $Ahr^{d}$ allele occurs in AKR, DBA/2, and 129 strains.[31,32] AHR-responsive C57BL/6 mice have a receptor with an affinity for agonists approximately 10 to 20 times higher than that of the AHR in the nonresponsive DBA/2J strains.[33] The Ah receptors in these two strains differ by an Ala-to-Val mutation at amino acid 375.[31,34] The equivalent position in the human AHR is at Val-381, which, when mutated to Asp, completely abolishes ligand-binding activity of the human receptor.[35] In humans, differences in AHR ligand-binding affinity due to polymorphic variations may result in individual differences in dioxin-mediated susceptibility to disease.[35,36]

Because the effects of dioxin on gene regulation occur as a result of binding of AHR–ARNT complexes to *cis*-acting promoter sequences, researchers have focused on the temporal and spatial expression of the AHR to gain a perspective on its functions. Expression of the gene encoding the AHR occurs in a tissue- and developmentally specific manner.[37] The promoter of the AHR gene is G–C rich and contains no TATA or CAAT boxes. Analyses of the promoter sequence have revealed elements that are present in many housekeeping genes, including several binding sites for the transcription factor Sp-1, a cAMP response element, an AP-1 binding site, E-box sites, and two elements demonstrated in other genes to confer placenta-specific expression.[38,39] In addition, the AHR regulates its own expression. The ligand-bound AHR activates the transcription of a novel repressor gene, encoding the AHR repressor (AHRR), which interacts with ARNT and represses transcription of the AHR gene. The AHR and AHRR form a regulatory circuit in the xenobiotic signal transduction pathway which suggests a novel mechanism for regulation of AHR function that may determine tissue-specific sensitivity to environmental pollutants.[40]

Some of the highest levels of AHR gene expression are found in the lung, while moderate levels are found in liver, thymus, placenta, brain, heart, and spleen, and low levels are found in skeletal muscle.[41] AHR gene expression can be altered by TCDD, phenobarbital, serum and mitogens, retinoic acid, and transforming growth factor β (TGF-β).[42–46] AHR gene expression is also dependent on chromatin structure, as indicated by treatment of tissue culture cells with chemicals that affect histone deacetylase (HDAC) activity. HDAC inhibitors, such as sodium butyrate and trichostatin A, increase the constitutive activation of the AHR gene promoter. Blocking histone acetylation with the E1A oncoprotein, a negative regulator of the CBP/p300 histone acetylator complex, results in a decrease in AHR promoter activity.[47]

Despite countless reports of exogenous ligands binding to the AHR, an endogenous, physiologic ligand has yet to be identified. Several studies have demonstrated that endogenous AHR ligands may exist. Putative ligands include the tryptophan derivative, indole[3,2-b]carbazole, biliverdin and bilirubin, and lipoxin A, an arachidonic acid metabolite.[48–50] Cell-specific activation of the AHR in the absence of exogenous ligands has also been known for quite some time. Singh et al.[51] found that the AHR/ARNT complex is localized to the nucleus in HeLa cells in the absence of exogenous ligands, and similar findings were observed by Chang and Puga,[52] who reported that the endogenous AHR ligand is a likely substrate for *CYP1A1* and *CYP1A2*. The search continues with the hope that once the putative ligands are found, questions can be answered as to the role of the AHR, independent of its ability to regulate xenobiotic metabolism.

Our laboratory, as well as others, has found that the AHR interacts with the retinoblastoma (RB) protein and that this interaction is critical for the transcriptional regulation of E2F-dependent and cell-cycle-specific genes (Figure 8.1).[53–56] We can ask questions based on these data that relate not only to AHR-dependent signaling mechanisms for xenobiotic detoxification, but also to transcriptional responses from complex protein interactions involving the AHR. Through the use of microarray analysis, these questions can be addressed in a relatively short time and at a low cost.

One of the difficulties in interpreting gene expression changes following a TCDD challenge is that the outcomes exhibit cell-type and cell-cycle dependence. For example, in murine hepatoma Hepa1c1c7 cells, constitutive *CYP1A1* mRNA levels are reduced by more than 45% in TCDD-challenged cells arrested at $G_2/M$ of the cell cycle by tubulin disruption and stabilization relative to TCDD-challenged asynchronous cultures.[57] Accumulation of the dioxin-inducible gene, NAD(P)H quinone oxidoreductase-1 (NQO-1), was also reduced in tubulin-destabilized $G_2/M$ synchronized cultures challenged with dioxin.[57] Of particular interest regarding the role of the AHR in cell-cycle control is that a decline in *CYP1A1* mRNA is independent of both AHR protein levels and nuclear translocation, but TCDD responsiveness is restored when these cells are released from $G_2/M$. Furthermore, dioxin-altered *CYP1A1* expression levels during $G_1/S$ are much greater than in early $G_1$ or in $G_2/M$.[57] It should be noted, however, that only two dioxin-inducible genes were examined in this study, and the observations may not be universal.

The effect of AHR on cell-cycle events was observed for other phase I CYPs as well. Induction of *CYP1B1* or *CYP1A1* transcription in vascular smooth muscle cells (vSMCs) is less responsive to TCDD than to BaP.[58] Vascular endothelial cells (vECs) preferentially express inducible *CYP1A1*, whereas, vSMCs express inducible and constitutive *CYP1B1*.[58,59] TCDD exposure does not result in increased PAH metabolic capacity as determined by aryl hydrocarbon hydroxylase (AHH) activity in vSMCs; however, AHH activity is increased in $Ahr^{-/-}$, as well as $Ahr^{+/+}$ vSMCs as a result of BaP challenge in a cell-cycle-dependent manner.[58] Dioxin increased levels of *CYP1A1* and *CYP1B1* mRNAs in the human mammary carcinoma MCF-7 cell line and uterine cancer line RL95-2, but not in lymph node cancer of the prostate (LNCaP) cells.[60] The differences in expression of these two dioxin-sensitive genes appear to be related to cell-type-specific methylation patterns of the AHRE in the 5′ promoter regions.[60] Methylase inhibitors restore *CYP1A1* expression in LNCaP cells, but do not change expression in the aforementioned cell types.[60] Furthermore, TCDD caused LNCaP cells to block testosterone-mediated transcriptional activity, and testosterone inhibits TCDD-mediated expression of *CYP1A1*.[60,61]

## 8.3 DIOXIN AND THE DISEASE PROCESS

Most epidemiological data on the human health effects of dioxin have been derived from accidental industrial exposures, particularly the high-profile cases at Seveso, Italy, in 1976 and at Hamburg, Germany, in 1953.[62,63] Employees and people in surrounding areas were exposed to relatively high levels of TCDD for prolonged periods, resulting in increased incidence of thyroid disease, appendicitis, infectious diseases of the intestine and upper respiratory tract, peripheral nervous system disorders, and chronic liver disease.[63] These accidental exposures were not limited to Europe. In

Verona, Missouri, and Newark, New Jersey, workers exposed to trichlorophenol and 2,4,5-trichlorophenoxyacetic ester exhibited elevated gamma-glutamyl transferase enzyme and high-density lipoprotein levels, as well as elevated testosterone in males and increased luteinizing and follicle-stimulating hormones in females.[63,64] Despite obvious adverse health effects in humans exposed to dioxin and dioxin-like congeners, laboratory animal models develop adverse effects at doses that are seemingly benign in humans. In some cases, human body burdens of dioxin range from 13 to 7000 ng/kg, a range that is associated with developmental toxicity in laboratory animals exposed *in utero*.[65] TCDD levels of 25 μg/kg of body mass have been reported in humans, an extremely high and ultimately lethal level in laboratory animals.[66,67] By comparison, TCDD levels for individuals exposed in Seveso were approximately 10 μg/kg, which translates into 56 μg/kg in lipids circulating in the blood in adults.[66]

As in humans, rats and mice exposed to TCDD display a variety of adverse biological effects, including wasting syndrome, thymic involution, immune system dysfunction, reproductive and developmental effects, teratogenesis, epithelial hyperplasia and metastasis, and cardiovascular disease.[68–70] In contrast, TCDD has been found to be chemoprotective in that exposure reduces the incidence of breast cancer in laboratory animals.[71,72] Epidemiological data from Seveso indicate a similar protective effect in humans.[73]

Additional lines of evidence suggest that dioxin has the ability to inhibit tumor growth, as well as act as a tumor promoter. TCDD inhibits the proliferation and metastasis of tumor cells *in vitro* in an AHR-dependent manner,[72,74,75] most likely by altering apoptotic and cell-cycle regulatory events.[53–56,76] These findings have been eclipsed by the obvious tumor-promoting ability of dioxin.[77,78] This effect also results from the ability of dioxin to alter gene expression patterns and/or interfere with cell-cycle regulatory events.[54–56,76] The ability of TCDD to act as a tumor promoter has been directly attributed to its ability to inhibit apoptosis, thereby reducing elimination of "initiated" cells in several *in vitro* and *in vivo* systems, an effect that is dependent on the AHR.[79] The effect of dioxin on apoptosis has been attributed to several distinct but incompletely understood mechanisms, including modulation of p53 levels and downregulation of TGF-β transcription.[80] The AHR may also modulate the cellular response to apoptotic stimuli via its role in the RB pathway, particularly regarding the E2F transcription factor, a known mediator of apoptosis under conditions of cell-cycle deregulation (Figure 8.1).[81] The AHR has some ability to alter the transactivation activity of E2F in the absence of RB,[56] the outcome of which is currently unknown. The above scenarios of simultaneous tumor growth inhibition and tumor promotion resulting from TCDD exposure, while seemingly contradictory, are likely attributable to the cell- and tissue-specific effects of dioxin discussed earlier in this chapter.

Although absolute causality of dioxin-associated disorders is not well established, it should be acknowledged that individuals in the human population exposed to dioxin differ in their responses to this toxicant. A major focus in our laboratory is the effect of dioxin and PAHs on gene expression patterns, a probable key to understanding differential susceptibilities and cell-type-specific effects of dioxin exposure. Included in the range of responses to dioxin is the development of cardiovascular disease. TCDD is associated with cardiac abnormalities, such as dilated cardiomyopathy, and with symptoms of congestive heart failure.[82] Exposure to dioxin and dioxin-like compounds may play an important role in the onset and progression of ischemic heart disease. This hypothesis is gaining strong support, owing to the similarities in origin and progression of the atherogenic and carcinogenic processes.[83]

Dioxin likely affects multiple components of the cardiovascular system. The ability of TCDD to alter lipid metabolism was first reported nearly 20 years ago. Lovati et al.[84] found that 20 μg/kg TCDD resulted in a significant increase in plasma triglyceride levels, specifically the LDL fraction, in New Zealand rabbits fed both a normal or a 0.5% cholesterol diet. This work suggested that TCDD might block the degradation or uptake of triglycerides, resulting in hypertriglyceridemia. Recently, our laboratory reported that *Apoe*$^{+/+}$ and *Apoe*$^{-/-}$ mice challenged with TCDD demonstrated increased serum triglyceride levels, particularly of the LDL fraction.[70] An increase in atherosclerotic lesions was also observed in *Apoe*$^{-/-}$ mice challenged with 150 ng/kg TCDD three times a week for 7 to 26 weeks, with no lesions reported in *Apoe*$^{+/+}$ mice. Vascular smooth muscle cells, the key cellular component in atherosclerotic plaques, challenged with 1 nM TCDD exhibit increased expression of caspase-1, p21$^{waf/cip1}$, cyclin D1, mdm2, p15, and p57 mRNA by more than twofold. Coordinately, expression of the proteins encoded by these genes was also altered; however, the temporal patterns did not always coincide with mRNA expression, indicating that TCDD may control the regulation of these critical proteins at both the translational and transcriptional levels. These data indicate that TCDD may increase the incidence of ischemic heart disease through dual mechanisms by: (1) increasing lipid metabolism, and (2) altering gene expression patterns in the vasculature.[70]

## 8.4 CONSIDERATIONS IN THE USE OF MICROARRAYS TO STUDY DIOXIN

Toxicologists are primarily concerned with the adverse response of environmental chemicals on an exposed organism. Altered gene expression patterns observed in microarray experiments are often thought to be the result of a direct toxic response. This is a common misconception, owing to the implicit expectation that an exposure to a toxicant must be deleterious, though this is not always the case. This is similar to the concept of hazard vs. risk in risk assessment. Although an individual is exposed to a toxic compound, this does not imply a toxic response will take place, as the intricate workings of the whole organism can adapt until the threshold of the exposure is reached. As a case in point, altered gene expression patterns observed in TCDD-treated cells or tissues may be the result of an adaptive response that has little to do with a toxic outcome. The difficulty lies in determining in which category to place the response.

Functional assays of cell proliferation, toxicity, and apoptosis serve to complement gene expression analyses by connecting altered genes that at first appear to have unrelated or contradictory functions. It may be that the changes observed for many of the genes in which expression is altered by a toxic exposure, such as heat shock proteins, apoptotic genes, and cell-cycle inhibitors, have an adaptive purpose that may be misconstrued as a toxic response. Clustering or the generation of hierarchies for sets of genes from microarray datasets that share similar patterns of expression is a critical step in identifying regulatory and functional pathways, thus resulting in logical and testable hypotheses. Those genes that share an expression profile are placed in a given cluster, and it is assumed the genes in a cluster are coregulated by a common regulatory mechanism. Gene clusters give the first clues as to possible cellular and molecular mechanisms of a toxic response. The construction of gene hierarchies based on mechanisms of transcription, function, location, systemic effects, and disease processes may facilitate gene grouping (Figure 8.2). The use of gene arrays as a toxicological method is unique in that the data obtained permit numerous interpretations of the results, rather than a single outcome from one assay. Thus, functional assays must accompany and complement microarray experiments to allow for logical interpretation of the numerous and often seemingly unrelated functions of the altered genes.

Understanding the mechanism of a toxic response, acute or otherwise, requires knowledge of the biology of the exposed organism, and hypotheses must be proposed that incorporate mechanisms based on this knowledge. Microarray technology becomes extremely useful in this pursuit for two major reasons. First, once a gene expression profile — a transcriptional signature — has been generated for a particular toxic agent, the information becomes the starting point for the development of new hypotheses to characterize the mechanisms of toxicity. For example, if we observe that a given toxicant activates a receptor tyrosine kinase signal transduction pathway, we may expect that it will alter the expression of various proto-oncogenes, such as *RAS*, *RAF*, *FOS*, and *JUN*, a proposal that can be experimentally tested. Second, the transcriptional signature of the known toxic substance can be used as a reference point to compare the signatures of numerous other toxicants (Figure 8.2).

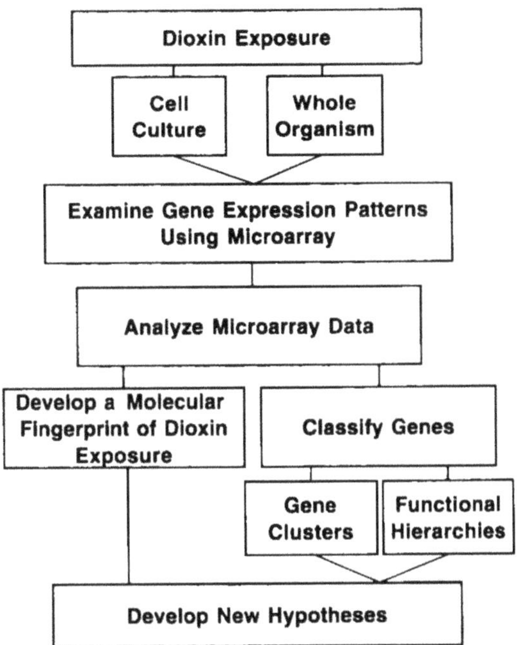


**FIGURE 8.2** Flow chart for examining dioxin-mediated global gene expression patterns. RNA isolated from dioxin-challenged cells or organisms is used to examine global gene expression patterns using microarray hybridization. These data are analyzed and subsequently used to develop a molecular fingerprint of dioxin exposure or lead to gene clustering and the development of functional hierarchies. The ultimate goal is to use the new data to develop new hypotheses that can lead to further understanding of the effects of dioxin.

Thus, toxicity testing in animal models may not have to be conducted with more than one agent (or, at most, a few agents) exhibiting similar transcriptional signatures.

The transcriptional response to a toxicant may be initiated through a variety of signaling routes. Signals originating at the cell surface may be transmitted to secondary messengers through kinases or redox reactions that can activate cytoplasmic receptors and mitogen-activated protein (MAP) kinase cascades, resulting in a multivariate transcriptional response that can be detected by microarray analysis. Clusters and hierarchies, as mentioned above, facilitate gene groupings based presumably on common mechanisms of transcriptional regulation, function, and cell location, as well as systemic effects and disease processes. However, researchers must recognize and heed the limitations imposed by the current lack of knowledge of gene regulatory networks required to interpret gene array results properly.

In recent years, the quest to discover novel pathways of dioxin-mediated gene regulation has resulted in three well-documented studies.[85–87] The first of these was reported by our laboratory, in which we used microarrays to establish a database of TCDD-responsive genes in HepG2 cells. These results revealed many clusters of related gene functions and pointed to the possibility that interactions among these clusters may provide an explanation for the multiple effects of TCCD, in particular, and of AHR activation, in general.[85] Expression of more than 300 genes was affected by dioxin treatment, including 114 that were upregulated and 196 that were downregulated by a magnitude of 2.1 or greater. Several hierarchical groups of genes altered by TCDD treatment were identified in these experiments, as well as seven gene groups, or clusters, that corresponded to well-defined cellular or biological processes. Of these clusters, four were of particular interest to cardiovascular biology, as they included genes involved in Ras/MAP kinase signaling pathways, calcium regulation, cardiovascular and pulmonary functions, and cell-cycle regulation and apoptosis. In addition, we identified genes involved in development, cell adhesion, cancer and metastasis, drug metabolism and DNA stability, protein trafficking and membrane integrity, receptor-associated kinases and phosphatases, and transcriptional regulation.[85] Similar results have been obtained with BaP, another AHR ligand. Like TCDD, BaP treatment in HepG2 cells altered the expression of genes critical for vascular functions, such as endothelial NO synthase, PAI-1, VEGF, FGL-2, troponin, and calmodulin-binding proteins, as well as genes involved in the regulation of cell cycle, signal transduction, and apoptosis.[85] Signal transduction occurs via amplification of a stimulus that originates either at the cell surface or in the cytoplasm using secondary messengers, such as kinases, phosphatases, or redox reactions that activate cytoplasmic receptors and kinase cascades. The ultimate outcome is a transcriptional response that is observed as a "signature" of gene expression. The change in expression of numerous genes resulting from TCDD treatment was inhibited by the protein synthesis inhibitor cyclohexamide, suggesting that the effect of dioxin on these genes is secondary, rather than a primary transcriptional response. However, the effects of dioxin on other known targets, such as the genes for PAI-2 and IL-1β,[88] *FOS* and *JUN* immediate-early gene families,[89–90] *COX-1* and *COX-2*,[91–97] and tumor necrosis factor α (TNF-α),[98–101] could be the result of a primary, secondary, or even higher order response resulting from cascade signaling events.

The interest in dioxin-mediated gene regulation is the main focus in numerous laboratories. In a study closely resembling those described above, Frueh and coworkers[87] examined gene expression changes in HepG2 cells challenged with 10 nM TCDD for 18 hours (compared to 4 and 8 hours in the study described previously).[85] Of the more than 12,000 genes analyzed, 85 were upregulated and 27 were downregulated. These researchers confirmed the expression of a handful of these genes by northern blot analysis and reverse transcriptase–polymerase chain reaction (RT-PCR), including genes coding for CYP1A1, cot kinase, XMP, HM74, human enhancer of filamentation 1 (HEF-1), metallothionein, PAI-1, and HM74.[87] The addition of cyclohexamide completely blocked TCDD-mediated expression of XMP and metallothionein mRNA, indicating that, for these genes, dioxin acts through a nontranscriptional mechanism, possibly through additional transactivating proteins or secondary messengers. Many of the same genes were identified in both studies. More important are the unique genes that were identified in the two studies. These differences powerfully reveal the need for common protocols, resources, and, most importantly, reference standards, in order to allow comparability among microarray data sets. Microarray use has become widespread; hence, the proper means for comparing data arising from different laboratories is a key consideration. The lack of a common set of guidelines for data analysis is a major problem, and, at present, a standardized system for data analysis has yet to be developed to address this problem.

The use of cell cultures and whole animals is essential for understanding the cell- and tissue-specific effects of dioxin. Data obtained from *in vitro* studies should be confirmed by *in vivo* studies, and vice versa. The effects of TCDD *in vivo* using C57Bl/6J mice were recently examined by Thomas and coworkers.[86] The focus of their study was the identification of common mouse gene sets that can be used to predict responses to different toxicants. RNA isolated from livers of TCDD-treated mice was compared to liver RNA from vehicle-treated controls. These data were compared with other toxicants insults, including non-coplanar PCBs (PCB-153, Arochlor 1260, phenobarbital), peroxisome proliferators (Cipro, Wy-16,463), inflammatory molecules (TNF-α, lipopolysaccharide, IL-6), and hypoxia-inducible agents (cobalt, phenylhydrazine) to generate gene expression profiles based on structural, as well as functional, differences in the chemicals. A set of 12 predictive genes for the various classes of chemicals were identified: CYP1A2, FMO5, CYP4A14, IL-18, CYP2B10, CYP4A10, mouse BMMT, CYP2C29, CYP1A1, SAA112, and two unknown transcripts. As would be expected, the two genes with the highest expression levels after TCDD challenge were *CYP1A1* and *CYP1A2*, which were expressed at comparable levels in a time-dependent manner. One important aspect of this study that differed significantly from those discusses above was the use of whole mouse livers treated *in vivo*, rather than the use of the transformed human liver HepG2 cells. Cell lines in general are presumed to exhibit responses that result from mutations or epigenetic events that have accumulated during the process of *in vitro* propagation. Certainly, future experiments with transgenic and gene knockout mice with defined genetic backgrounds will address the relevant issue of organ-specific gene expression profiles induced by TCDD or other AHR ligands and provide a complete picture of the molecular signature of dioxin exposure, both *in vivo* and *in vitro*, that can be used as toxicological markers.

## 8.5 INTERPRETING TOXICOGENOMIC DATA FROM DIOXIN STUDIES

A major concern in interpreting microarray data is the application of appropriate statistical analyses, a problem that we have recently addressed in relation to our own microarray studies.[85] Before any analysis is carried out, the significance of the numerical results must be determined. In dual-labeled microarray experiments, the fluorescent precursors Cy3- and Cy5-labeled dCTP or dUTP are incorporated into target cDNA using total RNA from control and experimental groups. By convention, DNA printed on the microarray slide is referred to as the *probe* and the fluorescent cDNA, made from the test mRNA, is called the *target.*[102] The hybridization reaction is often referred to as *competitive*, although the probe is usually in large molar excess relative to either target, making hybridization essentially noncompetitive. The microarray slide is scanned by lasers set at specific wavelengths to excite the dyes, and the emitted fluorescence of the hybridized Cy3- and Cy5-labeled cDNA targets is recorded as the sum total of the pixel intensities for each gridded feature, each feature representing a different gene in the array.

Differential gene expression is dictated by the ratio of Cy3/Cy5 fluorescence in each feature. The critical criterion for inclusion of a cDNA in a group of either induced or repressed expression is that the balanced differential expression (i.e., the normalized Cy3/Cy5 pixel ratio for that grid feature) must be greater (if induced) or smaller (if repressed) than a certain arbitrary value at either tail of the distribution of all the values. This value must be determined for each experiment, because it depends on the variance of the experimental distribution of all the points in any given experiment, and it cannot be assumed that all experiments will have a homogeneous variance. If the value is set, for example, at Cy3/Cy5 $\geq 2$ and this corresponds to a statistical likelihood of $p < 0.05$, then those genes are considered to be repressed by the treatment if Cy3/Cy5 $< 0.5$ and induced if Cy3/Cy5 $> 2$. However, when dealing with large numbers of genes, as is the case in microarray experiments, $p < 0.05$ may still correspond to a very large number of outliers, and its choice may lead to incorrect conclusions. Thus, for a 10,000-clone library, $p < 0.05$ will include roughly 500 genes at either end of the statistical distribution. These genes will score as up- or downregulated, but in fact are the result of chance alone.

One way to reduce the number of apparently deregulated genes is to set the threshold at a higher value (for example, at $p < 0.01$), but this may force the corresponding Cy3/Cy5 ratios to unreasonable levels, such that only ratios $> 4$ or $< 0.25$ will score as induced or repressed, respectively. The preferred alternative solution is to include replicate microarrays for each experiment. At the same threshold level of $p < 0.05$, the probability that the *same* 500 genes would show up as outliers by chance alone a second time would be $0.05^2$, or 0.0025%. Hence, with a single repetition, the number of genes erroneously included in a group is reduced from 500 to 25. A second repetition further reduces this number to just one gene. Experimental evidence from commercial microarrays indicates that, in a good experiment, 99% of the microarray spots display <1.4 or >0.7 differential expression when compared with themselves; thus, in experimental conditions, genes may be considered to be differentially expressed when they change by >1.4 or <0.7. Others have chosen a ratio greater than 1.7-fold as their criteria.[103] Obviously, the ratio will depend on the tightness (variance) of the overall distribution and cannot be set *a priori*. In our experiments with microarrays containing over 8000 sequences, we determined that less than 6% of the genes were induced or repressed by more than 2.1-fold. Thus, in three repetitions of the experiment, 28 of the 310 genes that we found to be deregulated by TCDD in HepG2 cells may have shown expression changes by chance alone.

Repetition is the best approach to improve the reliability and statistical significance of microarray data. By increasing to three the number of microarray slides used for any one set of samples, outliers occurring by chance alone are effectively eliminated, increasing the overall statistical power of the analysis. It is also advisable to switch fluorescent labels in each repetition, such that control cDNA is labeled twice with Cy3 and once with Cy5, and the experimental cDNA is labeled twice with Cy5 and once with Cy3. The logic behind this stems from the observation that, at least for direct labeling procedures, the reverse transcriptase enzyme incorporates less effectively the longer-side chain substituted Cy5-dUTP (or dCTP) than the shorter Cy3-dUTP (or dCTP), resulting in an apparent under-representation of Cy5-labeled targets, regardless of source (control or experimental). Therefore, indirect methods for labeling cDNA with Cy3 and Cy5 dyes (see http://www.microarray.org) essentially eliminate dye incorporation bias and are the most widely used and recommended. Nonetheless, switching the dye labeling for the control and experimental RNAs is still advisable to uncover all possible variability.

It is essential to verify the results of microarray experiments by other quantitative means. It is a common observation that the level of differential gene expression determined from microarray hybridization results is not always in agreement with other means of expression analyses. In our laboratory, we use real-time RT-PCR (Real-time RT-PCR) amplification for this purpose. We design primers specific for the mRNAs of the genes detected by microarray hybridization and use these primers in conjunction with SYBR-green staining to quantify amplified DNA products. Real-time PCR allows one to complete a typical PCR experiment in less than 30 minutes, and the built-in fluorimetric detection system for *in silico* quantification of amplified products provides accurate quantitation of transcript abundance. A further advantage of this system is that several (up to four) different transcripts may be analyzed in the same reaction, thus allowing for the simultaneous quantitation of relative expression levels for different genes.

Determining the changes in accumulated mRNA levels is required in order to fully understand the regulatory mechanisms that take place as a consequence of toxicant exposure. However, it is often the case that changes in mRNA levels are not accompanied by parallel changes in levels of the corresponding protein.[104] Again, to fully understand the cellular response to a toxicant, it is necessary to determine the levels of the proteins encoded by the genes identified by the microarray data, as ultimately it is the changes in protein levels that dictate cell physiology. Consequently, an essential corollary to any microarray experiment is to prepare protein extracts from control and toxicant-exposed cells or tissues for further analysis.

Interactive and higher order levels of regulatory complexity are beyond our current capabilities of dissection and are often overlooked when analyzing microarray data. For example, high-order regulatory complexity is a particularly difficult problem when whole-organ homogenates are used as the source of RNA in microarray experiments. Gene expression changes identified in whole-organ arrays are likely the result of alterations that have taken place in a subset of cells or tissue. The identity of the cells, how they interact with other cells, and how the genes that they express affect the physiology of the whole organ cannot be known using these sorts of methods. If technically feasible, the cells in question should be cultured along with the different cell types from the same organ for additional microarray analysis that will complement total organ microarray analysis. Alternatively, specific groups of cells may be removed by laser capture microdissection from discrete tissue sections for further gene profiling analyses. This may provide a clearer insight into the significance of any expression profile changes. For example, vascular endothelial cells relay vital signals to vascular smooth muscle cells via cytokines that alter gene expression patterns and ultimately growth responses.[83] Gene expression patterns of neither total aortic homogenates nor of independent cultures of endothelial and smooth muscle cells provide a complete picture of the transcriptional responses in the arterial wall. Co-culture experiments in trans-well chambers that allow physically separated cells to pass diffusible agents from one cell to another may help define which genes respond to primary signals and which to secondary signals resulting from cellular cross-talk.

## 8.6 SUMMARY

The experiments discussed in this chapter demonstrate the diversity among microarrays and the challenges associated with examining the effects of TCDD, a clear example of the complexity of global gene expression patterns associated with dioxin exposure. Researchers have a long road ahead before a common protocol for performing microarray analysis is established. With the numerous micro- and macro-arrays commercially available, it will likely be years before this takes place. Once we have achieved our goal of unifying gene expression techniques, what do we hope to gain from our newfound knowledge? First, we, as toxicologists, can begin to utilize the available gene knockout and transgenic mice to their full potential to elucidate molecular targets of TCDD in the presence, as well the absence, of AHR, *CYP1A1*, *CYP1A2*, and *CYP1B1*, to name just a few. Second, we can develop disease- and toxicant-specific arrays, a process that is currently underway in some laboratories. But, this future may not be as near as we would hope as the gene expression profiles for dioxin, as with most toxicants, vary within each cell type, organ, and animal. Therefore, utmost care must be given to assure reproducibility and reliability of the data. What is most interesting, and disturbing at the same time, is that for all of the attention given to dioxin and its role in gene expression, the number of papers on the subject is limited.[85–87] Given all of these facts, the future holds endless possibilities, and perhaps one day we will be able to analyze the toxic, as well as pharmacologic, response for thousands of chemicals with extreme accuracy and minimal time.

## ACKNOWLEDGMENTS

Preparation of this review and the cited research performed in the authors' laboratories was supported in part by National Institutes of Health Grants R01 ES06273, ES10807, and P30 ES06096 (A.P., C.R.T.). J.K.K. is supported by National Institute of Environmental Health Sciences National Research Service Award F32 ES11250 and by a Society of Toxicology/Colgate-Palmolive Postdoctoral Fellowship Award. J.L.M. is supported by a University of Cincinnati Predoctoral Functional Genomics Fellowship.

## REFERENCES

1. Van den, B.M. et al., The toxicokinetics and metabolism of polychlorinated dibenzo-*p*-dioxins (PCDDs) and dibenzofurans (PCDFs) and their relevance for toxicity, *CRC Crit. Rev. Toxicol.*, 24, 1, 1994.
2. Kaiser, J., Toxicology. Panel backs EPA dioxin assessment, *Science*, 290, 1071, 2000.
3. Hoffman, E.C. et al., Cloning of a factor required for activity of the Ah (dioxin) receptor, *Science*, 252, 954, 1991.
4. Crews, S.T., Thomas, J.B., and Goodman, C.S., The *Drosophila* single-minded gene encodes a nuclear protein with sequence similarity to the per gene product, *Cell*, 52, 143, 1988.
5. Citri, Y. et al., A family of unusually spliced biologically active transcripts encoded by a *Drosophila* clock gene, *Nature*, 326, 42, 1987.
6. Burbach, K.M., Poland, A., and Bradfield, C.A., Cloning of the Ah-receptor cDNA reveals a distinctive ligand-activated transcription factor, *Proc. Natl. Acad. Sci. USA*, 89, 8185, 1992.
7. Perdew, G.H., Association of the Ah receptor with the 90-kDa heat shock protein, *J. Biol. Chem.*, 263, 13802, 1988.
8. Chen, H.S. and Perdew, G.H., Subunit composition of the heteromeric cytosolic aryl hydrocarbon receptor complex, *J. Biol. Chem.*, 269, 27554, 1994.
9. Carver, L.A. and Bradfield, C.A., Ligand-dependent interaction of the aryl hydrocarbon receptor with a novel immunophilin homolog *in vivo*, *J. Biol. Chem.*, 272, 11452, 1997.

10. Ma, Q. and Whitlock, J.P., Jr., A novel cytoplasmic protein that interacts with the Ah receptor, contains tetratricopeptide repeat motifs, and augments the transcriptional response to 2,3,7,8-tetrachlorodibenzo-*p*-dioxin, *J. Biol. Chem.*, 272, 8878, 1997.
11. Wilhelmsson, A. et al., The specific DNA-binding activity of the dioxin receptor is modulated by the 90 kD heat shock protein, *EMBO J.*, 9, 69, 1990.
12. Reyes, H., Reiz-Porszasz, S., and Hankinson, O., Identification of the Ah receptor nuclear translocator protein (ARNT) as a component of the DNA binding form of the Ah receptor, *Science*, 256, 1193, 1992.
13. Wang, G.L. and Semenza, G.L., Purification and characterization of hypoxia-inducible factor 1, *J. Biol. Chem.*, 270, 1230, 1995.
14. Nebert, D.W. and Gonzalez, F.J., P450 genes. Structure, evolution and regulation, *Annu. Rev. Biochem.*, 56, 945, 1987.
15. Tukey, R.H. and Nebert, D.W., Regulation of mouse cytochrome P3-450 by the Ah receptor. Studies with a P3-450 cDNA clone, *Biochemistry*, 23, 6003, 1984.
16. Savas, U. et al., Mouse cytochrome P-450EF, representative of a new 1B subfamily of cytochrome P-450s. Cloning, sequence determination, and tissue expression, *J. Biol. Chem.*, 269, 14905, 1994.
17. Paulson, K.E. et al., Analysis of the upstream elements of the xenobiotic compound-inducible and positionally regulated glutathione S-transferase Ya gene, *Mol. Cell Biol.*, 10, 1841, 1990.
18. Hempel, J., Harper, K., and Lindahl, R., Inducible (class 3) aldehyde dehydrogenase from rat hepatocellular carcinoma and 2,3,7,8-tetrachlorodibenzo-*p*-dioxin-treated liver: distant relationship to the class 1 and 2 enzymes from mammalian liver cytosol/mitochondria, *Biochemistry*, 28, 1160, 1989.
19. Nebert, D.W. et al., Role of the aromatic hydrocarbon receptor and [Ah] gene battery in the oxidative stress response, cell cycle control, and apoptosis, *Biochem. Pharmacol.*, 59, 65, 2000.
20. Wilson, C.L. and Safe, S., Mechanisms of ligand-induced aryl hydrocarbon receptor-mediated biochemical and toxic responses, *Toxicol. Pathol.*, 26, 657, 1998.
21. Fernandez-Salguero, P. et al., Immune system impairment and hepatic fibrosis in mice lacking the dioxin-binding Ah receptor, *Science*, 268, 722, 1995.
22. Schmidt, J.V. et al., Characterization of a murine *Ahr* null allele: involvement of the Ah receptor in hepatic growth and development, *Proc. Natl. Acad. Sci. USA*, 93, 6731, 1996.
23. Mimura, J. et al., Loss of teratogenic response to 2,3,7,8-tetrachlorodibenzo-*p*-dioxin (TCDD) in mice lacking the Ah (dioxin) receptor, *Genes Cells*, 2, 645, 1997.
24. Dalton, T.P. et al., Targeted knockout of Cyp1a1 gene does not alter hepatic constitutive expression of other genes in the mouse [Ah] battery, *Biochem. Biophys. Res. Commun.*, 267, 184, 2000.
25. Liang, H.-C. et al., *Cyp1A2(–/–)* null mutant mice develop normally, but show deficient drug metabolism, *Proc. Natl. Acad. Sci. USA*, 93, 1671, 1996.
26. Buters, J.T. et al., Cytochrome P450 CYP1B1 determines susceptibility to 7,12-dimethylbenz[*a*]anthracene-induced lymphomas, *Proc. Natl. Acad. Sci. USA*, 96, 1977, 1999.
27. Hildebrand, C.E. et al., Regional linkage analysis of the dioxin-inducible P-450 gene family on mouse chromosome 9, *Biochem. Biophys. Res. Commun.*, 130, 396, 1985.
28. Corchero, J. et al., Organization of the CYP1A cluster on human chromosome 15: implications for gene regulation, *Pharmacogenetics*, 11, 1, 2001.
29. Fernandez-Salguero, P. et al., Aryl hydrocarbon receptor-deficient mice are resistant to 2,3,7,8-tetrachlorodibenzo-*p*-dioxin-induced toxicity, *Toxicol. Appl. Pharmacol.*, 140, 173, 1996.
30. Shimizu, Y. et al., Benzo[*a*]pyrene carcinogenicity is lost in mice lacking the aryl hydrocarbon receptor, *Proc. Natl. Acad. Sci. USA*, 97, 779, 2000.
31. Poland, A., Palen, D., and Glover, E., Analysis of the four alleles of the murine aryl hydrocarbon receptor, *Mol. Pharmacol.*, 46, 915, 1994.
32. Poland, A. and Glover, E., Characterization and strain distribution pattern of the murine Ah receptor specified by the $Ah_d$ and $Ah_{b-3}$ alleles, *Mol. Pharmacol.*, 38, 306, 1990.
33. Poland, A. et al., Genetic expression of aryl hydrocarbon hydroxylase activity. Induction of monooxygenase activities and cytochrome P1-450 formation by 2,3,7,8-tetrachlorodibenzo-*p*-dioxin in mice genetically "non-responsive" to other aromatic hydrocarbons, *J. Biol. Chem.*, 249, 5599, 1974.
34. Chang, C.-Y. et al., Ten nucleotide differences, five of which cause amino acid changes, are associated with the Ah receptor locus polymorphism of C57BL/6 and DBA/2 mice, *Pharmacogenetics*, 3, 312, 1993.

35. Ema, M. et al., Dioxin binding activities of polymorphic forms of mouse and human arylhydrocarbon receptors, *J. Biol. Chem.*, 269, 27337, 1994.
36. Wanner, R. et al., Polymorphism at codon 554 of the human Ah receptor: different allelic frequencies in Caucasians and Japanese and no correlation with severity of TCDD induced chloracne in chemical workers, *Pharmacogenetics*, 9, 777, 1999.
37. Abbott, B.D., Birnbaum, L.S., and Perdew, G.H., Developmental expression of two members of a new class of transcription factors: I. Expression of aryl hydrocarbon receptor in the C57BL/6N mouse embryo, *Dev. Dynamics*, 204, 133, 1995.
38. Schmidt, J.V., Carver, L.A., and Bradfield, C.A., Molecular characterization of the murine Ahr gene. Organization, promoter analysis, and chromosomal assignment, *J. Biol. Chem.*, 268, 22203, 1993.
39. Fitzgerald, C.T., Nebert, D.W., and Puga, A., Regulation of mouse Ah receptor (Ahr) gene basal expression by members of the Sp family of transcription factors, *DNA Cell Biol.*, 17, 811, 1998.
40. Mimura, J. et al., Identification of a novel mechanism of regulation of Ah (dioxin) receptor function, *Genes Dev.*, 13, 20, 1999.
41. Li, W. et al., Ah receptor in different tissues of C57BL/6J and DBA/2J mice: use of competitive polymerase chain reaction to measure Ah-receptor mRNA expression, *Arch. Biochem. Biophys.*, 315, 279, 1994.
42. Pollenz, R.S., The aryl-hydrocarbon receptor, but not the aryl-hydrocarbon receptor nuclear translocator protein, is rapidly depleted in hepatic and non-hepatic culture cells exposed to 2,3,7,8-tetrachlorodibenzo-*p*-dioxin, *Mol. Pharmacol.*, 49, 391, 1996.
43. Okey, A.B. and Vella, L.M., Elevated binding of 2,3,7,8-tetrachlorodibenzo-*p*-dioxin and 3-methylcholanthrene to the Ah receptor in hepatic cytosols from phenobarbital-treated rats and mice, *Biochem. Pharmacol.*, 33, 531, 1984.
44. Vaziri, C. et al., Expression of the aryl hydrocarbon receptor is regulated by serum and mitogenic growth factors in murine 3T3 fibroblasts, *J. Biol. Chem.*, 271, 25921, 1996.
45. Wanner, R. et al., The differentiation-related upregulation of aryl hydrocarbon receptor transcript levels is suppressed by retinoic acid, *Biochem. Biophys. Res. Commun.*, 209, 706, 1995.
46. Dohr, O. et al., Effect of transforming growth factor-beta1 on expression of aryl hydrocarbon receptor and genes of Ah gene battery: clues for independent down-regulation in A549 cells, *Mol. Pharmacol.*, 51, 703, 1997.
47. Garrison, P.M. et al., Effects of histone deacetylase inhibitors on the Ah receptor gene promoter, *Arch. Biochem. Biophys.*, 374, 161, 2000.
48. Rannug, A. et al., Certain photooxidized derivatives of tryptophan bind with very high affinity to the Ah receptor and are likely to be endogenous signal substances, *J. Biol. Chem.*, 262, 15422, 1987.
49. Phelan, D. et al., Activation of the Ah receptor signal transduction pathway by bilirubin and biliverdin, *Arch. Biochem. Biophys.*, 357, 155, 1998.
50. Schaldach, C.M., Riby, J., and Bjeldanes, L.F., Lipoxin A4: a new class of ligand for the Ah receptor, *Biochemistry*, 38, 7594, 1999.
51. Singh, S.S., Hord, N.G., and Perdew, G.H., Characterization of the activated form of the aryl hydrocarbon receptor in the nucleus of HeLa cells in the absence of exogenous ligand, *Arch. Biochem. Biophys.*, 329, 47, 1996.
52. Chang, C.-Y. and Puga, A., Constitutive activation of the aromatic hydrocarbon receptor, *Mol. Cell. Biol.*, 18, 525, 1998.
53. Elferink, C.J., Ge, N.L., and Levine, A., Maximal aryl hydrocarbon receptor activity depends on an interaction with the retinoblastoma protein, *Mol. Pharmacol.*, 59, 664, 2001.
54. Ge, N.-L. and Elferink, C.J., A direct interaction between the aryl hydrocarbon receptor and retinoblatoma protein, *J. Biol. Chem.*, 273, 22708, 1998.
55. Strobeck, M.W. et al., Restoration of retinoblastoma mediated signaling to Cdk2 results in cell cycle arrest, *Oncogene*, 19, 1857, 2000.
56. Puga, A. et al., Aromatic hydrocarbon receptor interaction with the retinoblastoma protein potentiates repression of E2F-dependent transcription and cell cycle arrest, *J. Biol. Chem.*, 275, 2943, 2000.
57. Santini, R.P. et al., Regulation of Cyp1a1 induction by dioxin as a function of cell cycle phase, *J. Pharmacol. Exp. Ther.*, 299, 718, 2001.
58. Kerzee, J.K. and Ramos, K.S., Constitutive and inducible expression of Cyp1a1 and Cyp1b1 in vascular smooth muscle cells: role of the Ahr bHLH/PAS transcription factor, *Circ. Res.*, 89, 573, 2001.

59. Zhao, W., Parrish, A.R., and Ramos, K.S., Constitutive and inducible expression of cytochrome P450IA1 and P450IB1 in human vascular endothelial and smooth muscle cells [letter], *In Vitro Cell Dev. Biol. Anim.*, 34, 671, 1998.
60. Jana, N.R. et al., Comparative effects of 2,3,7,8-tetrachlorodibenzo-*p*-dioxin on MCF-7, RL95–2, and LNCaP cells: role of target steroid hormones in cellular responsiveness to CYP1A1 induction, *Mol. Cell Biol. Res. Commun.*, 4, 174, 2000.
61. Jana, N.R. et al., Cross-talk between 2,3,7,8-tetrachlorodibenzo-p-dioxin and testosterone signal transduction pathways in LNCaP prostate cancer cells, *Biochem. Biophys. Res. Commun.*, 256, 462, 1999.
62. Needham, L.L. et al., Exposure assessment: serum levels of TCDD in Seveso, Italy, *Environ. Res.*, 80, S200, 1999.
63. Zober, A., Ott, M.G., and Messerer, P., Morbidity follow up study of BASF employees exposed to 2,3,7,8-tetrachlorodibenzo-*p*-dioxin (TCDD) after a 1953 chemical reactor incident, *Occup. Environ. Med.*, 51, 479, 1994.
64. Sweeney, M.H. et al., Review and update of the results of the NIOSH medical study of workers exposed to chemicals contaminated with 2,3,7,8-tetrachlorodibenzodioxin, *Teratog. Carcinog. Mutagen.*, 17, 241, 1997.
65. DeVito, M.J. et al., Comparisons of estimated human body burdens of dioxinlike chemicals and TCDD body burdens in experimentally exposed animals [review], *Environ. Health Perspect.*, 103, 820, 1995.
66. Bertazzi, P.A. et al., The Seveso studies on early and long-term effects of dioxin exposure: a review, *Environ. Health Perspect.*, 106(suppl. 2), 625, 1998.
67. Geusau, A. et al., Severe 2,3,7,8-tetrachlorodibenzo-*p*-dioxin (TCDD) intoxication: clinical and laboratory effects, *Environ. Health Perspect.*, 109, 865, 2001.
68. Greenlee, W.F., Sutter, T.R., and Marcus, C., Molecular basis of dioxin actions on rodent and human target tissues [review], *Prog. Clin. Biol. Res.*, 387, 47, 1994.
69. Vanden Heuvel, J.P. and Lucier, G., Environmental toxicology of polychlorinated dibenzo-*p*-dioxins and polychlorinated dibenzofurans., *Environ. Health Perspect.*, 100, 189, 1993.
70. Dalton, T.P. et al., Dioxin exposure is an environmental risk factor for ischemic heart disease, *Cardiovasc. Toxicol.*, 1, 285, 2002.
71. Kociba, R.J. et al., Results of a two-year chronic toxicity and oncogenicity study of 2,3,7,8-tetrachlorodibenzo-*p*-dioxin in rats, *Toxicol. Appl. Pharmacol.*, 46, 279, 1978.
72. Safe, S. et al., 2,3,7,8-Tetrachlorodibenzo-*p*-dioxin (TCDD) and related compounds as antioestrogens: characterization and mechanism of action [review], *Pharmacol. Toxicol.*, 69, 400, 1991.
73. Bertazzi, P.A. et al., Cancer incidence in a population accidentally exposed to 2,3,7,8-tetrachlorodibenzo-*para*-dioxin [see comments], *Epidemiology*, 4, 398, 1993.
74. Osborne, R. and Greenlee, W.F., 2,3,7,8-Tetrachlorodibenzo-*p*-dioxin (TCDD) enhances terminal differentiation of cultured human epidermal cells, *Toxicol. Appl. Pharmacol.*, 77, 434, 1985.
75. Hushka, D.R. and Greenlee, W.F., 2,3,7,8-Tetrachlorodibenzo-*p*-dioxin inhibits DNA synthesis in rat primary hepatocytes, *Mutat. Res.*, 333, 89, 1995.
76. Kolluri, S.K. et al., $p27_{kip1}$ induction and inhibition of proliferation by the intracellular Ah receptor in developing thymus and hepatoma cells, *Genes Dev.*, 13, 1742, 1999.
77. Pitot, H.C. et al., Quantitative evaluation of the promotion by 2,3,7,8-tetrachlorodibenzo-*p*-dioxin of hepatocarcinogenesis, *Cancer Res.*, 40, 3616, 1980.
78. Knutson, J.C. and Poland, A., Keratinization of mouse teratoma cell line XB produced by 2,3,7,8-tetrachlorodibenzo-*p*-dioxin: an *in vitro* model of toxicity, *Cell*, 22, 27, 1980.
79. Schwarz, M. et al., Ah receptor ligands and tumor promotion: survival of neoplastic cells, *Toxicol. Lett.*, 112–113, 69, 2000.
80. Worner, W. and Schrenk, D., Influence of liver tumor promoters on apoptosis in rat hepatocytes induced by 2-acetylaminofluorene, ultraviolet light, or transforming growth factor beta 1, *Cancer Res.*, 56, 1272, 1996.
81. Johnson, D.G., The paradox of E2F1: oncogene and tumor suppressor gene, *Mol. Carcinog.*, 27, 151, 2000.
82. Walker, M.K. and Catron, T.F., Characterization of cardiotoxicity induced by 2,3,7,8-tetrachlorodibenzo-*p*-dioxin and related chemicals during early chick embryo development, *Toxicol. Appl. Pharmacol.*, 167, 210, 2000.

83. Ross, R., The pathogenesis of atherosclerosis: a perspective for the 1990s, *Nature*, 362, 801, 1993.
84. Lovati, M.R. et al., Increased plasma and aortic triglycerides in rabbits after acute administration of 2,3,7,8-tetrachlorodibenzo-*p*-dioxin, *Toxicol. Appl. Pharmacol.*, 75, 91, 1984.
85. Puga, A., Maier, A., and Medvedovic, M., The transcriptional signature of dioxin in human hepatoma HepG2 cells, *Biochem. Pharmacol.*, 60, 1129, 2000.
86. Thomas, R.S. et al., Identification of toxicologically predictive gene sets using cDNA microarrays, *Mol. Pharmacol.*, 60, 1189, 2001.
87. Frueh, F.W. et al., Use of cDNA microarrays to analyze dioxin-induced changes in human liver gene expression, *Toxicol. Lett.*, 122, 189, 2001.
88. Sutter, T.R. et al., Targets for dioxin: genes for plasminogen activator inhibitor-2 and interleukin-1 beta, *Science*, 254, 415, 1991.
89. Hoffer, A., Chang, C.-Y., and Puga, A., Dioxin induces *fos* and *jun* gene expression by Ah receptor-dependent and -independent pathways, *Toxicol. Appl. Pharmacol.*, 141, 238, 1996.
90. Puga, A., Nebert, D.W., and Carrier, F., Dioxin induces expression of c-*fos* and c-*jun* proto-oncogenes and a large increase in transcription factor AP-1, *DNA Cell Biol.*, 11, 269, 1992.
91. Puga, A. et al., Sustained increase in intracellular free calcium and activation of cyclooxygenase-2 expression in mouse hepatoma cells treated with dioxin, *Biochem. Pharmacol.*, 54, 1287, 1997.
92. Kraemer, S.A. et al., Regulation of prostaglandin endoperoxide H synthase-2 expression by 2,3,7,8-tetrachlorodibenzo-*p*-dioxin, *Arch. Biochem. Biophys.*, 330, 319, 1996.
93. Olnes, M.J., Verma, M., and Kurl, R.N., 2,3,7,8-tetrachlorodibenzo-*p*-dioxin modulates expression of the prostaglandin G/H synthase-2 gene in rat thymocytes, *J. Pharmacol. Exp. Ther.*, 279, 1566, 1996.
94. Liu, Y., Levy, G.N., and Weber, W.W., Induction of human prostaglandin endoperoxide H synthase-2 (PHS-2) mRNA by TCDD, *Prostaglandins*, 53, 1, 1997.
95. Vogel, C. et al., Modulation of prostaglandin H synthase 2 mRNA expression by 2,3,7,8-tetrachlorodibenzo-*p*-dioxin in mice, *Arch. Biochem. Biophys.*, 351, 265, 1998.
96. Wolfle, D. et al., Induction of cyclooxygenase expression and enhancement of malignant cell transformation by 2,3,7,8-tetrachlorodibenzo-*p*-dioxin, *Carcinogenesis*, 21, 15, 2000.
97. Lee, C.A. et al., 2,3,7,8-Tetrachlorodibenzo-*p*-dioxin induction of cytochrome P450-dependent arachidonic acid metabolism in mouse liver microsomes: evidence for species-specific differences in responses, *Toxicol. Appl. Pharmacol.*, 153, 1, 1998.
98. Connor, M.J., Nanthur, J., and Puhvel, S.M., Influence of 2,3,7,8-tetrachlorodibenzo-*p*-dioxin (TCDD) on TNF-alpha levels in the skin of congenic haired and hairless mice, *Toxicol. Appl. Pharmacol.*, 129, 12, 1994.
99. Vogel, C. and Abel, J., Effect of 2,3,7,8-tetrachlorodibenzo-*p*-dioxin on growth factor expression in the human breast cancer cell line MCF-7, *Arch. Toxicol.*, 69, 259, 1995.
100. Fan, F. et al., Cytokines (IL-1beta and TNFalpha) in relation to biochemical and immunological effects of 2,3,7,8-tetrachlorodibenzo-*p*-dioxin (TCDD) in rats, *Toxicology*, 116, 9, 1997.
101. Moos, A.B., Oughton, J.A., and Kerkvliet, N.I., The effects of 2,3,7,8-tetrachlorodibenzo-*p*-dioxin (TCDD) on tumor necrosis factor (TNF) production by peritoneal cells, *Toxicol. Lett.*, 90, 145, 1997.
102. Duggan, D.J. et al., Expression profiling using cDNA microarrays, *Nat. Genet.*, 21, 10, 1999.
103. Braxton, S. and Bedilion, T., The integration of microarray information in the drug development process, *Curr. Opin. Biotechnol.*, 9, 643, 1998.
104. Haynes, P.A. et al., Proteome analysis: biological assay or data archive?, *Electrophoresis*, 19, 1862, 1998.

# 9 Toxicogenomic Dissection of the Antioxidant Response

*Jiang Li and Jeffrey A. Johnson*

## CONTENTS



## 9.1 INTRODUCTION

Transcriptional activation of many antioxidant genes is mediated through a *cis*-acting enhancer named the antioxidant-responsive element (ARE). AREs have been detected in the promoters of the following phase II detoxifying enzymes: rat and mouse glutathione S-transferase (GST) Ya, rat GST P, rat and human NAD(P)H:quinone oxidoreductase (NQO1), and murine heme oxygenase-1 (HO1).[1,2] Several other phase II detoxifying enzymes and antioxidant genes such as γ-glutamylcysteine ligase catalytic (GCLC) and regulatory (GCLR) subunits, as well as ferritin heavy and light chains, are also suspected to be upregulated through ARE activation.[3,4] AREs share a consensus motif (TGAC/TNNNGC) originally identified by mutational analysis of the rat GST-Ya ARE.[5,6] Thus, identification of the ARE was an initial step leading to elucidation of the molecular mechanism for antioxidant response.

A metabolite of the widely used food antioxidant butylated hydroxyanisole (BHA), *tert*-butylhydroquinone (tBHQ) is a known inducer of phase II enzymes in various model systems, including IMR-32 human neuroblastoma cells.[7] Increasing evidence indicates that tBHQ induces these phase II detoxifying enzymes through the ARE and that the transcription factor Nrf2 (NF-E2 p45-related factor 2) is essential for ARE-mediated induction of these genes.[1–4] Recently, our laboratory showed that activation of the human NQO1 ARE is dependent on Nrf2 and that tBHQ dramatically induces Nrf2 nuclear translocation in human neuroblastoma cells.[8] The binding of Nrf2 to the ARE leads to transcriptional activation of a score of genes such as NQO1, HO1, multiple forms of GST, glutathione reductase (GR), and thioredoxin reductase (TR).[9–16] Because the *cis*-regulatory motif acts as a binding site for Nrf2 to control expression, coexpressed genes sharing the same regulatory

0-8493-1334-1/03/$0.00+$1.50

motifs are likely to be networked under the same regulatory control and could potentially be identified by high-throughput microarray analysis.

## 9.2 TIME-DEPENDENT CHANGES IN ARE-DRIVEN GENE EXPRESSION

In the initial part of our study, we monitored the expression level of thousands of genes in parallel and tried to identify and characterize the time-dependent changes in gene expression associated with tBHQ treatment in IMR-32 human neuroblastoma cells using HuU95A Affymetrix oligonucleotide arrays.

### 9.2.1 Deduction of the Magnitude of Gene Expression

IMR-32 cells were treated with tBHQ (10 μM) or vehicle (0.01% EtOH) for 24 hr initially, and four pairs of samples were generated over a 4-month period. All samples were analyzed using the human U95A arrays (Affymetrix). As shown in Figure 9.1A, dramatic variations were observed in the number of genes called different (increased or decreased), even between the pair-matched comparisons (bold in Figure 9.1A). Affymetrix gene expression software uses the difference call (see Chapter 3) as a basic parameter for evaluating transcriptional up- or downregulation. We employed an in-house approach and defined increased, decreased, or no-change levels of expression for individual genes based on ranking the difference calls from six 2×2, four 3×3, and one 4×4 matrix comparisons. Briefly, in this system we binned genes into five categories: no change = 0, marginal increase = 1, increase = 2, marginal decrease = –1, and decrease = –2. The cutoff value for increase or decrease was set as $\pm n^2$ ($n \geq 2$) because of the presence of marginal calls. For example, in a 2×2 matrix that generates four datasets, genes called marginal increase (I)/decrease (–I) would have to be called in each comparison to pass the criterion of $n^2$ = 4/–4; therefore, the cutoff in the 3×3 matrix would be $n^2$ = 9/–9 and that in the 4×4 matrix would be $n^2$ = 16/–16. Analysis of the data from the 8-hr time point by a Latin-square comparison is shown in Figure 9.1B. A significant decrease in the number of genes passing the rank analysis was observed when comparing the 2×2 matrix with the 3×3 matrix for both increased and decreased genes. Variability appeared to be greater in the number of increased genes than that of decreased genes (Figure 9.1B). This variability in gene number, however, does not reflect a consistent change in the same gene, thereby accounting for the low number of decreased genes following tBHQ treatment. These data indicate that a minimum of three independent samples should be run to determine differential gene expression between control and treatment groups. Application of this 3×3 matrix analysis to the microarray data at each time point (4, 8, 24, and 48 hr) led to the identification of approximately 200 genes for which the expression was consistently increased (196) or decreased (4).

### 9.2.2 Functional Categorization

Evaluation of the reproducibility of microarray data was based on the coefficient of variation (CV) (standard deviation divided by the mean) for average difference changes (ADC). Previsously identified ARE-driven genes such as NQO1, HO1, TR, GR, and GCLR were used to determine a rational cutoff value for CV. The CVs of these genes were all below 1.0; therefore, we considered this value as our cutoff. Only those changes complying with this criterion were considered further for functional categorization and clustering analysis. Of the 200 genes, 101 (1% of 9670 genes in U95Av2 chips) genes made the final list. These genes are listed by functional categorization in Table 9.1. Interestingly, there were no decreased genes, suggesting that tBHQ is a selective and potent activator of the ARE. Transcriptional upregulation peaked at 8 hr with more than 50% of listed genes showing increased expression at this time point.

A

| Group | Jan-tBHQ | Feb-tBHQ | Mar-tBHQ | Apr-tBHQ |
|---|---|---|---|---|
| Jan-Control | **171/283** | 607/526 | 294/852 | 1225/879 |
| Feb-Control | 503/1633 | **171/151** | 248/1703 | 540/1290 |
| Mar-Control | 1164/493 | 1878/368 | **597/300** | 1709/253 |
| Apr-Control | 1071/510 | 1891/340 | 529/345 | **1658/252** |

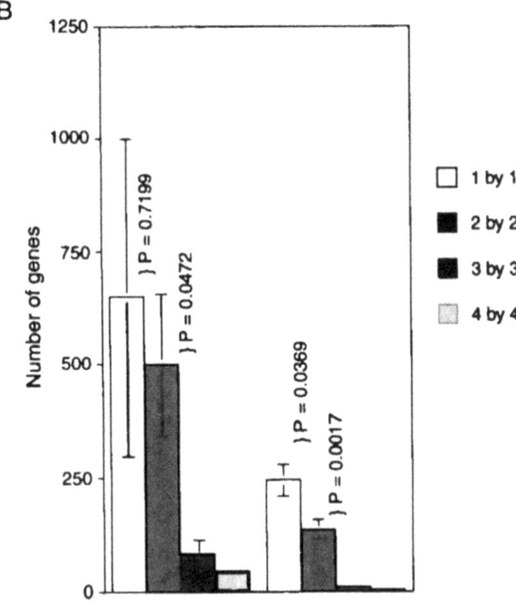


**FIGURE 9.1** (A) Matrix comparisons ($n \times n$) plus ranking analysis. By means of cross comparisons between treatment and control groups, genes are labeled as increased (I) or decreased (D). Data gathered from pair-wise comparisons are **bold**. (B) Histograms of the number of genes labeled I or D. In addition to four pair-matched 1×1 comparisons (Jan., Feb., Mar., and Apr.), a random selection was conducted to generate six 2×2 (Jan.+Feb., Jan.+Mar., Jan.+Apr., Feb.+Mar., Feb.+Apr., and Mar.+Apr.), four 3×3 (Jan.+Feb.+Mar., Jan.+Feb.+Apr., Feb.+Mar.+Apr., and Jan.+Mar.+Apr.), and one 4×4 (Jan+Feb+Mar+Apr) matrix comparisons. Statistical analysis was conducted by one-tailed, unequal variance $t$-test. Significant $p$ values ($p < 0.05$) and the comparisons are shown in the graph. (From Li, J. and Johnson J.A., *Physiol. Genomics*, 9, 137, 2002. With permission.)

One of the major clusters of transcriptionally activated genes in IMR-32 cells was the phase II detoxification enzymes. Within this category of genes were early-response genes (NQO1, HO1, GR, GSTM3, GCLR, and TR) and late-response genes (ferritin heavy and light chain). Interestingly, unlike other phase II detoxification enzymes, HO1 was transiently upregulated at 8 hr of treatment and not changed at 24 or 48 hr. Other antioxidative systems such as hepatic dihydrodiol dehydrogenase and its isoform, KIAA0119, malate NADP oxidoreductase, and breast cancer cytosolic NADP(+)-dependent malic enzyme were also found to be upregulated by tBHQ (Table 9.1). All of these genes ranked higher than other genes induced by tBHQ, suggesting their importance in the tBHQ-mediated antioxidant effect. These findings are consistent with previous reports by Burczynski et al.,[17] suggesting that certain members of the human dihydrodiol dehydogenases are regulated by an ARE-type mechanism, even though no functional AREs have been found within the regulatory regions of these genes to date. The tBHQ-inducible human dihydrodiol dehydrogenase has also been shown to metabolize 4-hydroxynonenal and other products of lipid peroxidation, suggesting a physiologically relevant role for this inducible enzyme in the cellular response to oxidative stress.[18]

**TABLE 9.1**
**Overview of Transcriptional Upregulation Induced by tBHQ in IMR-32 Cells**

| Category of Genes | GenBank Accession Number | SOM Clustering (3×4) | Description of Genes |
|---|---|---|---|
| Cell death/apoptosis | D83699 | C11 | Brain 3UTR of mRNA for neuronal death protein (harakiri) |
| | N/A | C8 | Bax delta |
| Chaperones/heat shock proteins | N/A | C0 | Heat shock protein Hsp40 homolog |
| | X87949 | C1 | BiP protein (a member of Hsp70 family of chaperones) |
| | W28493 | C2 | Heat shock 70-kDa protein 8 |
| CNS-specific function | U79299 | C3 | Neuronal olfactomedin-related ER localized protein |
| | AF009674 | C3 | Axin (AXIN) |
| | U40572 | C1 | β2-syntrophin (SNT β2) |
| | M25756 | C5 | Secretogranin II gene |
| | Z48054 | C3 | Peroxisomal targeting signal 1 |
| | U73304 | C4 | CB1 cannabinoid receptor (CNR1) |
| | AB023209 | C4 | KIAA0992 (palladin) |
| | N/A | C11 | Neurofibromatosis 2 tumor suppressor |
| | L27745 | C9 | Voltage-operated calcium channel, α-1 subunit |
| | AB012851 | C8 | Musashi (RNA-binding protein) |
| Cytoskeleton | AB002323 | C7 | Dynein, cytoplasmic, heavy polypeptide 1 (KIAA0325) |
| | W26631 | C4 | Microtuble-associated protein 1A |
| | X15306 | C4 | NF-H gene, exon 1 |
| | S67247 | C1 | Smooth muscle myosin heavy-chain isoform (Smemb) |
| | M22299 | C4 | T-plastin polypeptide |
| Detoxification and antioxidative stress | M81600 | C7 | NAD(P)H-quinone oxidoreductase |
| | X15722 | C6 | Glutathione reductase |
| | N/A | C3 | Glutathione transferase M3 (GSTM3) |
| | Z82244 | C2 | Heme oxygenase 1 |
| | N/A | C5 | Breast cancer cytosolic NADP(+)-dependent malic enzyme |
| | AL049699 | C4 | Malic enzyme 1, soluble (NADP-dependent malic enzyme, malate oxidoreductase) |
| | D17793 | C6 | KIAA0119 (aldo-keto reductase family 1) |
| | U05861 | C7 | Hepatic dihydrodiol dehydrogenase |
| | X91247 | C5 | Thioredoxin reductase |
| | L35546 | C2 | γ-Glutamylcysteine ligase regulatory subunit |
| | AL031670 | C8 | Ferritin, light polypeptide-like 1 |
| | J04755 | C8 | Ferritin H processed pseudogene |
| | L20941 | C7 | Ferritin heavy chain |

*(continued)*

**TABLE 9.1 (CONTINUED)**
**Overview of Transcriptional Upregulation Induced by tBHQ in IMR-32 Cells**

| Category of Genes | GenBank Accession Number | SOM Clustering (3×4) | Description of Genes |
|---|---|---|---|
| DNA repair | N/A | C2 | Human growth arrest and DNA-damage-inducible protein (gadd45) mRNA |
| | N/A | C4 | ERCC5 excision repair protein |
| Extracellular matrix | AL050138 | C1 | Elastin microfibril interface located protein |
| | M92642 | C8 | Alpha-1 type XVI collagen |
| Glycoprocess | U84007 | C1 | Glycogen debranching enzyme isoform 1 (AGL) |
| | U84011 | C5 | Glycogen debranching enzyme isoform 6 (AGL) |
| | L12711 | C7 | Transketolase (tk) |
| Immunosystem | D32129 | C7 | HLA class-I (HLA-A26) heavy chain |
| RNA processing/modification | L22009 | C11 | hnRNP H |
| | AL03168 | C8 | Splicing factor, arginine/serine-rich 6 (SRP55–2) (isoform 2) |
| Signaling | L20861 | C5 | Wnt-5a |
| | N/A | C3 | Guanine nucleotide-binding protein Rap2, Ras-oncogene related |
| | AJ011679 | C3 | Rab6 GTPase activating protein |
| | D79990 | C2 | Ras association domain family 2 (KIAA0168) |
| | L36870 | C3 | MAP kinase kinase 4 (MKK4) |
| | U35113 | C4 | Metastasis-associated mta1 |
| | M88714 | C8 | Bradykinin receptor (BK-2) |
| | U46751 | C0 | Phosphotyrosine-independent ligand p62 for the Lck SH2 domain |
| | Y13493 | C2 | Protein kinase; Dyrk2 |
| | Z85986 | C10 | Clone 108K11 on chromosome 6p21 contains SRP20 (protein serine/threnione kinase) |
| | N/A | C11 | Interferon-inducible RNA-dependent protein kinase (Pkr) |
| | X59656 | C11 | Crk-like gene CRKL (protein tyrosine kinase) |
| | N/A | C9 | Ptdins 4-kinase (PI4Kb) |
| | L35594 | C4 | Autotaxin (ectonucleotide pyrophosphatase/phosphodiesterase) |
| Transcription regulation | D50922 | C3 | KIAA0132 (kelch-like ECH-associated protein 1) |
| | U66561 | C1 | Kruppel-related zinc finger protein (ZNF184) |
| | X78992 | C3 | ERF-2 |
| | AF078096 | C0 | Forkhead/winged helix-like transcription factor 7 (FKHL7) |
| | AF040963 | C5 | Mad4 homolog (Mad4) |
| | AF096870 | C8 | Estrogen-responsive B box protein (EBBP) |
| | X87838 | C11 | β-Catenin |
| | U19969 | C3 | Two-handed zinc finger protein ZEB |
| | N/A | C9 | DNA-binding protein (APRF) |
| | D88827 | C0 | Zinc finger protein FPM3 |

*(continued)*

**TABLE 9.1 (CONTINUED)**
**Overview of Transcriptional Upregulation Induced by tBHQ in IMR-32 Cells**

| Category of Genes | GenBank Accession Number | SOM Clustering (3×4) | Description of Genes |
|---|---|---|---|
| Transcription regulation *(cont'd)* | U10324 | C0 | NF-90 |
| | X96381 | C6 | erm |
| Translation/posttranslation regulation | U20180 | C10 | Iron-regulatory protein 2 (IRP2) |
| | D26600 | C8 | Proteasome subunit HsN3 |
| Others | U29332 | C4 | Heart protein (FHL-2) |
| | X64728 | C1 | CHML |
| | AF055001 | C0 | Homocysteine-inducible, endoplasmic reticulum stress inducible, ubiquitin-like domain member 1 |
| | AB007865 | C3 | Fibronectin leucine rich transmembrane protein 2 (KIAA0405) |
| | X54232 | C0 | Heparan sulfate proteaglycan (glypican) |
| | Z25535 | C2 | Nuclear pore complex protein hnup153 |
| | D19878 | C1 | Transmembrane protein |
| | U23070 | C9 | Putative transmembrane protein (nma) |
| | AL031781 | C3 | Human ortholog of zebrafish Quaking protein homolog ZKQ-1 (isoform 1) |
| | AF070598 | C6 | Clone 24410 ABC transporter |
| | U60644 | C7 | HU-K4 (phospholipase D) |
| | AJ131581 | C9 | Latrophilin-2 |
| | AF061573 | C0 | Protocadherin (PCDH8) |
| Unknown | AL039458 | C4 | DKFZp434N0910 |
| | AL080143 | C3 | DKFZp434N043 |
| | AL050261 | C0 | DKFZp547E2110 |
| | AL096729 | C8 | DKFZp434D044 |
| | AB007962 | C3 | KIAA0493 |
| | D38522 | C1 | KIAA0080 |
| | D26069 | C3 | KIAA0041 |
| | AB020685 | C6 | KIAA0878 |
| | D87443 | C3 | KIAA0254 |
| | AB011169 | C3 | KIAA0597 |
| | D87437 | C6 | KIAA0250 |
| | AB018285 | C4 | KIAA0742 |
| | AB002351 | C6 | KIAA0353 |
| | AI768188 | C1 | *Homo sapiens* cDNA |
| | AW024285 | C3 | *Homo sapiens* cDNA |
| | N92920 | C3 | *Homo sapiens* cDNA |
| | AC004382 | C9 | Chromosome 16 BAC clone CIT987SK-A-152E5 |

*Note:* The genes upregulated by tBHQ are functionally categorized.. Self-organizing maps (SOMs) were used to group the identified genes into clusters, and the cluster number that each individual gene belongs to is provided (C0–C11).

*Source:* Li, J. and Johnson, J.A., *Physiol. Genomics*, 9, 137, 2002. With permission.

A

$n = 3$

| Gene Name | $T_{8h}$ | $C_{8h}$ | $T_{24h}$ | $C_{24h}$ |
|---|---|---|---|---|
| NQO1 | 10535.6±1074.0* | 1758.4±249.2 | 14069.3±1355.1* | 3754.3±847.4 |
| GCLR | 15498.5±1819.9* | 1529.5±284.8 | 8331.7±1506.5* | 2004.1±31.1 |
| GCLC | 5796.6±91.4 | 4024.8±734.4 | 5635.7±932.9 | 4348.5±169.0 |
| HO1 | 1507.1±386.4* | -722.8±209.1 | -1138.9±932.3 | -1124.9±557.6 |

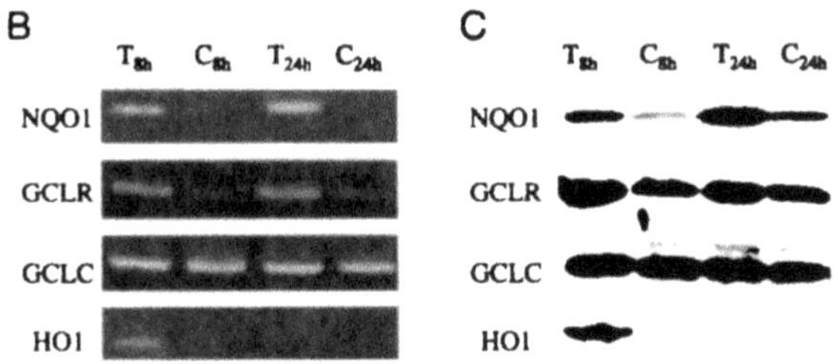


**FIGURE 9.2** Verification of upregulated gene expression in IMR-32 cells treated by tBHQ. After treatment with tBHQ for 8 and 24 hr, total RNA (1 μg) from IMR-32 cells was extracted for (A) microarray analysis and (B) RT-PCR with specific primers for NQO1, HO1, GCLC, and GCLR. Statistical analysis was conducted by one-tailed, unequal variance *t*-test; "*" represents a significant difference of the average difference between control (C) and treatment (T) groups ($p < 0.05$). (C) Protein lysates from IMR-32 cells were prepared and subjected to western blot with specific antibodies toward NQO1, GCLC, GCLR, and HO1. (From Li, J. and Johnson J.A., *Physiol. Genomics*, 9, 137, 2002. With permission.)

Other categories of genes include those that enhance the growth and differentiation of neurons (neuronal olfactomedin-related ER localized protein, axin, NF-H), function as chaperone proteins or respond to heat shock proteins (hsp40 homolog, Bip protein), participate in various signaling pathways (Wnt-5a, MAPKK4, phosphotyrosine-independent ligand p62 for the Lck SH2 domain), and modulate transcription (KIAA0132 and ERF-2). Consistent with the microarray data (Figure 9.2A), reverse transcription–polymerase chain reaction (RT-PCR) and western blot confirmed the increase in mRNA and protein levels of several detoxifying enzymes (NQO1, GCLR, and HO1) (Figure 9.2B).

Nuclear transcription factors including c-Jun, c-Fos, Jun-B, Jun-D, Fra-1, Fra-2, ATF-3, ATF4, NF-κB, and small Maf proteins (MafK and MafF) have been reported to possibly influence the Nrf2–ARE interaction. However, the expression levels of these genes did not change at the mRNA level after treatment with tBHQ. Of particular note, KIAA0132, the human homolog of mouse Keap1, was increased. Keap1 is a cytosolic chaperone of Nrf2. Exposure of cells to inducers disrupts the Keap1–Nrf2 complex and allows Nrf2 to translocate into the nucleus, where it binds to the ARE and stimulates transcription.[19] We propose that the transcriptional elevation of Keap1 mRNA likely serves as a transcriptionally mediated, negative-feedback mechanism to control the expression of ARE-driven genes. However, little is understood regarding the mechanism by which the cytosolic Keap1–Nrf2 complex functions as a cytoplasmic sensor for oxidative stress.

Our laboratory and others have demonstrated that exposure to tBHQ triggers nuclear accumulation of the transcription factor Nrf2, which binds to the ARE.[7,19] These data provide evidence for a pathway of signal transduction in response to oxidative stress via the ARE. Comparative analysis of the promoter sequences of the genes with similar expression profiles should provide a strategy for unraveling common regulatory sequences and overlapping gene expression networks modulating ARE-driven genes. A search for potential Nrf2 binding sites within the 5′-flanking regions of a selected gene set was performed using MatInspector v.2.2 based on TRANSFAC 4.0 (http://transfac.gbf.de) to determine whether their expression profiles could be predicted by the presence or absence of Nrf2 binding sites. Indeed, the gene containing no Nrf2 binding site (GCLC) was not increased by tBHQ, whereas the transcription of genes whose promoter regions contain Nrf2 binding sites (e.g., NQO1, GCLR, Wnt5a, TR, and Malate NADP oxidoreductase) were markedly increased

**TABLE 9.2**
**Analysis of the Potential Transcription Factor Binding Sites in the ARE-Driven Genes**

| Name of Genes | Source of Sequence (GenBank Accession Number) | Expression Profile by Microarray Analysis | Transcription Factor Binding Site | | | |
|---|---|---|---|---|---|---|
| | | | Nrf 2 | NFE2 | NF-κB | AP1 |
| NADPH:quinone oxidoreductase | 5′ flanking region (M81596 J05348) | Increase | 3 | 3 | 4 | 27 |
| Dimeric dihydrodiol dehydrogenase | Chromosome 19 working draft (NT_011190) | Increase | 1 | 0 | 4 | 14 |
| γ-Glutamylcysteine ligase catalytic subunit | 5′ flanking region (L39773) | No change | 0 | 0 | 1 | 48 |
| | Promoter region (S79432) | No change | 0 | 0 | 1 | 36 |
| γ-Glutamylcysteine ligase regulatory subunit | 5′ flanking region (U72210) | Increase | 2 | 1 | 1 | 48 |
| | 5′ flanking region (AF028815) | Increase | 1 | 1 | 0 | 24 |
| Malate NADP oxidoreductase | 5′ flanking region (L34809) | Increase | 1 | 1 | 3 | 47 |
| Heme oxygenase 1 | Promoter region (AF145047) | Increase | 2 | 0 | 1 | 45 |
| Wnt 5a | Promoter region (U39837) | Increase | 6 | 0 | 11 | 22 |
| KIAA0132 (a human homolog of Keap1) | 5′ flanking region (AF361887) | Increase | 2 | 1 | 2 | 17 |
| Thioredoxin reductase | Chromosome 12 working draft (NT_024383) | Increase | 1 | 2 | 1 | 40 |

*Note:* The potential transcription factor (TF) binding sites were analyzed by MatInspector v.2.2 on selected genes. Sequences selected were located within 3000 bp upstream of the coding region. The transcription factor binding sites were identified with ≥95% homology to their consensus sequences in the TRANSFAC database. Core similarity ≥ 0.95; matrix similarity ≥0.85.

by tBHQ exposure (Table 9.2). It should be noted that AP-1 and NF-κB binding sites are distributed abundantly throughout the 5′-flanking region of all these genes, including GCLC, suggesting that these transcription factors do not play a quintessential role in transcriptional activation by tBHQ. The presence of Nrf2 binding sites thus appears to be directly correlated with the tBHQ inducibility of these genes.

### 9.2.3 SOM Clustering

Self-organizing maps (SOMs; see Chapter 4) based on unsupervised neural network algorithms were applied to cluster and analyze gene expression patterns in this study. In contrast to (1) the rigid structure of hierarchical clustering aproaches, (2) the strong prior hypotheses necessary in Bayesian clustering, or (3) the relative lack of structure in *k*-means clustering, SOMs are ideally suited to exploratory data analysis by allowing one to impose partial structure on the clusters which facilitates easy visualization and interpretation.[20,21] This analysis assigns genes to the single group or cluster of genes that most closely share related expression patterns across samples. This approach is also expected to yield insights into biological relevance, because coordinate regulation of groups of genes often signifies a role in common processes or pathways.[22]

In this study, SOM automatically and quickly extracted the gene expression profiles from among the most prominent features of the data (Figure 9.3). Most genes previously identified in the literature as tBHQ-inducible were grouped within different clusters, although some of them were grouped together as depicted in Table 9.1. The expression profiles induced by a specific agent, therefore, were far more complicated than expected. In addition, these genes grouped into 12 distinct clusters with strikingly consistent patterns, yet no apparent functional correlation existed among the clustered genes (Table 9.1). For example, the genes grouped in cluster 7 are involved in detoxification and cellular antioxidant defense (NQO1, ferritin heavy chain, and hepatic dihydrodiol dehydrogenase), cytoskeleton construction (Dynein heavy polypeptide), immune response (HLA class 1 [HLA-A26] heavy chain), and energy metabolism (Transketolase). The SOM results showed that some of the potential ARE-driven genes were clustered together based on their gene expression similarity. However, there are also examples of ARE-driven genes that did not co-cluster with other known ARE-related genes, suggesting that multiple pathway and/or transcription factors may be involved in controlling the expression level of each individual gene.

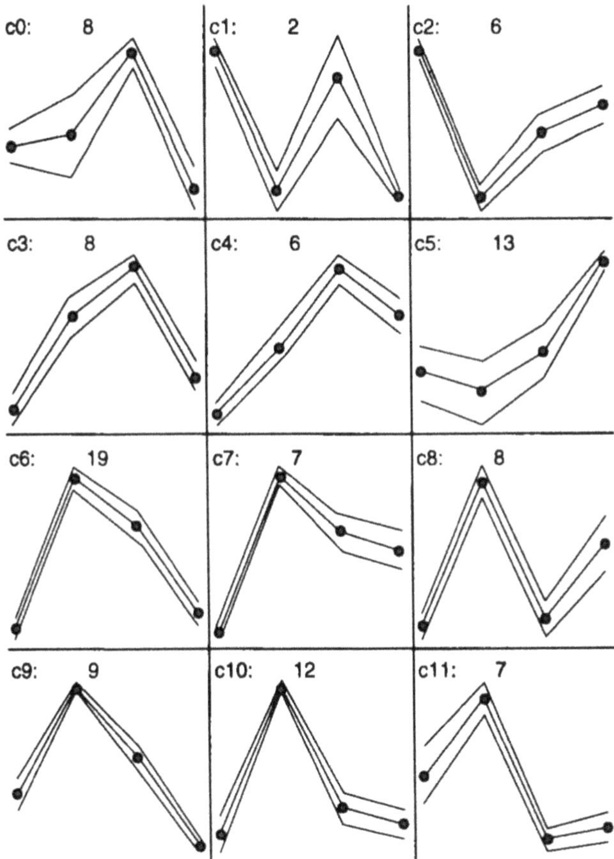


**FIGURE 9.3** Cluster analysis of probe sets on human U95A arrays revealed a gene-expression wave in IMR-32 cells after treatment with tBHQ. Self-organizing maps (SOMs) were used to group the 101 identified genes into clusters based on similar expression dynamics over the four time points. The SOM algorithm grouped them into 12 discrete clusters. The label in the upper left corner of each inset represents the cluster number (c0–c11). The number at the top center of each inset represents the number of genes in that particular cluster. Each time point (4, 8, 24, and 48 hr) is represented by black dots on the graphs. Thick lines indicate the mean expression values, and thin lines indicate standard deviation. (From Li, J. and Johnson J.A., *Physiol. Genomics*, 9, 137, 2002. With permission.)

## 9.3 PI3-KINASE REGULATION OF THE ARE IN IMR-32 HUMAN NEUROBLASTOMA CELLS

Although the role of Nrf2 in ARE activation seems to be evident, the upstream regulatory mechanisms by which ARE activating signals are linked to Nrf2 and how this transcription factor is released from the Nrf2-Keap1 complex remain to be elucidated. Phosphatidylinositol 3-kinase (PI3-kinase) phosphorylates phosphatidylinositol at the D-3 position and has been shown to form a heterodimer consisting of 85-kDa (adaptor protein) and 110-kDa (catalytic) subunits.[23,24] The role of PI3-kinase in intracellular signaling has been underscored by its implication in a plethora of biological responses such as cell growth, differentiation, apoptosis, calcium signaling, and insulin signaling.[24–28] Among the downstream targets of PI3-kinase are phospholipase C and the serine/threonine kinase Akt (protein kinase B).[25,27,29–30] Akt, one of the most well-known downstream targets of PI3-kinases, protects cells from apoptosis by phosphorylation and inhibition of Bad protein.[31,32] Based on the protective effects of PI3-kinase and the observation that induction of phase II enzymes is thought to be a protective response in cells, we were interested in determining whether PI3-kinase was involved in ARE regulation.

We found evidence that activation of the human NQO1 ARE (hNQO1–ARE) by tBHQ is mediated by PI3-kinase in IMR-32 human neuroblastoma cells. Pretreatment with LY 294002 (selective PI3-kinase inhibitor) inhibited both hNQO1–ARE–luciferase expression and endogenous NQO1 protein induction (Figure 9.4). In addition, transfection of IMR-32 cells with constitutively active PI3-kinase selectively activated the ARE in a dose-dependent manner that was completely inhibited by treatment with LY 294002 (Figure 9.5). Pretreatment of cells with the PI3-kinase inhibitors, LY 294002, and wortmannin significantly decreased Nrf2 nuclear translocation induced by tBHQ (Figure 9.6), and ARE activation by constitutively active PI3-kinase was completely blocked by dominant-negative Nrf2. Taken together, these data clearly show that ARE activation by tBHQ is dependent on PI3-kinase, which lies upstream of Nrf2.[7]

## 9.4 MICROARRAY ANALYSIS OF A PI3-KINASE-SENSITIVE ARE-DRIVEN GENE SET CONFERRING PROTECTION FROM OXIDATIVE STRESS APOPTOSIS IN IMR-32 CELLS

Others have shown that treating cells with tBHQ, a strong inducer of phase II detoxification enzymes via activation of the ARE, can protect cells from oxidative stress.[33–35] Induction of NQO1 in N18-RE-105 neuronal cells by tBHQ prior to glutamate treatment was correlated with a significant decrease in glutamate toxicity. Glutamate toxicity in these cells is not due to *N*-methyl-D-aspartate receptor activation and calcium influx. Rather, it is a result of competitive inhibition of cysteine uptake, depletion of GSH, increased oxidative stress, and apoptosis.[36–38] Subsequent studies by Murphy and colleagues[35] using $H_2O_2$ and dopamine to induce oxidative stress, however, demonstrated that N18-RE-105 cells over-expressing NQO1 were not resistant to cytotoxicity. These data suggest that the protective effect conferred by tBHQ may not simply be due to an increase in one gene but the coordinated upregulation of many genes. We hypothesize that it is this combined effect of tBHQ-mediated activation on multiple genes regulated by the ARE that is a principal component generating the protective response.

In follow-up studies, $H_2O_2$ was used to generate oxidative injury and subsequent apoptosis in cultured IMR-32 cells. Our results clearly show that $H_2O_2$ induces IMR-32 cell apoptosis through a caspase 3-dependent pathway. The apoptotic process and development of cell injury in $H_2O_2$-treated IMR-32 cells was prevented and/or delayed by strengthening the antioxidant capacity of these cells through the transcriptional activation of ARE-driven genes in response to treatment with tBHQ (Figure 9.7). Using the MTS cell viability assay, we showed that preincubation with the PI3-kinase inhibitor LY294002 (20 μM) for 30 min prior to tBHQ treatment completely reversed the protective effect of tBHQ (Table 9.3). These data suggest that PI3-kinase dependent, ARE-driven

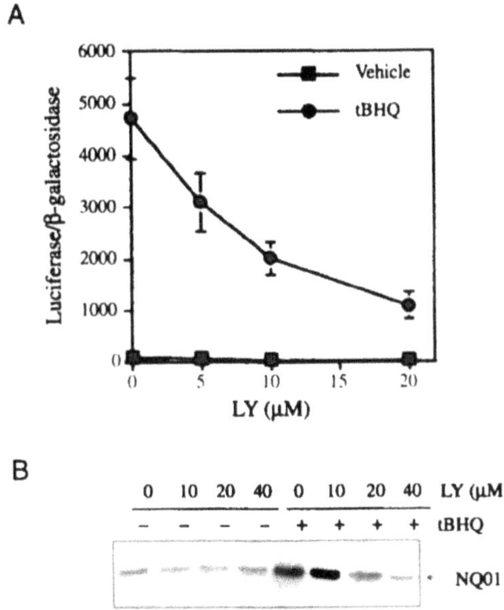


**FIGURE 9.4** Effect of LY 294002 on ARE activation and induction of NQO1. (A) IMR-32 cells were transfected with hNQO1-ARE-luciferase reporter construct (80 ng/well) and CMV-β-galactosidase (20 ng/well). After 24 hr, cells were pretreated with increasing doses of LY 294002. After 30 min of pretreatment, cells were treated with vehicle (■) or tBHQ (●, 10 μM). After 24 hr of treatment, cells were lysed, and luciferase and galactosidase activities were determined as described in the experimental procedures. Data are expressed as the ratio of luciferase to β-galactosidase activity. Each data point represents the mean ± SE ($n$ = 6). (B) IMR-32 cells were pretreated with increasing doses of LY 294002, followed by vehicle (–) or tBHQ (+, 10 μM) treatment. After 24 hr of treatment, whole cell extracts were prepared, and 50 μg of protein was resolved by SDS-PAGE. Transferred membrane was probed with NQO1 antibody. (From Lee, J.M. et al., *J. Biol. Chem.*, 276, 20011, 2001. With permission.)

genes are responsible for the cytoprotective effect manifested by tBHQ. Oligonucleotide microarray analysis identified the genes that changed in parallel with these treatments. When the protective effects of tBHQ pretreatment were reversed by inhibition of PI3-kinase, 46 of the 63 genes increased by tBHQ were significantly blocked (Table 9.4). Although we have not screened the entire human genome, the genes identified here represent a first approximation of the set of ARE-inducible antiapoptotic genes encoding proteins that can directly or indirectly attenuate the apoptotic process.

## 9.5 SUMMARY

In conclusion, oxidative stress can be defined as an imbalance in which free radicals and their products exceed the capacity of cellular antioxidant defense mechanisms. A gain in product formation or a loss in protective mechanisms can disturb this equilibrium, leading to programmed cell death (PCD). Numerous publications have demonstrated a correlation between direct supplementation of medium with chemical antioxidants and a decreased apoptotic rate in cell lines.[39–40] Induction of multiple antioxidant-regulated genes may also provide a way to increase an antioxidative potential and protect cells from apoptosis, as demonstrated in this chapter. We refer to this process as programmed cell life, or PCL. An equilibrium exists such that any increase in the forces that drive PCD, therefore, are balanced by increasing the forces that drive PCL. The set of ARE-driven survival genes identified in these microarray studies represents the first set of genes shown to modulate PCD and confirms that the PI3-kinase–Nrf2–ARE pathway is crucial for the transcription of these PCL genes.

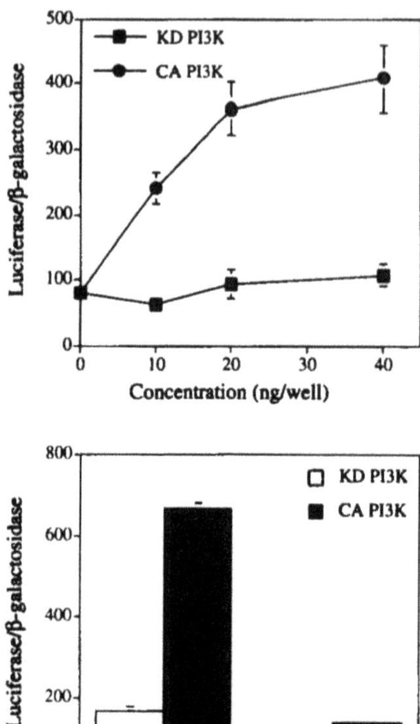


**FIGURE 9.5** Constitutively active PI3-kinase activates the ARE. (A) IMR-32 cells were cotransfected with hNQO1–ARE–luciferase reporter construct (80 ng/well), CMV-β-galactosidase (20 ng/well), and the indicated amount of PI3-kinase plasmid. After 24 hr, cells were harvested in lysis buffer. Luciferase and galactosidase activities were measured as described in the experimental procedures. Each data point represents the mean ± SE ($n = 6$). (B) Cells were cotransfected with hNQO1–ARE–luciferase reporter construct (80 ng/well), CMV-β-galactosidase (20 ng/well), and PI3-kinase plasmid (40 ng/well). After 24 hr, cells were treated with vehicle or LY 294002 (20 μM) for another 24 hr. Luciferase and β-galactosidase activities were determined as described in experimental procedures. Each bar represents the mean ± SE ($n = 6$). (From Lee, J.M. et al., *J. Biol. Chem.*, 276, 20011, 2001. With permission.)

A

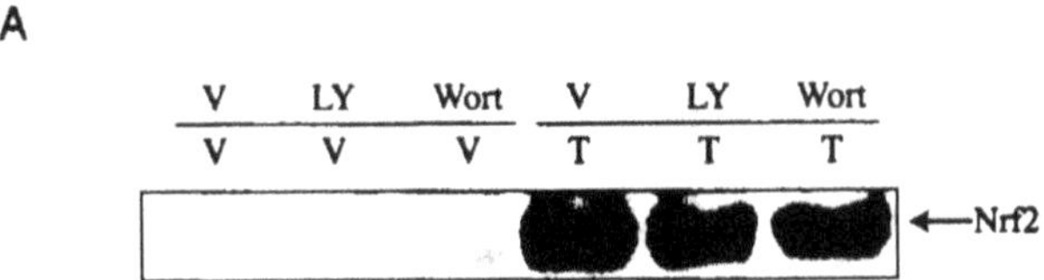


B

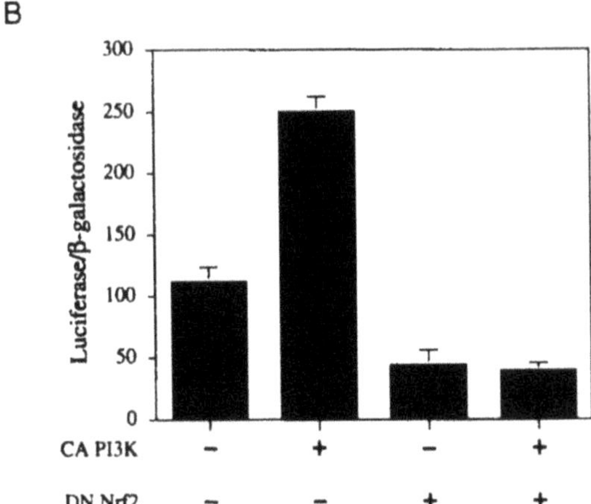


**FIGURE 9.6** PI3-kinase is linked to Nrf2. (A) IMR-32 cells were pretreated with vehicle (V), LY 294002 (LY, 20 μM), or wortmannin (Wort, 1 μM) for 30 min and treated with vehicle (V) or tBHQ (T, 10 μM). After 6 hr, nuclear extracts were isolated and resolved by SDS-PAGE. Transferred membrane was probed with Nrf2 antibody. (B) Cells were cotransfected with hNQO1–ARE–luciferase reporter construct (80 ng/well), CMV-β-galactosidase (20 ng/well), constitutively active PI3-kinase p110 (CA PI3-kinase, 40 ng/well), and dominant-negative Nrf2 (DN Nrf2, 5 ng/well). After 24 hr, cells were harvested, and luciferase and β-galactosidase activities were determined as described in the experimental procedures. Each data bar represents the mean ± SE ($n$ = 6). (From Lee, J.M. et al., *J. Biol. Chem.*, 276, 20011, 2001. With permission.)

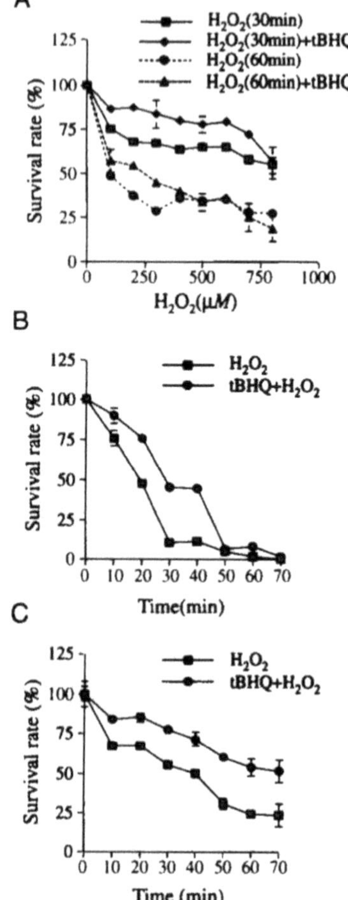


**FIGURE 9.7** Time and concentration-dependent $H_2O_2$ cytotoxicity in IMR-32 cells and the protective effect of tBHQ. Cell viability was determined by the MTS assay. The survival rate was calculated. The final concentration of $H_2O_2$ varied from 100 μM to 800 μM, and cells were pretreated with tBHQ (10 μM) for 24 hr. (A) Treatment with varied concentration of $H_2O_2$ for 30 or 60 min. (B) Time-dependent cytotoxicity and protection at low cell density (40–50% confluence). (C) Time-dependent cytotoxicity and protection at high cell density (80–90% confluence). Each point represents mean ± SEM ($n = 3$). (From Li, J., Lee, J.M., and Johnson, J.A., *J. Biol. Chem.*, 277, 388, 2002. With permission.)

**TABLE 9.3**
**Inhibition of PI3-Kinase Completely Reversed the Protective Effect of tBHQ against $H_2O_2$-Induced Cytotoxicity**

| Time (min) | $H_2O_2$ (%) | tBHQ + $H_2O_2$ (%) | LY294002 + tBHQ + $H_2O_2$ (%) | LY294002 + tBHQ (%) |
|---|---|---|---|---|
| 0 | 100.0 ± 3.5 | 100.0 ± 6.8 | 100.0 ± 3.2 | 93.9 ± 0.9 |
| 10 | 66.9 ± 1.2 | 83.8 ± 0.7* | 63.9 ± 2.0 | 96.5 ± 3.2* |
| 30 | 54.9 ± 2.1 | 77.1 ± 1.9* | 56.3 ± 2.8 | 99.2 ± 3.9* |
| 50 | 30.7 ± 2.6 | 60.0 ± 1.4* | 33.2 ± 1.5 | 106.1 ± 1.0* |

*Note:* IMR-32 cells were pretreated with LY294002 (20 μM) for 30 min before tBHQ treatment for 24 hr. MTS assay was performed after $H_2O_2$ (200 μM) treatment for 30 min. Cell survival rates are calculated and represented as the mean ± SEM ($n = 3$). Statistic analysis was conducted by one-tailed, unequal variance *t*-test and "*" represents a significant difference compared to LY294002 + tBHQ + $H_2O_2$ group ($p < 0.05$).

*Source:* Li, J., Lee, J.M., and Johnson, J.A., *J. Biol. Chem.*, 277, 388, 2002. With permission.

**TABLE 9.4**
**Effect of LY294002 on the Transcriptional Upregulation of Genes Induced by tBHQ**

| Category of Genes | Description of Genes | 8-hr (3×3) – LY294002 ADC (mean ± SEM) | FC (mean ± SEM) | R | 8-hr (3×3) + LY294002 ADC (mean ± SEM) | FC (mean ± SEM) | R | Sig |
|---|---|---|---|---|---|---|---|---|
| Chaperones/heat shock proteins | **Heat shock protein hsp40 homolog** | 592 ± 150 | 2.07 ± 0.56 | 12 | –61 ± 42 | –0.40 ± 0.38 | 0 | HS |
| | **BiP protein (a member of Hsp70 family of chaperones)** | 5975 ± 1409 | 1.89 ± 0.21 | 10 | 354 ± 625 | –0.29 ± 0.42 | 0 | HS |
| | Heat shock 70-kDa protein 8 (W28493) | 1551 ± 734 | 3.28 ± 1.43 | 12 | 300 ± 151 | 0.68 ± 0.60 | 3 | NS |
| CNS specific function | **Neuronal olfactomedin-related ER localized protein** | 12,825 ± 1907 | 1.60 ± 0.09 | 12 | 5149 ± 413 | 1.26 ± 0.02 | 2 | HS |
| | **Axin (AXIN)** | 2002 ± 319 | 6.38 ± 1.58 | 9 | 509 ± 307 | 2.54 ± 1.61 | 0 | HS |
| | **NF-H gene, exon 1** | 8972 ± 1411 | 2.04 ± 0.16 | 18 | 3495 ± 474 | 1.69 ± 0.10 | 8 | HS |
| | **β2-Syntrophin (SNT β2)** | 1008 ± 228 | 3.27 ± 0.71 | 10 | 245 ± 200 | 0.50 ± 0.43 | 0 | HS |
| | **Secretogranin II gene** | 2512 ± 1461 | 2.12 ± 1.00 | 10 | –1604 ± 775 | –0.81 ± 0.38 | –4 | S |
| | **Peroxisomal targeting signal 1** | 900 ± 110 | 1.97 ± 0.11 | 12 | 463 ± 58 | 1.27 ± 0.03 | 0 | HS |
| | **CB1 cannabinoid receptor (CNR1)** | 468 ± 106 | 1.90 ± 0.48 | 10 | –25 ± 21 | –0.62 ± 0.34 | 0 | HS |
| | Microtuble-associated protein 1A (W26631) | 6634 ± 1702 | 1.91 ± 0.24 | 10 | 5881 ± 1199 | 2.36 ± 0.46 | 2 | NS |
| | KIAA0992 (palladin) | 2004 ± 443 | 1.54 ± 0.13 | 11 | 1264 ± 201 | 1.30 ± 0.06 | 4 | NS |
| Detoxification and antioxidative stress | **NAD(P)H-quinone oxidoreductase** | 8685 ± 972 | 6.42 ± 0.38 | 18 | 2075 ± 156 | 2.49 ± 0.12 | 18 | HS |
| | **Glutathione reductase** | 941 ± 120 | 2.70 ± 0.15 | 16 | 331 ± 32 | 1.96 ± 0.17 | | HS |
| | **Glutathione transferase M3 (GSTM3)** | 979 ± 138 | 2.54 ± 0.31 | 14 | 329 ± 59 | 1.37 ± 0.09 | 2 | HS |
| | **Heme oxygenase 1** | 2025 ± 300 | 6.11 ± 1.30 | 18 | 499 ± 26 | 3.08 ± 0.12 | 8 | HS |
| | **Breast cancer cytosolic NADP(+)-dependent malic enzyme** | 5465 ± 700 | 4.21 ± 0.26 | 17 | 2051 ± 119 | 3.20 ± 0.15 | 14 | HS |
| | **Malic enzyme 1, soluble** | 2484 ± 379 | 7.00 ± 1.59 | 18 | 780 ± 107 | 4.12 ± 0.56 | 14 | HS |
| | **KIAA0119 (aldo-keto reductase family 1)** | 3374 ± 1151 | 7.96 ± 2.13 | 18 | 206 ± 37 | 1.67 ± 0.37 | 0 | S |
| | **Hepatic dihydrodiol dehydrogenase** | 3717 ± 1194 | 5.87 ± 1.82 | 14 | 581 ± 93 | 1.89 ± 0.20 | 6 | S |
| | Thioredoxin reductase | 20747 ± 2324 | 2.66 ± 0.11 | 18 | 19560 ± 1478 | 2.19 ± 0.09 | 18 | NS |

*(continued)*

**TABLE 9.4 (CONTINUED)**
**Effect of LY294002 on the Transcriptional Upregulation of Genes Induced by tBHQ**

| Category of Genes | Description of Genes | 8-hr (3×3) – LY294002 | | | 8-hr (3×3) + LY294002 | | | |
|---|---|---|---|---|---|---|---|---|
| | | ADC (mean ± SEM) | FC (mean ± SEM) | R | ADC (mean ± SEM) | FC (mean ± SEM) | R | Sig |
| | Gamma-glutamylcysteine ligase regulatory subunit | 12026 ± 1417 | 9.52 ± 0.99 | 18 | 12101 ± 1185 | 9.43 ± 1.14 | 18 | NS |
| DNA repair | Human growth arrest and DNA-damage-inducible protein (gadd45) mRNA | 4593 ± 1076 | 1.58 ± 0.15 | 10 | 3342 ± 858 | 1.43 ± 0.12 | 6 | NS |
| | ERCC5 excision repair protein | 519 ± 258 | 1.82 ± 1.25 | 9 | 23 ± 339 | –0.36 ± 0.50 | 2 | NS |
| Extracellular matrix | **Elastin microfibril interface located protein** | 2015 ± 371 | 5.59 ± 0.94 | 13 | –167 ± 213 | –1.69 ± 0.94 | 0 | HS |
| Glycoprocess | **Glycogen debranching enzyme isoform 1 (AGL)** | 2406 ± 527 | 2.29 ± 0.28 | 12 | 262 ± 328 | 0.41 ± 0.44 | 2 | HS |
| | **Glycogen debranching enzyme isoform 6 (AGL)** | 6103 ± 1777 | 1.76 ± 0.51 | 10 | 1009 ± 1743 | 0.66 ± 0.43 | 4 | S |
| Signaling | **Wnt-5a** | 1425 ± 161 | 4.46 ± 0.71 | 18 | 644 ± 53 | 3.67 ± 0.28 | 15 | HS |
| | **Guanine nucleotide-binding protein Rap2, Ras-oncogene related** | 1324 ± 177 | 4.14 ± 0.62 | 11 | 122 ± 100 | 0.89 ± 0.75 | 0 | HS |
| | **Rab6 GTPase-activating protein** | 1658 ± 520 | 1.57 ± 0.43 | 10 | –14 ± 278 | –0.33 ± 0.37 | 0 | HS |
| | **KIAA0168 (Ras association domain family 2)** | 500 ± 100 | 1.98 ± 0.17 | 10 | –31 ± 62 | –0.24 ± 0.54 | 0 | HS |
| | **MAP kinase kinase 4** | 792 ± 179 | 2.73 ± 0.74 | 9 | –113 ± 56 | –1.06 ± 0.51 | 0 | HS |
| | **Metastasis-associated mta1** | 3393 ± 927 | 1.53 ± 0.40 | 9 | –262 ± 522 | –0.34 ± 0.37 | 0 | HS |
| | **Phosphotyrosine independent ligand p62 for the Lck SH2 domain** | 3990 ± 616 | 3.88 ± 0.72 | 10 | 2668 ± 199 | 1.80 ± 0.07 | 6 | S |
| | Protein kinase; Dyrk2 | 2605 ± 551 | 1.63 ± 0.13 | 10 | 2172 ± 631 | 1.33 ± 0.09 | 6 | NS |
| Transcription | **KIAA0132 (kelch-like ECH-associated protein 1)** | 3466 ± 623 | 2.8 ± 0.44 | 12 | 1284 ± 145 | 1.5 ± 0.06 | 5 | S |
| | **Kruppel-related zinc finger protein (ZNF184)** | 760 ± 214 | 1.76 ± 0.44 | 10 | 1 ± 192 | –0.39 ± 0.49 | 2 | HS |
| | **ERF-2** | 1728 ± 246 | 2.02 ± 0.14 | 12 | 857 ± 317 | 1.13 ± 0.47 | 6 | S |
| | Forkhead/winged helix-like transcription factor 7 (FKHL7) | 1457 ± 347 | 1.78 ± 0.21 | 9 | 2258 ± 985 | 1.58 ± 0.65 | 6 | NS |
| | Mad4 homolog (Mad4) | 1443 ± 390 | 2.83 ± 0.80 | 11 | 1178 ± 169 | 1.68 ± 0.12 | 2 | NS |
| Others | **Heart protein (FHL-2)** | 753 ± 117 | 1.54 ± 0.08 | 9 | 150 ± 135 | 0.21 ± 0.42 | 0 | HS |
| | **Smooth muscle myosin heavy chain isoform** | 2319 ± 558 | 1.69 ± 0.19 | 10 | –348 ± 482 | –0.38 ± 0.42 | 0 | HS |
| | **CHML** | 780 ± 179 | 2.34 ± 0.35 | 12 | 39 ± 147 | –0.05 ± 0.57 | 4 | HS |

| | | | | | | | |
|---|---|---|---|---|---|---|---|
| | **Homocysteine-inducible, endoplasmic reticulum ubiquitin-like domain member 1** | 951 ± 283 | 1.54 ± 0.41 | 10 | 117 ± 58 | 0.42 ± 0.36 | 0 | HS |
| | **KIAA0405 (fibronectin leucine-rich transmembrane protein 2)** | 1293 ± 434 | 2.07 ± 0.71 | 12 | 285 ± 88 | 1.13 ± 0.04 | 0 | HS |
| | **Heparan sulfate proteaglycan (glypican)** | 3220 ± 603 | 1.61 ± 0.08 | 10 | 1058 ± 82 | 1.31 ± 0.03 | 0 | HS |
| | **Nuclear pore complex protein hnup153** | 6456 ± 1641 | 1.91 ± 0.24 | 14 | 2700 ± 959 | 0.87 ± 0.39 | 4 | S |
| | Transmembrane protein | 2149 ± 673 | 1.70 ± 0.23 | 10 | 838 ± 904 | −0.20 ± 0.52 | 6 | NS |
| | Human ortholog of zebrafish Quaking protein homolog ZKQ-1 (isoform 1) | 3402 ± 1474 | 3.87 ± 1.70 | 10 | 1124 ± 561 | 0.50 ± 0.39 | 0 | NS |
| | Protocadherin (PCDH8) | 779 ± 410 | 2.09 ± 0.93 | 9 | 256 ± 200 | 0.36 ± 0.54 | 6 | NS |
| Unknown | **DKFZp434N0910** | 631 ± 81 | 2.46 ± 0.25 | 10 | 171 ± 60 | 1.30 ± 0.51 | 3 | HS |
| | **DKFZp434N043** | 951 ± 150 | 1.73 ± 0.13 | 9 | 167 ± 106 | 0.42 ± 0.37 | 2 | HS |
| | **DKFZp547E2110** | 4518 ± 1084 | 2.72 ± 0.75 | 12 | 2089 ± 380 | 1.52 ± 0.10 | 6 | S |
| | **KIAA1180** | 4058 ± 851 | 1.90 ± 0.28 | 10 | 1272 ± 758 | 0.77 ± 0.40 | 2 | S |
| | **KIAA0493** | 535 ± 86 | 2.31 ± 0.34 | 14 | 178 ± 42 | 1.13 ± 0.29 | 2 | HS |
| | **KIAA0080** | 1449 ± 245 | 3.00 ± 0.50 | 10 | 62 ± 292 | −0.17 ± 0.52 | 0 | HS |
| | **KIAA0041** | 717 ± 196 | 2.46 ± 0.94 | 12 | -45 ± 40 | −0.56 ± 0.31 | 0 | HS |
| | **KIAA0254** | 744 ± 203 | 1.6 ± 0.40 | 10 | 194 ± 86 | 0.47 ± 0.40 | 0 | S |
| | ***Homo sapiens* cDNA (AI768188)** | 4197 ± 932 | 2.19 ± 0.38 | 12 | 732 ± 486 | 0.71 ± 0.39 | 0 | HS |
| | KIAA0597 | 5239 ± 1409 | 1.83 ± 0.50 | 10 | 4018 ± 1255 | 1.36 ± 0.11 | 2 | NS |
| | KIAA0742 | 1283 ± 384 | 1.51 ± 0.54 | 9 | 599 ± 307 | 0.69 ± 0.34 | 0 | NS |
| | KIAA0353 | 617 ± 189 | 1.62 ± 0.62 | 12 | 363 ± 97 | 1.27 ± 0.32 | 0 | NS |
| | *Homo sapiens* cDNA (N92920) | 1514 ± 394 | 3.97 ± 1.47 | 10 | 677 ± 437 | 0.36 ± 1.03 | 2 | NS |

*Note:* The genes upregluated by tBHQ were functionally categorized, and genes with FC over 1.5 are listed. R represents the values calculated by ranking analysis based on *different call*, and a value ranging from 9 to 18 is considered to be a significantly increased gene expression. Sig represents significance by one-tailed, unequal-variance *t*-test based on ADC gathered from 3 × 3 groups without LY294002 compared to the corresponding value in the presence of LY294002. S, significant ($p < 0.05$); HS, highly significant ($p < 0.01$); NS, not significant. The names of genes for which upregulation was blocked by LY294002 are shown in **bold**.

*Source:* Li, J., Lee, J.M., and Johnson, J.A., *J. Biol. Chem.*, 277, 388, 2002. With permission.

## REFERENCES


1. Itoh, K. et al., Regulatory mechanisms of cellular response to oxidative stress, *Free Radic. Res.*, 31, 319, 1999.
2. Jaiswal, A.K., Regulation of genes encoding NAD(P)H:quinone oxidoreductases, *Free Radic. Biol. Med.*, 29, 254, 2000.
3. Talalay, P., Chemoprotection against cancer by induction of phase 2 enzymes, *Biofactors*, 12, 5, 2000.
4. Tamayo, P. et al., Interpreting patterns of gene expression with self-organizing maps: methods and application to hematopoietic differentiation, *Proc. Natl. Acad. Sci. USA*, 96, 2907–2912, 1999.
5. Friling, R.S. et al., Xenobiotic-inducible expression of murine glutathione S-transferase Ya subunit gene is controlled by an electrophile-responsive element, *Proc. Natl. Acad. Sci. USA*, 87, 6258, 1990.
6. Paulson, K.E. et al., Analysis of the upstream elements of the xenobiotic compound-inducible and positionally regulated glutathione S-transferase Ya gene, *Mol. Cell. Biol.*, 10, 1841, 1990.
7. Lee, J.M. et al., Phosphatidylinositol 3-kinase, not extracellular signal-regulated kinase, regulates activation of the antioxidant-responsive element in IMR-32 human neuroblastoma cells, *J. Biol. Chem.*, 276, 20011, 2001.
8. Lee, J.M. et al., Nrf2-dependent activation of the antioxidant responsive element by *tert*-butylhydroquinone is independent of oxidative stress in IMR-32 human neuroblastoma cells, *Biochem. Biophys. Res. Commun.*, 280, 286, 2001.
9. Alam, J. et al., Nrf2, a Cap'n'Collar transcription factor, regulates induction of the heme oxygenase-1 gene, *J. Biol. Chem.*, 274, 26071, 1999.
10. Chan, J.Y. and Kwong, M., Impaired expression of glutathione synthetic enzyme genes in mice with targeted deletion of the Nrf2 basic-leucine zipper protein, *Biochim. Biophys. Acta*, 1517, 19, 2000.
11. Eftekharpour, E., Holmgren, A., and Juurlink, B.H., Thioredoxin reductase and glutathione synthesis is upregulated by *t*-butylhydroquinone in cortical astrocytes but not in cortical neurons, *Glia*, 31, 241, 2000.
12. Jeyapaul, J. and Jaiswal, A.K., Nrf2 and c-Jun regulation of antioxidant response element (ARE)-mediated expression and induction of gamma-glutamylcysteine synthetase heavy subunit gene, *Biochem. Pharmacol.*, 59, 1433, 2000.
13. Kang, K.W. et al., Activation of phosphatidylinositol 3-kinase and Akt by *tert*-butylhydroquinone is responsible for antioxidant response element-mediated rGSTA2 induction in H4IIE cells, *Mol. Pharmacol.*, 59, 1147, 2001.
14. Kwak, M.K. et al., Role of transcription factor Nrf2 in the induction of hepatic phase 2 and antioxidative enzymes *in vivo* by the cancer chemoprotective agent, 3H-1, 2-dimethiole-3-thione, *Mol. Med.*, 7, 135, 2000.
15. Li, N. et al., Induction of heme oxygenase-1 expression in macrophages by diesel exhaust particle chemicals and quinones via the antioxidant-responsive element, *J. Immunol.*, 165, 3393, 2000.
16. McMahon, M. et al., The Cap'n'Collar basic leucine zipper transcription factor Nrf2 (NF-E2 p45-related factor 2) controls both constitutive and inducible expression of intestinal detoxification and glutathione biosynthetic enzymes, *Cancer Res.*, 61, 3299, 2001.
17. Burczynski, M.E., Lin, H.K., and Penning, T.M., Isoform-specific induction of a human aldo-keto reductase by polycyclic aromatic hydrocarbons (PAHs), electrophiles and oxidative stress: implications for the alternative pathway of PAH activation catalyzed by human dihydrodiol dehydrogenases, *Cancer Res.*, 59, 607–614, 1999.
18. Burczynski, M.E., Sridhar, G.R., Palackal, N.T., and Pening, T.M., The reactive oxygen species- and Michael acceptor-inducible human aldo-keto reductase AKR1C1 reduces the α,β-unsaturated aldehyde 4-hydroxy-2-nonenal to 1,4-dihydroxy-2-nonene, *J. Biol. Chem.*, 276, 2890–2897, 2001.
19. Dhakshinamoorthy, S. and Jaiswal A.K., Functional characterization and role of Inrf2 in antioxidant response element-mediated expression and antioxidant induction of NAD(P)H:quinone oxidoreductase 1 gene, *Oncogene*, 20, 3906, 2001.
20. Butte, A.J. et al., Discovering functional relationships between RNA expression and chemotherapeutic susceptibility using relevance networks, *Proc. Natl. Acad. Sci. USA*, 97, 12182, 2000.
21. Dieckgraefe, B.K. et al., Analysis of mucosal gene expression in inflammatory bowel disease by parallel oligonucleotide arrays, *Physiol. Genomics*, 4, 1, 2000.

22. Eisen, M.B. et al., Cluster analysis and display of genome-wide expression patterns, *Proc. Natl. Acad. Sci. USA*, 95, 14863, 1998.
23. Klippel, A., Escobedo, J.A., Hu, Q., and Williams, L.T., A region of the 85-kilodalton (kDa) subunit of phosphatidylinositol 3-kinase binds the 110-kDa catalytic subunit *in vivo*, *Mol. Cell. Biol.*, 13, 5560, 1993.
24. Shepherd, P.R., Withers, D.J., and Siddle, K., Phosphoinositide 3-kinase: the key switch mechanism in insulin signalling, *Biochem. J.*, 333, 471, 1998.
25. Franke, T.F., Kaplan, D.R., and Cantley, L.C., PI3K: downstream AKTion blocks apoptosis, *Cell*, 88, 435, 1997.
26. Jiang, B.H. et al., Myogenic signaling of phosphatidylinositol 3-kinase requires the serine–threonine kinase Akt/protein kinase B, *Proc. Natl. Acad. Sci. USA*, 96, 2077, 1999.
27. Rameh, L.E. et al., Phosphoinositide 3-kinase regulates phospholipase Cgamma-mediated calcium signaling, *J. Biol. Chem.*, 273, 23750, 1998.
28. Sabbatini, P. and McCormick, F., Phosphoinositide 3-OH kinase (PI3K) and PKB/Akt delay the onset of p53-mediated, transcriptionally dependent apoptosis, *J. Biol. Chem.*, 274, 24263, 1999.
29. Falasca, M. et al., Activation of phospholipase C gamma by PI3-kinase-induced PH domain-mediated membrane targeting, *EMBO J.*, 17, 414, 1998.
30. Le Good, J.A. et al., Protein kinase C isotypes controlled by phosphoinositide 3-kinase through the protein kinase PDK1, *Science*, 281, 2042, 1998.
31. Datta, S.R. et al., Akt phosphorylation of BAD couples survival signals to the cell-intrinsic death machinery, *Cell*, 91, 231, 1997.
32. Dudek, H. et al., Regulation of neuronal survival by the serine–threonine protein kinase Akt, *Science*, 275, 661, 1997.
33. Duffy, S., So, A., and Murphy, T.H., Activation of endogenous antioxidant defenses in neuronal cells prevents free radical-mediated damage, *J. Neurochem.*, 71, 69, 1998.
34. Murphy, T.H., DeLong M.J., and Coyle, J.T., Enhanced NAD(P)H:quinone reductase activity prevents glutamate toxicity produced by oxidative stress, *J. Neurochem.*, 56, 990, 1991.
35. Murphy, T.H. et al., Glutamate toxicity in a neuronal cell lone involves inhibition of cystine transport leading to oxidative stress, *Neuron*, 2, 1547, 1989.
36. Budihardjo, I. et al., Biochemical pathways of caspase activation during apoptosis, *Annu. Rev. Cell. Dev. Biol.*, 15, 269, 1999.
37. Li, P. et al., Cytochrome c and dATP-dependent formation of Apaf-1/caspase-9 complex initiates an apoptotic protease cascade, *Cell*, 91, 479, 1997.
38. Sugano, N., Ito, K., and Murai, S., Cyclosporin A inhibits $H_2O_2$-induced apoptosis of human fibroblasts, *FEBS Lett.*, 447, 274, 1999.
39. Lee, M.H. et al., Effect of proteasome inhibition on cellular oxidative damage, antioxidant defences and nitric oxide production, *J. Neurochem.*, 78, 32, 2001.
40. Qin, F. et al., Antioxidant vitamins prevent cardiomyocyte apoptosis produced by norepinephrine infusion in ferrets, *Cardiovasc. Res.*, 51, 736, 2001.

# 10 An Overview of Mechanistic Toxicogenomic Studies

*Julia Scheel, Marie-Charlotte von Brevern, and Thorsten Storck*

## CONTENTS



## 10.1 INTRODUCTION

In the previous chapters, we have seen how specific expression profiling studies have provided insight into new patterns and mechanisms of regulation induced by prototypical enzyme inducers (Chapters 6 and 7), dioxin (Chapter 8), or antioxidants like tBHQ (Chapter 9). These types of studies, as well as experiments examining the patterns of response to DNA-damaging agents, metal toxicants, and endocrine disruptors, are in the focus of toxicological research and, therefore, naturally represent "hot spots" for the application of the newly developed transcription profiling technologies. Results obtained within these areas are discussed in the following chapter; other areas of mechanistic toxicogenomic research, such as methamphetamine-induced neurotoxicity,[1–6] acetamphetamine toxicity,[7–11] studies on particles and fibers,[12–16] and studies dealing with various other topics,[17–33] are not discussed in this chapter, but are summarized in Tables 10.1 to 10.4.

## 10.2 REVIEW OF PUBLISHED STUDIES

### 10.2.1 PPAR Pathways/Peroxisome Proliferators

A number of toxicogenomic studies are concerned with compounds that act via peroxisome proliferator activated receptors (PPARs). Although substantial research has been done to characterize the corresponding pathways, the related mechanisms are still not fully understood, and it is hoped that toxicogenomic studies might shed some light on the remaining questions. PPARs are a family of ligand-activated nuclear transcription factors belonging to the steroid receptor superfamily regulating the expression of a variety of genes involved in lipid metabolism, energy balance, eicosanoid signaling, cell differentiation, and tumorigenesis.[34]

0-8493-1334-1/03/$0.00+$1.50

**TABLE 10.1**
**Overview of DNA Array-Based Toxicogenomic Studies in Rats**

| Tissue/Cells | Strain, Gender | Compounds Analyzed | Type of Array | Number of Probes | Organization or Company | Ref. |
|---|---|---|---|---|---|---|
| Liver | Sprague–Dawley, male | Wy-14,643, clofibrate, gemfibrozil, phenobarbital | cDNA | ~1700 | NIEHS; Boehringer-Ingelheim Pharmaceuticals | Hamadeh et al.[39] |
| | | Wy-14,643, clofibrate, gemfibrozil, phenobarbital, phenytoin, diethylhexyl phthalate (DEHP), hexobarbital | cDNA | ~1700 | NIEHS; Boehringer-Ingelheim Pharmaceuticals | Hamadeh et al.[40] |
| | | 3-Methylcholanthrene, phenobarbital, dexamethasone, clofibrate | Oligonucleotide | 1443 | Merck | Gerhold et al.[42] |
| | | 15 known hepatotoxicants: carbon tetrachloride, allyl alcohol, aroclor 1245, methotrexate, diquat, carbamazepine, methapyrilene, arsenic, diethylnitrosamine, monocrotaline, dimethyl-formamide, amiodarone, indomethacin, etoposide, 3-methylcholanthrene | Oligonucleotide | ~1000 | Abbott; Rosetta | Waring et al.[78] |
| | Wistar, male | Microcystin-LR, phenobarbital, lipopolysaccharide, carbon tetrachloride, thioacetamide, cyproterone acetate | Oligonucleotide | 1600 | Pfizer | Bulera et al.[77] |
| Hepatocytes | Sprague–Dawley, male | Carbon tetrachloride, allyl alcohol, aroclor 1245, methotrexate, diquat,carbamazepine, methapyrilene, arsenic, diethylnitrosamine, monocrotaline, dimethyl-fomamide, amiodarone, indomethacin, etoposide, 3-methylcholanthrene | Oligonucleotide | ~1000 | Abbott | Waring et al.[79] |
| | | Changes over time | Oligonucleotide | 8700/~1000 | Eli Lilly | Baker et al.[17] |
| Liver, epithelial cell line TRL1215 | Originally derived from Fischer F344 rats | Arsenic | cDNA | 588 | NCI, NIEHS; University of Kansas | Liu et al.[107] |
| | | | cDNA | 588/207 | NIEHS | Chen et al.[110] |
| Lung, liver | Sprague–Dawley, male | Hexavalent chromium [Cr(VI)] | cDNA | 207 | University of Genoa | D'Agostini et al.[112] |

| | | | | | | |
|---|---|---|---|---|---|---|
| Lung | Sprague–Dawley, male | Diesel exhaust particles | cDNA | 207 | Cardiff University | Reynolds and Richards[12] |
| | Fischer F344, male | Diesel exhaust particles | cDNA | 588/207 | National Institute for Environmental Studies (Japan); Chiba University; Aomori University | Sato et al.[13] |
| Alveolar macrophages | | Diesel exhaust particles | cDNA | 465 | University of Tsukuba | Koike et al.[14] |
| Fibroblasts over-expressing (Rat1-myc) or lacking Myc (myc-null), WT (TGR-1, Rat1) | | VP-16 | cDNA | 6500 | NCI/NIH | Yu et al.[18] |
| Retinal neurons | Sprague–Dawley, female | *N*-methyl-D-aspartate (NMDA) | cDNA | 588 | University of Louisville | Laabich et al.[19] |
| Omenta | Wistar | Asbestos | cDNA | 588 | University of Düsseldorf | Sandhu et al.[15] |
| Uterus | Wistar, ovariectomized, female | Tamoxifen, estradiol 17-beta | cDNA | 512 | MRC | Green et al.[85] |
| Prostate gland | Noble rats, male | Testosterone, estradiol 17-beta | cDNA | 588 | University of Hong Kong | Ouyang et al.[88] |
| Liver, (kidney, brain, spleen, pancreas) | Sprague–Dawley, male | Acetaminophen, benzo(*a*)pyrene, clofibrate | cDNA | 7704 | Incyte | Cunningham et al.[11] |
| Developing reproductive tract | Long Evans, fetuses, female | 2,3,7,8-Tetrachlorodibenzo-*p*-dioxin (TCDD) | cDNA | 5147 | University of North Carolina | Hurst et al.[69] |
| Immortalized neuronal cell line CSM 14.1, derived from fetal mesencephalon | | Methamphetamine, dopamine | cDNA | 1176 | NIH | Cai et al.[2] |
| Fetal testes | Sprague–Dawley, female | Di(n-butyl) phthalate | cDNA | 588 | Sanofi-Synthelabo | Shultz et al.[20] |

**TABLE 10.2**
**Overview of DNA Array-Based Toxicogenomic Studies in Mice**

| Tissue/Cells | Strain, Gender | Compounds Analyzed | Type of Array | Number of Probes | Organization or Company | Ref. |
|---|---|---|---|---|---|---|
| Liver | Swiss-Webster, male | Cadmium chloride, benzo(a)pyrene, trichloroethylene | cDNA | 148 | University of California (Davis); Molecular Dynamics | Bartosiewicz et al.[109] |
| | | Beta-naphthoflavone (BNF) | cDNA | 148 | University of California (Davis); Molecular Dynamics | Bartosiewicz et al.[75] |
| | 129/SV, male | Sodium arsenite, sodium arsenate | cDNA | 140 | NIEHS; University of Kansas | Liu et al.[108] |
| | Crl:CD-1, male | $O^2$-Vinyl 1-(Pyrrolidin-1-yl)diazen-1-ium-1,2-diolate (V-PYRRO/NO)/D-Galactosamine/lipopolysaccharide (GlaN/LPS) | cDNA | 140/1176 | NIEHS, NCI | Liu et al.[21] |
| | C57Bl/6x129/Ola hybrid mice, Cox+/+, Cox-1-/+ and Cox-2 -/+ | Acetaminophen | Oligonucleotide | >11,000 | NIH | Reilly et al.[8] |
| | C57Bl/6x129/Ola hybrid mice | Acetaminophen | Oligonucleotide | >11,000 | NHLBI; NCI/NIH | Reilly et al.[7] |
| | CD-1, male | Acetaminophen | cDNA | 450/18,378 | AstraZeneca | Ruepp et al.[9] |
| | MT-null mice/WT, male | Cadmium | cDNA | 140 | NCI, NIEHS; University of Kansas | Liu et al.[105] |
| | PPAR-alpha null mice/pure-bred WT | Wy-14,643 | Oligonucleotide | 6500 | University of Lausanne; Pfizer; NCI | Kersten et al.[45] |
| | C57BL/6J, wt and AOX-/-, PPAR-alpha-/-, PPAR-alpha-/-AOX-/- (DKO) | Wy-14,643 | cDNA | 7483 | Northwestern University (Chicago) | Cherkaoui-Malki et al.[44] |
| | C75BL/6N; CD-1 | Wy-14,643, fenofibrate | Oligonucleotide | 12,000 | Eisai | Yamazaki et al.[43] |

| | | | | | | |
|---|---|---|---|---|---|---|
| | WT, CAR -null | Phenobarbital | cDNA | 8736 | Laboratory of Reproductive and Developmental Toxicology, NIEHS | Ueda et al.[61] |
| | CD-1: TTR-HFH-11 (tg)/wt | Carbon tetrachloride | cDNA | 1200 | University of Illinois (Chicago) | Wang et al.[22] |
| | C57BL/6J, male | Aroclor 1260, beta-naphthoflavone (BNF), ciprofibrate, cobalt chloride, TCDD, IL-6, LPS, PCB-153, phenobarbital, phenylhydrazine, TNF-alpha, Wy-16,463 | cDNA | 1200 | University of Wisconsin; Aeomica; University of Helsinki; Northwestern University (Chicago); Molecular Dynamics | Thomas et al.[41] |
| Kidney | MT-null mice 129-Mt1tm/Bri, Mt2tm/Bri 129/SvPCJ background and MT-null cells; WT | Lead | cDNA | 1178 | NCI, NIEHS | Qu et al.[113] |
| Testes | 129/Sv mice, male | Cadmium | cDNA | 588 | University of Kansas | Liu et al.[104] |
| Uterus, ovary | Female | Estradiol 17-beta/2.2-bis(p-hydroxy-phenyl)-1.1.1-trichloroethane (HPTE), flutamide | cDNA | 588 | CIIT | Waters et al.[87] |
| Uterus | Female | Tamoxifen, toremifene, raloxifen, estradiol | cDNA | 1176 | Medical Research Council Toxicology Unit (UK) | Parrott et al.[86] |
| Dopamine neurons, ventral midbrain | B6D2F1xB6C3F1, transgenic (TH-lacZ) | Methamphetamine | cDNA | 288 mouse/ 1100 human | NIA, NIDA, NIH; Johns Hopkins School of Medicine (Baltimore) | Barrett et al.[4] |
| Frontal cortex | CD-1, male | Methamphetamine | cDNA | 588 | NIH/NIDA | Cadet et al.[1] |
| | | Methamphetamine | cDNA | 588 | NIH/NIDA | Jayanthi et al.[5] |
| Striatum | c-Fos knockout/wt | Methamphetamine | cDNA | 1176 | NIH/NIDA | Cadet et al.[3] |
| Ventral midbrain | Swiss-Webster, albino, male | Methamphetamine | cDNA | ~1600 | Johns Hopkins University (Baltimore); NIA/NIH | Xie et al.[6] |
| Lymph nodes | Balb/c, female | Toluene diisocyanate, oxazolone, nonanoic acid | Oligonucleotide | 6519 | NIOSH | He et al.[23] |
| Three tumor cell lines | Nude mice, injected with transformed BALB/c-3T3 cells | Beryllium sulfate | cDNA | 1176 | NIOHS | Joseph et al.[24] |

**TABLE 10.3**
**Overview of DNA Array-Based Toxicogenomic Studies in Humans**

| Tissue/Cells | Compounds Analyzed | Type of Array | Number of Probes | Organization/Company | Ref. |
|---|---|---|---|---|---|
| Liver (biopsies) | Arsenic | cDNA | 588 | Guiyang Medical College Hospital; NCI; University of North Carolina; Southwest Endemic Prevention Station | Lu et al.[106] |
| Primary hepatocytes | Rifampin | cDNA/oligo-nucleotide | >200/>1000 | Georgetown University | Rae et al.[25] |
| Hepatoma cell line HepG2 | 100 compounds, data shown for cisplatin, transplatin, diflunisal, flufenamic acid | cDNA | 250 | R.W. Johnson PRI | Burczynski et al.[162] |
| | 2,3,7,8-tetrachlorodibenzo-*p*-dioxin (TCDD) | cDNA | 12,412 | Stanford University | Frueh et al.[68] |
| | | cDNA | 9182 | University of Cincinnati | Puga et al.[67] |
| | Ethanol, carbon tetrachloride | cDNA | 588 | Unilever | Harries et al.[158] |
| | Diethyl malate (DEM) | cDNA | 767 | GlaxoSmithKline | Casey et al.[26] |
| | Acetaminophen, caffeine, thioacetamide+C16 | cDNA | ~5000 | Affymetrix Research Institute | Gore et al.[10] |
| 60 cancer cell lines | 1400 compounds | cDNA | 8000 | NIH/NCI | Scherf et al.[135] |
| | Expression profiles before drug treatment, selection for chemosensitivity | Oligonucleotide | 6817 | NIH/NCI; MIT; Dana-Farber Cancer Institute; Harvard Medical School | Staunton et al.[137] |
| 39 cancer cell lines | Correlation of expression profiles with chemosensitivity profiles | cDNA | 9216 | Japanese Foundation for Cancer Research; University of Tokyo; RIKEN | Dan et al.[136] |
| MERAkt cells, MERAkt/NIH3T3 cells | Tamoxifen, raloxifen, ICI-182780, ZK955 | cDNA | | Berlex Biosciences | Kuhn et al.[27] |
| ZR75-1 breast cancer cells | Estradiol 17-beta, 4-hydroxytamoxifen, raloxifen, faslodex | Oligonucleotide | 5600 | Imperial Cancer Research Fund (UK) | Soulez and Parker[93] |
| MCF7 breast cancer cell line | Tamoxifen, raloxifen, ICI-182780, ZK955 | cDNA | 1901 | NIEHS | Lobenhofer et al.[90] |
| | Tamoxifen, estradiol 17-beta | cDNA | 23,000/44,000 | University of Kentucky; University of California/Los Angeles; Research Genetics; Applied Genomics; Stanford University; University of North Carolina | Finlin et al.[91] |
| | 4 estrogenic compounds | cDNA | 600 | Syngenta | Pennie et al.[92] |
| MDA-MB-231 cancer cell line | 15dPGJ2 | cDNA | 1176 | Wake Forest University Baptist Medical Center (North Carolina) | Clay et al.[58] |

| IMR-32 neuroblastoma cells | Tert-butylhydroquinone/LY294002 | Oligonucleotide | 12,000 | University of Wisconsin | Li et al.[28] |
|---|---|---|---|---|---|
| Astrocytes, non-small cell lung carcinoma PC-14 cells | TZT-1027 | cDNA | 588 | National Cancer Center (Tokyo) | Natsume et al.[29] |
| HT-29 colon cancer cells | Troglitazone, 15dPGJ2 | cDNA | 205 | Dokkyo University | Shimada et al.[56] |
| Colon carcinoma cell line MOSER S (M-S) | (DMSO), GW7845/rosiglitazone | Oligonucleotide | 5600/5184 | Vanderbilt University; Research Genetics; Veterans Affairs Medical Center (Nashville); Harvard Medical School; GlaxoSmithKline | Gupta et al.[57] |
| Mononuclear cells (hMNCs) | Amphotericin B | cDNA | 588 | University of Mississippi | Cleary et al.[30] |
| Coronary artery endothelial cells | Nicotine | cDNA | 4000 | University of Southampton | Zhang et al.[31] |
| EpiDerm™ skin model | Sodium lauryl sulphate | cDNA | ~3600 | Unilever; Center for Cutaneous Research; The Royal School of Medicine and Dentistry (UK) | Fletcher et al.[33] |
| Papillomavirus-immortalized bronchial epithelial cell line HBE1 | Smoke, hydrogen peroxide | cDNA | 9600 | University of California (Davis), National Taiwan University Hospital | Yoneda et al.[32] |
| HeLa cells | Cadmium | cDNA | 7075 | National Institute of Industrial Health (Japan) | Yamada and Koizumi[103] |
| Astrocytes | Lead | cDNA | 588 | Johns Hopkins University School of Medicine; Kennedy Krieger Research Institute | Hossain et al.[114] |
| HL-60 promonocytes | Iron | cDNA | 43 | University of Texas; South Texas Veterans Health Care System | Alcantara et al.[115] |
| Human papillomavirus (HPV18)-immortalized human bronchial epithelial (BEP2D) cells | Asbestos | cDNA | 588 | Columbia University | Zhao et al.[16] |
| Non-small cell lung carcinoma (NSCLC) SQ-5 | Troglitazone | cDNA | 160 | Gunma University | Satoh et al.[55] |
| Resting and proliferating peripheral blood lymphocytes from leukemia patients | Ionizing radiation | cDNA | 70 | German Cancer Research Center; Axaron Bioscience AG | Mayer et al.[131] |
| Panel of 12 human cancer cell lines, including MCF-7 and the myeloid cell line ML-1 | Ionizing radiation | cDNA | ~1300 | NIH/NCI; NHGRI | Amundson et al.[129] |
| Peripheral blood lymphocytes | Ionizing radiation | cDNA | 6728 | NIH/NCI; NHGRI | Amundson et al.[130] |
| TK6 lymphoblastoid cell line | Ionizing radiation | cDNA | 588/277 | Radiation Protection Bureau | Ford et al.[128] |
| Fibroblasts (HFW), derived from newborn foreskin | Sodium arsenite | cDNA | 568 | Academia Sinica; National Yang-Ming University Taipei | Yih et al.[111] |

**TABLE 10.4**
**Overview of Open Methods-Based Toxicogenomic Studies**

| Type of Method | Species | Strain, Gender | Tissue/Cells | Compounds Analyzed | Organization or Company | Ref. |
|---|---|---|---|---|---|---|
| Differential display | Human | *In vitro* | MRC-9, IMR-90 normal human embryonic lung cells; SBC-2, EBC-1 lung cancer cell line | NiO, Ni3S2 | Beijing Medical University; Toyama Medical and Pharmaceutical University; McMaster University | Mao et al.[126] |
| | Human, rat | *In vitro*; Fischer 344, female | Hepatocytes, human HepG2 liver cell line + rat liver | Ethinyl estradiol | Johns Hopkins School of Hygiene and Public Health | Chen et al.[98]; Chen et al.[99] |
| | Mouse | B6C3F1, female | Liver | Chloroform | CIIT | Kegelmeyer et al.[159] |
| | | | Liver | Dichloroacetic acid | NCTR | Thai et al.[50] |
| | | | Liver | Dimethylnitrosamine (DMN) | University of Minnesota | Bhattacharjee et al.[160] |
| | | C57B6, male | Liver | Di(2-ethylhexyl) phthalate (DEHP) | University of California; Riverside | Muhlenkamp et al.[49] |
| | | C57B1/10J, male | Liver | Phenobarbital | AstraZeneca | Garcia-Allan et al.[62] |
| | Mouse, rat | SV129, male; F344, male | Liver | Wy-14,643, phenobarbital/diethylnitrosamine (DEN/PB), di(2-ethylhexyl)phthalate (DEHP), di-n-butyl phthalate (DBP), gemfibrozil, PPAR-alpha null phenotype | CIIT; North Carolina State University | Anderson et al.[48] |
| | Rat | Wistar, male | Liver | Phenobarbital, Wy-14,643 | University of Surrey; Rhone-Poulenc | Rockett et al.[46] |

| | | | | | | |
|---|---|---|---|---|---|---|
| Differential display, representational difference analysis (RDA), SSH | Rat | F344, male | Rat liver + hepatocytes | Aflatoxin B1 | NCTR | Harris et al.[161] |
| GeneCalling® | Rat | Sprague–Dawley, male | Liver | GW9578 | CuraGen Corporation | Gould Rothberg et al.[51] |
| SSH | Human | *In vitro* | MCF-7 breast cancer cell line | Estradiol 17-beta | Stanford Universtiy | Ghosh et al.[96] |
| | | *In vitro* | MCF-7 breast cancer cell line | Estradiol 17-beta | A&M University | Chen et al.[94] |
| | Rat | *In vitro* | 5L hepatoma cell line | 2,3,7,8-tetrachlorodibenzo-p-dioxin (TCDD) | Forschungszentrum Karlsruhe | Kolluri et al.[70] |
| | | Wistar | Pleural mesothelium, mesothelial cell lines RZ 328, 44R.M.-4 | Crocidolite asbestos | Heinrich Heine University | Sandhu et al.[15] |
| | Rat, guinea pig | Wistar, male; Duncan-Hartley, male | Liver | Wy-14,643 | University of Surrey | Rockett et al.[47] |
| SABRE | Mouse | SV129 | Liver | Wy-14,643 (PPAR-alpha null phenotype) | University of Lausanne; Pfizer Global Research and Development; NCI | Kersten et al.[45] |
| SAGE | Human | *In vitro* | Breast cancer cell line | Estradiol 17-beta | University of Texas | Charpentier et al.[95] |
| | Mouse | C57B6, female | Liver | 2,3,7,8-tetrachlorodibenzo-p-dioxin (TCDD) | University of Kanazawa; University of Tokyo; National Institute for Environmental Studies Japan | Kurachi et al.[73] |
| EST quantitation | Human | *In vitro* | Prostate carcinoma cell line LNCaP | Synthetic androgen R1881 | Fred Hutchinson Cancer Research Center | Clegg et al.[100] |

PPARα is predominantly expressed in liver and is activated by cellular long-chain fatty acid derivatives, thus stimulating β-oxidation of fatty acids in peroxisomes. This receptor also mediates the effects of fibrates, including gemfibrozil and fenofibrate, a class of synthetic PPARα agonists that act as potent hypolipidemic drugs in the treatment of cardiovascular diseases.[35] By induction of the peroxisomal β-oxidation system of fatty acids in rodents, fibrates and other peroxisome proliferators such as fatty acids, certain plasticizers, and herbicides cause proliferation of peroxisomes, hepatomegaly, and eventually hepatocarcinoma.[36–38]

A number of studies have applied microarrays to elucidate the molecular basis for the pleiotropic effects of these nongenotoxic hepatocarcinogens in the rodent liver. Several reports have been published based on the rat liver as a model system. Hamadeh et al.[39,40] selected three peroxisome proliferators for analysis: Wy-14,643, the most prominent model ligand and a strong synthetic agonist for PPARα; clofibrate; and gemfibrozil, as well as the classical enzyme inducer phenobarbital. Their report[39] provides a detailed analysis of the findings, including a description of interesting molecular changes and pathway relationships associated with peroxisome proliferator exposure. In addtion, pattern recognition approaches, including hierarchical clustering, principal component analysis (PCA), and pairwise comparisons, were utilized that made it possible to distinguish between the different compounds and classes. In a related study,[40] these experiments provided a training set for the prediction of blinded samples, which had been treated with either a peroxisome proliferator (diethylhexyl phthalate [DEHP]) or enzyme inducers (phenytoin, hexobarbital). A similar study of 12 substances from five functional classes included two peroxisome proliferators, ciprofibrate and Wy-14,643, to identify gene sets predictive for their respective classes.[41] Another group reported on the effects of Wy-14,643 and three differently acting compounds and performed a careful mechanistic interpretation of array data to characterize compound-specific profiles.[42]

Other studies have used the mouse liver as a model system. Recently, Yamazaki et al.[43] analyzed the effects of Wy-14,643 and fenofibrate, a relatively weak PPARα agonist. Besides an expected regulation of genes involved in β-oxidation by both compounds, differences were also established; for instance, metallothionein 1 and 2 mRNA levels were regulated exclusively by Wy-14,643. In addition, a number of new target genes regulated by both agonists were identified, among these serum amyloid A (SAA) 2, for which expression was markedly decreased. Pretreatment of mice with fenofibrate inhibited acute-phase elevation of this gene, indicating that fenofibrate might reduce the plasma SAA concentration in patients with secondary amyloidosis. Cherkaoui-Malki et al.[44] also tested for alterations in gene expression induced by Wy-14,643 and identified approximately 700 genes with a more than fourfold deviation from their basal expression level, most of which had not been previously described in this context. Because the expression of a subset of this group of genes was unchanged in PPARα null mice, these functional genomic studies strongly suggest that the regulation of those genes is dependent on the action of PPARα. Kersten et al.[45] used an oligonucleotide array comprised of 6500 genes to screen for PPARα target genes combined with Northern blot and subtractive hybridization (selective amplification via biotin and restriction-mediated enrichment [SABRE]). Direct activation of PPARα through WY-14,643 led to a decrease of numerous genes involved in amino acid metabolism. Similar effects were also observed by comparison of fasted PPARα wild-type vs. null mice. These results indicate a key role of PPARα in the control of intermediary metabolism during fasting. Unfortunately, differences in study design and the heterogeneous amount and quality of information provided about the experimental data make it difficult to compare the resulting gene lists of the various studies.

In addition to DNA arrays, open transcription profiling technologies have also been used to resolve mechanistic questions on the mode of toxicity of peroxisome proliferators. Two papers by Rockett et al.[46,47] describe the use of differential display and suppression subtractive hybridization (SSH) to discover genes involved in peroxisome-proliferator-induced nongenotoxic carcinogenesis in rats. In the earlier report, they compared the response of a peroxisome proliferator to another tumor-promoting agent of a non-PP type (Wy-14,643 vs. phenobarbital) similar to the array-based study of Hamadeh et al.[39] In the second study, they employed a similar strategy to distinguish between carcinogenesis-specific and -unspecific effects of Wy-14,643 by comparing species with different susceptibilities toward peroxisome-proliferator-induced hepatocarcinogenesis (rat vs. guinea pig). Some of the genes identified in these studies were known targets, while others had not previously been shown to be inducible by peroxisome proliferators; the latter group of genes belonged to several cellular pathways (e.g., xenobiotic or amino acid metabolism). Generally, the comparison to guinea pig proved difficult because the genome is poorly characterized and thus only a few genes could be assigned to genes of known function.

Using differential display as well, Anderson et al.[48] compared hepatic adenomas with adjacent non-tumor liver tissue from rats fed with Wy-14,643 for 78 weeks and found an increase of specific acute-phase proteins in the tumors. This observation was used as a starting point to analyze the expression of these and other acute-phase protein genes by standard molecular biology techniques employing various treatment regimens and animal models to identify peroxisome-proliferator-specific effects. The authors observed a disregulation of hepatic acute-phase protein gene expression after PPARα activation by two other peroxisome proliferators in both rats and mice. This may indicate peroxisome-proliferator-induced changes in cytokine signaling networks that regulate both acute-phase protein gene expression and hepatocellular proliferation.

Two further studies using differential display examined the effects of other peroxisome-proliferating tumor promoters. Muhlenkamp and Gill[49] isolated potentially regulated clones that were induced after DEHP treatment in mouse liver. They focused on GRP58, a gene belonging to a family of stress-regulated molecular chaperones that is not inducible by clofibrate. In another study, the effects of dichloroacetic acid (DCA) on mouse liver were studied.[50] Six genes were found to be regulated, most of them involved in fatty acid metabolism.

Gould Rothberg et al.[51] applied their differential-display-based GeneCalling® technology in a thorough analysis of the response of rats to a novel synthetic peroxisome-proliferator-activated PPARα ligand indicated for lipid disorders. Following treatment, 2.4% of the assayed 9000 rat liver transcripts appeared to be regulated more than 1.5-fold. From this set, 50 distinctly modulated genes were selected and identified by sequencing. Most genes could be grouped into six discrete metabolic pathways. Of the 50 responsive genes, 26 were directly involved in fatty acid metabolism and therefore most likely related to the mode of action of the drug. Additionally, a number of genes were identified that did not fall into representative cellular pathways. Overall, this study was in agreement with previous findings published in the literature. The authors discussed possible implications of the genes identified regarding different mechanisms of drug action or toxicity, stressing the need for further follow-up studies to assess the relevance of the observed changes for drug response more thoroughly.

As mentioned earlier, most of the studies using either SSH or variations of differential display have suffered from high rates of false positives and, therefore, have resulted in relatively small numbers of truly regulated clones compared to the amount of effort invested. Additionally, lists of regulated genes from these types of studies are difficult to compare because they frequently show very little overlap. This may be due to the existing differences in experimental settings or the more problematic fact that the applied versions of differential display are quite often incapable of consistently isolating rare or weakly regulated transcripts. Therefore, studies using open methods such as SSH or differential display (DD) often only uncover a small fraction of those genes that have altered expression upon a toxic insult.

Another PPAR family member, PPARγ, is mainly expressed in adipose tissue and plays a crucial role as a transcription factor in adipocyte differentiation. While PPARγ regulates genes involved in lipogenesis and peripheral glucose utilization,[35] it is also presumed to have diverse functions beyond energy homeostasis.[52] PPARγ ligands have shown antineoplastic activities *in vitro*[53,54] and thus may represent interesting targets for cancer therapy. In particular, activation of PPARγ by the thiazolidinedione (TZD) class of antidiabetic drugs elicits growth inhibition in a variety of malignant tumors *in vitro* and *in vivo*.

Several analyses have concentrated on these PPARγ-induced growth inhibitory and apoptotic effects.[55,56–58] In a study by Satoh et al.,[55] non-small-cell lung carcinoma cells were stimulated with the synthetic TZD troglitazone. By activation of PPARγ, the growth-arrest and DNA-damage-inducible 153 gene (GADD153) was found to be induced using a subtraction cloning assay as well as a cDNA array of 160 apoptosis-related genes. Among four potential PPARγ-activated apoptosis genes that were regulated more than twofold on the cDNA array, two were identical to genes identified with the subtraction cloning assay. Using a broader human cDNA array, Clay et al.[58] identified p21 and p27 as candidate genes critical for the apoptotic effects of a cyclopentenone prostaglandin derivative of arachidonic acid, 15dPGJ2. Like TZDs, 15dPGJ2 is known to activate PPARγ and possesses a potent antiproliferative activity. Shimada et al.[56] took advantage of another model system, human colon cancer cells, to characterize the apoptotic processes mediated by 15dPGJ2 and troglitazone using cDNA microarrays. Downregulation of c-*myc* and upregulation of the genes for c-*jun* and for GADD153 were found to be associated with apoptosis.

### 10.2.2 Enzyme Inducers, TCDD, and PCBs

Mediation of enzyme induction is one of the central themes in toxicology, as drug-metabolizing enzymes are mainly responsible for xenobiotic functionalization, activation, and/or detoxification. In addition, many of these enzymes are upregulated by their substrates. A variety of enzyme inducers, both those discussed below and in previous chapters, are among the best understood toxicants. These studies are relevant for the validation of the new transcriptomics-based technologies in the field of toxicology, because a number of inducer-regulated genes that have already been isolated via other methods thus represent positive controls in microarray-based experiments involving the inducers of interest. These novel approaches are expected to contribute to the resolution of some of the remaining questions concerning the complex mechanisms of action of these compounds.

One of the best investigated enzyme-inducing compounds is phenobarbital (PB), a common sleep aid. PB induces various genes encoding xenobiotic (drug, steroid)-metabolizing enzymes such as cytochrome P450s (CYPs) and transferases. The nuclear orphan constitutive androstane receptor (CAR) is implicated in mediating PB responses.[59,60] Upon PB exposure, CAR translocates to the nucleus and heterodimerizes with the retinoid X receptor (RXR), leading to transcriptional activation of *CYP2B* and various other genes. cDNA microarray analyses were employed to compare liver gene expression in CAR-null mice and CAR wild-type mice, resulting in a list of 138 PB-responsive genes, about half of which were under the control of CAR.[61] Because a number of genes such as *CYP4A10* and *CYP4A14* were basally elevated in CAR-null mice, CAR may act as a transcriptional repressor for those genes. This broad spectrum of CAR-regulated genes suggests that CAR possesses a diversity of previously unrecognized functions in the regulation of hepatic genes.

In the papers by Hamadeh et al.,[39,40] PB was included as a prototype enzyme inducer and was the only representative of this functional class in the study. The authors provide a very comprehensive pathway map in which PB-regulated genes are highlighted, and they make a distinction between genes previously known or unknown to be regulated by PB as well as genes differentially induced between the two different treatment periods (24 hr, 2 wk). A study by Garcia-Allan et al.[62] used differential display to screen for genes regulated in mouse liver after stimulation with PB. Besides six novel genes differentially expressed, they identified an amine N-sulfotransferase that was significantly induced exclusively by PB but not by five other rodent nongenotoxic hepatocarcinogens.

The toxic actions mediated by the aryl hydrocarbon receptor (AHR) pathway have been attracting the interest of researchers over decades.[63–65] The environmental contaminant 2,3,7,8-tetrachlorodibenzo-*p*-dioxin (TCDD, or dioxin) is the prototypical AHR ligand, representing a broad group of halogenated aromatic hydrocarbons (HAHs), including polychlorinated dibenzodioxins (PCDDs), dibenzofurans (PCDFs), and biphenyls (PCBs). The process of activation is similar to that of PB and CAR/RXR. Upon ligand binding in the cytosol, AHR translocates to the nucleus, where it heterodimerizes with the aryl hydrocarbon nuclear translocator (ARNT) to build a transcriptionally active complex that regulates genes of the so-called AHR gene battery, including a number of phase I and phase II drug-metabolizing enzymes. TCDD is known to induce a wide range of diverse toxic and biochemical responses in laboratory animals and in humans which are not exclusively mediated by AHR. TCDD also acts as a potent endocrine disruptor.[66]

The effects of TCDD have been investigated in a number of microarray-based studies. As discussed in Chapter 8 and reviewed briefly here, treatment of human HepG2 cells with 10 nM TCDD for 8 hr led to more than twofold gene expression changes in 310 genes and 400 ESTs out of approximately 9000 probes on a cDNA array.[67] In this study, the protein synthesis inhibitor cycloheximide (CX) was added to distinguish between primary effects — still detectable with CX treatment — and secondary effects, which require protein synthesis. The authors grouped regulated genes according to their cellular functions. In a second study making use of HepG2 cells, the same dosing of TCDD was used, but exposure was significantly longer (18 hr).[68] Using another gene set on a different type of array, the authors came up with a list of 112 TCDD-regulated genes. The conclusion drawn in both studies, however, remained the same: the observed transcriptional changes by TCDD are much more complex than expected.

Taking into account the known teratogenic, reproductive, and developmental effects of TCDD, Hurst et al.[69] investigated its effects after *in utero* exposure during female fetal development, combining cDNA array analysis with histological, morphological, and immunohistochemistry data. Array analysis provided information on the induction of several genes at gestation days 18 and 19 that may be involved in the TCDD-induced vaginal dysmorphogenesis.

A good example for the value of open transcription profiling technologies for mechanistic analysis of toxicity is the study of Kolluri et al.[70] who examined the effect of TCDD exposure on the rat hepatoma cell line 5L by SSH to identify new target genes of the Ah receptor. They found 17 dioxin-inducible cDNA clones, 11 of which had not previously been reported to be regulated upon TCDD treatment. Out of a group of four genuinely novel clones, they assigned one clone to the N-myristoyltransferase (NMT) 2 gene. Confirming the induction of the NMT2 gene *in vivo* and of myristoyltransferase activity *in vitro*, the authors were able make a connection between dioxin exposure and protein myristoylation. This corresponds with earlier findings of elevated NMT protein levels and activity in certain tumor types.[71,72] Thus, induction of NMT2 may play a role in TCDD-mediated carcinogenicity.

One of the few SAGE (serial analysis of gene expression) studies in the field of toxicogenomics was presented by Kurachi et al.[73] Approximately 56,000 tags were analyzed from normal and TCDD-treated mouse liver libraries. This corresponded to approximately 15,000 different transcripts in each sample and 24,000 different transcripts in total, which produced only about a 30% overlap between the two samples. In general, about 95% of all transcripts were detected less than 10 times (tag-count), demonstrating that most of the genes identified were expressed at low levels, similar to previous findings using SAGE[74] and MPSS™ studies performed at Axaron. Treatment with TCDD changed the expression levels of 346 transcripts, including 94 ESTs ($p < 0.05$), and revealed a substantially broader and more complex response than previously reported. The TCDD-responsive genes belonged to different functional classes, such as signal transduction, protein trafficking, transcription factors, metabolic enzymes, stress response, plasma proteins, and detoxification. The above-mentioned studies on TCDD-treated human HepG2 cells using microarrays[67,68] showed only little overlap with the SAGE data. This may be due to the different model systems used but makes it rather difficult to build reasonable hypotheses regarding the effects of TCDD using comparisons of these datasets.

β-Naphtoflavone (BNF), 3-methylcholanthrene (3-MC), and benzo[*a*]pyrene belong to the same class of planar, hydrophobic, AHR-activating compounds as TCDD. BNF induces the *CYP1A1* and *CYP1A2* genes. The time course of BNF induction in mouse liver was analyzed in detail in a study employing a comprehensive range of dosages and treatment intervals using a cDNA array comprised of 148 probes selected from different toxicologically relevant functional classes.[75] Enzyme induction by 18 different known *CYP1A1* inducers (e.g., BNF, PCB-52, lansoprazol) was also investigated using microarrays in the soil nematode *Caenorhabditis elegans,* which contains 80 cytochrome P450 genes.[76]

Enzyme inducers and AHR ligands were part of several DNA array-based studies including other compound classes.[40–42,77–79] Among other compounds, Thomas et al.[41] worked on the AHR ligands TCDD and BNF and the enzyme inducers PB, PCB-153, and Aroclor-1260, as well as other mechanistic classes. Class-specific expression profiles could be distinguished between AHR ligands and other enzyme inducers as well as compared to other functional classes. Aroclor-1260, a PCB mixture that contains primarily non-coplanar, PB-type PCBs and a proportion of MC-like PCBs, did not cluster with TCDD. According to the hierarchical clustering performed in the studies of Waring et al.,[36,37] 3-MC and Aroclor-1254 fall into one and the same cluster. This may be due to the MC-type fraction of the Aroclor-1254 mixture or because no further enzyme inducers of the PB- or MC-type were included in the study. It appears, therefore, that the number of compounds belonging to a certain toxicological endpoint included in a database can influence the resolution of subsequent predictions.

### 10.2.3 Hormones and Endocrine Disruptors

The endocrine system is involved in processes that guide development, growth, reproduction, and behavior. In the field of endocrine research, interest is mostly concentrated on steroid hormone-receptor-mediated pathways, in particular actions via the estrogen receptor (ER). ER is an important pharmaceutical target for hormone replacement therapy in menopausal women and the prevention of breast cancer.[80] Similar to CAR and AHR, ER is a ligand-activated receptor acting as a transcriptional activator of target genes, with estrogen (17-β estradiol, estradiol, $E_2$) being the natural ligand. Estrogens are proven human carcinogens that exert some of their effects by stimulating cell proliferation in target organs.[81] Major goals in current toxicogenomic studies are the investigation of ER target genes and determining the impacts of other xenobiotics on ER-mediated signaling. Transcription profiling is a particularly valuable tool in the field of steroid receptor signaling, as the primary effects of these pathways lead to altered expression of target genes.

Some xenobiotics (e.g., drugs, pesticides) interfere with the functions of the endocrine system and thus are referred to as ***endocrine disruptors***. Endocrine disruptors may act as ER agonists or antagonists or as selective ER modulators (SERMs) which can have either agonistic or antagonistic function, depending on the tissue.[82] This is the reason why some SERMs have beneficial and undesirable effects at the same time, depending on the tissue in which they exert their effects. For example, both tamoxifen and raloxifen function as antagonists in the breast and as agonists in bone, but differ in their effects in the endometrium, where tamoxifen is an agonist and raloxifen an antagonist.[83] In case of an agonistic action, the resulting proliferative effect might contribute to the development of endometrial cancers.[84]

Estradiol has been proven to exert reproductive and developmental effects in rodents, and a number of toxicogenomic studies have been perfomed in this area.[85–87] Using DNA microarrays, Waters et al.[87] characterized the gene expression profiles of $E_2$ and 2,2-*bis*(*p*-hydroxyphenyl)-1,1,1-trichloroethane (HPTE), a metabolite of methoxychlor (MCX) with estrogenic activity (xenoestrogen), as well as the effects of the antiandrogen flutamide in the uteri and ovaries of mice. HPTE is a selective ER-α agonist but possesses a much weaker binding capacity to ER-α than $E_2$. HPTE is also an ER-β and androgen receptor (AR) antagonist. The authors found overlapping expression profiles of the two estrogenic compounds, as well as gene expression changes linked exclusively to one or the other. In general, transcription profiles and selective binding properties of the compounds were found to be well correlated. Discrepancies occurred between microarray and reverse transcription–polymerase chain reaction (RT-PCR) data. It should be noted that, in contrast to RT-PCR, the DNA array analysis was performed with only one animal per treatment group, which might be the reason for the poor correlation between RT-PCR and DNA array results.

Immunohistochemical data would be helpful to further clarify the mechanism. Green et al.[85] compared the actions of $E_2$, tamoxifen, toremifene, and raloxifen in the uteri of ovariectomized rats. The study design combined the use of semiquantitative RT-PCR, cDNA microarrays, immunohistochemistry, and enzymatic assays. Fold decreases of ER mRNAs (ER-α, -β, and -β2) were found to be equivalent after treatment with all compounds, indicating that these effects are not the primary cause for the observed differences in efficacy of induction of ornithine decarboxylase (ODC) and creatine kinase brain type (CK-BB) by these compounds.

Another study analyzed the effects of estrogen and testosterone on prostate cancer.[88] These hormones play an important role in prostate cancer. Changes in the ratio of testosterone and estrogens with increasing age are one of the potential risk factors for prostate cancer in men — the most frequently diagnosed malignant cancer in the Western world.[89] Using the rat animal model developed by Noble and colleagues, the authors observed upregulation of several genes during sex-hormone-induced prostate carcinogenesis, including elevations in testosterone-repressed prostatic message-2 (TRPM-2), matrix metalloproteinase-7 (MMP-7), and inhibitor of differentiation (ID-1), as detected by microarray analysis and further confirmed by semiquantitative RT-PCR, western blotting, and immunohistochemical studies.

Several studies on human breast cancer cell lines have been conducted to characterize relations of estrogenic actions and breast cancer.[90–96] Using microarrays, Finlin et al.[91] identified a *ras*-like gene whose expression was induced rapidly in estrogen-stimulated MCF-7 cells and suppressed with tamoxifen.[91] Because the new gene, termed RERG (*ras*-related and estrogen-regulated growth inhibitor), was found to be expressed at a low level in a significant percentage of primary breast tumors with poor clinical prognosis, it is possible that downregulation of RERG may contribute to breast tumorigenesis. Based upon the same model system, changes in gene expression were monitored in another study with cDNA microarrays at six different time points ranging from 1 to 48 hr after estrogen stimulation of MCF-7 cells.[90] At time points with an increased number of cells in the S-phase, a considerable number of genes associated with DNA replication were found to be induced, further elucidating the mechanism of mitogenic activity of estrogen. In addition to known ER-responsive genes, Soulez and Parker[93] identified a considerable number of novel target genes in human ZR75-1 cells. The most pronounced effect was an over 100-fold increase in *CYP2B* expression over a 24-hr period. Interestingly, *CYP2B* was not detected in MCF-7 cells.

The three other studies on the same model system used open approaches to analyze the effects of estrogen on gene expression.[94–96] In two of the studies, SSH was the method of choice; in the context of the inhibitory effects of AHR agonists on the estrogen receptor pathway, 35 genes associated with this receptor cross-talk were identified as being induced by estrogen and inhibited by TCDD. Many of the regulated genes were involved in cell proliferation, possibly contributing to the inhibition of estrogen-induced growth by AHR agonists.[94] In the other SSH study, Ghosh et al.[96] isolated 14 estrogen-responsive genes, two of which (PDZK1 and GREB1) were also significantly over-expressed in ER-positive primary breast cancers, representing potential new biomarkers for estrogen-responsive tumors. An in-depth view of the gene expression changes associated with estrogen action was attempted by another SAGE study[95] in which the transcript levels of 12,550 genes were monitored. Only about 0.4% of those genes showed an induction of at least threefold following 3 hr of treatment, including five novel E2-inducible genes (E2IG1–E2IG5) as well as several interesting genes potentially involved in cell-cycle progression.

In order to screen for possible endocrine-disrupting substances in groundwater, Custodia et al.[97] developed a high-throughput assay based on a *Caenorhabditis elegans* DNA microarray. The authors exposed *C. elegans* to a number of vertebrate steroids and promote this model system as being a useful screening tool for identifying putative endocrine disruptors.

The mitosuppressive effect of ethinyl estradiol (EE), a potent hepatic tumor promoter in rodents, was investigated in the livers of female rats by Chen et al.[98,99] using differential display. Several of the differentially expressed transcripts were mitochondrial genome-encoded genes that were also inducible after treatment of human HepG2 cells with EE, $E_2$, and certain estradiol catechol metabolites. The induction of those genes was accompanied by increased mitochondrial superoxide production and therefore seems to reflect increased respiratory chain activity.

In a recent study, Clegg and colleagues[100] compared the transcriptomes of a prostate cancer cell line in the presence and absence of androgen stimulation using the technique of EST quantitation. From the specific cDNA libraries they generated a total of 4400 expressed sequence tags (ESTs) representing 2486 distinct transcripts, 336 of which were expressed in both populations. After androgen treatment, 21 genes were detected as being significantly up- or downregulated. However, when comparing their EST quantitation results to Northern blot, they could confirm regulation for only 6 out of 10 genes. In order to identify additional androgen-responsive genes, they compared their digital EST results to SAGE profiling experiments examining androgen-regulated gene expression in the same cell line, which they could obtain from the SAGEmap website at the NCBI, a publicly available database (http://www.ncbi.nlm.nih.gov/SAGE/). A thorough analysis of the two datasets as well as other previously published literature on androgen-responsive genes was performed. SAGE and the EST quantitation approach did detect only a subset of all genes previously reported to be androgen-responsive in humans, and in no case did statistically solid regulation events agree across all datasets, showing the limitations of digital expression profiling in cases of insufficient overall sampling size (i.e., number of tags generated). Theoretical and empirical data suggest that approximately 650,000 transcripts must be sampled in order to identify all but very rare mRNAs in a cell.[101] This number is currently reached (and exceeded) only by the MPSS™ technology, which allows the analysis of up to 1 million transcripts in parallel.

### 10.2.4 Metal-Induced Toxicity

Metals and metalloids cause a broad range of toxicological problems and many exert carcinogenic effects; consequently, these compounds have attracted the interest of toxicologists for many years. Human exposure is partly due to their pharmacological use (for instance, in the use of arsenic against infectious diseases) and is becoming increasingly important in modern environmental toxicology.[102] In the past 2 years, microarray analysis has also been applied to this challenging field.[78,79,103–115]

Cadmium (Cd) is a heavy metal, occurring in the environment, that exerts toxic effects on renal, hepatic, respiratory, skeletal, and reproductive systems.[116] Cadmium has been classified as a human carcinogen.[117] Bartosiewicz et al.[109] performed a combined analysis of three compounds, cadmium chloride, benzo[*a*]pyrene, and trichloroethylene, using a small array of 150 genes. Ten genes were induced significantly, in addition to a number of significantly repressed genes. The induction of methallothionein (MT) I and II genes and several of the heat shock and early-response genes in mouse liver were consistent with previous results reported in the literature.

Authors studying the effects of acute cadmium exposure on stress-related gene expression in the liver of wild-type (wt) and MT-I/II-null (MT-null) mice observed a marked increase of mRNAs coding for heat shock and stress-response proteins.[105] Among the genes found to be suppressed were cytochrome P450s (CYPs), UDP-glucuronosyltransferases (UGTs), and manganese (Mn) superoxide dismutase. MT-null mice proved to be more sensitive to Cd-induced stress-related gene expression compared to wt mice, indicating a crucial role for MT. MT is also involved in lead-induced toxicity and, in analogy to its role in inhibiting cadmium-induced toxicity, MT-null mice show an increased sensitivity toward lead exposure.[113] In addition to microarray analysis, free radical production was analyzed in the study of Liu et al.[105] in order to quantify the well-known effects of Cd-induced oxidative stress. Although the mechanism is still unknown, the weak redox potential of Cd supports the hypothesis that indirect effects of Cd may ultimately be responsible for oxidative stress. Expression analysis of Cd-treated samples revealed a suppression of Mn-superoxide dismutase, supporting the hypothesis that Cd may disturb expression of components in the antioxidant defense system. Marked induction of *hsp60* and DT-diaphorase, phospholipase A2, and *CYP3A25* all further support a possible involvement of mitochondrial disruption and activation of oxidases contributing to oxidative stress. Similarly, inductions of stress response genes (including those for methallothionein and other antioxidant genes) were observed in Cd-treated human HeLa cells.[103] The mentioned studies provide evidence for the existence of well-characterized, reproducible transcriptional effects resulting from acute Cd exposure. Those effects were unrelated to Cd-induced testicular injury observed in mice.[104]

The metalloid arsenic is another important toxic agent that has been under toxicogenomic investigation. Arsenic is an environmental toxicant with a proven carcinogenic potential in humans, causing cancers of the skin, lung, urinary bladder, and liver.[118] A number of microarray-based studies are trying to unravel the molecular mechanisms of arsenic-induced injury which are still poorly understood.[79,80,106–108,110,111] To our knowledge, only one toxicogenomic study has reported on human *in vivo* exposure to arsenic.[106] The authors received liver biopsy samples from a Chinese population that had been exposed to arsenic for 6 to 10 years and showed accompanying degenerative liver lesions. RNA from six of the livers was compared to six normal livers (obtained from a U.S. hospital) by using a cDNA array comprised of 588 cancer-related genes. 60 genes were found to be differentially expressed, including genes involved in apoptosis, cell-cycle regulation, and DNA damage response. However, because ethnicity represents a confounding variable in these studies, revealing the origin of these transcriptional differences (arsenic exposure or simple ethnic diversity) is still not possible. With that caveat in mind. however, some of the differentially regulated genes in this study were also identified in a microarray analysis using rodent liver cells. Similar to the results in humans, dramatic increases in c-*myc*, PCNA, and cyclin D1 were observed in chronic arsenite-transformed rat liver TRL1215 cells.[110] In this study, 80 differentially expressed genes were identified, including stress-response genes such as *hsp70*, which was found to be regulated in several other studies (e.g., in a study using arsenic-treated human fibroblasts as model system)[111] and in a study testing for acute effects of arsenic in mice.[108] Comparison of acute and chronic or subchronic studies may enable the distinction of acute-phase and stress genes from genes specifically associated with long-term arsenic exposure. Again, an overall comparison of the results is difficult due to the use of different DNA arrays and nonstandardized study designs.

Arsenic was also among the 15 compounds included in the predictive study of Waring et al.[78] Depending on the clustering algorithm used, arsenic clustered together with various compounds, including methapyrilene. A common underlying mechanism between these compounds might lie in their common induction of periportal necrosis in the rat liver, which is one of several forms of hepatotoxicity exerted by arsenic.[119]

Epidemiological studies have revealed that exposure to certain nickel compounds is strongly correlated to the incidence of lung and nasal carcinomas.[120] In addition to the initiation of lipid peroxidation and reactive oxygen species formation,[121–122] nickel has been found to induce chromosomal aberrations and DNA–DNA and DNA–protein crosslinks,[123–125] all of which can lead to abnormal gene expression. In a study performed to elucidate the mechanism of nickel-induced cell transformation, Mao and colleagues[126] analyzed two nickel-transformed human embryonic lung cell lines by differential display. They detected very specific responses to individual nickel compounds: NiO stimulated the expression of a different set of genes than did $Ni_3S_2$. Two of the clones identified were also differentially regulated in certain lung tumor cell lines, pointing to a possible role for these genes in the process of lung carcinogenesis. One of these clones showed high homology to the Mena gene, whose protein product is probably involved in microfilament assembly.[127] It is thus formally possible that Ni-induced disregulation of this gene could disturb the process of chromosomal distribution during mitosis and thus contribute to some of the chromosomal aberrations observed after nickel exposure.

### 10.2.5 DNA Damage through Ionizing Radiation

Various studies have employed microarrays to investigate the mechanisms of DNA damage induced by ionizing irradiation in human cells.[128–131] Ionizing radiation randomly damages cellular components and induces a variety of DNA lesions,[132–133] thus inducing several cellular DNA repair mechanisms.[134] Because the ability of a cell to repair DNA damage may be dependent on the proliferative activity of the cell, an important question is whether quiescent or dividing cells should be studied. Two reports have addressed this problem.[130–131] In the study of Mayer et al.[131] in collaboration with our group, a DNA repair-focused microarray was used to analyze the expression profiles of resting and stimulated (proliferating) peripheral blood lymphocytes (PBLs). Gamma-radiation-induced DNA damage and repair were assessed by the Comet assay and compared to gene expression profiles in the PBLs. Differences in the extent of radiation-induced DNA damage or DNA repair capacity could not be observed among the gamma-irradiated resting and proliferating lymphocytes, whereas significantly elevated background damage was observed in stimulated cells. Twelve genes were found to be regulated in response to the mitogenic stimulus, and most of these are indeed involved in DNA replication. Amundson et al.[130] also searched for potential markers of radiation exposure in peripheral blood lymphocytes and demonstrated that stimulation did not appear to enhance the expression of DNA damage markers compared to irradiated quiescent cells. An induction of several markers, including DNA-damage-binding protein 2 (DDB2), Xeroderma pigmentosum group C complementing protein, XPC), and cyclin-dependent kinase inhibitor 1A (CDKN1A), could be reproduced in *ex vivo*-irradiated PBLs from multiple independent donors. DDB2 encodes for a subunit of DDB, a protein known to be involved in ultraviolet repair. In a previous study, the effects of ionizing radiation on the human myeloid cell line ML-1 were investigated, resulting in the identification of 48 differentially regulated genes.[129] The expression of a restricted number of these putative marker genes was systematically examined in a panel of 12 different human cell lines, revealing significant differences in expression between the cell lines investigated.

Because inter-individual differences in response to radiation are well known, it would be very helpful to have a screening tool at hand to easily predict the susceptibility of individuals to radiation. In order to take a first step toward the selection of suitable markers of radiation response, Ford et al.[128] irradiated lymphoblastoid cells with x-rays and identified 19 regulated genes, with changes ranging from 10-fold repression to a 12-fold induction. Some of the radiation-responsive genes (e.g., serotonin receptor, growth inhibitory factor) encoded for proteins for which expression had not been previously reported in lymphocytes. The 19 putative marker genes for radiation response may aid in further attempts to establish individual risk assessment on the basis of gene expression changes.

### 10.2.6 Chemotherapeutic Agents

Quite often no clear line can be drawn between adverse and beneficial or physiological, pharmacological, and toxic effects of drugs. This is especially true for the cytotoxic effects of compounds used in anticancer therapy. On the one hand, the cytotoxic effects on cancer cells are desired; on the other hand, adverse effects on normal cells are of major concern. Three microarray-based studies provide an extensive analysis of a larger number of human cancer cell lines in an effort to correlate chemosensitivity profiles with gene expression profiles of the unexposed cell lines. The analysis of direct correlations between drug action and gene expression is not provided.[135–137] Two of these studies[135,137] are based on the same panel of 60 U.S. National Cancer Institute (NCI) cancer cell lines that have been used by the NCI Developmental Therapeutics Program.[138–141] In both studies, transcription profiles of cell lines tend to cluster according to their tissue of origin, instead of clustering according to their chemosensitivity class. Applying a combination of data analysis approaches including a leave-one-out cross-validation procedure, training sets were used to generate tissue-independent, gene-expression-based classifiers for each of the 232 compounds tested. The predictive power of these classifiers was significantly higher than would be expected by chance alone.[137] A third study used a smaller number of 39 human cancer cell lines, 9 of which are not included in the NCI panel.[136]

The aim of these laborious studies was to improve the development of individualized cancer chemotherapies. For a first level of analysis it seems reasonable to examine gene expression and drug sensitivity relationships separately. The studies provide evidence that it is possible to screen samples for genetic determinants of differences in the sensitivity toward drugs, representing an important advance toward individualized treatment regimens for patients with different types of cancers. While putative marker genes for chemosensitivity have been identified in these studies, additional analyses of treated cells are necessary to provide further insights into the exact mode of action of these and other anticancer drugs. The number of transcription profiling studies dealing with these questions is too large for adequate representation in this chapter; however, we included a selection of studies in the references.[142–157]

## 10.3 SUMMARY

The transcriptomics-based studies surveyed here were conducted to improve our knowledge of toxicological responses. They have already generated an enormous amount of data, much of which still remains to be exploited. It is anticipated that in the coming years more rigorous standards of microarray data generation and analysis adopted by multiple laboratories will begin to more accurately define the true sets of genes induced by various toxicants.

By now, transcription profiling can provide evidence in favor of, or in contradiction to, certain mechanistic hypotheses and can provide new hints on the mechanism of toxicity of a compound; however, toxicogenomic data alone are insufficient to deduce complete sequences of toxic events. The studies discussed in this chapter show that we still have only a few pieces of the puzzle at hand which serve as starting points for further analysis. Supplementary information derived from methods such as histopathology, immunohistochemistry, proteomics, or metabolic studies are likely to further improve the interpretation of gene expression data. Ultimately, these ever-increasingly complex mechanistic studies will result in a better understanding of the cellular and molecular effects of toxicants in target and nontarget tissues. It is likely that the sum total of results from basic mechanistic toxicogenomic studies such as those presented here will eventually result in practical advances for society, ranging from early detection diagnostic assays to better strategies for therapeutic intervention following toxicant exposures.

## REFERENCES


1. Cadet, J.L., Jayanthi, S., McCoy, M.T., Vawter, M., and Ladenheim, B., Temporal profiling of methamphetamine-induced changes in gene expression in the mouse brain: evidence from cDNA array, *Synapse*, 41(1), 40, 2001.
2. Cai, N.S., McCoy, M.T., and Cadet, J.L., Profile of genes showing increased expression following dopamine, and methamphetamine exposure in an immortalized neuronal cell line, *Restor. Neurol. Neurosci.*, 18(2–3), 57, 2001.
3. Cadet, J.L., McCoy, M.T., and Ladenheim, B., Distinct gene expression signatures in the striata of wild-type and heterozygous c-*fos* knockout mice following methamphetamine administration: evidence from cDNA array analyses, *Synapse*, 44(4), 211, 2002.
4. Barrett, T., Xie, T., Piao, Y., Dillon-Carter, O., Kargul, G.J., Lim, M.K., Chrest, F.J., Wersto, R., Rowley, D.L., Juhaszova, M., Zhou, L., Vawter, M.P., Becker, K.G., Cheadle, C., Wood, W.H., 3rd, McCann, U.D., Freed, W.J., Ko, M.S., Ricaurte, G.A., and Donovan, D.M., A murine dopamine neuron-specific cDNA library and microarray: increased COX1 expression during methamphetamine neurotoxicity, *Neurobiol. Dis.*, 8(5), 822, 2001.
5. Jayanthi, S., McCoy, M.T., Ladenheim, B., and Cadet, J.L., Methamphetamine causes coordinate regulation of SRC, Cas, Crk, and the Jun N-terminal kinase-Jun pathway, *Mol. Pharmacol.*, 61(5), 1124, 2002.

6. Xie, T., Tong, L., Barrett, T., Yuan, J., Hatzidimitriou, G., McCann, U.D., Becker, K.G., Donovan, D.M., and Ricaurte, G.A., Changes in gene expression linked to methamphetamine-induced dopaminergic neurotoxicity, *J. Neurosci.*, 22(1), 274, 2002.
7. Reilly, T.P., Bourdi, M., Brady, J.N., Pise-Masison, C.A., Radonovich, M.F., George, J.W., and Pohl, L.R., Expression profiling of acetaminophen liver toxicity in mice using microarray technology, *Biochem. Biophys. Res. Commun.*, 282(1), 321, 2001.
8. Reilly, T.P., Brady, J.N., Marchick, M.R., Bourdi, M., George, J.W., Radonovich, M.F., Pise-Masison, C.A., and Pohl, L.R., A protective role for cyclooxygenase-2 in drug-induced liver injury in mice, *Chem. Res. Toxicol.*, 14(12), 1620, 2001.
9. Ruepp, S.U., Tonge, R.P., Shaw, J., Wallis, N., and Pognan, F., Genomics and proteomics analysis of acetaminophen toxicity in mouse liver, *Toxicol. Sci.*, 65(1), 135, 2002.
10. Gore, M.A., Morshedi, M.M., and Reidhaar-Olson, J.F., Gene expression changes associated with cytotoxicity identified using cDNA arrays, *Funct. Integr. Genomics*, 1(2), 114, 2000.
11. Cunningham, M.J., Liang, S., Fuhrman, S., Seilhamer, J.J., and Somogyi, R., Gene expression microarray data analysis for toxicology profiling, *Ann. N.Y. Acad. Sci.*, 919, 52, 2000.
12. Reynolds, L.J. and Richards, R.J., Can toxicogenomics provide information on the bioreactivity of diesel exhaust particles?, *Toxicology*, 165(2–3), 145, 2001.
13. Sato, H., Sagai, M., Suzuki, K.T., and Aoki, Y., Identification, by cDNA microarray, of A-raf, and proliferating cell nuclear antigen as genes induced in rat lung by exposure to diesel exhaust, *Res. Commun Mol. Pathol. Pharmacol.*, 105(1–2), 77, 1999.
14. Koike, E., Hirano, S., Shimojo, N., and Kobayashi, T., cDNA microarray analysis of gene expression in rat alveolar macrophages in response to organic extract of diesel exhaust particles, *Toxicol. Sci.*, 67(2), 241, 2002.
15. Sandhu, H., Dehnen, W., Roller, M., Abel, J., and Unfried, K., mRNA expression patterns in different stages of asbestos-induced carcinogenesis in rats, *Carcinogenesis*, 21(5), 1023, 2000.
16. Zhao, Y.L., Piao, C.Q., Wu, L.J., Suzuki, M., and Hei, T.K., Differentially expressed genes in asbestos-induced tumorigenic human bronchial epithelial cells: implication for mechanism, *Carcinogenesis*, 21(11), 2005, 2000.
17. Baker, T.K., Carfagna, M.A., Gao, H., Dow, E.R., Li, Q., Searfoss, G.H., and Ryan, T.P., Temporal gene expression analysis of monolayer cultured rat hepatocytes, *Chem. Res. Toxicol.*, 14(9), 1218, 2001.
18. Yu, Q., He, M., Lee, N.H., and Liu, E.T., Identification of Myc-mediated death response pathways by microarray analysis, *J. Biol. Chem.*, 277(15), 13059, 2002.
19. Laabich, A., Li, G., and Cooper, N.G., Characterization of apoptosis-genes associated with NMDA mediated cell death in the adult rat retina, *Brain Res. Mol. Brain Res.*, 91(1–2), 34, 2001.
20. Shultz, V.D., Phillips, S., Sar, M., Foster, P.M., and Gaido, K.W., Altered gene profiles in fetal rat testes after *in utero* exposure to di(*n*-butyl) phthalate, *Toxicol. Sci.*, 64(2), 233, 2001.
21. Liu, J., Saavedra, J.E., Lu, T., Song, J.G., Clark, J., Waalkes, M.P., and Keefer, L.K., O(2)-vinyl 1-(pyrrolidin-1-yl)diazen-1-ium-1,2-diolate protection against D-galactosamine/endotoxin-induced hepatotoxicity in mice: genomic analysis using microarrays, *J. Pharmacol. Exp. Ther.*, 300(1), 18, 2002.
22. Wang, X., Hung, N.J., and Costa, R.H., Earlier expression of the transcription factor HFH-11B diminishes induction of p21(CIP1/WAF1) levels and accelerates mouse hepatocyte entry into S-phase following carbon tetrachloride liver injury, *Hepatology*, 33(6), 1404, 2001.
23. He, L.Z., Tolentino, T., Grayson, P., Zhong, S., Warrell, R.P., Jr., Rifkind, R.A., Marks, P.A., Richon, V.M., and Pandolfi, P.P., Histone deacetylase inhibitors induce remission in transgenic models of therapy-resistant acute promyelocytic leukemia, *J. Clin. Invest.*, 108(9), 1321, 2001.
24. Joseph, P., Muchnok, T., and Ong, T., Gene expression profile in BALB/c-3T3 cells transformed with beryllium sulfate, *Mol. Carcinog.*, 32(1), 28, 2001.
25. Rae, J.M., Johnson, M.D., Lippman, M.E., and Flockhart, D.A., Rifampin is a selective, pleiotropic inducer of drug metabolism genes in human hepatocytes: studies with cDNA and oligonucleotide expression arrays, *J. Pharmacol. Exp. Ther.*, 299(3), 849, 2001.
26. Casey, W., Anderson, S., Fox, T., Dold, K., Colton, H., and Morgan, K., Transcriptional and physiological responses of HepG2 cells exposed to diethyl maleate: time course analysis, *Physiol. Genomics*, 8(2), 115, 2002.

27. Kuhn, I., Bartholdi, M.F., Salamon, H., Feldman, R.I., Roth, R.A., and Johnson, P.H., Identification of AKT-regulated genes in inducible MERAkt cells, *Physiol. Genomics*, 7(2), 105, 2001.
28. Li, J., Lee, J.M., and Johnson, J.A., Microarray analysis reveals an antioxidant responsive element-driven gene set involved in conferring protection from an oxidative stress-induced apoptosis in IMR-32 cells, *J. Biol. Chem.*, 277(1), 388, 2002.
29. Natsume, T., Nakamura, T., Koh, Y., Kobayashi, M., Saijo, N., and Nishio, K., Gene expression profiling of exposure to TZT-1027, a novel microtubule-interfering agent, in non-small cell lung cancer PC-14 cells and astrocytes, *Invest. New Drugs*, 19(4), 293, 2001.
30. Cleary, J.D., Rogers, P.D., and Chapman, S.W., Differential transcription factor expression in human mononuclear cells in response to amphotericin B: identification with complementary DNA microarray technology, *Pharmacotherapy*, 21(9), 1046, 2001.
31. Zhang, S., Day, I.N., and Ye, S., Microarray analysis of nicotine-induced changes in gene expression in endothelial cells, *Physiol. Genomics*, 5(4), 187, 2001.
32. Yoneda, K., Peck, K., Chang, M.M., Chmiel, K., Sher, Y.P., Chen, J., Yang, P.C., Chen, Y., and Wu, R., Development of high-density DNA microarray membrane for profiling smoke- and hydrogen peroxide-induced genes in a human bronchial epithelial cell line, *Am. J. Respir. Crit. Care Med.*, 164(10, pt. 2), S85, 2001.
33. Fletcher, S.T., Baker, V.A., Fentem, J.H., Basketter, D.A., and Kelsell, D.P., Gene expression analysis of EpiDerm following exposure to SLS using cDNA microarrays, *Toxicol. In Vitro*, 15(4–5), 393, 2001.
34. Clarke, S.D., Thuillier, P., Baillie, R.A., and Sha, X., Peroxisome proliferator-activated receptors: a family of lipid-activated transcription factors, *Am. J. Clin. Nutr.*, 70(4), 566, 1999.
35. Kersten, S., Desvergne, B., and Wahli, W., Roles of PPARs in health, and disease, *Nature*, 405(6785), 421, 2000.
36. Lock, E.A., Mitchell, A.M., and Elcombe, C.R., Biochemical mechanisms of induction of hepatic peroxisome proliferation, *Annu. Rev. Pharmacol. Toxicol.*, 29, 145, 1989.
37. van den Bosch, H., Schutgens, R.B., Wanders, R.J., and Tager, J.M., Biochemistry of peroxisomes, *Annu. Rev. Biochem.*, 61, 157, 1992.
38. Reddy, J.K., Azarnoff, D.L., and Hignite, C.E., Hypolipidaemic hepatic peroxisome proliferators form a novel class of chemical carcinogens, *Nature*, 283(5745), 397, 1980.
39. Hamadeh, H.K., Bushel, P.R., Jayadev, S., Martin, K., DiSorbo, O., Sieber, S., Bennett, L., Tennant, R., Stoll, R., Barrett, J.C., Blanchard, K., Paules, R.S., and Afshari, C.A., Gene expression analysis reveals chemical-specific profiles, *Toxicol. Sci.*, 67(2), 219, 2002.
40. Hamadeh, H.K., Bushel, P.R., Jayadev, S., DiSorbo, O., Bennett, L., Li, L., Tennant, R., Stoll, R., Barrett, J.C., Paules, R.S., Blanchard, K., and Afshari, C.A., Prediction of compound signature using high density gene expression profiling, *Toxicol. Sci.*, 67(2), 232, 2002.
41. Thomas, R.S., Rank, D.R., Penn, S.G., Zastrow, G.M., Hayes, K.R., Pande, K., Glover, E., Silander, T., Craven, M.W., Reddy, J.K., Jovanovich, S.B., and Bradfield, C.A., Identification of toxicologically predictive gene sets using cDNA microarrays, *Mol. Pharmacol.*, 60(6), 1189, 2001.
42. Gerhold, D., Lu, M., Xu, J., Austin, C., Caskey, C.T., and Rushmore, T., Monitoring expression of genes involved in drug metabolism and toxicology using DNA microarrays, *Physiol. Genomics*, 5(4), 161, 2001.
43. Yamazaki, K., Kuromitsu, J., and Tanaka, I., Microarray analysis of gene expression changes in mouse liver induced by peroxisome proliferator-activated receptor alpha agonists, *Biochem. Biophys. Res. Commun.*, 290(3), 1114, 2002.
44. Cherkaoui-Malki, M., Meyer, K., Cao, W.Q., Latruffe, N., Yeldandi, A.V., Rao, M.S., Bradfield, C.A., and Reddy, J.K., Identification of novel peroxisome proliferator-activated receptor alpha (PPARalpha) target genes in mouse liver using cDNA microarray analysis, *Gene Expr.*, 9(6), 291, 2001.
45. Kersten, S., Mandard, S., Escher, P., Gonzalez, F.J., Tafuri, S., Desvergne, B., and Wahli, W., The peroxisome proliferator-activated receptor alpha regulates amino acid metabolism, *FASEB J.*, 15(11), 1971, 2001.
46. Rockett, J.C., Esdaile, D.J., and Gibson, G.G., Molecular profiling of non-genotoxic hepatocarcinogenesis using differential display reverse transcription–polymerase chain reaction (ddRT-PCR), *Eur. J. Drug Metab. Pharmacokinet.*, 22(4), 329, 1997.

47. Rockett, J.C., Swales, K.E., Esdaile, D.J., and Gibson, G.G., Use of suppression-PCR subtractive hybridisation to identify genes that demonstrate altered expression in male rat and guinea pig livers following exposure to Wy-14,643, a peroxisome proliferator and non-genotoxic hepatocarcinogen, *Toxicology*, 144(1–3), 13, 2000.
48. Anderson, S.P., Cattley, R.C., and Corton, J.C., Hepatic expression of acute-phase protein genes during carcinogenesis induced by peroxisome proliferators, *Mol. Carcinog.*, 26(4), 226, 1999.
49. Muhlenkamp, C.R. and Gill, S.S., A glucose-regulated protein, GRP58, is down-regulated in C57B6 mouse liver after diethylhexyl phthalate exposure, *Toxicol. Appl. Pharmacol.*, 148(1), 101, 1998.
50. Thai, S.F., Allen, J.W., DeAngelo, A.B., George, M.H., and Fuscoe, J.C., Detection of early gene expression changes by differential display in the livers of mice exposed to dichloroacetic acid, *Carcinogenesis*, 22(8), 1317, 2001.
51. Gould Rothberg, B.E., Sundseth, S.S., DiPippo, V.A., Brown, P.J., Winegar, D.A., Gottshalk, W.K., Shenoy, S.G., and Rothberg, J.M., The characterization of PPAR alpha ligand drug action in an *in vivo* model by comprehensive differential gene expression profiling, *Funct. Integr. Genomics*, 1(5), 294, 2001.
52. Kliewer, S.A. and Willson, T.M., The nuclear receptor PPARgamma — bigger than fat, *Curr. Opin. Genet. Dev.*, 8(5), 576, 1998.
53. Sarraf, P., Mueller, E., Jones, D., King, F.J., DeAngelo, D.J., Partridge, J.B., Holden, S.A., Chen, L.B., Singer, S., Fletcher, C., and Spiegelman, B.M., Differentiation and reversal of malignant changes in colon cancer through PPARgamma, *Nat. Med.*, 4(9), 1046, 1998.
54. Brockman, J.A., Gupta, R.A., and Dubois, R.N., Activation of PPARgamma leads to inhibition of anchorage-independent growth of human colorectal cancer cells, *Gastroenterology*, 115(5), 1049, 1998.
55. Satoh, T., Toyoda, M., Hoshino, H., Monden, T., Yamada, M., Shimizu, H., Miyamoto, K., and Mori, M., Activation of peroxisome proliferator-activated receptor-gamma stimulates the growth arrest and DNA-damage inducible 153 gene in non-small cell lung carcinoma cells, *Oncogene*, 21(14), 2171, 2002.
56. Shimada, T., Kojima, K., Yoshiura, K., Hiraishi, H., and Terano, A., Characteristics of the peroxisome proliferator activated receptor gamma (PPARgamma) ligand induced apoptosis in colon cancer cells, *Gut*, 50(5), 658, 2002.
57. Gupta, R.A., Brockman, J.A., Sarraf, P., Willson, T.M., and DuBois, R.N., Target genes of peroxisome proliferator-activated receptor gamma in colorectal cancer cells, *J. Biol. Chem.*, 276(32), 29681, 2001.
58. Clay, C.E., Atsumi, G.I., High, K.P., and Chilton, F.H., Early *de novo* gene expression is required for 15-deoxy-delta 12,14-prostaglandin J2-induced apoptosis in breast cancer cells, *J. Biol. Chem.*, 276(50), 47131, 2001.
59. Sueyoshi, T. and Negishi, M., Phenobarbital response elements of cytochrome P450 genes and nuclear receptors, *Annu. Rev. Pharmacol. Toxicol.*, 41, 123, 2001.
60. Zelko, I. and Negishi, M., Phenobarbital-elicited activation of nuclear receptor CAR in induction of cytochrome P450 genes, *Biochem. Biophys. Res. Commun.*, 277(1), 1, 2000.
61. Ueda, A., Hamadeh, H.K., Webb, H.K., Yamamoto, Y., Sueyoshi, T., Afshari, C.A., Lehmann, J.M., and Negishi, M., Diverse roles of the nuclear orphan receptor CAR in regulating hepatic genes in response to phenobarbital, *Mol. Pharmacol.*, 61(1), 1, 2002.
62. Garcia-Allan, C., Lord, P.G., Loughlin, J.M., Orton, T.C., and Sidaway, J.E., Identification of phenobarbitone-modulated genes in mouse liver by differential display, *J. Biochem. Mol. Toxicol.*, 14(2), 65, 2000.
63. Schmidt, J.V. and Bradfield, C.A., Ah receptor signaling pathways, *Annu. Rev. Cell Dev. Biol.*, 12, 55, 1996.
64. Rowlands, J.C. and Gustafsson, J.A., Aryl hydrocarbon receptor-mediated signal transduction, *Crit. Rev. Toxicol.*, 27(2), 109, 1997.
65. Schrenk, D., Impact of dioxin-type induction of drug-metabolizing enzymes on the metabolism of endo- and xenobiotics, *Biochem. Pharmacol.*, 55(8), 1155, 1998.
66. Safe, S., Wang, F., Porter, W., Duan, R., and McDougal, A., Ah receptor agonists as endocrine disruptors: antiestrogenic activity and mechanisms, *Toxicol. Lett.*, 102–103, 343, 1998.
67. Puga, A., Maier, A., and Medvedovic, M., The transcriptional signature of dioxin in human hepatoma HepG2 cells, *Biochem. Pharmacol.*, 60(8), 1129, 2000.

68. Frueh, F.W., Hayashibara, K.C., Brown, P.O., and Whitlock, J.P., Use of cDNA microarrays to analyze dioxin-induced changes in human liver gene expression, *Toxicol. Lett.*, 122(3), 189, 2001.
69. Hurst, C.H., Abbott, B., Schmid, J.E., and Birnbaum, L.S., 2,3,7,8-tetrachlorodibenzo-*p*-dioxin (TCDD) disrupts early morphogenetic events that form the lower reproductive tract in female rat fetuses, *Toxicol. Sci.*, 65(1), 87, 2002.
70. Kolluri, S.K., Balduf, C., Hofmann, M., and Gottlicher, M., Novel target genes of the Ah (dioxin) receptor: transcriptional induction of N-myristoyltransferase 2, *Cancer Res.*, 61(23), 8534, 2001.
71. Raju, R.V., Moyana, T.N., and Sharma, R.K., N-myristoyltransferase overexpression in human colorectal adenocarcinomas, *Exp. Cell Res.*, 235(1), 145, 1997.
72. Rajala, R.V., Radhi, J.M., Kakkar, R., Datla, R.S., and Sharma, R.K., Increased expression of N-myristoyltransferase in gallbladder carcinomas, *Cancer*, 88(9), 1992, 2000.
73. Kurachi, M., Hashimoto, S., Obata, A., Nagai, S., Nagahata, T., Inadera, H., Sone, H., Tohyama, C., Kaneko, S., Kobayashi, K., and Matsushima, K., Identification of 2,3,7,8-tetrachlorodibenzo-*p*-dioxin-responsive genes in mouse liver by serial analysis of gene expression, *Biochem. Biophys. Res. Commun.*, 292(2), 368, 2002.
74. Zhang, L., Zhou, W., Velculescu, V.E., Kern, S.E., Hruban, R.H., Hamilton, S.R., Vogelstein, B., and Kinzler, K.W., Gene expression profiles in normal and cancer cells, *Science*, 276(5316), 1268, 1997.
75. Bartosiewicz, M., Trounstine, M., Barker, D., Johnston, R., and Buckpitt, A., Development of a toxicological gene array and quantitative assessment of this technology, *Arch. Biochem. Biophys.*, 376(1), 66, 2000.
76. Menzel, R., Bogaert, T., and Achazi, R., A systematic gene expression screen of *Caenorhabditis elegans* cytochrome P450 genes reveals CYP35 as strongly xenobiotic inducible, *Arch. Biochem. Biophys.*, 395(2), 158, 2001.
77. Bulera, S.J., Eddy, S.M., Ferguson, E., Jatkoe, T.A., Reindel, J.F., Bleavins, M.R., and De La Iglesia, F.A., RNA expression in the early characterization of hepatotoxicants in Wistar rats by high-density DNA microarrays, *Hepatology*, 33(5), 1239, 2001
78. Waring, J.F., Jolly, R.A., Ciurlionis, R., Lum, P.Y., Praestgaard, J.T., Morfitt, D.C., Buratto, B., Roberts, C., Schadt, E., and Ulrich, R.G., Clustering of hepatotoxins based on mechanism of toxicity using gene expression profiles, *Toxicol. Appl. Pharmacol.*, 175(1), 28, 2001.
79. Waring, J.F., Ciurlionis, R., Jolly, R.A., Heindel, M., and Ulrich, R.G., Microarray analysis of hepatotoxins *in vitro* reveals a correlation between gene expression profiles and mechanisms of toxicity, *Toxicol. Lett.*, 120(1–3), 359, 2001.
80. Jordan, V.C. and Morrow, M., Tamoxifen, raloxifene, and the prevention of breast cancer, *Endocr. Rev.*, 20(3), 253, 1999.
81. Marquardt, H., Chemical carcinogenesis, in *Toxicology*, Marquardt, H., Schäfer, S.G., McClellan, R., and Welsch, F., Eds., Academic Press, New York, 1999, p. 151.
82. Cohen, I.R., Sims, M.L., Robbins, M.R., Lakshmanan, M.C., Francis, P.C., and Long, G.G., The reversible effects of raloxifene on luteinizing hormone levels and ovarian morphology in mice, *Reprod. Toxicol.*, 14(1), 37, 2000.
83. Levenson, A.S. and Jordan, V.C., Selective oestrogen receptor modulation: molecular pharmacology for the millennium, *Eur. J. Cancer*, 35(12), 1628, 1999.
84. Fisher, B., Costantino, J.P., Redmond, C.K., Fisher, E.R., Wickerham, D.L., and Cronin, W.M., Endometrial cancer in tamoxifen-treated breast cancer patients: findings from the National Surgical Adjuvant Breast and Bowel Project (NSABP) B-14, *J. Natl. Cancer Inst.*, 86(7), 527, 1994.
85. Green, A.R., Parrott, E.L., Butterworth, M., Jones, P.S., Greaves, P., and White, I.N., Comparisons of the effects of tamoxifen, toremifene, and raloxifene on enzyme induction, and gene expression in the ovariectomised rat uterus, *J. Endocrinol.*, 170(3), 555, 2001.
86. Parrott, E., Butterworth, M., Green, A., White, I.N., and Greaves, P., Adenomyosis — a result of disordered stromal differentiation, *Am. J. Pathol.*, 159(2), 623, 2001.
87. Waters, K.M., Safe, S., and Gaido, K.W., Differential gene expression in response to methoxychlor and estradiol through ERalpha, ERbeta, and AR in reproductive tissues of female mice, *Toxicol. Sci.*, 63(1), 47, 2001.
88. Ouyang, X.S., Wang, X., Lee, D.T., Tsao, S.W., and Wong, Y.C., Up-regulation of TRPM-2, MMP-7, and ID-1 during sex hormone-induced prostate carcinogenesis in the Noble rat, *Carcinogenesis*, 22(6), 965, 2001.

89. Landis, S.H., Murray, T., Bolden, S., and Wingo, P.A., Cancer statistics, 1998, *CA Cancer J. Clin.*, 48(1), 6, 1998.
90. Lobenhofer, E.K., Bennett, L., Cable, P.L., Li, L., Bushel, P.R., and Afshari, C.A., Regulation of DNA replication fork genes by 17beta-estradiol, *Mol. Endocrinol.*, 16(6), 1215, 2002.
91. Finlin, B.S., Gau, C.L., Murphy, G.A., Shao, H., Kimel, T., Seitz, R.S., Chiu, Y.F., Botstein, D., Brown, P.O., Der, C.J., Tamanoi, F., Andres, D.A., and Perou, C.M., RERG is a novel *ras*-related, estrogen-regulated, and growth-inhibitory gene in breast cancer, *J. Biol. Chem.*, 276(45), 42259, 2001.
92. Pennie, W.D., Woodyatt, N.J., Aldridge, T.C., and Orphanides, G., Application of genomics to the definition of the molecular basis for toxicity, *Toxicol. Lett.*, 120(1–3), 353, 2001.
93. Soulez, M. and Parker, M.G., Identification of novel oestrogen receptor target genes in human ZR75–1 breast cancer cells by expression profiling, *J. Mol. Endocrinol.*, 27(3), 259, 2001.
94. Chen, I., Hsieh, T., Thomas, T., and Safe, S., Identification of estrogen-induced genes downregulated by AHR agonists in MCF-7 breast cancer cells using suppression subtractive hybridization, *Gene*, 262(1–2), 207, 2001.
95. Charpentier, A.H., Bednarek, A.K., Daniel, R.L., Hawkins, K.A., Laflin, K.J., Gaddis, S., MacLeod, M.C., and Aldaz, C.M., Effects of estrogen on global gene expression: identification of novel targets of estrogen action, *Cancer Res.*, 60(21), 5977, 2000.
96. Ghosh, M.G., Thompson, D.A., and Weigel, R.J., PDZK1 and GREB1 are estrogen-regulated genes expressed in hormone-responsive breast cancer, *Cancer Res.*, 60(22), 6367, 2000.
97. Custodia, N., Won, S.J., Novillo, A., Wieland, M., Li, C., and Callard, I.P., *Caenorhabditis elegans* as an environmental monitor using DNA microarray analysis, *Ann. N.Y. Acad. Sci.*, 948, 32, 2001.
98. Chen, J., Schwartz, D.A., Young, T.A., Norris, J.S., and Yager, J.D., Identification of genes whose expression is altered during mitosuppression in livers of ethinyl estradiol-treated female rats, *Carcinogenesis*, 17(12), 2783, 1996.
99. Chen, J., Gokhale, M., Li, Y., Trush, M.A., and Yager, J.D., Enhanced levels of several mitochondrial mRNA transcripts and mitochondrial superoxide production during ethinyl estradiol-induced hepatocarcinogenesis and after estrogen treatment of HepG2 cells, *Carcinogenesis*, 19(12), 2187, 1998.
100. Clegg, N., Eroglu, B., Ferguson, C., Arnold, H., Moorman, A., and Nelson, P.S., Digital expression profiles of the prostate androgen-response program, *J. Steroid Biochem. Mol. Biol.*, 80(1), 13, 2002.
101. Velculescu, V.E., Madden, S.L., Zhang, L., Lash, A.E., Yu, J., Rago, C., Lal, A., Wang, C.J., Beaudry, G.A., Ciriello, K.M., Cook, B.P., Dufault, M.R., Ferguson, A.T., Gao, Y., He, T.C., Hermeking, H., Hiraldo, S.K., Hwang, P.M., Lopez, M.A., Luderer, H.F., Mathews, B., Petroziello, J.M., Polyak, K., Zawel, L., Kinzler, K.W. et al., Analysis of human transcriptomes, *Nat. Genet.*, 23(4), 387, 1999.
102. Henschler, D., Wichtige Gifte und Vergiftungen, in *Allgemeine und spezielle Pharmakologie und Toxikologie*, Forth, W., Henschler, D., Rummel, W., and Starke, K., Eds., Spektrum Verlag, Heidelberg, 1996, p. 815.
103. Yamada, H. and Koizumi, S., DNA microarray analysis of human gene expression induced by a non-lethal dose of cadmium, *Ind. Health*, 40(2), 159, 2002.
104. Liu, J., Corton, C., Dix, D.J., Liu, Y., Waalkes, M.P., and Klaassen, C.D., Genetic background but not metallothionein phenotype dictates sensitivity to cadmium-induced testicular injury in mice, *Toxicol. Appl. Pharmacol.*, 176(1), 1, 2001.
105. Liu, J., Kadiiska, M.B., Corton, J.C., Qu, W., Waalkes, M.P., Mason, R.P., Liu, Y., and Klaassen, C.D., Acute cadmium exposure induces stress-related gene expression in wild-type and metallothionein-I/II-null mice, *Free Radic. Biol. Med.*, 32(6), 525, 2002.
106. Lu, T., Liu, J., LeCluyse, E.L., Zhou, Y.S., Cheng, M.L., and Waalkes, M.P., Application of cDNA microarray to the study of arsenic-induced liver diseases in the population of Guizhou, China, *Toxicol. Sci.*, 59(1), 185, 2001.
107. Liu, J., Chen, H., Miller, D.S., Saavedra, J.E., Keefer, L.K., Johnson, D.R., Klaassen, C.D., and Waalkes, M.P., Overexpression of glutathione S-transferase II and multidrug resistance transport proteins is associated with acquired tolerance to inorganic arsenic, *Mol. Pharmacol.*, 60(2), 302, 2001.
108. Liu, J., Kadiiska, M.B., Liu, Y., Lu, T., Qu, W., and Waalkes, M.P., Stress-related gene expression in mice treated with inorganic arsenicals, *Toxicol. Sci.*, 61(2), 314, 2001.
109. Bartosiewicz, M., Penn, S., and Buckpitt, A., Applications of gene arrays in environmental toxicology: fingerprints of gene regulation associated with cadmium chloride, benzo(*a*)pyrene, and trichloroethylene, *Environ. Health Perspect.*, 109(1), 71, 2001.

110. Chen, H., Liu, J., Merrick, B.A., and Waalkes, M.P., Genetic events associated with arsenic-induced malignant transformation: applications of cDNA microarray technology, *Mol. Carcinog.*, 30(2), 79, 2001.
111. Yih, L.H., Peck, K., and Lee, T.C., Changes in gene expression profiles of human fibroblasts in response to sodium arsenite treatment, *Carcinogenesis*, 23(5), 867, 2002.
112. D'Agostini, F., Izzotti, A., Bennicelli, C., Camoirano, A., Tampa, E., and De Flora, S., Induction of apoptosis in the lung but not in the liver of rats receiving intra-tracheal instillations of chromium (VI), *Carcinogenesis*, 23(4), 587, 2002.
113. Qu, W., Diwan, B.A., Liu, J., Goyer, R.A., Dawson, T., Horton, J.L., Cherian, M.G., and Waalkes, M.P., The metallothionein-null phenotype is associated with heightened sensitivity to lead toxicity and an inability to form inclusion bodies, *Am. J. Pathol.*, 160(3), 1047, 2002.
114. Hossain, M.A., Bouton, C.M., Pevsner, J., and Laterra, J., Induction of vascular endothelial growth factor in human astrocytes by lead. Involvement of a protein kinase C/activator protein-1 complex-dependent and hypoxia-inducible factor 1-independent signaling pathway, *J. Biol. Chem.*, 275(36), 27874, 2000.
115. Alcantara, O., Kalidas, M., Baltathakis, I., and Boldt, D.H., Expression of multiple genes regulating cell cycle and apoptosis in differentiating hematopoietic cells is dependent on iron, *Exp. Hematol.*, 29(9), 1060, 2001.
116. Goering, P.L., Waalkes, M.P., and Klaassen, C.D., Toxicology of cadmium, in *Toxicology of Metals: Biochemical Aspects, Handbook of Experimental Pharmacology*, Vol. 115, Goyer, R.A. and Cherian, M.G., Eds., Springer-Verlag, New York, 1995, p. 189.
117. IARC, *Cancer Monographs on the Evaluation of the Carcinogenic Risk to Humans*, Vol. 58, International Agency for Research on Cancer, Lyon, France, 1993.
118. Abernathy, C.O., Liu, Y.P., Longfellow, D., Aposhian, H.V., Beck, B., Fowler, B., Goyer, R., Menzer, R., Rossman, T., Thompson, C., and Waalkes, M., Arsenic: health effects, mechanisms of actions, and research issues, *Environ. Health Perspect.*, 107(7), 593, 1999.
119. Zimmermann, H.J., *Hepatotoxicity*, Lippincott/Williams & Wilkins, Philadelphia, 1999.
120. Anttila, A., Pukkala, E., Aitio, A., Rantanen, T., and Karjalainen, S., Update of cancer incidence among workers at a copper/nickel smelter and nickel refinery, *Int. Arch. Occup. Environ. Health*, 71(4), 245, 1998.
121. Huang, X., Zhuang, Z., Frenkel, K., Klein, C.B., and Costa, M., The role of nickel and nickel-mediated reactive oxygen species in the mechanism of nickel carcinogenesis, *Environ. Health Perspect.*, 102(suppl. 3), 281, 1994.
122. Misra, M., Rodriguez, R.E., and Kasprzak, K.S., Nickel induced lipid peroxidation in the rat: correlation with nickel effect on antioxidant defense systems, *Toxicology*, 64(1), 1, 1990.
123. Conway, K. and Costa, M., Nonrandom chromosomal alterations in nickel-transformed Chinese hamster embryo cells, *Cancer Res.*, 49(21), 6032, 1989.
124. Patierno, S.R. and Costa, M., DNA-protein cross-links induced by nickel compounds in intact cultured mammalian cells, *Chem. Biol. Interact.*, 55(1–2), 75, 1985.
125. Shen, H.M. and Zhang, Q.F., Risk assessment of nickel carcinogenicity and occupational lung cancer, *Environ. Health Perspect.*, 102(suppl. 1), 275, 1994.
126. Mao, X., Kashii, T., Hayashi, R., Sassa, K., Fujishita, T., Maruyama, M., Kobayashi, M., and Liu, S., Cloning of differentially expressed sequence tags from nickel-transformed human embryonic lung cells, *Cancer Lett.*, 161(1), 57, 2000.
127. Gertler, F.B., Niebuhr, K., Reinhard, M., Wehland, J., and Soriano, P., Mena, a relative of VASP and *Drosophila* Enabled, is implicated in the control of microfilament dynamics, *Cell*, 87(2), 227, 1996.
128. Ford, B.N., Wilkinson, D., Thorleifson, E.M., and Tracy, B.L., Gene expression responses in lymphoblastoid cells after radiation exposure, *Radiat. Res.*, 156(5, pt. 2), 668, 2001.
129. Amundson, S.A., Bittner, M., Chen, Y., Trent, J., Meltzer, P., and Fornace, A.J., Jr., Fluorescent cDNA microarray hybridization reveals complexity and heterogeneity of cellular genotoxic stress responses, *Oncogene*, 18(24), 3666, 1999.
130. Amundson, S.A., Do, K.T., Shahab, S., Bittner, M., Meltzer, P., Trent, J., and Fornace, A.J., Jr., Identification of potential mRNA biomarkers in peripheral blood lymphocytes for human exposure to ionizing radiation, *Radiat. Res.*, 154(3), 342, 2000.

131. Mayer, C., Popanda, O., Zelezny, O., von Brevern, M.C., Bach, A., Bartsch, H., and Schmezer, P., DNA repair capacity after gamma-irradiation and expression profiles of DNA repair genes in resting and proliferating human periferal blood lymphocytes, *DNA Repair*, 1, 237, 2002.
132. Ward, J.F., DNA damage produced by ionizing radiation in mammalian cells: identities, mechanisms of formation, and reparability, *Prog. Nucleic Acid Res. Mol. Biol.*, 35, 95, 1988.
133. Hutchinson, F., Chemical changes induced in DNA by ionizing radiation, *Prog. Nucleic Acid Res. Mol. Biol.*, 32, 115, 1985.
134. Friedberg, E.C., Walker, G.C., and Siede, W., *DNA Repair and Mutagenesis*, American Society for Microbiology, Washington, D.C., 1995.
135. Scherf, U., Ross, D.T., Waltham, M., Smith, L.H., Lee, J.K., Tanabe, L., Kohn, K.W., Reinhold, W.C., Myers, T.G., Andrews, D.T., Scudiero, D.A., Eisen, M.B., Sausville, E.A., Pommier, Y., Botstein, D., Brown, P.O., and Weinstein, J.N., A gene expression database for the molecular pharmacology of cancer, *Nat. Genet*, 24(3), 236, 2000.
136. Dan, S., Tsunoda, T., Kitahara, O., Yanagawa, R., Zembutsu, H., Katagiri, T., Yamazaki, K., Nakamura, Y., and Yamori, T., An integrated database of chemosensitivity to 55 anticancer drugs and gene expression profiles of 39 human cancer cell lines, *Cancer Res.*, 62(4), 1139, 2002.
137. Staunton, J.E., Slonim, D.K., Coller, H.A., Tamayo, P., Angelo, M.J., Park, J., Scherf, U., Lee, J.K., Reinhold, W.O., Weinstein, J.N., Mesirov, J.P., Lander, E.S., and Golub, T.R., Chemosensitivity prediction by transcriptional profiling, *Proc. Natl. Acad. Sci. USA*, 98(19), 10787, 2001.
138. Grever, M.R., Schepartz, S.A., and Chabner, B.A., The National Cancer Institute: cancer drug discovery and development program, *Semin. Oncol.*, 19(6), 622, 1992.
139. Stinson, S.F., Alley, M.C., Kopp, W.C., Fiebig, H.H., Mullendore, L.A., Pittman, A.F., Kenney, S., Keller, J., and Boyd, M.R., Morphological and immunocytochemical characteristics of human tumor cell lines for use in a disease-oriented anticancer drug screen, *Anticancer Res.*, 12(4), 1035, 1992.
140. Monks, A., Scudiero, D.A., Johnson, G.S., Paull, K.D., and Sausville, E.A., The NCI anti-cancer drug screen: a smart screen to identify effectors of novel targets, *Anticancer Drug Des.*, 12(7), 533, 1997.
141. Weinstein, J.N., Myers, T.G., O'Connor, P.M., Friend, S.H., Fornace, A.J., Jr., Kohn, K.W., Fojo, T., Bates, S.E., Rubinstein, L.V., Anderson, N.L., Buolamwini, J.K., van Osdol, W.W., Monks, A.P., Scudiero, D.A., Sausville, E.A., Zaharevitz, D.W., Bunow, B., Viswanadhan, V.N., Johnson, G.S., Wittes, R.E., and Paull, K.D., An information-intensive approach to the molecular pharmacology of cancer, *Science*, 275(5298), 343, 1997.
142. Kaminski, N., Allard, J., and Heller, R.A., Use of oligonucleotide arrays to analyze drug toxicity, *Ann. N.Y. Acad. Sci.*, 919, 1, 2000.
143. Sun, Y., Identification and characterization of genes responsive to apoptosis: application of DNA chip technology and mRNA differential display, *Histol. Histopathol.*, 15(4), 1271, 2000.
144. Huang, Q., Dunn, R.T., 2nd, Jayadev, S., DiSorbo, O., Pack, F.D., Farr, S.B., Stoll, R.E., and Blanchard, K.T., Assessment of cisplatin-induced nephrotoxicity by microarray technology, *Toxicol. Sci.*, 63(2), 196, 2001.
145. Radaeva, S., Jaruga, B., Hong, F., Kim, W.H., Fan, S., Cai, H., Strom, S., Liu, Y., El-Assal, O., and Gao, B., Interferon-alpha activates multiple STAT signals and down-regulates c-Met in primary human hepatocytes, *Gastroenterology*, 122(4), 1020, 2002.
146. Qin, L.F., Lee, T.K., and Ng, I.O., Gene expression profiling by cDNA array in human hepatoma cell line in response to cisplatin treatment, *Life Sci.*, 70(14), 1677, 2002.
147. Zhou, Y., Gwadry, F.G., Reinhold, W.C., Miller, L.D., Smith, L.H., Scherf, U., Liu, E.T., Kohn, K.W., Pommier, Y., and Weinstein, J.N., Transcriptional regulation of mitotic genes by camptothecin-induced DNA damage: microarray analysis of dose- and time-dependent effects, *Cancer Res.*, 62(6), 1688, 2002.
148. Chang, B.D., Swift, M.E., Shen, M., Fang, J., Broude, E.V., and Roninson, I.B., Molecular determinants of terminal growth arrest induced in tumor cells by a chemotherapeutic agent, *Proc. Natl. Acad. Sci. USA*, 99(1), 389, 2002.
149. Wang, Y., Rea, T., Bian, J., Gray, S., and Sun, Y., Identification of the genes responsive to etoposide-induced apoptosis: application of DNA chip technology, *FEBS Lett.*, 445(2–3), 269, 1999.
150. Certa, U., Seiler, M., Padovan, E., and Spagnoli, G.C., High density oligonucleotide array analysis of interferon-alpha2a sensitivity and transcriptional response in melanoma cells, *Br. J. Cancer*, 85(1), 107, 2001.

151. Johnsson, A., Zeelenberg, I., Min, Y., Hilinski, J., Berry, C., Howell, S.B., and Los, G., Identification of genes differentially expressed in association with acquired cisplatin resistance, *Br. J. Cancer*, 83(8), 1047, 2000.
152. Johnsson, A., Byrne, P., de Bruin, R., Weiner, D., Wong, J., and Los, G., Identification of gene clusters differentially expressed during the cellular injury responses (CIR) to cisplatin, *Br. J. Cancer*, 85(8), 1206, 2001.
153. Dan, S., and Yamori, T., Repression of cyclin B1 expression after treatment with adriamycin, but not cisplatin in human lung cancer A549 cells, *Biochem. Biophys. Res. Commun.*, 280(3), 861, 2001.
154. Lam, L.T., Pickeral, O.K., Peng, A.C., Rosenwald, A., Hurt, E.M., Giltnane, J.M., Averett, L.M., Zhao, H., Davis, R.E., Sathyamoorthy, M., Wahl, L.M., Harris, E.D., Mikovits, J.A., Monks, A.P., Hollingshead, M.G., Sausville, E.A., and Staudt, L.M., Genomic-scale measurement of mRNA turnover and the mechanisms of action of the anti-cancer drug flavopiridol, *Genome Biol.*, 2(10), research0041, 2001.
155. Deliliers, G.L., Servida, F., Fracchiolla, N.S., Ricci, C., Borsotti, C., Colombo, G., and Soligo, D., Effect of inositol hexaphosphate (IP6) on human normal and leukaemic haematopoietic cells, *Br. J. Haematol.*, 117(3), 577, 2002.
156. Kruger, A., Soeltl, R., Sopov, I., Kopitz, C., Arlt, M., Magdolen, V., Harbeck, N., Gansbacher, B., and Schmitt, M., Hydroxamate-type matrix metalloproteinase inhibitor batimastat promotes liver metastasis, *Cancer Res.*, 61(4), 1272, 2001.
157. Kaminski, N., Allard, J.D., Pittet, J.F., Zuo, F., Griffiths, M.J., Morris, D., Huang, X., Sheppard, D., and Heller, R.A., Global analysis of gene expression in pulmonary fibrosis reveals distinct programs regulating lung inflammation and fibrosis, *Proc. Natl. Acad. Sci. USA*, 97(4), 1778, 2000.
158. Harries, H.M., Fletcher, S.T., Duggan, C.M., and Baker, V.A., The use of genomics technology to investigate gene expression changes in cultured human liver cells, *Toxicol. In Vitro*, 15(4–5), 399, 2001.
159. Kegelmeyer, A.E., Sprankle, C.S., Horesovsky, G.J., and Butterworth, B.E., Differential display identified changes in mRNA levels in regenerating livers from chloroform-treated mice, *Mol. Carcinog.*, 20(3), 288, 1997.
160. Bhattacharjee, A., Lappi, V.R., Rutherford, M.S., and Schook, L.B., Molecular dissection of dimethylnitrosamine (DMN)-induced hepatotoxicity by mRNA differential display, *Toxicol. Appl. Pharmacol.*, 150(1), 186, 1998.
161. Harris, A.J., Shaddock, J.G., Manjanatha, M.G., Lisenbey, J.A., and Casciano, D.A., Identification of differentially expressed genes in aflatoxin B1-treated cultured primary rat hepatocytes and Fischer 344 rats, *Carcinogenesis*, 19(8), 1451, 1998.
162. Burczynski, M.E., McMillian, M., Ciervo, J., Li, L., Parker, J.B., Dunn, R.T., 2nd, Hicken, S., Farr, S., and Johnson, M.D., Toxicogenomics-based discrimination of toxic mechanism in HepG2 human hepatoma cells, *Toxicol. Sci.*, 58(2), 399, 2000.

# Section 5

## Predictive Toxicogenomics

# 11 Gene Expression Profile Databases in Toxicity Testing

*Robert T. Dunn II and Kyle L. Kolaja*

## CONTENTS



## 11.1 INTRODUCTION

The past several years have seen the rapid application of genomics to toxicology. In fact, the term *toxicogenomics* has been derived[1,2] to refer broadly to the study of gene expression in virtually any given toxicity paradigm. Measurement of gene expression is also frequently referred to as transcript profiling.[3,4] A critical hypothesis of toxicogenomics is that altered gene expression is one of the earliest measurable responses to a toxic challenge and that the resulting transcript profile reveals mechanistic insight into the toxicity as well as how each cell type is programmed to respond. In order to derive the most knowledge and value from these data, sound toxicologic study design principles must be included, such as dose response, time response, and species specificity of effects. Additionally, both target and nontarget organ effects should be assessed at the molecular level. Together, these principles have the best potential for creating a reliable tool for toxicological decision making. Finally, the ultimate achievement of toxicogenomics will be the ability to use transcript profiling to predict toxic outcomes accurately and reliably, thus greatly enhancing the efficiency of drug candidate selection. In order to obtain this goal of predictive toxicology, experimental databases must be constructed that include the classical standards of toxicity evaluation, such as histopathology and serum chemistry, which can be correlated with differential expression of specific sets of genes that occur temporally *in advance* of the clinical observations of toxicity. In this chapter, we focus on the characteristics of useful and meaningful toxicogenomic databases.

What is meant by *predictive toxicology*? The rapid pace of toxicogenomics research has created a new niche vocabulary and a flurry of publications vying to establish research benchmarks for the field. Thus, for the purposes of this chapter, some definition is warranted. Simply stated, predictive

0-8493-1334-1/03/$0.00+$1.50

toxicology is the use of short-term toxicogenomic data (e.g., *in vitro* or short-term *in vivo* data) to accurately predict findings in longer duration studies (e.g., chronic and/or carcinogenicity studies). To do so, a database of compounds with known chronic toxicity must be developed. Such a database could be queried with novel gene expression profiles from uncharacterized compounds. Samples in this case are mRNA from animal tissues or cells that have been exposed to a specific compound or other toxic challenge.

Often, however, an alternative paradigm is utilized and represented as predictive toxicology. In this paradigm, an unknown or "blinded" sample is predicted to fall into a particular class of toxicants on the basis of similarity in expression pattern between the blinded sample and a set of well-characterized samples. Blinded toxicity profiling of this sort at the level of gene expression is not predictive toxicology, but rather represents characterization of a toxicity pattern at the molecular level. A further criticism of this type of toxicogenomic research is that the "pressure-testing" of the database is far too simplistic, (e.g., a barbiturate-driven database is used to identify another barbiturate). True predictive toxicology is the utilization of gene profiles or comprehensive expression patterns in order to *predict* the eventual onset of toxicity or pathology that is not yet manifested. To illustrate the point, take the example of the potent hepatocarcinogen, aflatoxin B1 ($AFB_1$). It is well documented that administration of aflatoxin will result in liver tumors.[5] Thus, if rats are treated with $AFB_1$, one would ask, "Is there information present in the hepatic gene expression profiles shortly after treatment (e.g., 1, 3, or 7 days) that will ultimately correlate with the presence of tumors?" That is, can one predict from the gene expression profile from a short-term study that rats dosed chronically will eventually get liver tumors? The next step is to repeat this database experiment to include a broad collection of carcinogens. Obviously, tumorigenesis is a multistage and complex process, and it is not yet clear whether or not such databases will be useful in identifying specific gene expression profiles that can be informative, but such an experiment does demonstrate the essence of predictive gene expression.

Toxicogenomics as a subdiscipline of toxicology has rapidly evolved from a period of valid concern for data quality, integrity, and reproducibility to incorporating sophisticated experimental designs into mechanistic-driven studies and, finally, to applying transcript profiling to predict toxicity. During this period of substantial growth, gene expression experiments have been roundly scrutinized due to high cost, lofty and unrealistic expectations, and bioinformatic issues dealing with the difficulty of coping with tremendous data outflow from these studies. The application of toxicogenomics can range from early target identification through to post-marketing evaluation of adverse events. In early target identification and proof of principle experiments, toxicogenomics can be applied to tissues from diseased animals to help identify critical pathways and genes involved in disease process. Countless databases have been developed where disease models are characterized and potential disease targets are identified. Once targets are identified and small-molecule drug candidates are screened for potency and selectivity, early toxicology experiments (e.g., *in vitro* and/or short term *in vivo* studies) can be conducted. Toxicogenomic studies have been performed for mechanistic understanding of certain *in vitro* and *in vivo* toxicity models,[6–8] in addition to toxicology-related gene discovery[9] and assembly of comprehensive databases.[10] Pharmaceutical companies and research organizations are intrigued by the potential of toxicogenomics, but precise, value-added applications of this new technology with regard to drug development are critical for the future of the technology itself. Ultimately, backed by sound scientific principles such as appropriate experimental design and statistical examination of data, the utility of genomics in toxicology will be responsible for accelerating new chemical entities into the market. It is clear that toxicogenomics will strengthen the ability of toxicologists to more accurately understand toxicity and apply accurate risk assessment. However, it is unlikely that transcript profiling will be a panacea that replaces all of traditional toxicology; rather, it will become an additional yet valuable tool. For example, toxicogenomics can already be used to identify hepatic enzyme induction, yet it is unclear the advantage transcript profiling conveys upon solving transient and most likely posttranslational events such as drug-induced Q-Tc prolongation. Thus, in specific toxicities, toxico-

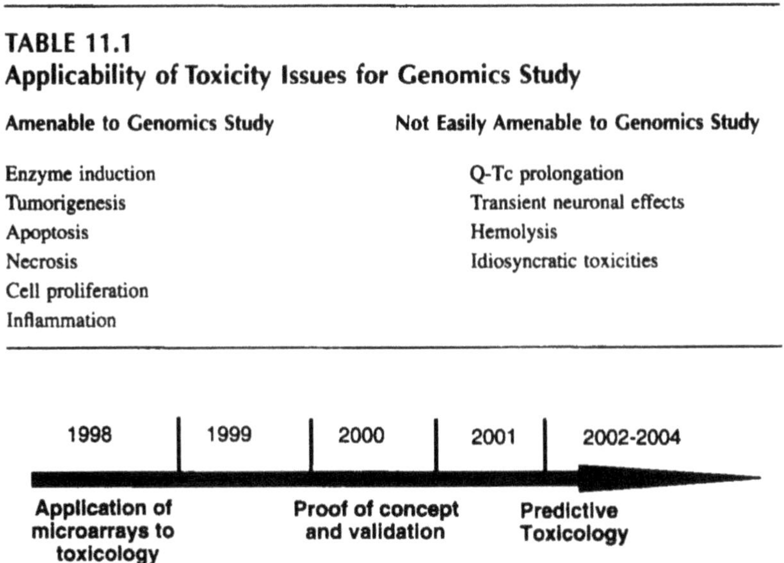

**TABLE 11.1**
**Applicability of Toxicity Issues for Genomics Study**

| Amenable to Genomics Study | Not Easily Amenable to Genomics Study |
|---|---|
| Enzyme induction | Q-Tc prolongation |
| Tumorigenesis | Transient neuronal effects |
| Apoptosis | Hemolysis |
| Necrosis | Idiosyncratic toxicities |
| Cell proliferation | |
| Inflammation | |



FIGURE 11.1 Toxicogenomics timeline. A relative timeline is depicted that marks the significant progress milestones of toxicogenomics over the past several years.

genomics will provide resolving power, while in other events it may not provide useful data (Table 11.1).

Over the past half decade, toxicogenomics has matured from simply measuring differentially expressed genes in toxicity studies, to solid, mechanistic proof of concept and validation, to the cusp of predictive toxicology (Figure 11.1). This chapter discusses the aspects of using toxicogenomics in assembling different types of reference databases and the considerations necessary for employing such databases for drug-development decision making.

## 11.2 TOXICOGENOMICS DATABASES

Currently, the term *database* is nearly as ubiquitous as it is vague and has become a source of confusion. Databases exist for the purpose of locating similarities or patterns for countless types of queries and/or to provide reference information for a given topic. Researchers are familiar with the National Center for Biotechnology Information (NCBI) PubMed website (http://www.ncbi.nlm.nih.gov), which is the user interface for a database of published literature citations in the biomedical sciences. When a text search query is entered into a dialog box on the PubMed homepage, the database is automatically searched and relevant information is retrieved and presented to the user. In the area of genomics, databases exist for vast sequences of nucleotides in order to find regions of similarity or uniqueness. Databases and tools available to the public for search purposes include GenBank (http://www.ncbi.nlm.nih.gov/Genbank/GenbankSearch.html) and the BLAST (basic local alignment search tool) (http://www.ncbi.nlm.nih.gov/BLAST/) algorithm.[11] Relevant to toxicology databases, a recent review by Fielden et al.[12] has summarized the various tools and strategies currently available for computer-based *in silico* toxicology. The databases mentioned previously are advantageous as they provide publicly available tools that permit early-stage research and hypothesis creation at minimal cost. These databases can also be queried with data from actual in-lab experiments. For example, if novel nucleotide sequences are identified in a set of experiments as being predictive for hepatocarcinogenicity, then these sequences can be compared to vast sequences built from a variety of species to identify the query sequences as novel or part of an existing genes or orthologs from another species. However, while providing a wealth

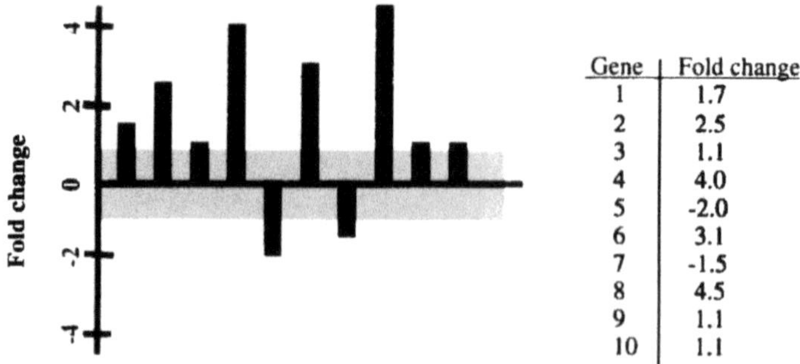


| Gene | Fold change |
|---|---|
| 1 | 1.7 |
| 2 | 2.5 |
| 3 | 1.1 |
| 4 | 4.0 |
| 5 | -2.0 |
| 6 | 3.1 |
| 7 | -1.5 |
| 8 | 4.5 |
| 9 | 1.1 |
| 10 | 1.1 |

**FIGURE 11.2** Diagram depicting a representative gene expression profile. Key features of the display are that genes are expressed in a ratio of treated to control. Note that in order for genes to have the same relative scale in the negative direction, the ratio values must be converted to the negative reciprocal. In addition, ratio-based expression eliminates any values in the range –1 to 1 (gray-shaded region).

of support, these approaches are best utilized as a framework from which hypothesis-driven data can be further explained, as these databases clearly cannot yet become a substitute for true scientific method-driven experimentation.

Fundamental differences exist between databases for toxicogenomics and sequence assemblies. Toxicogenomics databases consist of sets of empirically derived gene expression profiles. That is, a biological system is treated in some fashion (e.g., chemicals, disease process), mRNA is isolated, and relative expression changes in genes are determined and stored. Typically, gene changes are reported in a quantitative or qualitative fashion relative to some form of systematic and experimental controls. Quantitative gene expression measures are often presented as the ratio of treated to control such that a fold change in the gene of interest is recorded. For example, when gene A is expressed in a 2:1 ratio in the treated group relative to the control, this is called a twofold induction of gene A. Another method to report changes in gene expression is qualitative, such that genes expressed to a greater level than control are induced and genes expressed to a lower level than control are repressed. The fundamental principle of toxicogenomics databases is that the database essence is differential, quantitative, or semiquantitative data that can be seamlessly compared to other experimental results. Related information such as sequences and annotations are supportive of and electronically linked to the empirically derived data. In the ideal situation, the empirically derived gene expression data would be electronically linked to the NCBI annotation for a given gene, pathway information for the gene provided, coregulated genes identified, and some level of interpretation in a toxicological context available to the user.

A representation of a gene expression profile is depicted in Figure 11.2. In this example, genes are represented as individual columns across the $x$-axis, while the fold induction (or repression) scores are recorded at given values on the $y$-axis in response to a single treatment. Using this kind of view, the investigator can gain rapid insight into the expression of specific genes in response to a treatment. An alternative way to view changes in gene expression is to monitor the changes of a single gene across multiple experimental conditions. In this way, the response characteristics of a gene, including dose response, time response, compound specificity, and tissue distribution, can be fully characterized.

The term *database* for some can conjure up images of large and unwieldy datasets that can be managed only by bioinformaticists with high-powered computers and cutting-edge software. With user-defined databases and advances in software, this myth can be quickly dispelled. Databases can be derived from more limited datasets and can be as large or as small as the scientist needs. A database from a small focused study could be easily managed on a single computerized spreadsheet. Alternatively, databases can require multigigabyte storage areas but can be appropriately queried to display only the most relevant data. Toxicogenomics databases comprised of data, data management software, and data analysis applications are commercially available from a number of sources, including GeneLogic, Phase-1 Molecular Toxicology, and Iconix, among others (http://www.gene-chips.com/GeneChips.html#Corporate). The point is that the term *database* is independent of the size and manageability of the dataset and is more a function of being a storage, retrieval, and comparison system for data regardless of size, sophistication, and cost. Further, extremely powerful software applications require only brief training before the end user is ready to import and analyze data. The reality is that toxicogenomic databases are moving quickly toward being a ready-made, turn-key resource for all toxicologists, not just toxicogenomic specialists.

In order for toxicogenomics databases to have utility and ease of use, certain minimal components are required, including gene identifications and/or annotations, qualitative or quantitative information for each gene in response to a toxicant, and other forms of supportive information such as functional descriptions for genes and information about the experiment details (e.g., dose, exposure time, sex, species, strain, tissue type). Toxicogenomics databases as such form a reference framework for a variety of applications such as providing a means for comparing datasets or serving as a tool to generate new research hypotheses (e.g., linking expression information to other measures of toxicity). Value-added capabilities such as the ability to correlate transcript profiles to classical endpoints such as clinical chemistry and pathology are essential. Other more theoretical approaches to toxicogenomics are currently being explored as large sequence assemblies are studied and experiments are performed *in silico* as a means of generating new leads for research questions as well as possible functions for heretofore unidentified genes.[12] Finally, the ability to link genomics data to other emerging technologies such as proteomics and metabolic profiling (i.e., metabonomics) will result in synergistic characterization of toxicity issues and help advance the utility of all three emerging technologies.

## 11.3 DATABASE CONSIDERATIONS

Due to the current high-cost nature of genomics technology, the end goal and the specific characteristics of a toxicogenomics database must be clearly defined and fully understood prior to initiating experimentation. Basically, the fundamental question to be asked is: "What do I want the data to do?" If the database is to be used to find gene expression response characteristics for new chemical entities, the usefulness will be a function of the number and diversity of compounds that have been profiled relative to the compounds to be screened. If the database is to be used to generate new drug targets, then a multitude of disease tissues, including human and relevant preclinical models, should be used. If the database is to be used to generate new research hypotheses, then a database built on solid hypothesis-driven research is most appropriate. If the database is going to be used to support *in vitro* screening of compounds, then a database built on *in vitro* compound profiling is most appropriate. Due to the inherent differences between *in vitro* and *in vivo* models, the likelihood of being able to legitimately compare data from one system to the other is probably low. Thus, up-front decision making about the scope and application of a toxicogenomics database is critical.

The rationale for building a database can vary but may involve one of several paradigms. Databases are built in order to identify relatedness among compound effects at the level of gene expression. Alternatively, databases can be built to assess the toxicity of new compounds relative to more fully characterized prototype compounds. Finally, databases can be built for the purpose of identifying genes that are predictive of a toxic outcome. Once predictive genes are identified, the value of the entire database is decreased and the specific predictive genes become the high-value commodity.

### 11.3.1 Model Systems

Some database considerations are obvious while others are more transparent. First, the model system (rodent, canine, primate, cell line, etc.) for the biological phase of the database construction must be carefully selected. The decision about the most appropriate model takes into consideration such factors as how the genomic data will ultimately be used (e.g., early screening tool, preclinical lead selection/optimization, mechanistic insight), as well as the types of arrays (e.g., ADME-specific genes, Tox-specific genes, full genome, etc.) currently available to support the study.

If the intent of the genomics database is to profile large numbers of compounds (100 to >1000) rapidly, then an *in vitro* model system is clearly the best choice. In fact, one of the earliest studies in the field of toxicogenomics utilized the *in vitro* approach to screen compounds and characterize toxicity on the basis of gene expression profiling.[13] In a large-scale compound screening strategy, generating sufficient amounts of compound will likely be the rate-limiting step in database creation. Pharmaceutical compounds that are to be screened early in development are going to be in limited supply due to the difficulties associated with scale-up of combinatorial and medicinal chemistry processes, but milligram quantities of active pharmaceutical ingredient are feasible for an *in vitro* screening system. A compound database built on an *in vitro* strategy would be an internally relational database. That is, data will only be comparable within the selected *in vitro* system, and broad applicability to other *in vivo* systems as well as other *in vitro* systems would most likely be speculative.

Once the model has been selected, the most appropriate genomics platform must be employed. Currently, a variety of platforms are available, with new technologies constantly being created. Some questions that should be asked up front include: Will this particular array change significantly over time (i.e., will more genes be added or eliminated)? Will the data calculation algorithms remain fixed? How will improvements in technology affect the existing platform and results? Can data from this platform be compared to data derived from another platform? Are the genomic elements of the array relevant across species? How will raw data be stored? As the functions of unknown genes (ESTs) are established, how will the new knowledge be retrofitted to the existing database? These questions are addressed with consideration of the risk that previously generated data could be rendered obsolete as technology evolves. Clearly, one option to counter the current state of rapidly evolving technology is to archive the source samples used to generate the transcript profile data such that the same samples could be used to benchmark changes in the platforms and data handling methods. But, archiving samples is not without its own inherent costs, such as space requirements, in addition to the requirement to have a reference system in which data can be tracked back to the original sample, not to mention the costs of repeating experiments, all of which assumes that the original sample retains its integrity and characteristics over time. Another option is to build flexibility into the database such that recalculating and updating are inherently facile processes. Either way, the most important fact is to understand how changes in the database system affect the resultant data and whether the changes could affect previous conclusions drawn from those data.

### 11.3.2 Features of a Toxicogenomics Database

Databases are generally comprised of five major features: (1) data generation instruments and portal, (2) data, (3) storage, (4) data management (back end) and (5) analytical software (front end). One basic database layout concept, in a highly simplified diagram, is depicted in Figure 11.3. In this framework, data are generated by suitable instrumentation, and the data are stored in a centralized data warehouse (usually on a server) and organized in an efficient, yet non-user friendly, format (e.g., raw textual data tables). The data tables require external management by a database engine (software) that should be transparent to the user and is referred to as the *back end*. The database engines, many of which are commercially available, perform the steps necessary to merge and split data files as requested by the user operating at the interface or viewer. Data are viewed and analyzed

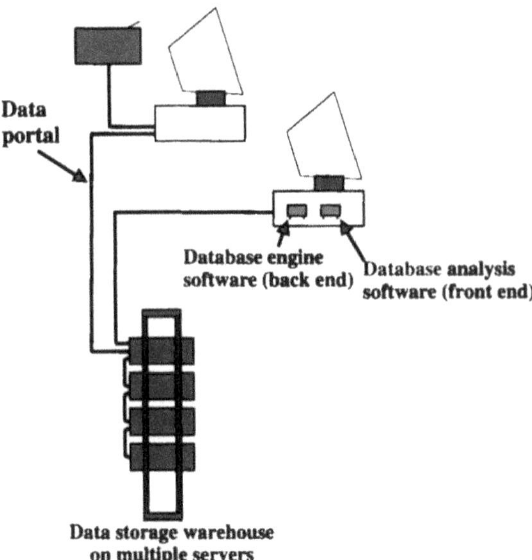


FIGURE 11.3 Conceptual layout of a toxicogenomics database system. In this diagram, raw data are produced at the data-generation workstation. Data are sent via the portal to a storage location on a server. Data are retrieved and managed by the database engine (back end) and manipulated and analyzed by the analysis software (front end).

by the data viewer, or *front end*. The goal of a viewer is to provide ease of use to proceed through logical steps in order to call in data, mathematically manipulate data, organize data, and perform numerous types of analysis.

A variety of front-end software packages are available including Rosetta Resolver®, Spotfire®, and GeneSpring®, among others. Software data analysis packages typically provide the user with a highly functional interface. Features of data analysis software are plentiful and more or less useful depending on the needs of the user. Commonly used data analyses include principle components analysis (a method of statistical data reduction to create visual images that describe the variability among individual sample populations), multidimensional scaling (a variant of principle components analysis), hierarchical clustering analysis (involving the formation of distinct clusters of data points that can be graphically displayed in scatterplots, histograms, or dendrograms), and the unequal variance $t$-test. One important concept in data analysis is that the end user must be aware of the strengths and limitations of the tools within each analysis. For example, comparison of gene expression profiles on the basis of simple correlation may yield statistically significant findings for genes with little to no biological relevance. Thus, statistical methods must be sensitive to the magnitude of change when deriving correlation coefficients. This will aid in avoiding gene expression profiles that differ substantially in magnitude yet appear similar on the basis of mere correlation. Understanding the strengths and limitations of an analysis method is fundamental yet often overlooked, resulting in incorrect or overstated conclusions.

## 11.4 DATABASE PARADIGMS

The following strategies are examples of the concepts behind the construction of toxicogenomics databases. Independent of cost and time, the conceptual framework of a database will determine its utility and applicability. At the present time, an initiative is ongoing to deal with the major issues of toxicogenomics. The International Life Sciences Institute (ILSI) is assembling gene expression data from some of the currently available gene expression technology platforms. The ILSI initiative has already begun to deliver useful information on the major toxicogenomics issues through the efforts of specific working groups within the ILSI Health and Environmental Sciences Institute.[14]

### 11.4.1 Chemical/Compound-Driven Databases

Compound-driven databases are conceived with the notion or hypothesis that structurally and chemically related compounds will behave similarly and have similar gene expression profiles within a range, in a selected biological system. For example, the environmental toxin TCDD (tetrachlorodibenzo-*p*-dioxin) elicits a number of its effects through the aryl hydrocarbon receptor (AHR). Thus, the hypothesis could be put forth that TCDD and its congeners would affect gene expression in a similar fashion. Likewise, compounds that activate the peroxisome-proliferator-activated receptor alpha (PPARα) (e.g., WY-14643, clofibrate, gemfibrozil) should induce changes in gene expression in much the same way across the class. Although the preceding examples are simplistic, they underlie the basic notion behind a compound-driven database. Indeed, experiments with peroxisome proliferators have been the basis for some of the recent "proof of concept" work in toxicogenomics.[15] Such compounds are useful for several reasons, including the existence of a large body of traditional toxicity information (dose–response and histology), inexpensive and readily available compounds, receptor-mediated effects, and a well-understood signaling pathway. A list of compounds and the receptors that mediate their effects are provided in Table 11.2. Ultimately, the goal of such a database is to provide a predictive framework for assessment of toxicity of new chemical entities or previously uncharacterized compounds. That is, if an unknown compound has a gene expression profile similar to an aromatic hydrocarbon or a peroxisome proliferator, certain assertions about the unknown compound can be made and tested experimentally.

While the notion of compound-driven toxicogenomics databases is pervasive, such an undertaking on a large scale (i.e., hundreds to thousands of compounds) is only feasible for *in vitro* systems at the present time. The reason for the reliance on an *in vitro* system in this case is simply due to the resource constraints of compound preparation. An *in vivo* approach to building a toxicogenomics database to mirror pharmacological structure–activity relationships (SARs) will not be widely applicable because demands for gram quantities of compounds for dosing animals may not be feasible without millions of dollars in funding. For a few cases of chemical classes, *in vivo* SAR datasets for toxicogenomics may be feasible, including peroxisome proliferators and aromatic hydrocarbons, but the applicability of such information to new drug design is minimal. The ultimate utility of the chemical-driven approach is to identify relationships between the structural features of diverse chemical molecules and how these features associate with observable

**TABLE 11.2**
**Compounds, Receptors, and Target Genes**

| Compound | Receptor | Regulated Genes |
|---|---|---|
| Estrogen, ethinyl estradiol | Estrogen receptor (ERα, ERβ) | *pS2* |
| Dihydrotestosterone | Androgen receptor (AR) | Spermidine synthase |
| Corticosterone, dexamethasone, | Glucocorticoid receptor (GR) | Tyrosine aminotransferase |
| WY-14643, clofibrate, gemfibrozil, diethylhexylphthalate | Peroxisome-proliferator-activated receptor α (PPARα) | Acyl-CoA oxidase |
| Pioglitazone, troglitazone, rosiglitazone | PPARγ | c-Cbl-associated protein (CAP) |
| Phenobarbital[a] | Constitutive androstane receptor (CAR) | *CYP2B1/2B2* |
| Pregnenolone-16α-carbonitrile | Pregnane X receptor (PXR) | *CYP3A* |
| TCDD, benzo[a]pyrene | Aryl hydrocarbon receptor (AHR) | *CYP1A* |

[a] Phenobarbital binds CAR with relatively low affinity, thus other mechanisms for *CYP2B1/2B2* upregulation are possible.[16]

pharmacology and toxicology in a biological system. Presumably, then, high-risk chemical features can be eliminated from subsequent molecules. However, identification of beneficial features that can be expanded upon are beyond the scope of toxicogenomics and thus will probably remain within the realm of validated high-throughput pharmacological assays. Currently, pharmacology groups within pharmaceutical companies use the SAR approach, but parallel toxicity SAR at the genomic level is still only conceptual.

Establishing the relationship between distinct chemical features and toxicogenomics is complicated by the fact that toxicity is often a multistep process. One possible way to address the problem of relating chemical information to toxicity is to employ the fundamental principles of toxicology, including dose and time response. A hypothesis that could be put forth is that gene expression changes at early time points are compound specific, while gene expression changes at later times are specific to the toxic response. Although logical, a comprehensive approach dramatically increases the amount of microarray work required.

### 11.4.2 Pathology-Driven Databases

A toxicogenomics database driven on the pathology concept relies on the ability of certain compounds to induce particular toxicologically relevant findings (Table 11.3). Endpoints in this context minimally include histopathological alterations and significant changes in serum chemistry parameters. The strength of such a database is based on the fact that these types of observations are very well documented and can be associated with toxicant-induced organ dysfunction. Thus, the pathology-driven database can be considered compound independent because structurally diverse compounds can cause similar toxicities. For example, both cisplatin and cephaloradine cause kidney tubular necrosis despite being structurally unrelated.[8,17] A gene expression profile database grouped according to the type of toxicological endpoint with which the profiles are associated would be useful in identifying the subclasses of expression profiles for different histopathological lesions. Importantly, the significance of being able to distinguish a particular lesion from a histopathologically similar lesion would be realized if an impact on risk assessment could be gained. New expression profiles can be compared to the database of pathology-associated profiles in order to classify and associate new data. Another advantage of the pathology-driven database is the inherent ability to leverage a large knowledge base of toxicity information. Historical literature contains information on classical toxicity endpoints for hundreds of compounds that can be mined and included in the database. Information such as target organ toxicity, dose–response relationships, solubility, carcinogenicity, and acute lethality ($LD_{50}$) enables accurate dosing and establishes correlative toxicity data. The correlation of traditional toxicity to gene expression forms the basis for connecting gene expression to more established measures of toxicity. In addition, if non-target tissues are used as a control, then mechanistic insight into target tissue toxicities is possible.

Gene expression changes, like other compound-mediated biological events, tend to follow distinctive time courses. Measurement of genes at a single time only provides a snapshot of the continuum of gene expression responses and may lead to incorrect conclusions. Interpreting single

**TABLE 11.3**
**Liver Toxicity Endpoints**

| Endpoint | Representative Toxin |
|---|---|
| Necrosis | Acetaminophen |
| Increased transaminases | Aflatoxin $B_1$ |
| Inflammation | Lipopolysaccharide |
| Hepatocyte proliferation | Phenobarbital |
| Peroxisome proliferation | Clofibrate |

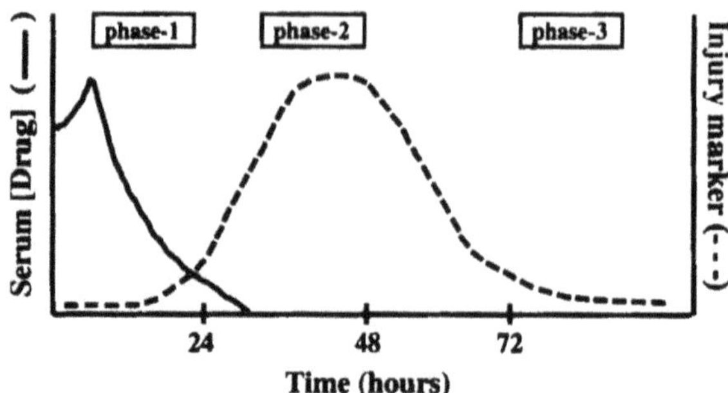


**FIGURE 11.4** Time–response measurement of genes relative to drug distribution and appearance of toxicity. The solid curve reflects the time course of drug presence in serum following a single dose. The dashed curve is the compound-associated toxicity. Each rectangular area represents distinct time frames for gene expression measurement.

time-point gene expression data is only minimally informative in regard to the underlying biological processes. A hypothetical example of time-course gene expression measurement, loosely based on hepatic injury models,[18,19] is depicted in Figure 11.4. In this example, the clinical manifestation of toxicity is delayed until ~24 hours after single-dose exposure. Depicted at the top of the figure are three phases of time in which gene expression changes could be assessed. During the first phase, the drug is undergoing distribution and is not yet at steady state; the initial effects of toxicity are occurring at the subcellular level with little or no accompanying tissue damage. During the second phase, the subcellular tissue damage process becomes severe enough to lead to overt damage of an increasingly greater proportion of cells in the hypothetical tissue, and the damage becomes detectable by traditional toxicity assays (e.g., transaminase assays, histological alterations). Finally, during the third phase, the tissue is undergoing repair processes necessary to restore normal organ homeostasis. Using this paradigm it is easy to understand the diversity of transcriptional responses that could take place. During the first phase of response, primary response genes (i.e., genes not likely to be dependent on synthesis of protein) are likely to be expressed. Genes in the primary response include c-*myc*[20] and c-*jun*[21] or interferon-inducible protein 10,[22] in addition to acute-phase-response genes induced in models of hepatic injury. The second phase of gene expression is indicative of the specific genomic response to toxicity. Genes in this category may include genes such as *waf-1*, *gadd45*, and *p21*.[23] Genes in the second phase control the cell cycle in order to limit permanent damage to critical cellular molecules such as DNA. Frequently, genes involved in inflammatory processes are also expressed during the second phase and may contribute to the resultant pathology.[24,25] Finally, in the third phase, the cells in the affected regions of tissue are commencing repair processes and initiating proliferation to replace cells lost to the toxic insult.[26,27] The preceding example, albeit hypothetical, provides a demonstration of how gene expression can be temporally connected with pathological observations of toxicity. The link between gene profiles and pathology provides mechanistic understanding of the underlying pathophysiological processes.

### 11.4.3 Hypothesis-Driven Databases

Creating a toxicogenomic database underneath a research hypothesis has an advantage in that critical scientific questions can be addressed while data are continually being added and the hypothesis evolves. In fact, hypothesis-driven databases may include features characteristic of the compound- and pathology-driven databases discussed earlier. However, a concise hypothesis will result in more datasets intended to answer focused research questions. In the end, a database built on hypothesis-driven research will likely be either an amalgamation of several different lines of parallel research or a single hypothesis taken down along a series of step-wise experiments.

Detailed descriptions of a hypothesis-driven database would be difficult to address due to the endless numbers of hypotheses that can be conceived; however, one concept worth considering is relating data across the "-omics" technologies. Genomics technologies have evolved sufficiently to allow for rapid, consistent, and reliable generation of data. Proteomics and metabonomics studies that are run in parallel with genomics are currently more limited in scope because the throughput rate of samples is still lower due to technological limitations. Yet, if history holds, both proteomics and metabonomics technologies will quickly overcome the current limitations and begin to yield data similar in scope to genomics. Thus, the study design for a hypothesis-driven genomics database should include the necessary sample collection protocols to enable future experiments to assess both protein and metabolite changes.

## 11.5 SUMMARY

The advancements in molecular biology that spawned the genomics revolution have resulted in a massive output of data that requires sophisticated software to be useful. Parallel advancements in miniaturization of genomics technologies and computer speed and storage capabilities have enabled the creation of incredibly vast databases of sequence and expression information. Toxicogenomics databases are unique in that the essential component is empirically derived differential gene expression data. Sequence and annotation or functional descriptions support the gene expression information and provide insight into toxic mechanisms. Finally, making the connection between gene expression profiles and classical toxic endpoints will result in the ability to predict the onset of toxicity earlier, in addition to identifying features of new chemical entities that have a toxic liability. Predictive toxicology promises new and accurate insight into toxicity, resulting in new and safer drugs being delivered to patients more quickly.

## REFERENCES

1. Nuwaysir E.F., Bittner M., Trent J., Barrett J.C., and Afshari C.A., Microarrays and toxicology: the advent of toxicogenomics, *Mol. Carcinog.*, 24, 153, 1999.
2. Farr, S.B. and Dunn, R.T., II, Concise review: gene expression applied to toxicology, *Toxicol Sci.*, 50, 1, 1999.
3. Kang, J.J., Kaysen, G.A., Jones, H., Jr., and Holland, M.J., Rat liver transcript profiling in normal and disease states using a kinetic polymerase chain reaction assay, *Methods*, 13, 437, 1997.
4. Shimkets, R.A., Lowe, D.G., Tai, J.T., Sehl, P., Jin, H., Yang, R., Predki, P.F., Rothberg, B.E., Murtha, M.T., Roth, M.E., Shenoy, S.G., Windemuth, A., Simpson, J.W., Simons, J.F., Daley, M.P., Gold, S.A., McKenna, M.P., Hillan, K., Went, G.T., and Rothberg, J.M., Gene expression analysis by transcript profiling coupled to a gene database query, *Nat. Biotechnol.*, 17, 798, 1999.
5. Pitot, H.C. and Dragon, Y.P., *Casarett and Doull's Toxicology*, 5th ed., McGraw-Hill, New York, 1996, chap. 8.
6. Waring, J.F., Ciurlionis, R., Jolly, R.A., Heindel, M., and Ulrich, R.G., Microarray analysis of hepatotoxins *in vitro* reveals a correlation between gene expression profiles and mechanisms of toxicity, *Toxicol. Lett.*, 31, 359, 2001.
7. Waring, J.F., Jolly, R.A., Ciurlionis, R., Lum, P.Y., Praestgaard, J.T., Morfitt, D.C., Buratto, B., Roberts, C., Schadt, E., and Ulrich, R.G., Clustering of hepatotoxins based on mechanism of toxicity using gene expression profiles, *Toxicol. Appl. Pharmacol.*, 175, 28, 2001.
8. Huang, Q., Dunn, R.T., II, Jayadev, S., DiSorbo, O., Pack, F.D., Farr, S.B., Stoll, R.E., and Blanchard, K.T., Assessment of cisplatin-induced nephrotoxicity by microarray technology, *Toxicol. Sci.*, 63, 196, 2001.
9. Rininger, J.A., DiPippo, V.A., and Gould-Rothberg, B.E., Differential gene expression technologies for identifying surrogate markers of drug efficacy and toxicity, *Drug Discov. Today*, 5, 560, 2000.
10. Furness, L.M., Analysis of gene and protein expression for drug mode of toxicity, *Curr. Opin. Drug Discov. Dev.*, 5, 98, 2002.

11. Altschul, S.F., Gish, W., Miller, W., Myers, E.W., and Lipman, D.J., Basic local alignment search tool, *J. Mol. Biol.*, 215, 403, 1990.
12. Fielden, M.R., Matthews, J.B., Fertuck, K.C., Halgren, R.G., and Zacharewski, T.R., *In silico* approaches to mechanistic and predictive toxicology: an introduction to bioinformatics for toxicologists, *CRC Crit. Rev. Toxicol.*, 32, 67, 2002.
13. Burczynski, M.E., McMillian, M., Ciervo, J., Li, L., Parker, J.B., Dunn, R.T., II, Hicken, S., Farr, S., and Johnson, M.D., Toxicogenomics-based discrimination of toxic mechanism in HepG2 human hepatoma cells, *Toxicol. Sci.*, 58, 399, 2000.
14. International Life Sciences Institute (ILSI) Health and Environmental Sciences Institute, *Newsletter*, 1, 1, 2002.
15. Hamadeh, H.K., Bushel, P.R., Jayadev, S., Martin, K., DiSorbo, O., Sieber, S., Bennett L., Tennant, R., Stoll, R., Barrett, J.C., Blanchard, K., Paules, R.S., and Afshari, C.A., Gene expression analysis reveals chemical-specific profiles, *Toxicol. Sci.*, 67, 219, 2002.
16. Moore, L.B., Parks, D.J., Jones, S.A., Bledsoe, R.K., Consler, T.G., Stimmel, J.B., Goodwin, B., Liddle, C., Blanchard, S.G., Willson, T.M., Collins, J.L., and Kliewer, S.A., Orphan nuclear receptors constitutive androstane receptor and pregnane X receptor share xenobiotic and steroid ligands, *J. Biol. Chem.*, 275, 15122, 2000.
17. Hori, R., Shimakura, M., Aramata, Y., Kizawa, K., Nozawa, I., Takahata, M., and Minami, S., Nephrotoxicity of piperacillin combined with furosemide in rats, *Jpn. J. Antibiot.*, 53, 582, 2000.
18. Bauer, I., Vollmar, B., Jaeschke, H., Rensing, H., Kraemer, T., Larsen, R., and Bauer, M., Transcriptional activation of heme oxygenase-1 and its functional significance in acetaminophen-induced hepatitis and hepatocellular injury in the rat, *J. Hepatol.*, 33, 395, 2000.
19. Schmiedeberg, P., Biempica, L., and Czaja, M.J., Timing of protooncogene expression varies in toxin-induced liver regeneration, *J. Cell Physiol.*, 154, 294, 1993.
20. Kulkarni, S.G., Harris, A.J., Casciano, D.A., and Mehendale, H.M., Differential protooncogene expression in Sprague–Dawley and Fischer 344 rats during 1,2-dichlorobenzene-induced hepatocellular regeneration, *Toxicology*, 139, 119, 1999.
21. Zhou, T., Zhou, G., Song, W., Eguchi, N., Lu, W., Lundin, E., Jin, T., and Nordberg, G., Cadmium-induced apoptosis and changes in expression of p53, c-jun and MT-I genes in testes and ventral prostate of rats, *Toxicology*, 142, 1, 1999.
22. Koniaris, L.G., Zimmers-Koniaris, T., Hsiao, E.C., Chavin, K., Sitzmann, J.V., and Farber, J.M., Cytokine-responsive gene-2/IFN-inducible protein-10 expression in multiple models of liver and bile duct injury suggests a role in tissue regeneration, *J. Immunol.*, 167, 399, 2001.
23. Wu, H., Wade, M., Krall, L., Grisham, J., Xiong, Y., and Van Dyke, T., Targeted *in vivo* expression of the cyclin-dependent kinase inhibitor p21 halts hepatocyte cell-cycle progression, postnatal liver development and regeneration, *Genes Dev.*, 10, 245, 1996.
24. Lalor, P.F., Shields, P., Grant, A., and Adams, D.H., Recruitment of lymphocytes to the human liver, *Immunol. Cell Biol.*, 80, 52, 2002.
25. Robertson, D.G., Reily, M.D., Sigler, R.E., Wells, D.F., Paterson, D.A., and Braden, T.K., Metabonomics: evaluation of nuclear magnetic resonance (NMR) and pattern recognition technology for rapid *in vivo* screening of liver and kidney toxicants, *Toxicol. Sci.*, 57, 326, 2000.
26. Diehl, A.M., Liver regeneration, *Front. Biosci.*, 7, E301, 2002.
27. Yin, L., Lynch, D., and Sell, S., Participation of different cell types in the restitutive response of the rat liver to periportal injury induced by allyl alcohol, *J. Hepatol.*, 31, 497, 1999.

# 12 Predictive Toxicogenomics

*Mark W. Porter, Arthur L. Castle, Michael S. Orr, and Donna L. Mendrick*

**CONTENTS**



0-8493-1334-1/03/$0.00+$1.50

## 12.1 INTRODUCTION

Toxicogenomics, the study of gene expression changes following toxicant exposure, could deliver insight into potential human toxicity earlier in the drug development process (Figure 12.1). This new technology may provide a platform for the identification and ranking of potential toxicities for new chemical entities (NCEs), thus increasing the likelihood of success in preclinical and clinical development of drugs. Traditional preclinical toxicology studies are time consuming and costly, may require high amounts of each compound under study, and may not accurately predict human toxicity. The limited number of endpoints evaluated in preclinical toxicology is estimated to be less than 100 in even the most comprehensive studies[1] and may not adequately measure the potential for toxicity of an NCE in humans. Transcriptome-wide gene expression studies targeted to understanding the impact a toxicant has across highly diversified, multifunctional gene sets and numerous pathways will revolutionize the manner in which toxicologists evaluate human safety issues.[1–6]

The field of toxicogenomics is growing quickly, as evidenced by the number of published articles in this field viewed over time (Figure 12.2). The majority of preliminary published articles were reviews describing the promise of this technology. Recent publications have focused on investigative pursuits through experimentation that attempt to characterize the applicability of toxicogenomics as a more sensitive measure for the classification of toxicity and its mechanisms.

The construction of a reference database and the implementation of several analytical and visualization techniques effective for toxicity discrimination and mechanism discovery are key components to a successful toxicogenomics program. All of these components are discussed in this chapter. Analysis techniques integral to a predictive program are reviewed, as are the parameters that affect the construction and utility of predictive toxicogenomics models. The ToxExpress™ (trademark of Gene Logic, Inc.; Gaithersburg, MD) program, Gene Logic's proprietary approach to toxicogenomics, is discussed, and, finally, specific applications of toxicogenomics using the ToxExpress reference database, analysis tools, and visualizations will be presented.

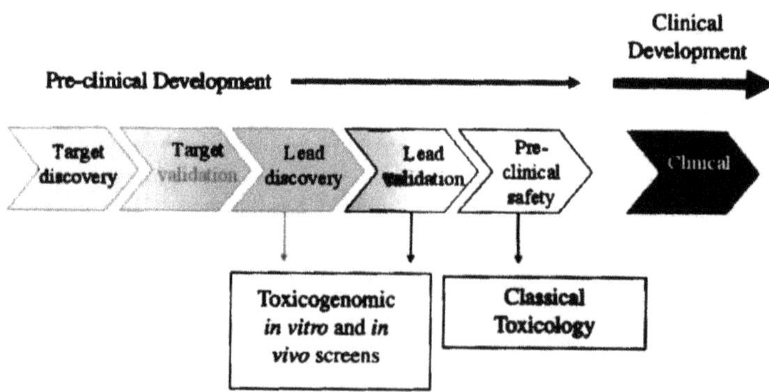


FIGURE 12.1 Toxicogenomics and drug development.

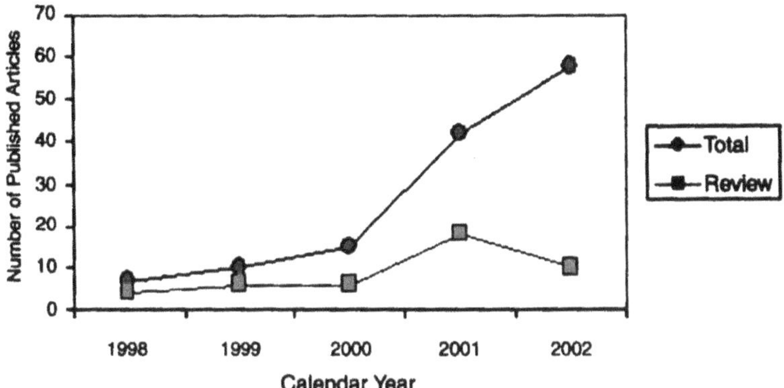


**FIGURE 12.2** The numbers of articles in 2002 are extrapolated from the first 6 months.

## 12.2 PREDICTIVE MODELING AND ANALYSIS USING LARGE GENE EXPRESS REFERENCE DATABASES

Predictive modeling is an application of statistical inference and does not inherently answer mechanism-based questions; however, the behavior of individual marker genes within predictive models can infer specific mechanisms in reference to the larger database. Therefore, several types of issues can be resolved using predictive modeling algorithms as a starting point, including: (1) early-stage compound toxicity ranking using *in vitro* or *in vivo* testing, (2) mechanistic-based analyses for toxicity, and (3) discovery of unique biomarkers that could be useful in evaluating preclinical and clinical studies. Contained within this section is a review of the steps in the construction and application of predictive models in general. The processes discussed include selection of gene markers, classification of samples and genes, validation of predictive models, and database construction and development.

### 12.2.1 Selection of Gene Markers

The advantage of microarray experiments in the classification of toxicity lies in their potential ability to measure gene expression levels across the entire transcriptome of an organism. A single chip experiment could yield expression information for up to tens of thousands of parameters, a much more informative situation than with the dozens of classical measures of toxicity. However, this advantage also presents multiple challenges, as, obviously, tens of thousands of parameters are more difficult to analyze than a few dozen. Whereas analysis of clinical chemistry and histopathology data can be performed inside the clinic by a trained observer with little more than a spreadsheet, the highly multidimensional space of microarray data quickly overwhelms the investigator. In addition, microarray platforms have been criticized in the industry due to the amount of inherent noise for any single measurement. It is difficult to separate the true measurement for any particular probe on a chip from the noise associated with that probe. Statistical validation along with subsequent biological validation is usually necessary for such an analysis.

To begin building predictive models using microarray experiment data, investigators must first identify marker genes that can classify the types of toxicity one is interested in assaying. For instance, the general toxic response can be theoretically defined as those genes for which expression levels are altered to approximately the same degree irrespective of the specific type of toxic effect the compound may induce. Likewise, a marker of toxicity that is specific to a type of pathology such as necrosis will be defined by its ability to be regulated in the same manner with all necrosis-causing agents. The potential for assays that yield both predictive results and mechanistic information in this context is great. Using predictive markers that characterize a general toxic response, specific pathologies, and even specific compounds of known toxicity, researchers can effectively fully characterize the toxicity of a compound and can concentrate only on those predictive markers that characterize the toxicity profile of interest.

Not all genes change with exposure to a particular toxicant. However, the definition of a "change" is not well characterized and is both platform and experiment specific. Microarray platforms provide for tens of thousands of gene expression measurements, where each of the genes being measured has an associated noise produced from the physical aspects of chip processing and from biological differences associated with individual specimens. By limiting the focus in predictive toxicology to only those genes that truly change due to toxicity, investigators can effectively increase the signal-to-noise ratio and improve the chance for successful classification of new compounds. Confounding biological factors include basic parameters such as pharmacological effects, time course effects for individual toxicants, and individual specimen susceptibility to the toxicant. Confounding physical microarray factors include basic parameters such as scanner variability, chip lot variability, and hybridization times. Both forms of variability are also influenced by more subtle factors that are too amorphous to identify. Although standardization of protocols at the chip processing and biological experiment levels will help to remedy these issues, the true solution, although costly, lies in the construction of a large reference database that contains many known toxicants and control compounds, multiple doses of those toxicants, several time points post-exposure, and biological replicates for each condition. Multiple doses of toxicants allow for the separation of the pharmacological effect from the toxic response. Time course data yield the obvious benefit of observable changes of the toxic response over time through gene expression and offer the additional benefit of improving the chances of observing a true toxic response compared to using a single time point. Finally, biological replicates are necessary for establishing significance measures of changes relevant to toxicity.

The wide variety of gene selection procedures all share the common goal of ranking genes with respect to distinct differences the analyst or biologist deems important. Ideally, ranking should be based on the probability that the expression differences are reproducibly different from a "normal" condition and not due to random noise.[7] This method of statistical inference depends on assumptions regarding the sample size and characteristics and adherence to the behavior of the universal population. All genes differ in their dynamic ranges of expression as well as in their overall biological variability, and each may be affected differently by microarray processing variability. A large reference database directly measures each one of these parameters, allowing for a reliable statistical inference that applies to each gene separately and adheres to the majority of parameters associated with the population.

Parametric and nonparametric techniques both have been applied to identify marker genes successfully for a number of different biological applications.[7–13] Normal distributions are optimal for gene selection procedures. In the event that a raw normal distribution of signal intensity for the majority of genes is not observed, numerous transformations may be applied to achieve this characteristic. Upon obtaining a normal distribution for the large majority of genes, various well-accepted, easily validated statistical tests of inference, including Student's *t*-test, analysis of variance (ANOVA), or the Z-test, can be applied to the dataset to rank genes.[7,12] Although the excessive numbers of variables prohibit the use of standard *p*-values as measures of significance, standard *post hoc* corrections, or bootstrapping and other randomization adjustments, can give reliable estimates of false positive probability.[7,11,12]

The more common situation for the researcher or analyst is the identification of markers based on a much smaller pool of data. In most cases, the number of conditions and the number of replicates for each of those conditions are limited. A fold-change threshold best serves as an *ad hoc* gene selection procedure in these cases.[14–16] It should be noted that the significance associated with a summary fold change value is only as powerful as the sample size available. Thus, adjustments to *p*-values and other significance measures become increasingly important. In previously published cases, the variability among all genes has been used to better estimate intra-gene variability.[10,16,17]

**TABLE 12.1**
**Multivariate Methods of Sample Classification**

| Method | How It Works |
|---|---|
| Linear discriminant analysis | Weights assigned to each gene based on distribution of known samples in the training set. Requires enough samples to estimate distribution adequately. Uses probability functions based on normal distributions and requires the assumption of normality. |
| Logistic regression | Weights assigned to each gene based on distribution in training set and requires enough samples to estimate distribution adequately. Estimates a linear transformation of the logit function for distribution values. Robust against departures from normal distributions. |
| Neural networks | Use functional units called *nodes* that calculate inputs, process relationships between input nodes, and calculate output. Weights assigned are learned from a training set. Requires many samples to accurately estimate weights. No assumptions on underlying distributions are needed. Can learn from new information. |
| Support vector machines | Draws an optimal complex boundary through multidimensional space that separates sample groups in the training set. Depends on each sample being good representatives of groups; highly sensitive to measurement errors or outliers. |
| Clustering approaches | Uses guilt-by-association method; samples are classified as to which they most closely resemble. Can be a useful method when the number of samples is small. Assumes the same measure of closeness can be applied equally to all genes; the measure of closeness (metric) can greatly influence results. |

### 12.2.2 Classification of Samples

The goal of predictive modeling is successful characterization of the toxicity of samples treated with compounds for which no or only a minimal amount of safety information is available. Classification success depends on the reliability that changes in known samples accurately reflect what will be seen in the population being tested. Classification as *normal* or *pathologic* must consider all possible normal states and abnormal pathologies; therefore, a large number of toxin-treated and control samples must be used to approximate the uniqueness of the markers to measuring the toxic events. The components of a predictive modeling approach are the marker genes themselves, a training set of samples, and a test set of samples with associated gene expression data. First, appropriate marker genes are selected and a predictive model is trained using the reference database. Measures of intensity, noise, and variation across the entire training set, including proper controls, are input into the system by way of the appropriate algorithm. Next, test samples are measured against the resulting model. Finally, proper algorithms are applied to the constructed model to assure the model is robust to the population outside the training set. Given the number of parameters in any microarray experiment, the risk of over-fitting any model is large, especially those built using nonlinear techniques. Cross-validation is discussed in the next section, but it should be noted that no predictive modeling exercise is complete without this type of assessment.

Many methods are actively used to classify samples described by a large numbers of parameters such as gene expression. Table 12.1 summarizes several methods, including logistic regression (LR), linear discriminant analysis (LDA), neural networks, support vector machines, and co-clustering approaches.[8,9,18–25] Each method classifies data using different formulas and rules, each has their own set of limitations, and each makes certain assumptions about data characteristics. The choice of the optimal method for a particular dataset depends on several factors, including sample size, data distribution, the intended capacity of the assay for sensitivity, and available processing power. Several gene expression datasets are available to the public for the comparison of these various methods.[26–29]

Logistic regression and linear discriminant analysis are well-described, basic methods of statistical inference that use linear weighted contributions from the signal intensity distribution of each gene to classify samples.[8,9,30] These methods generally assume a continuous, normal distribution and presuppose that relevant relationships between gene sets and samples necessary for prediction can be estimated with a linear or transformed linear relationship. It is especially important to justify these assumptions by using sample numbers on the order of at least dozens, hundreds, or even thousands, depending on the overall variability and noise associated with the dataset. LR estimates a linear transformation of the logit function based on the distribution values between groups, whereas LDA uses a linearly weighted probability function based on normal distributions.[8,9]

Neural networks were originally inspired by the architecture of the brain and its capacity to learn based on input and output interconnections of neurons. The process is based on the observations that strong associations between neurons are represented by stable synapses and that each neuron has an assigned threshold of activation. In artificial modeling terms, the neurons are replaced by intermediate input and output calculators called *nodes*, and the threshold is represented by an activation function specific for each node. As the model is trained, it increases or decreases the strength of node associations at the expense of others and constantly recalculates activation thresholds for individual nodes. The many advantages associated with neural networks include their ability to adapt to new data quickly and their ability to learn and identify extremely complex patterns; however, they require a large number of samples to accurately assign parameters to each node. Neural networks have been used increasingly for the general task of finding associations between genes given certain conditions. In addition, they have been used for the classification of samples once those associations have been defined.[31,30]

Clustering approaches use a guilt-by-association method to classify samples and have been applied to the classification of the toxic response.[18,19,21–23,32–34] Similarity matrices are used to group samples together that exhibit a similar overall gene expression pattern. These similarity matrices can be based on covariance, correlation, or other measures. Once a similarity metric is defined, these methods calculate signal intensity distances without the application of predetermined weights based on discriminative priority and in the absence of any formal training process. The result is a method that allows for a general classification that is easily represented in common visualizations; however, many relationships established by the method are insignificant yet will contribute to the final predictive outcome.[11,30,35] In some cases, clustering approaches are very sensitive to new samples being added to the training set. In the example of hierarchical clustering, an association tree is built with sample nodes hierarchically related to other sample nodes by distance measures, represented by linkage branches. The addition of a single sample to a training set originally consisting of 100 or more samples will completely restructure the tree to some extent, causing an inherent instability in thresholds for classification. Clustering approaches are commonly used for classification when few samples are available and more robust statistical inference and rule-based methods cannot be applied.

Many methods can be used to select target markers of toxicity and recognize and classify types of toxicity. All are limited, however, by how well the collected data resemble the true toxic gene expression profile population and by the fit of the collected data with the selected method. In order to establish a modeling approach, a method must be selected and applied uniformly to all samples and genes within the gene expression database of treated tissues.

### 12.2.3 Classification of Genes

The functional classification of genes can be achieved through classical methods of sequence analysis at the nucleotide and protein levels. Genes can also be classified by their overall observed ontology, or function, and by family membership. In addition, the functional classification of genes can be achieved through analysis of assays that measure the coregulation and coexpression of genes over multiple conditions (e.g., normal, diseased, treated, etc.). Gene expression is ideal for the assessment of common regulatory events that can lead to conclusions about the mechanisms of disease states and the mechanism of action of drugs, in both their pharmacological and toxic effects. The utility of gene classification in predictive genomics in general, and predictive toxicology specifically, is limited due to the fact that predictive toxicology is more concerned with the classification of samples. The classification of genes in this case is secondary to their ability to recognize the toxic response.

Clustering of genes is a widely used and accepted technique to find coregulation events; however, issues must be taken into account when applying clustering algorithms to these datasets. First, the same measure of similarity is used in calculating distances between genes, without regard for the dynamic range and variability associated with any one particular gene.[30] Therefore, no level of confidence can be assigned to the relationship between any two genes. Second, most clustering algorithms do not allow for membership in more than one cluster. This is not biologically accurate, as many genes can influence and reside in many different biological pathways. Finally, clustering algorithms, by definition, assign structure to the relationships between genes. In most cases, these structures are hierarchical in nature, assuming parent–child relationships that are not representative of a truly complex biological system. Despite these limitations, clustering is essential to the identification of unsupervised relationships between genes and allows for quick determinations about functional activities of genes in reference to the treatment or disease state being investigated.

Hierarchical clustering has been used extensively to categorize genes.[36–38] Each gene is related to others by means of an association tree that groups genes together and subsequently subgroups the resulting groups of genes. These trees can be built divisively by progressively splitting from one all-inclusive set of genes into smaller subgroups or by pairing samples together and iteratively building upon these pairs in an agglomerative approach.[18,35,38,39] The tree establishes the parent–child relationships between groups of genes and establishes metrics and distances between the groups.[35,38,40] Each group can be defined by an average expression profile across all of the component genes. Although the assumption is that all genes represented in the group will adhere to this pattern or something close to it, this is not always the case.[35,30] If the underlying system being measured is inherently hierarchical, then this structure will characterize the system well.[35] Despite the fact that biological systems do not truly exhibit hierarchical systems, visual relationships between genes are easily extracted. Thus both bench scientists and bioinformaticians often use the algorithm.

*K*-means clustering groups genes into a predetermined number of clusters.[39,41,42] All genes within a cluster are commonly related by their residence in that cluster. The method allows for the representation of nonhierarchical relationships and is useful for assays where only a few patterns of gene expression will dominate the system. A gene can only establish residency in one cluster, and the clusters have no inherent relationship to one another. Other methods, such as self-organizing maps and multidimensional scaling, map genes relative to each other in two- or three-dimensional space, but are not totally unsupervised methods.[8,22,39,43–45]

One of the potential pitfalls of clustering is that the algorithms do not remove genes based on their ability to conform to a desired outcome; therefore, if a gene does not belong to a cluster, or if it only has a tenuous relationship to another group of genes, it may still be clustered in relation to those genes. The majority of the results obtained through clustering may be driven solely by insignificant genes or by very few outlier genes. Proper marker gene selection, comparison of multiple clustering methods using a variety of metrics, and strong validation measures are suggested to help overcome these issues.[18,22,35,39,40,46] The addition of random noise to the system or cross-validation using randomization of sample classifiers will help to establish confidence for the entire clustering process.[11,23,29,47]

A variation of clustering involves the derivation of gene associations to build large networks of genetic associations. The multidimensional mapping of each gene to all other genes, either directly or through its interactions with other genes in the same network, essentially permits genes to appear in multiple clusters and best allows for determination of the true biological situation.[48–55] The simplest form of a genetic network is the Boolean network, where all genes are related to each other by an "on or off" pattern.[54,56] More complex algorithms include the modeling of a saturating S-shaped response, linear modeling of the function of each gene based on weighted effects of all other genes, and self-evolving algorithms.[37,49,50,55–57] Although these algorithms have the potential to aid the researcher in reverse engineering large-scale data into biological endpoints, they require large amounts of accurate data.[56]

### 12.2.4 Validation of Predictive Models

Once a predictive model has been built and success rates are established over the training set, it is important to further interrogate the model to assure its applicability to multiple test sets. The measure of predictive accuracy is no longer determined by its ability to predict the training set but is determined by the ability to predict multiple test sets relative to the original training set. This process allows for an overall assessment of the true error within the original model and will aid in the identification of modeling pitfalls such as data over-fitting, data assumptions that are not actually observed, and sample size issues. The process can also be used to determine thresholds at which model construction is halted, an especially useful concept in the building of neural network-based and classification tree-based models. Statistical cross-validation and resampling methods have been previously utilized in gene-expression-based predictive systems.[11,58]

The statistics discipline has produced a large number of manuscripts devoted solely to the exploration of various cross-validation and resampling procedures with no clear solutions to the biologist or bioinformaticist.[59–63] Many of these papers deal with relatively small intricacies that measure each cross-validation and resampling approach with respect to specific predictive modeling construction methods. For instance, the utility of a bootstrapping algorithm may result in quite different outcomes depending on whether the predictive model of choice is a neural network or a linear regression. These detailed comparative analyses of resampling and cross-validation methods have yet to make it into the microarray analysis domain to a significant extent. A relatively good cross-validation technique can be derived out of the available tools as long as reasonable precautions are taken not to overstate the robustness of a model instance. To that end, this overview does not take into account many of the technical or theoretical details that can be found in the above references.

In an attempt to estimate the overall error inherent to the modeling system, researchers must determine that the training set will apply to almost any incoming test set. The more exhaustive an approach of cross-validation and resampling that can validate the model, the more likely this hypothesis is to be true. The most basic cross-validation method involves removal of subsets of the training set in order to use this set as a test set of samples. Four important parameters need to be determined before beginning this process: (1) the number of samples to be removed from the training set as a percentage of the whole, (2) numbers of sample removal iterations, (3) sample replacement algorithm, and (4) marker gene reselection.

The number of samples to be removed from the training set is dependent on the original training set sample size. Resampling procedures are normally implemented by $k$-fold sample removal, where $k$ equals the number of different sets into which the original training set samples are divided. The lower the $k$ value, the more samples are to be removed from the training set and, in general, the more robust the model tends to be. For instance, in the case of a model composed of 600 training set samples, when $k = 10$ we remove 60 samples from the original training set, build the model on the remaining 540 samples, and establish classification success rates based on the 60 removed samples. Given the same original training set, when $k = 3$ we remove a total of 200 samples from the original training set and follow that same procedure. Obviously, much more information is removed upon each iteration with the $k = 3$ situation than with the $k = 10$ situation. If our success rates approach that of our original training set using both $k = 3$ and $k = 10$, then we have more confidence in our model based solely on the $k = 3$ result. Researchers can extend this to the popular method of leave one out (LOO). In this case, $k = 600$ and we remove only one sample out of the training set. Intuitively, our model will change only minimally and our new success rate will be calculated mainly on the numbers of outliers in the system. Given large sample sizes, LOO is insufficient for estimates of error, but it is a good technique for finding sample outliers and is a valuable determinant of error for very small sample sizes (e.g., <20).

An estimation of classification error inherent to the predictive model cannot be established from a single iteration of sample subset removal. Multiple iterations of sample removal allow for an estimation of both the deviation of a resampling-based error rate from the original rate and the error or variability associated with the original success rate. In general, the more subsets of samples that are removed in the overall process, the more accurate the validation procedure. A large determinant in the numbers of iterations desired for a modeling application is the estimated processing power required. This will be dependent on the type of model being implemented (e.g., linear vs. nonlinear), the numbers of samples in the model, and the numbers of gene variables being assayed. In addition to processing power, there may be a theoretical limitation on the number of iterations that can be implemented. Given the $k$-fold examples above, if the researcher decides not to reuse any samples from the subsets, then the maximum number of iterations will be $k$. However, if replacement of the samples is allowed so that a sample can be found randomly in more than one subset, then an exhaustive number of iterations are allowed and replacement of samples is necessary given a small $k$ relative to the total number of samples.

Gene selection is intimately tied to classification success rates. In the case of cross-validation or resampling procedures, the researcher is not only validating the success of the model relative to the universal population but is also validating the predictive gene set. Cross-validation procedures are especially prone to improper gene selection, as researchers may opt not to reselect a gene set once samples are removed. Upon removal of samples from the training set, predictive gene set changes should be noted upon reselection. Given multiple iterations of sample subset removal, those genes that seem to be fairly stable can be separated from those genes that fall in and out of the predictive gene set. Thus, the validation of the predictive gene set can aid in trimming of the final gene set on which the population will be measured. By not reselecting genes upon sample removal, a selection bias is generated that underestimates the true error of the model.[58]

No cross-validation or resampling procedure can be applied to all predictive models with the same success. The variations on all procedures are almost limitless, and the proper selection of one over another is based on many different parameters. We have outlined only a few here that seem to be the most important given our own modeling paradigm. A few points, however, are clear. First and foremost, some type of cross-validation and/or resampling procedure is required before any claims are made concerning the predictive accuracy of the system. Second, the constructed model is only as good as the validation approach the researcher has chosen, and claims should be limited to the observations from these results. Finally, a very real selection bias is associated with the absence of gene reselection upon removal of samples from the training set.

### 12.2.5 Reference Database Construction and Data Management

A complete toxicogenomics program should be applicable to several types of complex toxicogenomics-related analyses. The expanding field of toxicogenomics has focused on both well-defined and ill-defined issues. For instance, the ability to predict the potential human toxicity of drugs in a drug development pipeline is relatively easy to address in comparison to the ability to characterize novel pathways involved in the toxic response. The more complex and diverse the sample and compound treatment content is for a reference database, the more robust that database will be in addressing multiple issues. To this end, each database parameter requires a thorough investigation with respect to its intended utility and the effects it may have on several different analysis-related outcomes.

A reference database must answer the questions that researchers are currently investigating and must anticipate questions that will be investigated in the future. In the case of predictive toxicology, it is valuable to be able to identify toxicants and rank each compound relative to the entire drug development pipeline. It is also valuable to be able to characterize each compound by its propensity to induce different types of pathology, to separate rat-specific toxicity from toxicity that will eventually appear in humans, to find gene expression changes that precede observable pathology, and to find compound-specific effects that can only be measured by gene expression assays. Several commercial groups, including our own, are currently addressing these issues and others. Future issues are likely to deal with the ability of gene expression to identify yet uncharacterized pathways that are conduits for the toxic response, to correlate specific structural relationships to gene expression changes, and to more effectively link efficacy to toxicity potential.

It is rather difficult to cost-effectively build a reference database that will be able to answer all questions. In the case of structural relationship, it may require tens of thousands of compounds to link structure to expression data reliably, as the universe of compound structures is almost infinite. However, current knowledge of the toxic response allows for the identification of parameters that can be controlled. In the case of structural relationships, known toxicants are available that have isomers that exhibit no toxicity (see the cephalosporin examples at the end of this chapter). Although the relationship between structure and expression is still tenuous at this point, small numbers of these types of examples can be included in a reference database to assess mechanistic differences among multiple structural effects. A complete reference database not only will include those parameters directly related to the toxic response but will also at least contain proof-of-concept examples to further its growth into other areas.

A large reference database is necessarily constructed over a long period of time. Many parameters pertaining to animal treatment, sacrifice, and tissue and chip processing protocols will change over that period of time. The power of a reference database is in its overall ability to deal effectively with each of these changing parameters. By incorporating data generated from changes in these parameters, predictive models will become more and more insensitive to effects caused by them. In the case of predictive toxicology, only those markers will be selected that are unaffected by these variables. As an example, Gene Logic's ToxExpress reference database incorporates data generated from array scanners that were set to a high photomultiplier tube (PMT) level and data generated from scanners set to one tenth of that level, based on recommendations from the manufacturer, Affymetrix. Because we have incorporated these data into the database and build our predictive models based on that same data, we can effectively remove those genes that are affected by the scanner setting, thus making our models insensitive to this variation. Another application of this concept of removing variability in general is found in Section 12.4.

The database management system also plays an important role in the functionality of the reference database content. A relational database structure that links gene expression information to sample and gene annotations is necessary for an effective analysis. In addition, a reliable, visually appealing interface for a management system ensures that bench scientists and bioinformaticians alike can easily retrieve data and format it or export it for further analysis. Multiple commercial and publicly available visualizations and analysis tools are available and serve the purposes of each type of scientist.

## 12.3 THE ToxExpress PROGRAM

Gene Logic develops predictive gene expression-based models to test NCEs for potential toxicity. Gene Logic's extensive reference database contains gene expression profiles and associated classical toxicity endpoint data obtained from thousands of rats and from *in vitro* rat and human primary cells exposed to known toxicants or control compounds. With a thousand or more observations for every gene per organ or cell type in the reference database, direct examinations of gene expression distributions and gene-by-gene variability are possible, as is the selection of targets for determining both the presence of toxicity and its likely mechanism.

For each target organ or cell type, initial dose-ranging studies are completed to determine and verify toxic doses. The toxic dose is defined as one for which overt toxicity (e.g., change in serum aspartate amino transferase [AST] levels, pathological evidence of cholestasis, cytotoxicity in cell culture) is confirmed without undue stress to, or termination of, the animal. The compound being evaluated ultimately determines the study design. A typical study design includes three to four time points, two doses of toxicant, five replicates per time point and dose, and vehicle controls. Vehicle controls are absolutely essential to qualify a baseline for the entire reference database and to assess a baseline for each individual compound. No toxicogenomics program is complete without vehicle controls for each and every experiment, as no claims can be made about the gene expression response for the compounds in individual experiments without a baseline.

Well-characterized, known toxicants are selected based on target organ toxicity, compound class, and species specificity. Selected compounds exert toxic effects in (1) both humans and rats, (2) rats but not humans, or (3) humans but not rats. The use of these three compound types allows investigators to correlate gene expression changes between rats and humans and to help define *in vivo* and *in vitro* screens for proprietary drugs. Drug compounds for the reference database are also selected based on the type of toxicity within tissues (e.g., liver necrosis, steatosis, cholestasis) and the target organ distribution of toxic damage (e.g., liver only vs. liver and kidney). In addition, compounds are selected based on more interesting phenomena that may be specific to that compound. For instance, compounds that exert a heterogeneous response on general clinical and pathology parameters are included in the reference database, allowing investigators to conduct exhaustive studies for accurate toxicity prediction based on gene expression (see example in Section 12.4).

All tissue samples obtained from individual animals or cell culture are studied using the Affymetrix GeneChip® (registered trademark of Affymetrix, Inc.; Santa Clara, CA) Human Genome U133 set (a two-microarray set) or the Rat Genome U34 set (a three microarray set). Approximately 33,000 human known genes and expressed sequence tags (ESTs) are represented on the human U133 set. The rat U34 set contains approximately 26,000 fragments that represent an unspecified number of unique genes.

### 12.3.1 ToxExpress Data Management and Analysis

The Genesis Enterprise System™ (trademark of Gene Logic, Inc.; Gaithersburg, MD) administers and stores this enormous amount of data, serving as both a data management and analysis tool. Standard analysis operations such as fold change calculation, principal components analysis (PCA), and clustering are supported. To complement these operations, the Genesis system offers extensive functional annotations to characterize each fragment completely and to link these fragments to the metabolic and regulatory pathways in which they play a part. Links to public databases such as GenBank (gene information: ncbi.nlm.nih.gov/ntrez.index.html), UniGene (co-clustered gene information: ncbi.nlm.nih.gov/UniGene/), ExPASy (enzyme and protein information: http://us.expasy.org/), Swiss-Prot (protein information: expasy.ch.sprot.sprot-top.html), LocusLink (ncbi.nlm.nih.gov/LocusLink/), and BioCarta (exclusively licensed to Gene Logic) help refine and confirm findings. In addition, all relevant annotations from each of these public and proprietary resources are reported in line with gene expression data.

### 12.3.2 ToxExpress Predictive Modeling

ToxExpress predictive modeling integrates multiple LDA-based models to profile and rank lead compounds.[64] This system includes models based on common toxicity-responsive genes, models affected by compounds exerting similar pathologies or mechanisms of action, and models that define each compound uniquely. This allows users a detailed analysis of the compound under study. For example, a drug might be predicted to induce liver effects in a manner similar to that of

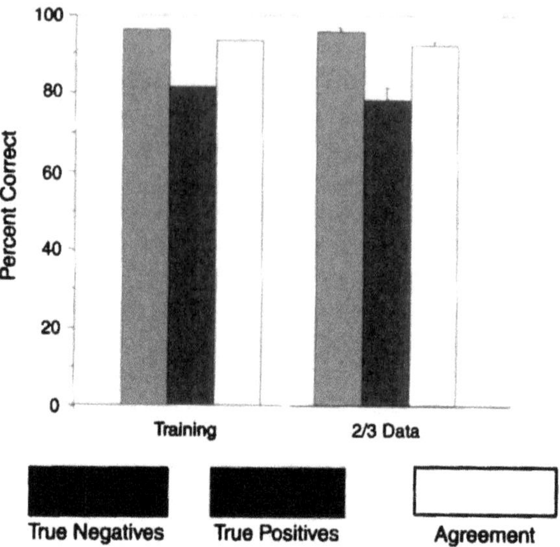


**FIGURE 12.3** Internal cross-validation results. Percent samples predicted correctly are plotted on the *y*-axis. Three categories are calculated: true negatives, true positives, and agreement, which is the summed average of all predictions. True negative rates are based on the correct prediction of "nontoxic" for all vehicles, all low doses of compounds, and high doses of non-hepatotoxicants. True positives are based on the correct prediction of "toxic" for high doses of hepatotoxicants. The bars on the left show rates using all training data. The bars on the right show rates using multiple iterations of modeling when different subsets of one third of the data have been removed. The small error bars and small deviation from the training set data show robustness of the predictive model and associated markers.

cyproterone acetate. As this is a relatively safe drug in humans, a company might choose to further develop the drug. Contrast this with the decision to be made if the drug matches a known strong toxicant such as alpha-nathylisothiocyanate (ANIT). Such detailed prediction is possible because of the massive amounts of gene expression data present in the ToxExpress reference database. Statistical inference ranks the probability that specific gene expression values are truly different and not the result of normal biological variation. High-quality gene expression data result in robust predictions based on the identity and magnitude of gene responses, not based solely on the number of gene expression changes. By adding gene-expression-based predictions to the existing knowledge base for each compound, a more comprehensive profile results.

### 12.3.3 Validation of the ToxExpress Predictive System for *In Vivo* Liver

Gene Logic's approach to validation of predictive models requires using both internal and external data sources. This approach assures that the models perform correctly and are robust enough to deliver results even when variability exists in the data tested. Gene Logic's validation of predictive models using internal data adheres to the concepts defined in the previous analysis section. This approach assesses the influence any one subset of data has on the modeling results. A strong influence suggests the reference database has not reached *convergence*, a theoretical state where the data within the database are sufficient to characterize the universe of compounds. Little or no influence suggests that the dataset is converged and that the database can stop growing for the purposes of prediction for the type of toxicity being measured. Each predictive model of general toxicity is validated by using $k = 3$, at least 10 iterations, sample replacement, and gene reselection upon each iteration. As shown in Figure 12.3, the *in vivo* liver model for the prediction of general toxicity is robust even when large amounts of data are removed. Examination of genes during the

**Using external data to validate the ToxExpress predictive system.**

| Samples | Customer Expectations | ToxExpress Predictive System | Interpretation |
|---|---|---|---|
| 126 | + | + | True positive (toxic) |
| 122 | – | – | True negatives (non-toxic) |
| 9 | – | + | False positive |
| 18 | + | – | False negative |
| 275 | | | |

*Note:* Agreement = 126 + 122/275 = 90%.

**FIGURE 12.4** Validating the *in vivo* liver predictive models using external data provides information relating to how well the system predicts toxicity against gene expression data generated at an external site. This data demonstrated that the ToxExpress predictive system has high concordance with the reference method. Every sample was scored as correct or incorrect. Samples were assigned a correct call when the customer agreed on the finding that was often supported by clinical data. An incorrect call was assigned when the customer disagreed with the finding, sometimes solely based on lack of clinical pathology.

validation process shows that the same genes are reselected upon each iteration, with very few deviations (results not shown).

The second validation approach assesses data from external sources. To date, seven companies have submitted Affymetrix GeneChip U34A data from over 270 rat liver samples. In all, 31 compounds were tested, 19 of which were not represented in the ToxExpress *in vivo* predictive models or the reference database. In addition, five strains of rats, multiple doses and time points (2 hr to 8 days), and data generated from multiple operators and sites were submitted. As shown in Figure 12.4, the ToxExpress predictive system demonstrated a high agreement between the customers' expectations and the predictive modeling results. Most importantly, of the three non-hepatotoxicant control compounds that were submitted by customers, all were correctly predicted as being nontoxicants.

Internal validation verifies that the ToxExpress predictive modeling for *in vivo* liver is robust, predicting and classifying the toxic response for individual compounds even when large amounts of data are removed from the training set. External validation verifies that the modeling can predict potential toxicity in compounds that are not in the database. Interestingly, the false positive and false negative rates for external data submitted by customers are approximately the same as that found by the internal cross-validation. This finding is important for two reasons: (1) it supports the hypothesis that the predictive markers of general toxicity apply to the universe of compounds with few exceptions, and (2) the ToxExpress reference database is sufficiently robust to overcome potential variability between sites of sample generation and chip processing.

## 12.4 APPLYING TOXICOGENOMICS

The following are examples of the utility of toxicogenomics and how researchers can apply gene expression data to address problems facing them during the development of drugs:

1. Structure–activity relationships (SARs)
2. *In silico* validation of gene dysregulation
3. Predicting a human-specific hepatotoxicant using gene expression dysregulation in rats
4. Accounting for confounding sources of variability in predictive models
5. Common toxicity marker expression in the presence of heterogeneous classical measure results
6. Insights into molecular mechanisms of toxicity at the pathway level

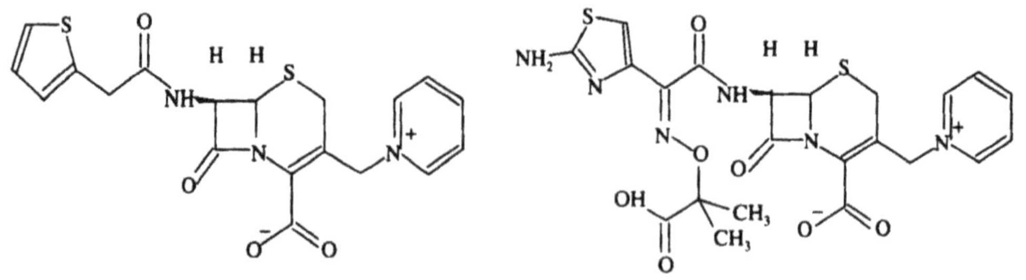

FIGURE 12.5 Structures of cephaloridine and ceftazidime, respectively.

### 12.4.1 Exploring SAR and the Mechanism of Toxicity for Two Closely Related Compounds

#### 12.4.1.1 Background

To improve the drug development pipeline in terms of safety, cost, and speed, investigators are working to build SAR models that will allow optimization of chemical structures. It is recognized that drugs of similar chemical structure can have divergent toxicities, and efforts are underway to develop models that utilize multiple parameters such as structure, mechanism of toxicity, and biotransformation to anticipate toxicity.[65–67] Such investigations into the mechanism of toxicity can be expanded to the level of gene dysregulation. One example of such an experiment is provided here.

Cephaloridine, a first-generation cephalosporin, is a renal toxicant, while ceftazidime, a third-generation cephalosporin, exhibits relatively little renal toxicity. These two compounds share a great deal of structural similarity (Figure 12.5). To understand why cephaloridine is toxic and ceftazidime is not, rats were exposed to one of these drugs, and the resultant gene expression data were collected from the kidneys and compared.

Normal adult Sprague–Dawley rats were treated intravenously at time 0 hr with a low or high dose (100 and 800 mg/kg, respectively) of cephaloridine, with 800 mg/kg of ceftazidime, or with saline (control). Rats were sacrificed 6, 24, 72, or 168 hr after this one dose (Table 12.2). The hemidissected right kidneys from each animal were analyzed using the Affymetrix GeneChip RGU34 rat genome microarray. At the time of sacrifice, serum was collected for clinical chemistry assessment, and one half of each hemidissected right and left kidney was submitted for routine light microscopic analysis. Selected results are shown in Table 12.2. Classical serum chemistry revealed minimal signs of renal toxicity at 24 hr in the high-dose cephaloridine group, as evidenced by a statistically significant, but modest (<twofold) increase in blood urea nitrogen (BUN). By 72 hr, renal toxicity was evident, as animals treated with high-dose cephaloridine exhibited >sixfold increases in both BUN and serum creatinine levels. BUN and serum creatinine showed no significant changes at any time after treatment with ceftazidime. Histological evaluation of the kidneys revealed minimal degeneration of proximal tubular epithelial cells in several high-dose cephaloridine-treated animals as early as 6 hr, with more extensive nephrosis by 24 hr post-dose. No signs of renal histopathological injury were observed in rats treated with ceftazidime.

To understand the differences underlying the mechanism of toxicity, a comparison of the early responses (6 hr post-treatment), those that may represent an early indication of future injury, to cephaloridine and ceftazidime was made using the GeneExpress® (registered trademark of Gene Logic, Inc.; Gaithersburg, MD) software system contrast analysis tool. Table 12.3 summarizes the number of genes that were significantly up- or downregulated after treatment with high doses of cephaloridine or ceftazidime when compared to animals treated with vehicle or a low dose of cephaloridine. All these genes showed statistically significant ($p < 0.05$) dysregulation 6 hr after treatment with high doses of cephaloridine or ceftazidime in comparison to animals treated with vehicle or a low dose of cephaloridine. An example of a stress-response gene (heat shock protein)

**TABLE 12.2**
**Summary of the *In Vivo* Cephalosporine Study Design and Clinical Chemistry Parameters**

| Compound | Dose (mg/kg) | Time Point (hr) | Blood Urea Nitrogen (BUN) | Creatinine |
|---|---|---|---|---|
| Saline (vehicle) | — | 6 | | |
| | | 24 | | |
| | | 72 | | |
| | | 168 | | |
| Cephaloridine | 100 | 6 | Normal | Normal |
| | | 24 | Normal | Normal |
| | | 72 | Normal | Normal |
| | | 168 | — | — |
| | 800 | 6 | Normal | Normal |
| | | 24 | Elevated[a] | Slightly elevated[a] |
| | | 72 | Elevated[a] | Elevated[a] |
| | | 168 | — | — |
| Ceftazidime | 800 | 6 | Normal | Normal |
| | | 24 | Normal | Normal |
| | | 72 | Normal | Normal |
| | | 168 | — | — |

[a] Statistically significant as compared to vehicle-treated rats sacrificed at the same time points

**TABLE 12.3**
**Numbers of Gene Expression Changes Observed across Structurally Similar Compounds[a]**

| Drug | Upregulated | Downregulated |
|---|---|---|
| High-dose cephaloridine | 44 | 2 |
| Ceftazidime | 2 | 0 |

[a] Data represented are from analysis of chip A of the Affymetrix GeneChip RGU34 rat genome microarray.

modified by cephaloridine, but not by ceftazidime, at both 6 and 24 hr is shown in Figure 12.6. The heat shock protein is only one of the 44 genes listed as upregulated by cephaloridine in Table 12.3. A gene-by-gene analysis can be performed on each of the 46 genes, providing a comprehensive picture.

Exporting the Affymetrix average difference values into a third-party tool (such as Microsoft Excel) enables additional visualization and analysis. For example, Figure 12.7 shows the same data as Figure 12.6, but Figure 12.7 uses summary statistics, such as mean and standard deviation, rather than showing individual sample information. The mean average difference values ($y$-axis) are given for each of the treatment conditions ($x$-axis), with standard deviations displayed as error bars for this heat shock protein.

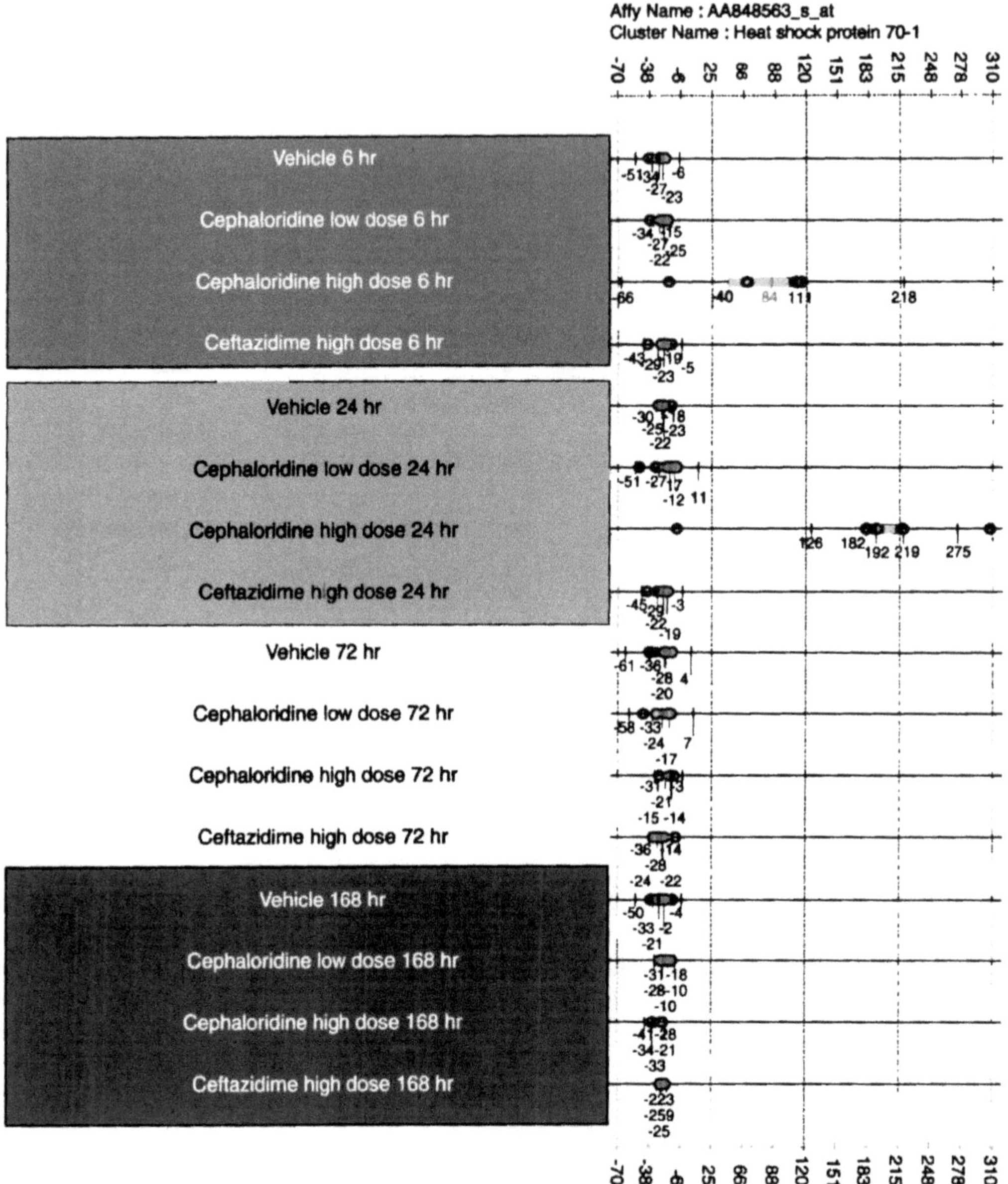


**FIGURE 12.6** SAR distinctions. A box plot generated with the GeneExpress® software system contrast analysis tool illustrates the average difference values for each treatment condition (circles). Each treatment condition (far left) contains multiple samples corresponding to individual animals within that group. A 95% CI and median are indicated for each set of samples. For example, treatment-condition cephaloridine, high dose 6 hr, has a CI between 64 and 139. The median is 116 for this same group and is indicated within the cluster of samples. (See color insert following page 112.)

### 12.4.1.2 Summary

These data demonstrate that at least one of the 44 identified genes shows an increase in gene expression at 6 and 24 hr after cephaloridine treatment while remaining unaffected after treatment with ceftazidime or vehicle. As a result, this gene may be a useful and potential toxicity marker when evaluating compounds similar in structure or with mechanisms of action similar to cephaloridine.

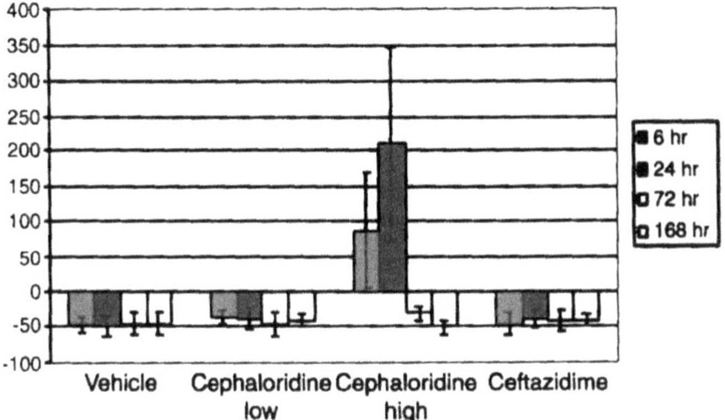


**FIGURE 12.7** Bar plot of the average difference values for heat shock protein calculated for vehicle, cephaloridine, and ceftazidime by using the GeneExpress® software system contrast analysis tool. The results shown are expressed as mean ± standard deviation. The observed increase in expression at 6 and 24 hr following treatment with high-dose cephaloridine is statistically significant ($p < 0.001$).

### 12.4.2 *In Silico* Validation of Gene Dysregulation Using *In Vivo* Rat Studies and Human Diseased Tissues

#### 12.4.2.1 Background

To elucidate new drug targets, drug discovery programs study differential gene expression in human normal and diseased tissues and/or in animals exposed to toxicants, the latter serving as a model for such human diseases. Gene Logic has developed a large reference database of gene expression derived from normal and diseased human tissue (BioExpress™), which can be used in combination with ToxExpress for *in silico* validation of drug targets. As an example, alpha-actinin 1 will be discussed here. This gene encodes a cross-linking protein involved in anchoring the visceral glomerular epithelial foot process to the renal glomerular basement membrane,[68] an interaction severely altered in glomerulosclerosis. Using the data in BioExpress, a contrast analysis determined that the alpha-actinin 1 gene was upregulated to a statistically significant degree ($p < 0.01$) in patients with glomerulosclerosis as compared to samples of normal human kidneys (Figure 12.8). To determine what toxicant might replicate such a gene dysregulation, the models in ToxExpress were examined using the same contrast analysis parameters.

Normal adult Sprague–Dawley rats were treated intravenously at time 0 hr with a low or high dose (10 and 150 mg/kg, respectively) of puromycin aminonucleoside (PAN) or with saline (control). Rats were sacrificed 6, 24, or 168 hr after this one dose (Table 12.4). The hemidissected right kidneys from each animal were submitted for gene expression analysis using the Affymetrix GeneChip RGU34 rat genome microarray. At the time of sacrifice, serum was collected for clinical chemistry assessment, and the other halves of the hemidissected kidneys were submitted for light microscopic analysis.

Animals treated with puromycin aminonucleoside (PAN) exhibited a statistically significant upregulation ($p < 0.01$) of alpha-actinin 1 at 168 hr following treatment with high-dose PAN (Figure 12.9). Administration of PAN to rats serves as a model of human minimal-change disease in that the glomerular visceral epithelial cells undergo morphological simplification, and this toxicant has been shown to induce glomerulosclerosis.[69] Classical toxicity parameters exhibited some possible signs of nephrotoxicity as early as 24 hr (e.g., <twofold elevation in BUN) without changes in urine analytes. The study director attributed this increase in BUN to dehydration. By 168 hr, obvious changes in BUN were present, as were alterations in urinary analytes. Histological evaluation of the kidney showed evidence of damage at only this time point, and this pathology was restricted

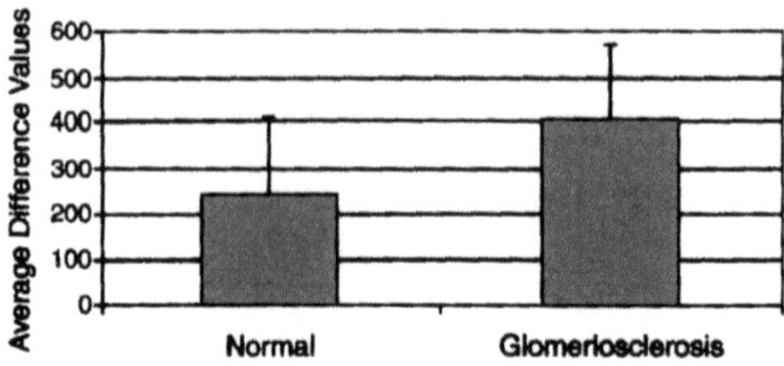


FIGURE 12.8 The expression of the mRNA that encodes human alpha-actinin 1 was examined in the human kidney samples present in the BioExpress™ (GeneLogic, Inc.; Gaithersburg, MD) database. (The data were generated on the Affymetrix GeneChip U133 human genome microarrays.) This analysis included 97 normal kidney samples and 6 patients with glomerulosclerosis. Results revealed a statistically significant ($p < 0.01$) upregulation of alpha-actinin 1 in patients with glomerulosclerosis, echoing what was seen in the ToxExpress data from the PAN rat study. The results shown are expressed as mean ± one standard deviation.

**TABLE 12.4**
**Summary of *In Vivo* Study Design, Clinical Chemistry, and Renal Histopathology Results**

| Compound | Dose (mg/kg) | Time Points (hr) | Blood Urea Nitrogen (BUN) | Serum Creatinine | Histopathology Diagnosis |
|---|---|---|---|---|---|
| Saline (vehicle) | — | 6 | Normal | Normal | Normal |
| | | 24 | Normal | Normal | Normal |
| | | 168 | Normal | Normal | Normal |
| Puromycin aminonucleoside (PAN) | 10 | 6 | Normal | Normal | Normal |
| | | 24 | Normal | Normal | Normal |
| | | 168 | Normal | Normal | Normal |
| | 150 | 6 | Normal | Normal | Normal |
| | | 24 | Elevated (<twofold)[a] | Normal | Normal |
| | | 168 | Elevated (fivefold)[a] | Normal | Nephrosis (moderately severe) |

[a] Statistically significant as compared to vehicle-treated rats sacrificed at the same time points.

to animals treated with the highest dose of PAN (Table 12.4). The literature reveals that immunohistochemical staining for alpha-actinin 1 has demonstrated an increase in this protein following treatment of rats with two agents that induce glomerular visceral epithelial cell injury: namely, antibodies to the glomerular basement membrane (nephrotoxic serum nephritis) and PAN.[70,71] These reports of protein alterations serve as some confirmation of the gene dysregulation seen in our studies.

#### 12.4.2.2 Summary

It has been reported that alpha-actinin 4 mutations can cause focal and segmental glomerulosclerosis in humans;[72] however, this case study may be the first indication that alpha-actinin 1 expression is altered in human glomerulosclerosis. These studies illustrate the value of gene expression data as

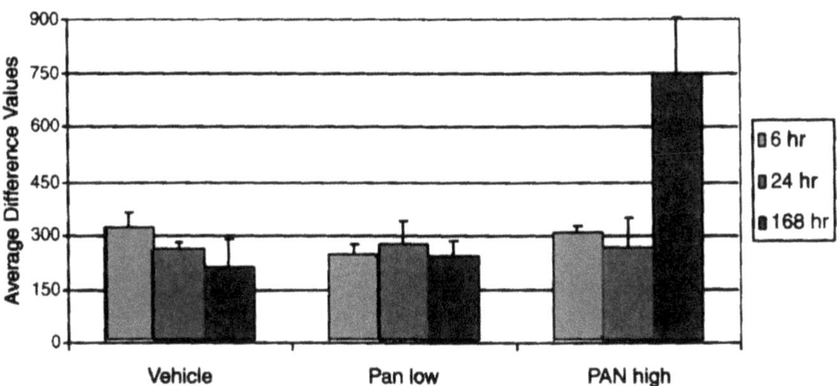


**FIGURE 12.9** Bar plot of the average difference values for alpha-actinin 1 calculated for vehicle and PAN low and high doses and obtained by using the GeneExpress® Software System contrast analysis tool. The results shown are expressed as mean ± standard deviation. The observed increase in the expression at 168 hr following treatment with high-dose PAN is statistically significant ($p < 0.01$).

a tool for examining toxicity and mechanism of action. The alpha-actinin 1 example demonstrates that, by using gene expression data obtained from human tissue (normal and diseased) along with toxicant-treated rat tissue, one may (1) identify possible new drug targets, (2) discover toxicity markers that relate to human disease states, (3) ascertain a potentially valuable animal system in which to screen drug compounds, and (4) offer *in silico* validation.

### 12.4.3 Prediction of a Human-Specific Toxicant Using Gene-Expression-Based Markers

#### 12.4.3.1 Background

Of the many potential advantages of gene-expression-based toxicity indication, one is based on the tenet that gene expression changes associated with a particular toxic response are exhibited despite the absence of any observable pathological change. Therefore, gene expression, as a more sensitive indicator of toxicity than classical measures, may be able to detect human-specific toxicity. The benefits of implementing such a system within a drug development pipeline are obvious. To determine if the ToxExpress predictive system can be applied to detect harmful drugs before they enter clinical trials, three acetylcholinesterase inhibitors were studied (Figure 12.10). Tacrine (Cognex®, a registered trademark of Pfizer; New York, NY) has been shown to cause liver dysfunction in about 30% of patients using it, despite a lack of observable hepatotoxicity during preclinical studies.[73,74] Donepezil (Aricept®, a registered trademark of Eisai Co., Ltd.; Tokyo, Japan) and physostigmine cause no hepatotoxicity in humans or rodents, yet exhibit a mechanism of action similar to that of tacrine.[75–77]

In this *in vivo* study, rats received one oral dose of vehicle or drug. Tacrine was employed at four dose levels, physostigmine at three, and Aricept at one. The study design is shown in Table 12.5. It should be noted that the highest doses of both tacrine and physostigmine caused similar levels of mortality due solely to the excessive pharmacological effect of acetylcholinesterase inhibition. No serum chemistry changes or liver alterations were observed, including histopathology, that would suggest hepatotoxicity over a 7-day post-dose time period (Figure 12.11). The gene expression levels of each 6- and 24-hr post-dose liver sample were measured on the Affymetrix GeneChip.

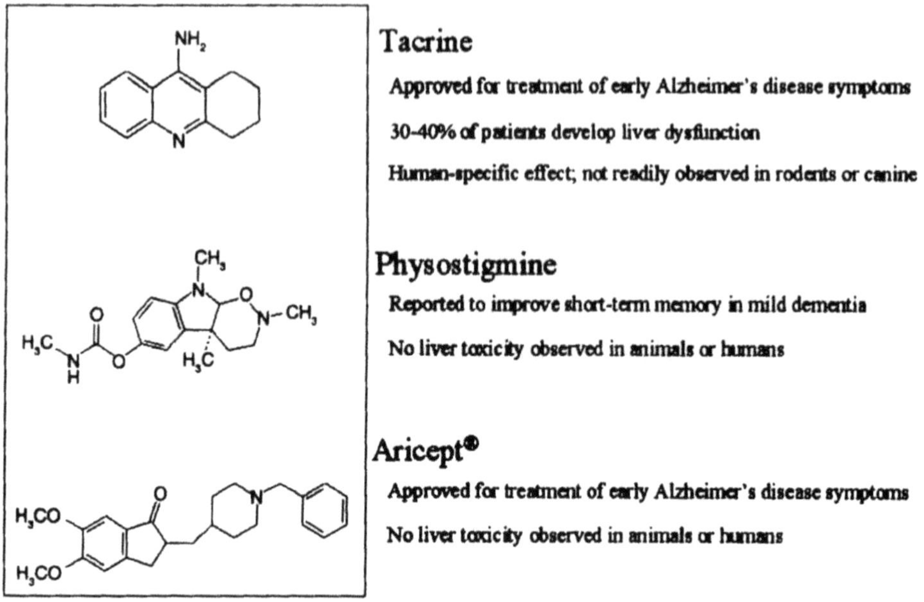

**Tacrine**

Approved for treatment of early Alzheimer's disease symptoms

30–40% of patients develop liver dysfunction

Human-specific effect; not readily observed in rodents or canine

**Physostigmine**

Reported to improve short-term memory in mild dementia

No liver toxicity observed in animals or humans

**Aricept®**

Approved for treatment of early Alzheimer's disease symptoms

No liver toxicity observed in animals or humans

**FIGURE 12.10** Three acetylcholinestase inhibitors used for the study.

**TABLE 12.5**
**Summary of *In Vivo* Study Design for Acetylcholinesterase Inhibitors**

| Compound | Dose (mg/kg) |
|---|---|
| Saline (vehicle) | — |
| Aricept | 1 |
| Physostigmine | 3 |
| | 12.5 |
| | 25 |
| Tacrine | 3 |
| | 10 |
| | 30 |
| | 40 |

*Note:* All animals were sacrificed at either 6 or 24 hr after administration.

As expected, the ToxExpress predictive system did not classify any of the vehicle-treated samples or those samples treated with low doses of the three inhibitors as toxin-exposed using only gene expression data.[78] The system successfully predicted a majority of the samples treated with a high dose of tacrine to be toxin exposed (Figure 12.12). Physostigmine and Aricept were correctly predicted as non-hepatotoxicants. The predictive ability of the ToxExpress predictive system is founded on several proprietary and public statistical algorithms that determine the identity and magnitude of gene responses relative to the entire reference database of known hepatotoxicants

**Tacrine**
No liver toxicity observed by classical methods (Clinical Chemistry/Histopathology) at any dose
Rats treated with 40 mg/kg showed clinical signs of exaggerated pharmacology
Several rats died

**Physostigmine**
No liver toxicity observed by classical methods at any dose
Rats treated with 25 mg/kg showed exaggerated pharmacology
Mortality rate at 25 mg/kg was on the same order as that of the 40 mg/kg Tacrine group

**Aricept®**
No liver toxicity observed by classical methods
No exaggerated pharmacology observed

**FIGURE 12.11** Observed effects in this study.

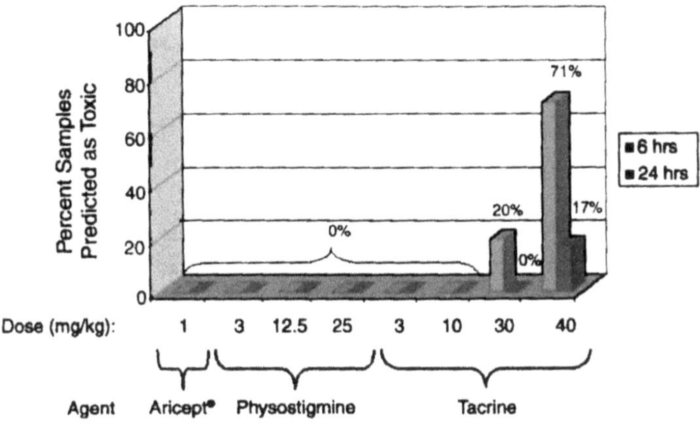


**FIGURE 12.12** Percent samples predicted as toxic by *in vivo* liver model are plotted on the *y*-axis with respect to each study group. Tacrine was predicted in the absence of observable rat toxicity by classical measures. The predictions are based on models that were built without any of the three compounds in the training set. Thus, each compound is a true test of the predictive model and associated markers. No other human-specific toxicant was used in the training set.

and is not established just on the number of changing genes. To illustrate this, gene expression levels from each treatment group were compared with those from control groups. Figure 12.13 shows the number of genes at each time point and dose that exceeded a twofold change in expression level as compared to control group gene expression with a significance level of $p < 0.01$ (using a two-tailed *t*-test of unequal variance). The study group treated with physostigmine at the highest dose exhibited significant changes in the expression levels of hundreds of genes, yet none of these samples was classified as toxin-exposed using predictive modeling. Samples treated with the highest dose of tacrine showed approximately the same numbers of significant changes greater than twofold, but these samples were classified correctly as toxin-exposed. There is less than a 2% overlap between the genes that change at the highest dose of physostigmine vs. the highest dose of tacrine (Figure 12.14). Thus, the toxicity-associated mechanism observed with tacrine is far different than the ambient mechanism observed with physostigmine.

Several examples of genes that are known to be associated with a general toxic response are shown (Figure 12.15 to Figure 12.17). In each case, two instances of probe fragments on the GeneChip are representative of each gene. Each gene is obviously upregulated by tacrine at 6 hr,

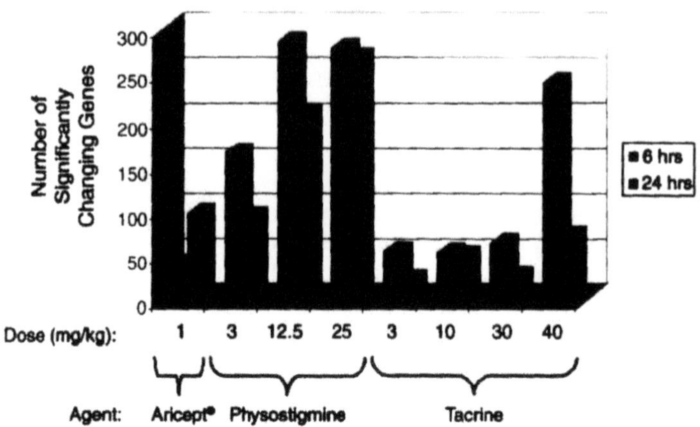


**FIGURE 12.13** Comparison of numbers of significantly changing genes over several time and dose points. Numbers of significantly changing genes over the study groups in the experiment are plotted on the *y*-axis. The numbers of changing genes with physostigmine are approximately the same as the number of changing genes with an early tacrine response. However, the two compounds are differentially predicted with respect to their toxic outcome. Significance is established as a fold change >2 relative to vehicles at same time point ($p < 0.01$; two-tailed *t*-test of unequal variance).

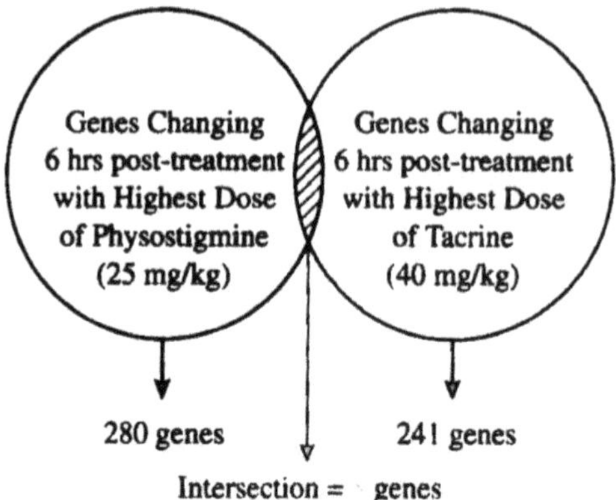


**FIGURE 12.14** Few significantly changing genes for tacrine and physostigmine overlap at the 6-hr time point. The genes changing with physostigmine administration are not the same genes changing with tacrine administration. The genes changing with tacrine are found in the predictive model for general liver toxicity, while the genes changing with physostigmine are not. Significance is established as a fold change >2 relative to vehicles at same time point ($p < 0.01$; two-tailed *t*-test of unequal variance).

the time at which we capture toxicity indication, but is not regulated by physostigmine or Aricept. It is important to note that these genes were determined to be general markers of toxicity using our entire reference database. In addition, no acetylcholinesterase inhibitors were used in building our predictive system. Therefore, the compounds we have used as examples in this case were not used to select these genes or the other genes we have determined to be predictive markers of toxicity.

### 12.4.3.2 Summary

In this predictive model, no other hepatotoxicant specific for humans was included in our training set, illustrating that statistical inference based on a large reference database of known hepatotoxicants is effective as a classification tool for compounds. The genes regulated upon tacrine administration are many of the same genes that have been identified from a large reference database as

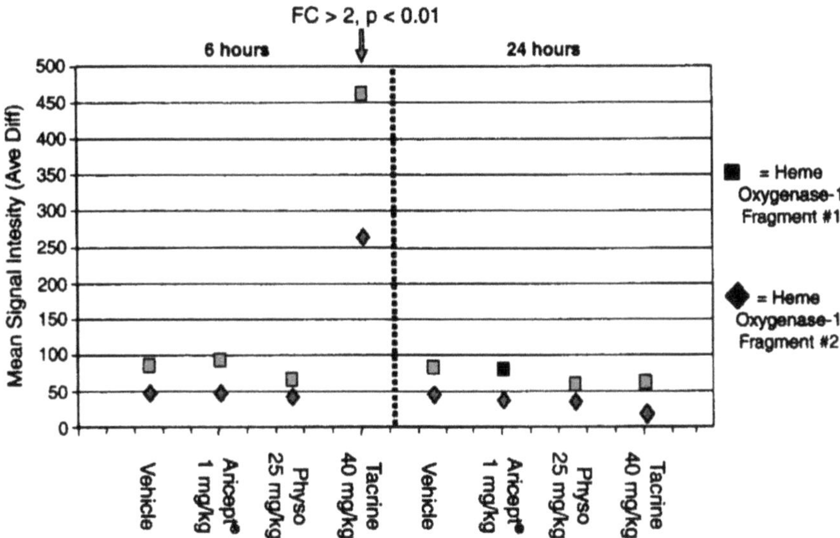


**FIGURE 12.15** Example of a predictive toxic marker gene (heme oxygenase-1) and its response to both drugs at both 6 and 24 hr. This is one of the genes responsible for the predictions of physostigmine and Aricept as "nontoxic" and tacrine as "toxic." Two fragment elements are tiled on the microarray for this gene. Both respond identically and both are found within the predictive model. NOTE: Physo = physostigmine.

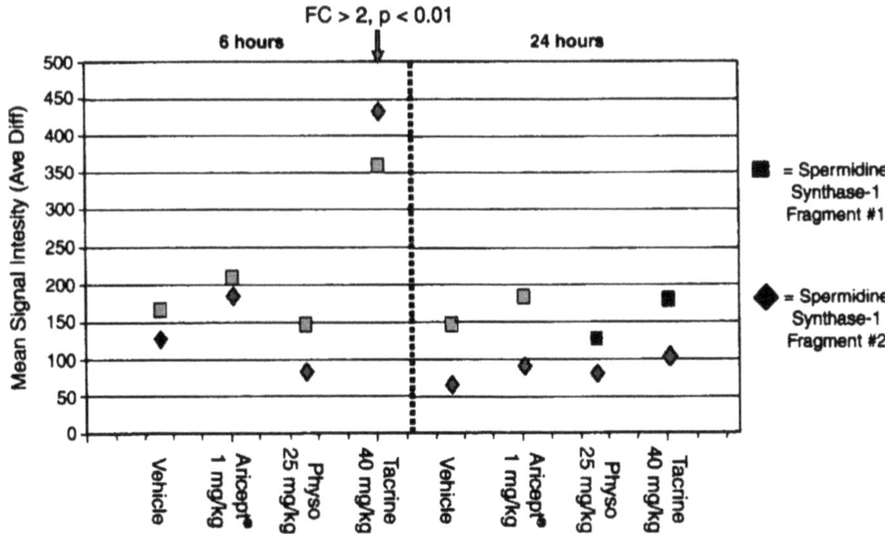


**FIGURE 12.16** Example of a predictive toxic marker gene (spermidine synthase) and its response to both drugs at both 6 and 24 hr. This is one of the genes responsible for the predictions of physostigmine and Aricept as "nontoxic" and tacrine as "toxic." Two fragment elements are tiled on the microarray for this gene. Both respond identically and both are found within the predictive model.

expression-based toxicity markers. This is not the case with physostigmine or Aricept. Toxicogenomic assessment in rats was able to predict that tacrine is a human toxicant, suggesting that such systems may be able to detect such toxicity early in the drug development pipeline before exposure to humans. It is equally important to recognize that the number of genes for which the expression level changes significantly upon exposure to a drug is insufficient to assign a toxicity classification,

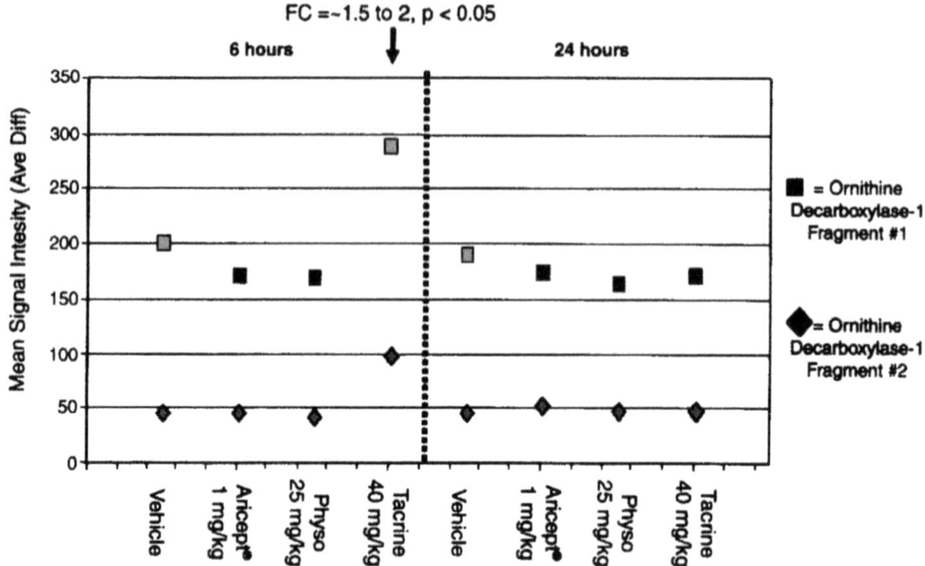


**FIGURE 12.17** Example of a predictive toxic marker gene (ornithine decarboxylase) and its response to both drugs at both 6 and 24 hr. This is one of the genes responsible for the predictions of physostigmine and Aricept as "nontoxic" and tacrine as "toxic." Two fragment elements are tiled on the microarray for this gene. Both respond identically and both are found within the predictive model.

as physostigmine, a non-hepatotoxicant, elicited the dysregulation of more genes in the liver than did tacrine.

### 12.4.4 Accounting for Confounding Sources of Variability in Predictive Models

#### 12.4.4.1 Background

Large reference database construction involves years of data collection. Parameters related to every aspect of the underlying process will change over that period. These parameters include multiple array processing protocols (e.g., PMT scanner setting, hybridization, and wash times), multiple biological conditions (e.g., strain of species, gender), and specific administration protocols that vary from compound to compound (e.g., types of vehicle, route of administration, feeding/fasting).

By incorporating different settings for each one of these parameters into the reference database, predictive modeling becomes somewhat insensitive to each change. In the most basic application of predictive modeling, the researcher would like to find those genes that are most applicable for the identification of a general toxic response. The marker selection process assays for those genes that fall into a certain distribution only when toxicity is observed by gene expression and for those that fall into a separate distribution when toxicity is not observed. Because toxicity is observed in conjunction with all other parameters that change from experiment to experiment and over time, in general, markers of toxicity should not be explained by these confounding factors to an appreciable degree. Therefore, confounding influences external to the toxic response can benefit the predictive model by making it more robust against these influences.

In an effort to assess the prevalence of confounding factors that may have been introduced over time into our set of predictive markers, we ran an analysis to determine if these markers could successfully predict general toxicity in the presence of such factors. In this example, carbon tetrachloride ($CCl_4$), a well-known hepatotoxicant, was administered to rats. The liver was extracted at various time points post-dose and run over the Affymetrix GeneChip RGU34 set. The study

**TABLE 12.6**
**Summary of *In Vivo* Study Design for $CCl_4$**

| Experiment | Compound | Time Points (hr) |
|---|---|---|
| Early | Corn oil (vehicle) | 3 |
| | | 6 |
| | | 24 |
| Early | $CCl_4$ | 3 |
| | | 6 |
| | | 24 |
| More recent | Corn oil (vehicle) | 3 |
| | | 6 |
| | | 24 |
| More recent | $CCl_4$ | 3 |
| | | 6 |
| | | 24 |

design can be found in Table 12.6. The experiment labeled as "early" was run approximately one year before the study labeled "more recent." In addition, the animal experiment was conducted at two separate and unaffiliated sites of sample generation. All tissue processing and chip processing took place at Gene Logic but were run by different personnel as part of Gene Logic's production processes. These protocols were known to change sufficiently over that period of time for several factors.

Figure 12.18 shows two principal components analysis (PCA) visualizations, where relative variability is shown on each axis. The highest contributing factor to the overall variability is shown on the $x$-axis and is known as the *first component*. The $y$-axis shows the second-highest component, and the $z$-axis shows the third-highest variability component. In Figure 12.18a, samples from both the early and more recent experiments are plotted on this visualization using all 26,202 non-control probe sets. The visualization shows two things of note: (1) the principal source of variability over all genes defines the time effect, and (2) the more recent experiment shows less within-experiment variability than the older experiment. Figure 12.18b shows the same samples plotted on a PCA visualization, this time measured only over the Gene Logic-determined general toxicity marker genes. In this visualization, toxicity is observed as the principal source of variability; therefore, confounding influences have been removed from the comparison.

The PCA visualization is only one of the many methods to assess confounding variables and their effects on the prediction and observation of toxic events at the gene expression level. It is only provided as an easy visualization where conclusions can be made on the general behavior of reliable predictive toxicity markers. The algorithm on which predictions are based should rely on distributions of the signal intensities of each marker and not on dimension reduction. It should be noted that our LDA predictive approach verifies the results provided in these PCA figures.

#### 12.4.4.2 Summary

The predictive markers for general toxicity identified by Gene Logic seem to be insensitive to most types of variability that may have been introduced into a historical reference database over the course of a year. This is mostly due to the fact that they are selected despite this variability within the database. As a reference database grows and focuses on additional variability parameters (e.g.,

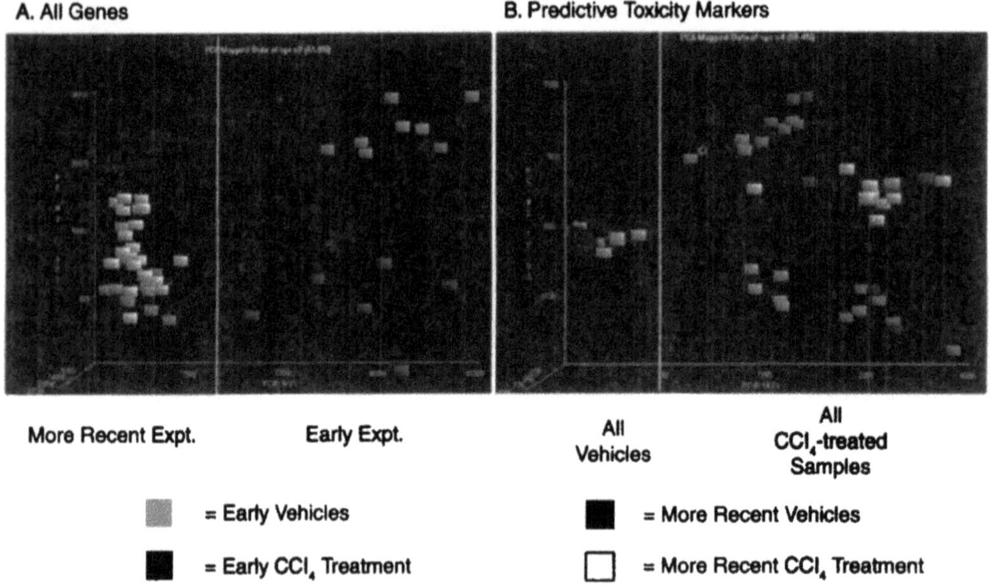


FIGURE 12.18 Two PCA visualizations show that gene selection is necessary for removing confounding factors of variability unrelated to toxicity. The left plot (A) shows the principal separation of samples due to the time the experiment was conducted when using all genes. The right plot (B) shows the principal separation of samples due to toxicity. In fact, there appears to be no observable differences between studies in the first three components. Within each plot, the highest contributing factor to the overall variability is shown on the $x$-axis and is known as the ***first component***. The $y$-axis shows the second-highest component, and the $z$-axis shows the third-highest variability component. (See color insert.)

gender, strain), the predictive markers should become insensitive to these as well. This is an advantage to the genome-wide assay approach.

### 12.4.5 Resolution of Heterogeneous Clinical Chemistry and Histopathology Results Using Gene Expression

#### 12.4.5.1 Background

Compounds are known to have differential toxic effects among individuals as measured by classical measures (i.e., clinical chemistry and histopathology). This heterogeneous response may be mediated at the gene, protein, enzyme, or metabolite level of the toxic mechanism. If this heterogeneity is determined by entities downstream from gene regulation, then gene expression assays may exhibit a more homogeneous response. To relate the heterogeneity of response between classical and gene expression measures, we orally administered acetaminophen (APAP) at pharmacological and toxic doses to rats. The animals were sacrificed at several time points post-dose, and the livers were extracted and processed over the Affymetrix GeneChip RGU34A, B, and C microarrays at Gene Logic's production processing facility. The study design is provided in Table 12.7.

All vehicle-treated rats and those treated with a low dose of APAP showed no clinical chemistry or pathology effects; however, the high-dose APAP treatment group exhibited a toxic response at 24 hr post-dose (Figure 12.19). The variability of clinical chemistry parameters such as alanine amino transferase (ALT) and AST levels and histopathology damage was high. A closer look at this group in our study showed the heterogeneity of the ALT and histopathological responses for individual rats (Figure 12.20). Some rats exhibited no evidence of liver necrosis or showed only minor ALT changes, whereas some rats showed more robust alterations in both classical toxicity endpoints. A few rats showed no increases in ALT levels, but did exhibit an observable pathology.

**TABLE 12.7**
**Summary of *In Vivo* Study Design for Acetaminophen**

| Compound | Dose (g/kg) | Time Points (hr) |
|---|---|---|
| Gum tragacanth (vehicle) | — | 3 |
| | | 6 |
| | | 24 |
| Acetaminophen | 0.1 | 3 |
| | | 6 |
| | | 24 |
| Acetaminophen | 2 | 3 |
| | | 6 |
| | | 24 |

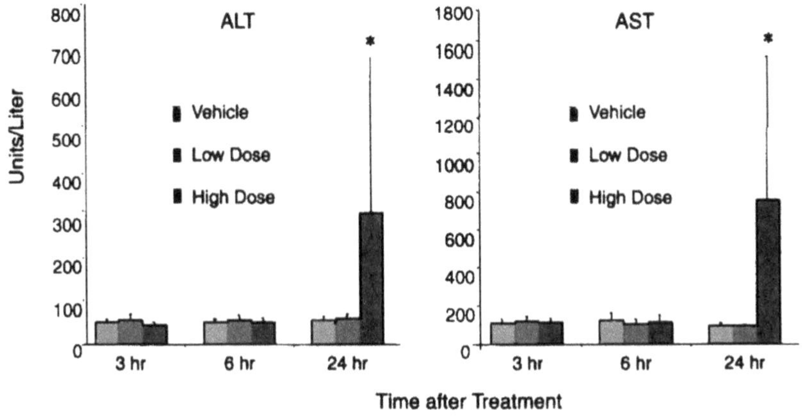

**FIGURE 12.19** ALT and AST levels plotted for each of the study groups. Significance is noted for the 24-hr high-dose group for APAP oral administration. Significance is established as a fold change >2 relative to vehicle-treated rats at the same time point ($p < 0.05$; two-tailed $t$-test of unequal variance).

This heterogeneity may be due to multiple causes, but it has been shown previously that APAP can cause this differential effect in rats when given orally, as opposed to the relative homogeneous response when the drug is administered by the intraperitoneal route.[79–81]

A large number of gene expression changes were observed for the highest dose treatment over all three time points relative to vehicle- and low-dose-treated samples. The rats sacrificed 24 hr post-dose showed many more gene expression changes than earlier time points, but it was noted that the same groups of genes seemed to be changing over all time points. This group of changing genes also showed a large amount of overlap with our predictive markers of general toxicity. Upon running each of the samples in the experiment through our models, this observation was verified (Figure 12.21). Each of the samples treated with a high dose of APAP was classified as "toxic" with the exception of one false negative observed at the 3-hr time point. Therefore, the gene expression changes and the classifications associated with those changes were successful in defining the toxic response before it was observed with classical measures.

A PCA was conducted using this data in order to observe the relative variability of each of the groups in the study (Figure 12.22). Relative variability is shown on each axis. The highest contributing factor to the overall variability is shown on the $x$-axis (the first component). The $y$-axis shows the second-highest component, and the $z$-axis shows the third highest variability component. Each

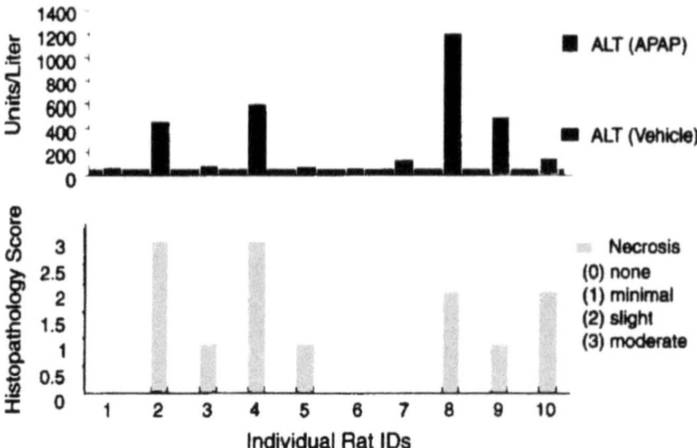


**FIGURE 12.20** ALT levels and necrosis severity levels for only the group sacrificed 24 hr post-dose with a high dose of APAP. A variable response of toxicity is observed for both measures.

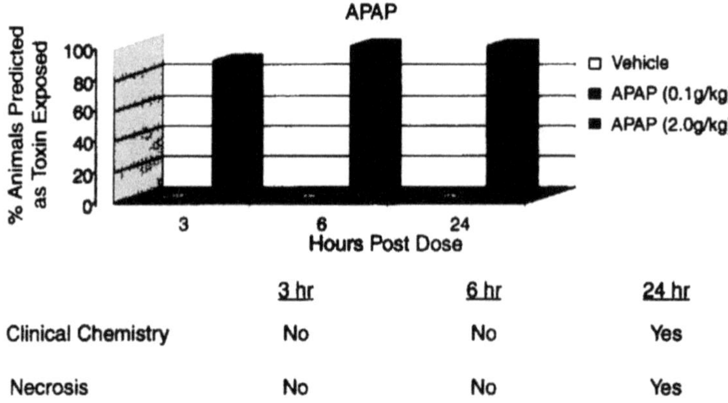


| | 3 hr | 6 hr | 24 hr |
|---|---|---|---|
| Clinical Chemistry | No | No | Yes |
| Necrosis | No | No | Yes |

**FIGURE 12.21** The percent animals predicted by markers of general toxicity as toxin-exposed for each study group by the treatment is shown. Also shown is a summary of overall clinical chemistry and histopathology for high dose-treatments across time. The homogeneous predictions are a contrast to the heterogeneity in classical measures shown in Figure 12.20. Also note that gene expression analysis identified potential toxicity before observable effects.

sample in the study is plotted in reference to all other samples. The genes used to define this variability were the genes that significantly changed at the highest dose administration relative to vehicle and low dose (ANOVA, $p < 0.01$). As all high-dose-treated samples fell to the left portion of the plot, we can assign the toxic response to the first principal source of variability, as expected from our gene selection parameters. In addition, time course can be assigned to the second most prevalent source of variability.

#### 12.4.5.2 Summary

The homogeneity of the gene response observed in the 24-hr high-dose-treated samples is in direct contrast to the heterogeneity of this same group measured by classical measures. Therefore, a group of genes exists that can identify the toxic response based on treatment, rather than heterogeneous observed outcomes. It has been difficult to find those genes that define the "high responders" category from the "no responders" category internal to that group. Thus, mediation of certain clinical and pathological outcomes may be dependent on events downstream from transcription. This assay may be a more sensitive indicator of potential toxicity in this regard.

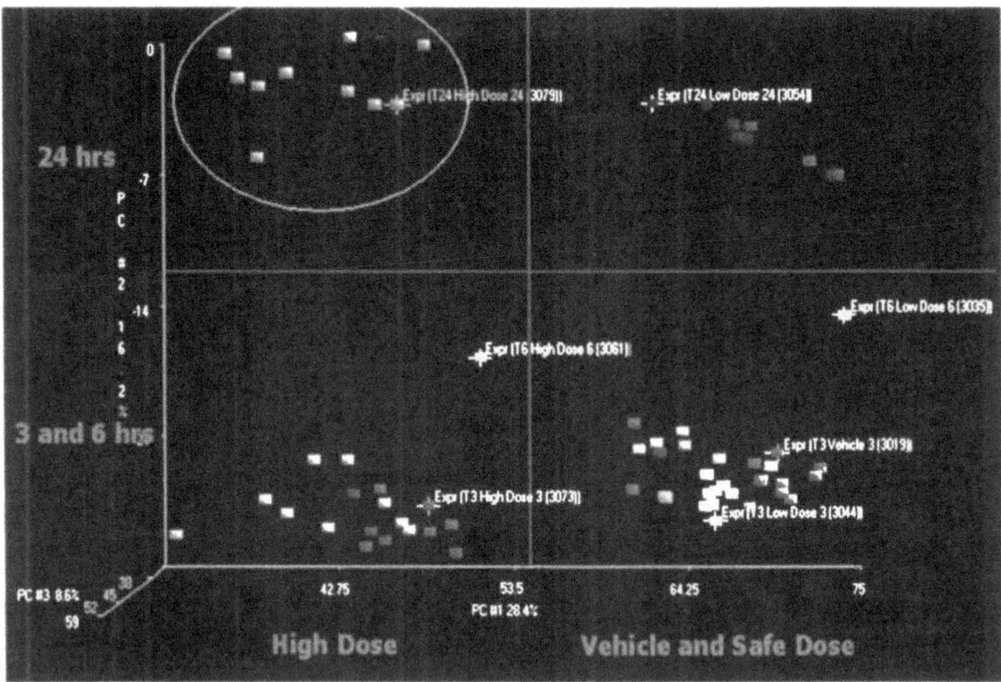


FIGURE 12.22 Principal components analysis (PCA) of APAP data is visualized for all animals in the study using those genes that change due to APAP administration at the highest dose. The homogeneity of the gene expression response is characterized by both a toxic response and a toxic response related to sacrifice time post-dose. (See color insert.)

## 12.4.6 Insights into the Molecular Mechanisms of Toxicity at the Pathway Level

### 12.4.6.1 Background

The advent of microarray technology allows the analysis of thousands of genes and multiple pathways from one experiment, providing a more thorough dissection of the molecular mechanisms of toxicity at the transcript level. The response of the rat liver to two classical liver toxicants, APAP and carbon tetrachloride, were investigated by microarray analysis, as their respective mechanisms of liver toxicity have been extensively studied using classic toxicological methods.[82] An overview of the chemical structures and types of liver toxicity induced by these toxicants are listed in Figure 12.23. This particular study focused on evaluating alterations in the glutathione pathway, as reduced glutathione plays a pivotal role in preventing overt toxicity by APAP.

When therapeutic doses of APAP are administered, APAP is metabolized by sulfation and glucoronidation, the major pathways of APAP metabolism.[83] When the sulfation and glucoronidation metabolic pathways become saturated by high doses of APAP, APAP is then metabolized by the cytochrome P450 system, specifically *CYP2E1* and *CYP1A2*, producing the *N*-acetyl-*p*-benzoquinone imine (NAPQI) reactive metabolites.[84,85] The NAPQI reactive metabolites are capable of covalently binding to proteins and inducing glutathione depletion in both the cytoplasmic and mitochondrial compartments. This is followed by necrotic cell death.[86,87] Glutathione detoxifies the NAPQI reactive metabolite by both nonenzymatic and enzymatic reactions catalyzed by glutathione transferases.[88–90] As glutathione is one of the key components for the detoxification of the NAPQI reactive intermediate, it was postulated that multiple genes in the glutathione synthesis pathway would be over-expressed following administration of toxic doses of APAP. It was also theorized

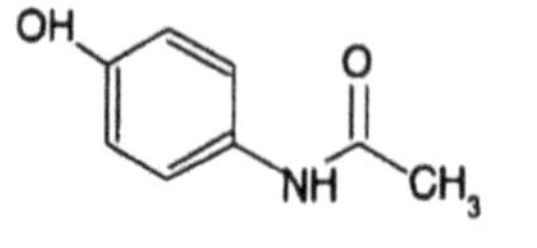

APAP
Drug (Acetaminophen; Paracetamol)
Centrilobular necrosis of the liver

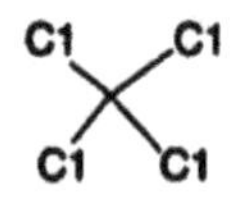

$CCl_4$
Solvent
Necrosis and steatosis (fatty liver)

**FIGURE 12.23** Compound structures and types of liver damage. (Images reproduced from The Merck Index.)

that the genes would be over-expressed in an attempt by the cell to increase the intracellular levels of reduced glutathione. The elevated levels of reduced glutathione could then be utilized by the cells for the detoxification of the NAPQI reactive intermediates, enhancing the ability of the cell to eliminate the toxic insult.

In this study, Sprague–Dawley rats were treated with APAP or $CCl_4$ (see Tables 12.7 and 12.6, respectively, for study design). RNA was isolated from the liver and evaluated by microarray analysis using the Affymetrix GeneChip RGU34. Elevations of ALT levels and signs of necrosis by histopathological examination confirmed that both APAP and $CCl_4$ induced liver damage at 24 hr. The well-characterized glutathione pathway was evaluated for time- or compound-specific differences in gene expression following exposure to either of the toxicants. In Figure 12.24, gene expression changes were determined, and the fold-change results were linked to the *Kyoto Encyclopedia of Genes and Genomes* (KEGG) illustration of the glutathione pathway. This visualization tool allows a toxicologist to determine which genes are being modulated by the toxicants in a single pathway of interest and provides a framework for evaluating multiple pathways for gene expression changes. Analysis of the glutathione pathway revealed that multiple genes were being over-expressed 24 hr post-treatment in response to APAP (2 g/kg) exposure, while the same genes responded within 3 hr of $CCl_4$ exposure. Importantly, the rate-limiting enzyme in the synthetic pathway of glutathione (Figure 12.25), γ-glutamylcysteine synthetase catalytic subunit, was over-expressed, a finding that has been previously documented in mice exposed to APAP.[91,92] In addition, two other key genes involved in glutathione synthesis were over-expressed: glutathione synthetase and glucose 6-phosphate dehydrogenase (Figure 12.24). A number of glutathione S-transferases were over-expressed as well (data not shown). Overall, the pathway indicates that the cells are attempting to increase levels of key genes needed for the synthesis of reduced glutathione, an essential component required to detoxify the NAPQI reactive intermediate. The global gene expression changes captured by the microarray analysis provides a more complete snapshot of the molecular mechanisms being invoked by the cells following exposure to toxic levels of APAP.

The results shown in Figure 12.24 indicate that the temporal regulation of the glutathione pathway for a number of important genes is quite different for $CCl_4$ and APAP, as the $CCl_4$ altered the expression levels within 3 hr of exposure (early) while APAP lacked differential expression at 3 and 6 hr but displayed significant alterations at 24 hr (late).

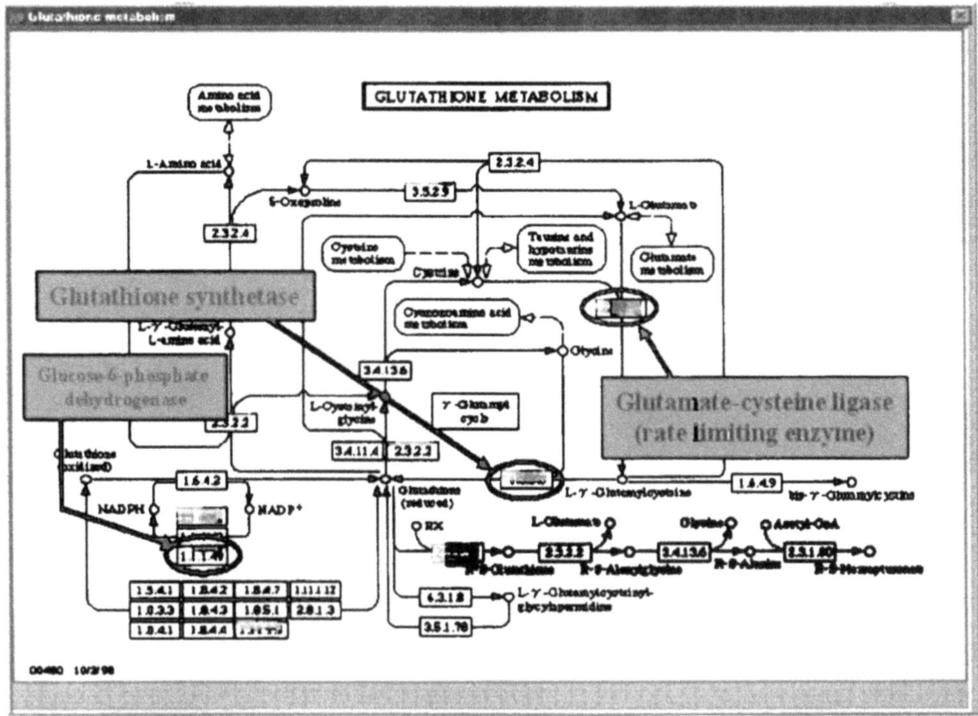


**FIGURE 12.24** Glutathione pathway. The fold-change values for genes that encode proteins in the glutathione pathway are illustrated in the context of the KEGG glutathione metabolism pathway. Overexpressed genes are brown, and the underexpressed genes are green. (See color insert.)

### 12.4.6.2 Summary

These results provide evidence that biologically relevant molecular mechanisms of toxicity can be elucidated from toxicogenomic studies. Analysis of the glutathione pathway revealed that both APAP and $CCl_4$ modulated the expression levels for a number of key genes required for maintaining glutathione levels, such as γ-glutamylcysteine synthetase (rate-limiting enzyme), glutathione synthetase, and glutathione 6-phosphate dehydrogenase In addition, these results indicate that similar expression changes are observed in the glutathione pathway for either $CCl_4$ or APAP in the rat, yet the time course for differential gene expression is distinct.

## 12.5 SUMMARY

As this chapter demonstrated, toxicogenomics has the potential to impact, in a new and exciting way, current in-house drug development efforts and toxicology assessments by:

1. Providing specific information about the mechanism of toxicity with simultaneous analysis of all perturbed pathways earlier in the drug development cycle than convention methods
2. Providing a way of screening and ranking compounds by developing and using predictive models (*in vivo* and *in vitro*) and by applying appropriate statistical methods to discriminate toxicity markers from nontoxic markers and incurred pathologies from a wide range of reference compounds
3. Profiling proprietary compounds through gene expression and defining unique compound-specific signatures

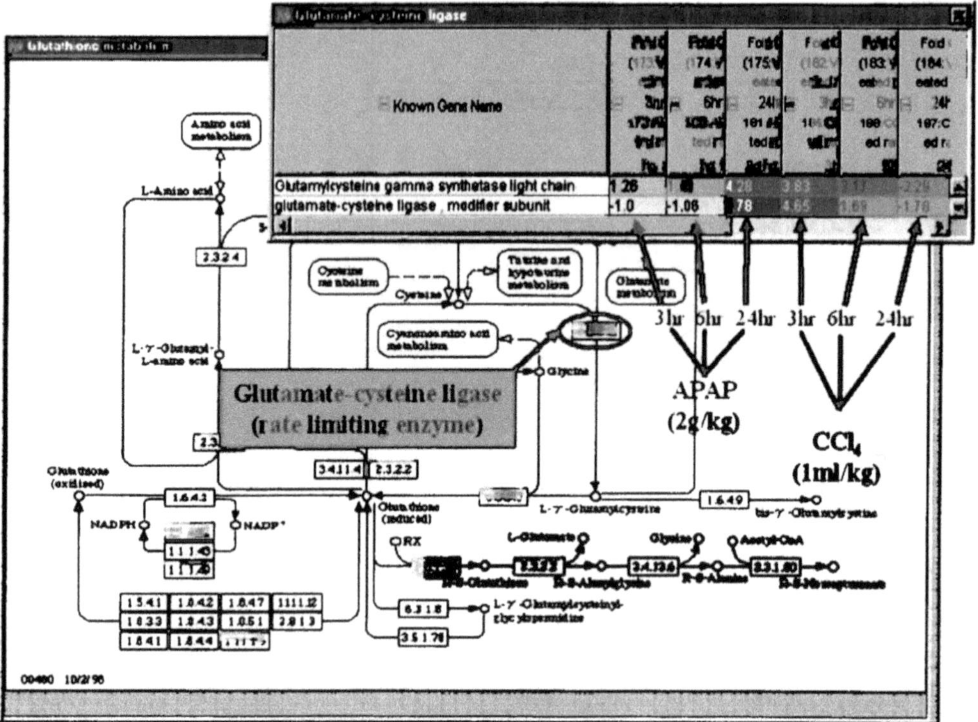


FIGURE 12.25 The rate-limiting enzyme for the glutathione pathway, glutamine-cysteine ligase, is upregulated upon APAP treatment at 24 hr and by $CCl_4$ treatment at 3 hr, as shown in the upper right-hand box. (See color insert.)

4. Identifying biomarkers to track potential toxicity in subchronic or chronic studies as well as in clinical development
5. Allowing a rapid assessment of toxicity of in-licensed compounds

The future of toxicogenomics will ultimately depend on how each researcher applies the technology, the degree of standardization and the underlying understanding of the complexity of gene expression analysis. This chapter provided a predictive-based perspective on the scientific issues surrounding the generation and successful use of toxicogenomics in hopes of continuing and refining the debate surrounding the use of gene expression data.

## REFERENCES

1. Farr, S. and Dunn, R.T., II, Concise review: gene expression applied to toxicology, *Toxicol. Sci.*, 50, 1, 1999.
2. Waring, J.F. and Ulrich, R.G., The impact of genomics-based technologies on drug safety evaluations, *Annu. Rev. Pharmacol. Toxicol.*, 40, 335, 2000.
3. Rockett, J.C. and Dix, D.J., Application of DNA arrays to toxicology, *Environ. Health Perspect.*, 107, 681, 1999.
4. Afshari, C.A. et al., Application of complementary DNA microarray technology to carcinogen identification, toxicology, and drug safety evaluation, *Cancer Res.*, 59, 4759, 1999.
5. Lovett, R.A., Toxicologists brace for genomics revolution, *Science*, 289, 536, 2000.
6. Pennie, W.D. et al., The principles and practice of toxicogenomics: applications and opportunities, *Toxicol. Sci.*, 54, 277, 2000.

7. Kerr, M.K. and Churchill, G.A., Statistical design and the analysis of gene expression microarray data, *Genet. Res.*, 77, 123, 2001.
8. Golub, T.R. et al., Molecular classification of cancer: class discovery and class prediction by gene expression monitoring, *Science*, 286, 531, 1999.
9. Raychaudhuri, S. et al., Pattern recognition of genomic features with microarrays: site typing of *Mycobacterium tuberculosis* strains, *Proc. Int. Conf. Intell. Syst. Mol. Biol.*, 8, 286, 2000.
10. Park, P.J. et al., A nonparametric scoring algorithm for identifying informative genes from microarray data, *Pac. Symp. Biocomput.*, 52, 2001.
11. Kerr, M.K. and Churchill, G.A., Bootstrapping cluster analysis: assessing the reliability of conclusions from microarray experiments, *Proc. Natl. Acad. Sci. USA*, 98, 8961, 2001.
12. Callow, M.J. et al., Microarray expression profiling identifies genes with altered expression in HDL-deficient mice, *Genome Res.*, 10, 2022, 2000.
13. Hamadeh, H.K. et al., Prediction of compound signature using high density gene expression profiling, *Toxicol. Sci.*, 67, 232, 2002.
14. Butte, A.J. et al., Determining significant fold differences in gene expression analysis, *Pac. Symp. Biocomput.*, 6, 2001.
15. Tsien, C.L. et al., On reporting fold differences, *Pac. Symp. Biocomput.*, 496, 2001.
16. Manduchi, E. et al., Generation of patterns from gene expression data by assigning confidence to differentially expressed genes, *Bioinformatics*, 16, 685, 2000.
17. Claverie, J.M., Computational methods for the identification of differential and coordinated gene expression, *Hum. Mol. Genet.*, 8, 1821, 1999.
18. Alon, U. et al., Broad patterns of gene expression revealed by clustering analysis of tumor and normal colon tissues probed by oligonucleotide arrays, *Proc. Natl. Acad. Sci. USA*, 96, 6745, 1999.
19. Perou, C.M. et al., Distinctive gene expression patterns in human mammary epithelial cells and breast cancers, *Proc. Natl. Acad. Sci. USA*, 96, 9212, 1999.
20. Furey, T.S. et al., Support vector machine classification and validation of cancer tissue samples using microarray expression data, *Bioinformatics*, 16, 906, 2000.
21. Alizadeh, A.A. et al., Towards a novel classification of human malignancies based on gene expression patterns, *J. Pathol.*, 195, 41, 2001.
22. Bittner, M. et al., Molecular classification of cutaneous malignant melanoma by gene expression profiling, *Nature*, 406, 536, 2000.
23. Califano, A. et al., Analysis of gene expression microarrays for phenotype classification, *Proc. Int. Conf. Intell. Syst. Mol. Biol.*, 8, 75, 2000.
24. Raychaudhuri, S. et al., Principal components analysis to summarize microarray experiments: application to sporulation time series, *Pac. Symp. Biocomput.*, 455, 2000.
25. Brown, M.P. et al., Knowledge-based analysis of microarray gene expression data by using support vector machines, *Proc. Natl. Acad. Sci. USA*, 97, 262, 2000.
26. DeRisi, J.L. et al., Exploring the metabolic and genetic control of gene expression on a genomic scale, *Science*, 278, 680, 1997.
27. Cho, R.J. et al., A genome-wide transcriptional analysis of the mitotic cell cycle, *Mol. Cell*, 2, 65, 1998.
28. Spellman, P.T. et al., Comprehensive identification of cell cycle-regulated genes of the yeast *Saccharomyces cerevisiae* by microarray hybridization, *Mol. Biol. Cell*, 9, 3273, 1998.
29. Alizadeh, A.A. et al., Distinct types of diffuse large B-cell lymphoma identified by gene expression profiling, *Nature*, 403, 503, 2000.
30. Raychaudhuri, S. et al., Basic microarray analysis: grouping and feature reduction, *Trends Biotechnol.*, 19, 189, 2001.
31. Khan, J. et al., Classification and diagnostic prediction of cancers using gene expression profiling and artificial neural networks, *Nat. Med.*, 7, 673, 2001.
32. Cheng, Y. and Church, G.M., Biclustering of expression data, *Proc. Int. Conf. Intell. Syst. Mol. Biol.*, 8, 93, 2000.
33. Sorlie, T. et al., Gene expression patterns of breast carcinomas distinguish tumor subclasses with clinical implications, *Proc. Natl. Acad. Sci. USA*, 98, 10869, 2001.
34. Waring, J.F. et al., Clustering of hepatotoxins based on mechanism of toxicity using gene expression profiles, *Toxicol. Appl. Pharmacol.*, 175, 28, 2001.

35. Planet, P.J. et al., Systematic analysis of DNA microarray data: ordering and interpreting patterns of gene expression, *Genome Res.*, 11, 1149, 2001.
36. Eisen, M.B. et al., Cluster analysis and display of genome-wide expression patterns, *Proc. Natl. Acad. Sci. USA*, 95, 14863, 1998.
37. Michaels, G.S. et al., Cluster analysis and data visualization of large-scale gene expression data, *Pac. Symp. Biocomput.*, 42, 1998.
38. Eisen, M.B. and Brown, P.O., DNA arrays for analysis of gene expression, *Methods Enzymol.*, 303, 179, 1999.
39. Sherlock, G., Analysis of large-scale gene expression data, *Curr. Opin. Immunol.*, 12, 201, 2000.
40. Heyer, L.J. et al., Exploring expression data: identification and analysis of coexpressed genes, *Genome Res.*, 9, 1106, 1999.
41. Aronow, B.J. et al., Microarray analysis of trophoblast differentiation: gene expression reprogramming in key gene function categories, *Physiol. Genomics*, 6, 105, 2001.
42. Huang, J. et al., Effects of ischemia on gene expression, *J. Surg. Res.*, 99, 222, 2001.
43. Dysvik, B. and Jonassen, I., J-Express: exploring gene expression data using JAVA, *Bioinformatics*, 17, 369, 2001.
44. Tamayo, P. et al., Interpreting patterns of gene expression with self-organizing maps: methods and application to hematopoietic differentiation, *Proc. Natl. Acad. Sci. USA*, 96, 2907, 1999.
45. Toronen, P. et al., Analysis of gene expression data using self-organizing maps, *FEBS Lett.*, 451, 142, 1999.
46. Altman, R.B. and Raychaudhuri, S., Whole-genome expression analysis: challenges beyond clustering, *Curr. Opin. Struct. Biol.*, 11, 340, 2001.
47. Yeung, K.Y. et al., Validating clustering for gene expression data, *Bioinformatics*, 17, 309, 2001.
48. McAdams, H.H. and Shapiro, L., Circuit simulation of genetic networks, *Science*, 269, 650, 1995.
49. Maki, Y. et al., Development of a system for the inference of large scale genetic networks, *Pac. Symp. Biocomput.*, 446, 2001.
50. van Someren, E.P. et al., Linear modeling of genetic networks from experimental data, *Proc. Int. Conf. Intell. Syst. Mol. Biol.*, 8, 355, 2000.
51. Hartemink, A.J. et al., Using graphical models and genomic expression data to statistically validate models of genetic regulatory networks, *Pac. Symp. Biocomput.*, 422, 2001.
52. D'Haeseleer, P. et al., Linear modeling of mRNA expression levels during CNS development and injury, *Pac. Symp. Biocomput.*, 41, 1999.
53. Wessels, L.F. et al., A comparison of genetic network models, *Pac. Symp. Biocomput.*, 508, 2001.
54. Akutsu, T. et al., Algorithms for identifying Boolean networks and related biological networks based on matrix multiplication and fingerprint function, *J. Comput. Biol.*, 7, 331, 2000.
55. Akutsu, T. et al., Inferring qualitative relations in genetic networks and metabolic pathways, *Bioinformatics*, 16, 727, 2000.
56. Liang, S. et al., Reveal, a general reverse engineering algorithm for inference of genetic network architectures, *Pac. Symp. Biocomput.*, 18, 1998.
57. Koza, J.R. et al., Reverse engineering of metabolic pathways from observed data using genetic programming, *Pac. Symp. Biocomput.*, 434, 2001.
58. Ambroise, C. and McLachlan, G.J., Selection bias in gene extraction on the basis of microarray gene-expression data, *Proc. Natl. Acad. Sci. USA*, 99, 6562, 2002.
59. Ronchetti, R. et al., Robust linear model selection by cross-validation, *JASA*, 92, 1017, 1997.
60. Shao, J., Linear model selection by cross-validation, *JASA*, 88, 486, 1993.
61. Efron, B. and Tibshirani, R., *An Introduction to the Bootstrap*, Chapman & Hall, London, 1993.
62. Tibshirani, R., A comparison of some error estimates for neural network models, *Neural Comput.*, 8, 152, 1996.
63. Plutowski, M.E., Survey: Cross-Validation in Theory and in Practice, www.emotivate.com/CvSurvey.doc, 1996.
64. Castle, A.L. et al., Predictive modeling of hepatotoxicants using microarrays and a linear discriminant modeling approach, *Toxicol. Sci.*, 66(1-S), 274, 2002.
65. Johnson, D.E. and Wolfgang, G.H., Assessing the potential toxicity of new pharmaceuticals, *Curr. Top. Med. Chem.*, 1, 233, 2001.

66. Soffers, A.E. et al., Computer-modeling-based QSARs for analyzing experimental data on biotransformation and toxicity, *Toxicol. In Vitro*, 15, 539, 2001.
67. Rosenkranz, H.S. et al., Development, characterization and application of predictive-toxicology models, *SAR QSAR Environ. Res.*, 10, 277, 1999.
68. Smoyer, W.E. and Mundel, P., Regulation of podocyte structure during the development of nephrotic syndrome, *J. Mol. Med.*, 76, 172, 1998.
69. Diamond, J.R. and Karnovsky, M.J., Focal and segmental glomerulosclerosis following a simgle intravenous dose of puromycin aminonucleoside, *Am. J. Pathol.*, 122, 481, 1986.
70. Smoyer, W.E. et al., Podocyte alpha-actinin induction precedes foot process effacement in experimental nephrotic syndrome, *Am. J. Physiol.*, 273, F150–F157, 1997.
71. Shirato, I. et al., Cytoskeletal changes in podocytes associated with foot process effacement in Masugi nephritis, *Am. J. Pathol.*, 148, 1283, 1996.
72. Kaplan, J.M. et al., Mutations in ACTN4, encoding alpha-actinin-4, cause familial focal segmental glomerulosclerosis, *Nat. Genet.*, 24, 251, 2000.
73. Woolf, T.F. et al., Bioactivation and irreversible binding of the cognition activator tacrine using human and rat liver microsomal preparations: species difference, *Drug Metab. Dispos.*, 21, 874, 1993.
74. Fitten, L.J. et al., Long-term oral administration of memory-enhancing doses of tacrine in mice: a study of potential toxicity and side effects, *J. Gerontol.*, 42, 681, 1987.
75. Mayeux, R. and Sano, M., Treatment of Alzheimer's disease, *N. Engl. J. Med.*, 341, 1670, 1999.
76. Nordberg, A. and Svensson, A.L., Cholinesterase inhibitors in the treatment of Alzheimer's disease. A comparison of tolerability and pharmacology, *Drug Saf.*, 19, 465, 1998.
77. Taylor, P., Anticholinesterase agents, in *Goodman and Gilman's Pharmacologic Basis of Therapeutics*, Hardman, J.G., Ed., McGraw-Hill, New York, 1996, chap. 8.
78. Porter, M.W. et al., Identification through gene expression of a toxic response in the rat for a compound that exhibits overt toxicity in humans, but not in rats, *Toxicol. Sci.*, 66 (1-S), 295, 2002.
79. Hessel, G. et al., Correlation between the severity of acute hepatic necrosis induced by acetaminophen and serum aminotransferase levels in fasted and sucrose-fed rats, *Braz. J. Med. Biol. Res.*, 29, 793, 1996.
80. Buttar, H.S. et al., Serum enzyme activities and hepatic triglyceride levels in acute and subacute acetaminophen-treated rats, *Toxicology*, 6, 9, 1976.
81. Tarloff, J.B. et al., Sex- and age-dependent acetaminophen hepato- and nephrotoxicity in Sprague–Dawley rats: role of tissue accumulation, nonprotein sulfhydryl depletion, and covalent binding, *Fund. Appl. Toxicol.*, 30, 13, 1996.
82. Klaasen, C.D., *Casarett and Doull's Toxicology: The Basic Science of Poisons*, Fifth ed., McGraw-Hill, New York, 1996, p. 1.
83. Vermeulen, N.P. et al., Molecular aspects of paracetamol-induced hepatotoxicity and its mechanism-based prevention, *Drug Metab. Rev.*, 24, 367, 1992.
84. Zaher, H. et al., Protection against acetaminophen toxicity in CYP1A2 and CYP2E1 double-null mice, *Toxicol. Appl. Pharmacol.*, 152, 193, 1998.
85. Mitchell, J.R. et al., Acetaminophen-induced hepatic necrosis. I. Role of drug metabolism, *J. Pharmacol. Exp. Ther.*, 187, 185, 1973.
86. Dahlin, D.C. et al., N-acetyl-*p*-benzoquinone imine: a cytochrome P-450-mediated oxidation product of acetaminophen, *Proc. Natl. Acad. Sci. USA*, 81, 1327, 1984.
87. Davidson, D.G. and Eastham, W.N., Acute liver necrosis following overdose of paracetamol, *Br. Med. J.*, 5512, 497, 1966.
88. Coles, B. et al., The spontaneous and enzymatic reaction of N-acetyl-*p*-benzoquinonimine with glutathione: a stopped-flow kinetic study, *Arch. Biochem. Biophys.*, 264, 253, 1988.
89. Rollins, D.E. and Buckpitt, A.R., Liver cytosol catalyzed conjugation of reduced glutathione with a reactive metabolite of acetaminophen, *Toxicol. Appl. Pharmacol.*, 47, 331, 1979.
90. Hinson, J.A. et al., 3-(glutathion-S-yl)acetaminophen: a biliary metabolite of acetaminophen, *Drug Metab. Dispos.*, 10, 47, 1982.
91. Deneke, S.M. and Fanburg, B.L., Regulation of cellular glutathione, *Am. J. Physiol.*, 257, L163–L173, 1989.
92. Chan, K. et al., An important function of Nrf2 in combating oxidative stress: detoxification of acetaminophen, *Proc. Natl. Acad. Sci. USA*, 98, 4611, 2001.

# 13 Unsupervised Hierarchical Clustering of Toxicants

*Jeffrey F. Waring and Roger G. Ulrich*

CONTENTS



## 13.1 INTRODUCTION

The application of genomics techniques to the field of toxicology has created a new science subdiscipline, *toxicogenomics*. Toxicogenomics is the study of the structure and transcriptional output of the entire genome as it relates and responds to adverse xenobiotic exposure. One of the main tools utilized in toxicogenomics research is microarray analysis. This technology allows one to monitor the expression of thousands of genes at the same time by quantitating the expression levels of messenger RNA (mRNA). The expression of virtually every gene in the entire genome can be determined for a single time point in one experiment, thus yielding an unprecedented amount of information regarding the physiological state of the cell. However, inherent in microarray experiments is the difficulty in extracting knowledge from such a vast amount of data. For example, consider the implications of using microarrays in a simple toxicology experiment looking at the effects of one compound on one tissue at three dose levels plus a control group at a single time point. If the study has 5 animals per treatment group, this would involve running microarray analysis on 20 samples. Assuming the microarray chip has 20,000 genes present, this study would yield 400,000 data points to analyze. If one wishes to look at different time points, comparator compounds, or tissues, the number of data points quickly exceeds 1 million.

Faced with such a plethora of data, how can one extract the most useful information that can aid in determining the safety of a new compound? One method that is often used is simply to look at the changes for the set of genes of interest. If a compound is suspected to result in cell death,

0-8493-1334-1/03/$0.00+$1.50

for instance, one could concentrate on gene changes involved in proliferation and cell-cycle regulation. Likewise, if a compound is suspected to be a mitochondrial damaging agent, one could concentrate on the regulation of genes involved in mitochondrial function. Certainly, information gained from such analysis can be useful. However, performing this kind of analysis presupposes that one has an idea of the type of toxicity to expect. In addition, concentrating on only a subset of genes limits the usefulness of microarray analysis. Examining the expression profiles of the entire genome, not just a subset, allows one to gain information on all of the changes in the cell resulting from treatment with a compound, changes that come from both intended (on-target) and untoward (off-target) effects. Such observations, in turn, allow one to learn valuable information regarding potential toxic liabilities without any presupposed knowledge of the compound.

For toxicogenomic studies, it is beneficial to look at the regulation of all of the genes present on a chip in order to identify any changes that may indicate potential toxicity resulting from treatment with the compound, including those changes not previously suspected and included in a candidate list. Along with data concerning off-target effects, though, are data describing the intended effect along with gene changes related to normal biological fluctuation and thus unrelated to specific chemical exposure. The challenge lies in how to manage the huge volume of data generated from microarray studies to extract the necessary and relevant information. While many different methods are available to do this, one that is being employed widely in the field of toxicogenomics is a statistical method known as *unsupervised learning*. The application of this method to the analysis of toxicogenomic data stems from the hypothesis that expression profiles reflect the molecular phenotype of a cell in a certain state. Breast cancer cells, for instance, can be distinguished from normal breast cells based on their gene expression profiles. Similarly, cells exposed to a toxin can be distinguished from control (normal) cells, and the resultant gene expression pattern is both characteristic for the particular compound or chemical class and contains information regarding the mechanism of action. Thus, by comparing the gene expression profile of an unknown compound against a reference database, it may be possible to determine if the compound has potential toxic liabilities.

The difference between supervised and unsupervised learning is that with supervised learning, the user has previous knowledge from outside the microarray experiment concerning the output value (gene expression) or the input variable (cell type or treatment condition). This information is used to direct the analysis of the microarray experimental results. For instance, in the case of a microarray experiment comparing normal tissue to breast cancer tissue, if the user knows the identification of the samples, this information can be used to group the expression profiles and thus identify genes that are regulated in the diseased state.

As introduced in Chapter 4, in unsupervised learning the user has no data concerning either the output value or the input variable. Basically, the task of unsupervised learning is to cluster or group the data by finding patterns or linear relationships. Because no training set is used, it is difficult to judge the accuracy or degree of error that may result from using unsupervised clustering. As a result, a variety of methods are used for unsupervised clustering, all of which are somewhat subject to interpretation. Examples of unsupervised clustering are agglomerative, divisive, and *k*-means clustering.

Previous studies have used unsupervised clustering to show that expression profiles obtained from microarrays can be used to classify cells from various tissues or in different physiological states.[1] In one particular study, gene expression signatures were used to cluster gene profiles from acute myeloid leukemia (AML) and acute lymphoblastic leukemia (ALL) cells. In this study, the authors were able to classify 29 out of 34 bone marrow or peripheral blood samples as being either AML or ALL with 100% accuracy using gene expression profiles; in fact, in at least one case the expression signatures proved to be a more sensitive and accurate method than traditional histopathology for classifying the tumor cells.[2] A similar analysis was performed by Ross et al.,[3] who used unsupervised cluster analysis to classify 60 cell lines from diverse tumor tissues. The cell lines clustered based on their tissues of origin. Microarrays and cluster analysis have also been used to classify changes in gene expression in cell-cycle control and pheromone signaling in *Saccharomyces cerevisiae*.[4] Previous work has also shown that cell responses to xenobiotic treatment can be indicators of the mechanism of action of the agent. Scherf et al.[5] used microarray and cluster analysis to study cell responses to 118 cancer drugs with known mechanisms of action. The studies showed that the cell responses to the cancer drugs grouped together according to the mechanism of action of the drug, regardless of the type of cell used.

This chapter focuses on one study that applied unsupervised clustering to expression profiles from rats treated with 15 different known hepatotoxins: allyl alcohol, amiodarone, Aroclor 1254, arsenic, carbamazepine, carbon tetrachloride, diethylnitrosamine, dimethylformamide, diquat, etoposide, indomethacin, methapyrilene, methotrexate, monocrotaline, and 3-methylcholanthrene. These agents have been shown to cause a variety of toxic hepatic effects, including hepatocellular hypertrophy, DNA damage, necrosis, steatosis, and cholestasis, among others. Gene expression analysis was done on RNA from the livers of treated rats and was compared against vehicle-treated controls. The gene expression results were clustered using different methods of unsupervised clustering, and the results were compared to histopathology, clinical chemistry, and historical data concerning mechanisms of toxicity.

## 13.2 MATERIALS AND METHODS

### 13.2.1 In-Life Studies

Rats were treated with the 15 hepatotoxins at levels that have previously been shown to be hepatotoxic but not lethal. Male Sprague–Dawley rats, approximately 6 to 12 weeks of age and weighing between 225 and 275 g, were injected intraperitoneally with vehicle (corn oil or saline), allyl alcohol (40 mg/kg/day),[6] amiodarone (100 mg/kg/day),[7] Aroclor 1254 (400 mg/kg/day),[8] arsenic (20 mg/kg/day),[9] carbamazepine (250 mg/kg/day),[10] carbon tetrachloride (1000 mg/kg/day),[8] diethylnitrosamine (100 mg/kg/day),[11] dimethylformamide (1000 mg/kg/day),[12] diquat (17.2 mg/kg/day),[13] etoposide (50 mg/kg/day),[14] indomethacin (20 mg/kg/day),[15] methapyrilene (50 mg/kg/day),[16] methotrexate (250 mg/kg/day),[17] monocrotaline (50 mg/kg/day),[18] or 3-methylcholanthrene (100 mg/kg/day).[19]

Allyl alcohol, Aroclor 1254, 3-methylcholanthrene, and carbon tetrachloride were suspended or dissolved in corn oil, while all other compounds were suspended or dissolved in saline. Each treatment group had three rats, which were dosed daily for three days and sacrificed on the fourth. Blood samples drawn from the animals at necropsy were used to measure serum concentrations or activities of blood urea nitrogen (BUN), creatinine, alanine amino transferase (ALT), aspartate amino transferase (AST), γ-glutamyl transferase (GGT), alkaline phosphatase (ALP), cholesterol, triglycerides, bilirubin, glucose, total protein, albumin, and globulins using an Abbott Aeroset clinical chemistry analyzer. The terminal body weights and liver weights were recorded. The left lateral lobe of the liver was processed for histopathologic evaluation. Approximately 100 mg from each liver was placed into TRIzol® reagent and immediately homogenized using a Turrax tissue grinder. The remaining portion of the liver was retained frozen for future study.

### 13.2.2 Microarray Analysis

RNA preparation and analysis was done according to the Affymetrix (Santa Clara, CA) protocol. Briefly, RNA from the three rats in each group was pooled using equal amounts from each rat to make a total of 200 μg of RNA for each treatment group. The integrity of the RNA from the pooled samples was determined using an Agilent 2100 Bioanalyzer. Following this, mRNA was extracted from the total RNA samples using the Qiagen Oligotex mRNA Midi Kit (Cat. No. 70042; Hilden, Germany). Complementary DNA (cDNA) was prepared from 1 μg of mRNA using the Superscript Choice system from Gibco-BRL Life Technologies (Cat. No. 18090-019; Gaithersburg, MD). The

Gibco protocol was followed exactly, with the exception that the primer used for the reverse transcription reaction was a modified T7 primer with 24 thymidines at the 5′ end. The sequence was:

5′–GGCCAGTGAATTGTAATACGACTCACTATAGGGAGGCGG-$(dT)_{24}$–3′

Following this, labeled cRNA was synthesized from the cDNA using the Enzo RNA Transcript Labeling Kit (Cat. No. 900182; New York, NY) according to the manufacturer's instructions. Approximately 20 µg of cRNA was then fragmented in a solution of 40 mM Tris-acetate, pH 8.1, 100 mM KOAc, and 30 mM MgOAc at 94°C for 35 minutes.

Labeled cRNA was hybridized to the Affymetrix GeneChip Test 2 Array to verify the quality of labeled cRNA. Following this, cRNA was hybridized to the Affymetrix Rat Toxicology U34 Array (Cat. No. 900252). The cRNA was hybridized overnight at 45°C.

### 13.2.3 Data Analysis and Clustering Algorithm

The microarray data was analyzed using the Rosetta Resolver™ v.2.0 Expression Data Analysis System. Sequences on Affymetrix chips are generally represented by an average of 20 perfect match (PM) probe sets to the sequence and corresponding mismatch (MM) sequences to control for background and nonspecific hybridization. Absolute intensity statistics are based on feature intensities and corresponding standard deviations provided in the GeneChip-generated CEL files. The absolute intensity statistics were computed by averaging the PM and corresponding MM feature intensity differences after removal of outliers; for a given probe set, the error-weighted PM/MM intensity difference was considered an outlier if it was more than 3 standard deviations away from the mean PM/MM intensity difference for that probe set. The *p*-values for the average of the $\log_{10}$ PM/MM differences were computed by pooling the standard deviations associated with the $\log_{10}$ of the PM/MM differences, weighting the mean of the $\log_{10}$ PM/MM differences using the pooled standard deviation and treating the resulting statistic as a normally distributed random variable with mean 0 and standard deviation 1.

Expression changes between two arrays were quantified as $\log_{10}$ (expression ratio), where the expression ratio was taken to be the ratio between normalized, error-weighted PM/MM difference intensities. We have observed, as have others,[20] significant spatial variation of control probes across the surface of an array, where the control probes should all be of the same intensity. Therefore, in addition to correcting for these regional gain biases, normalization between two arrays was performed by subdividing the arrays into 16 regions and normalizing the arrays region by region. An error model for the log ratio was applied to quantify the significance of expression changes between two samples. This error model assumed the log ratio statistic followed a standard normal distribution. The main purpose of the error model was to generate *p*-values for the log ratio statistic so that genes could be rank-ordered according to the significance of this statistic. Using *p*-values to rank-order genes, as opposed to using *p*-values in a classic hypothesis-testing context, is not sensitive to departures from normality.

The color displays given in Figure 13.3 show the $\log_{10}$ (expression ratio) as (1) red, when the treatment sample is upregulated relative to the control sample; (2) green, when the treatment sample is downregulated relative to the control sample; (3) black, when the $\log_{10}$ (expression ratio) is close to zero; and (4) gray, when data from one or both of the samples for a given probe set are unreliable. Clustering results presented in Figure 13.4 were generated by reordering genes and treatments according to a hierarchical clustering algorithm, as previously described.[21] Other clustering methods used were the divisive hierarchical clustering algorithm,[22] the *k*-means partitioning clustering algorithm,[23] and the self-organizing map (SOM)-based clustering algorithm.[24] For a review of these methods, the reader is referred to Chapter 4 in this textbook.

### 13.2.4 Statistical Analysis

The correlations of the clinical chemistry parameters to gene expression changes were determined using eight of the clinical chemistry variables: ALT, AST, ALP, cholesterol, bilirubin, glucose, GGT, and BUN. The log of the ratio of the treated and control clinical chemistry measurements was regressed onto the error-weighted log ratio values for each gene across all 15 treatments, yielding a correlation coefficient for each clinical chemistry variable/gene pair. For a given gene, a treatment was included in the least-squares fit, giving rise to the correlation coefficient only if the $p$-value for the error-weighted log ratio for gene expression values was $<0.01$. Correlation coefficients were computed only if four or more treatments met this $p$-value condition. If the correlation coefficient resulting from the least-squares fit was $>0.60$, the clinical chemistry variable/gene pair was determined to be significantly correlated.

## 13.3 RESULTS

### 13.3.1 Histopathology and Clinical Chemistry Results

With allyl alcohol treatment, two of three rats showed slightly elevated serum ALP activities compared with controls. All three had elevated serum ALT activities, ranging from 11 to 33 times the mean value of the corn oil control group. AST activities were similarly elevated, ranging from 6 to 18 times the control mean. The mean relative liver weight was significantly greater than control. Microscopic change was pronounced in all three rats. Hepatocellular necrosis was present in periportal regions, and both bile duct proliferation and peribiliary inflammatory infiltrates and fibrosis were increased above control. The numbers of mitotic figures derived from both hepatocytes and biliary epithelial cells exceeded control levels; hepatocellular figures were especially numerous. Minor amounts of hemorrhage and mineralization were also present (Figure 13.1).

Treatment of rats with Aroclor 1254 resulted in a twofold increase in average serum triglyceride concentration and a 140% increase in relative liver weight over the vehicle-treated controls. Mild midzonal hepatocellular fatty change was evident in all treated animals. In addition, diffuse mild hepatocytomegaly and increased hepatocellular mitotic activity were observed (Figure 13.1). Hepatocellular lipid (triglyceride) accumulation suggests an imbalance between clearance of triglycerides from the serum and subsequent secretion of very-low-density lipoprotein (VLDL).

Two of three animals treated with carbon tetrachloride showed mild to moderate increases in serum ALP activities relative to control. Serum ALT activities were roughly four- to sixfold greater, and serum AST activities were three- to fourfold greater than control. Two of the three rats had mildly higher serum triglyceride concentrations relative to control, and the mean relative liver weights increased significantly above control. Mild to moderate centrilobular hepatocellular vacuolar degeneration and necrosis were present in all three rats. Mild to marked increases in the number of hepatocellular mitotic figures were also evident in these three animals, and minimal to mild infiltrates of mononuclear cells were present in centrilobular areas of all treated animals (Figure 13.1).

All three of the livers from rats treated with diethylnitrosamine revealed severe, diffuse hepatic necrosis with attendant mild mononuclear inflammatory infiltrate. The eosinophilic necrosis was particularly centrilobular but extended out toward portal areas as well (Figure 13.2).

Two livers from rats treated with monocrotaline revealed minimal expressions of hepatocyte karyomegaly and single-cell necrosis. The liver from the third rat revealed a severe and uniformly widespread, diffuse, centrilobular necrosis. A sparse population of mixed inflammatory cells was present in the necrotic zones (Figure 13.2).

All of the livers from rats treated with methapyriline had a minimal expression of single-cell necrosis with a minimal mononuclear infiltrate. Several small portal areas had minimal expressions of oval cell hyperplasia and minimal mononuclear inflammation (Figure 13.2).

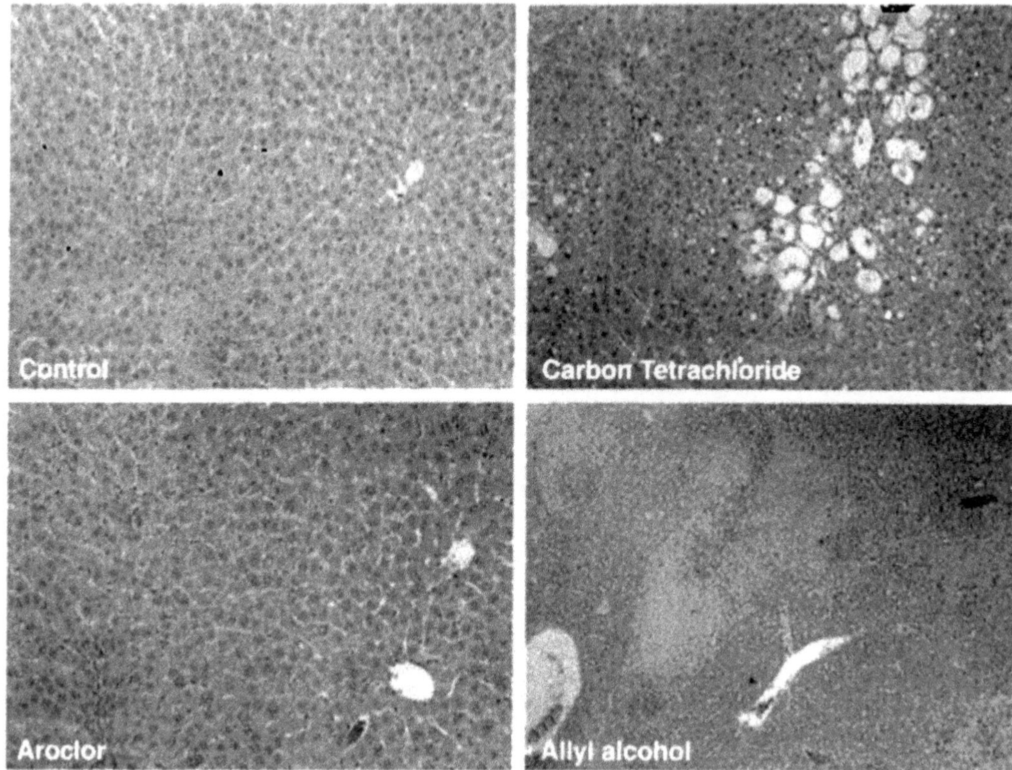


**FIGURE 13.1** Slide showing liver sections from control or rats treated with carbon tetrachloride, Aroclor 1254, or allyl alcohol. All sections are shown at 100× magnification except for allyl alcohol, which is shown at 50×. The central vein is located to the right in each slide. (See color insert following page 112.)

Rats treated with dimethylformamide showed varying degrees of diffuse and multifocal centrilobular necrosis. The livers from rats treated with indomethacin revealed mild to moderate numbers of microfoci of extramedullary hematopoiesis scattered about the liver. Rats treated with 3-methylcholanthrene showed an increase in relative liver weights of 132% over the vehicle-treated control group. Beyond the increase in liver size, no notable histological observations outside of normal parameters were seen with rats treated with 3-methylcholanthrene.

Two of three rats treated with diquat and one rat treated with methotrexate showed higher-than-expected serum blood urea nitrogen values (one rat treated with diquat did not survive and was not sampled). Two of three rats treated with carbamazepine showed slightly lower serum cholesterol concentrations. Rats treated with diquat, carbamazepine, methotrexate, arsenic, etoposide, and amiodarone showed no meaningful microscopic changes. Although the dose levels that we used have been shown to be hepatotoxic in rats, it is possible that three days of treatment was not long enough to observe the cellular changes reported in other studies.

### 13.3.2 Gene Expression Changes

RNA was harvested from the livers of treated rats. The RNA integrity of the pooled RNA was determined using an Agilent 2100 Bioanalyzer. The results showed that all of the RNA was intact (data not shown). Microarray analysis was done to determine differences in hepatic gene expression between hepatotoxin-treated and vehicle-treated rats. For each of the approximately 1000 genes present on the Affymetrix rat toxicology chip (RTU34), induction or repression values were calculated using the Rosetta Resolver™ v.2.0 Expression Data Analysis System. In order to ascertain

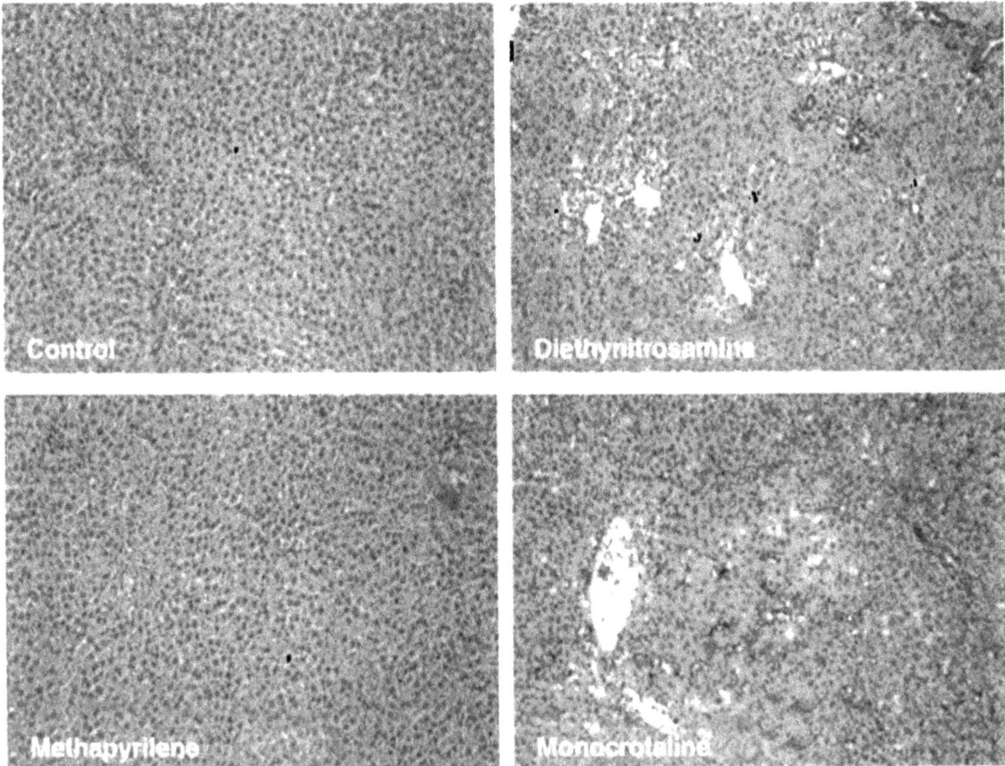


**FIGURE 13.2** Slide showing liver sections from control or rats treated with diethylnitrosamine, methapyrilene, or monocrotaline. All sections are shown at 100× magnification.

the extent of variablility resulting from RNA preparation, we processed RNA from the same samples and performed microarray analysis on these samples. We found extremely high correlation between the signature profiles of different RNA preparations (weighted correlation coefficient of common signature = 0.98; data not shown). We therefore concluded that single RNA preparations were sufficient for analysis.

RNA was pooled from the treatment groups and hybridized to the RTU34 microarray chips. The gene expression changes resulting from treatment with the 15 hepatotoxins are represented by the two-dimensional graphs shown in Figure 13.3. The individual genes are represented on the $x$-axis, and the different experiments are shown on the $y$-axis. The changes in gene expression are represented colorimetrically as described in the Methods section. Because it is possible to obtain different results depending on the clustering method used, we performed three separate clustering operations: (1) agglomerative, (2) divisive, and (3) $k$-means (for a review of these methods, see Chapter 4).[25] As noted earlier in this book, compounds that cluster together regardless of the clustering method used are likely to be more significant. The results show that some clusters are constant regardless of the statistical method used, while other clusters vary. With all three clustering methods, Aroclor 1254 clustered with 3-methylcholanthrene and allyl alcohol clustered with carbon tetrachloride. In addition, both agglomerative and $k$-means cluster analyses paired monocrotaline with diethylnitrosamine. Some clusters varied, depending on the method used. For instance, using agglomerative clustering, diquat clustered with methotrexate, while with divisive clustering diquat clustered with etoposide (compare Figures 13.3A and 13.5B).

By pooling the RNA from individually treated animals, we obtained signature expression profiles that should represent an average of the three animals. Potentially, the hepatotoxins may have clustered differently if we had obtained expression profiles from individual animals instead

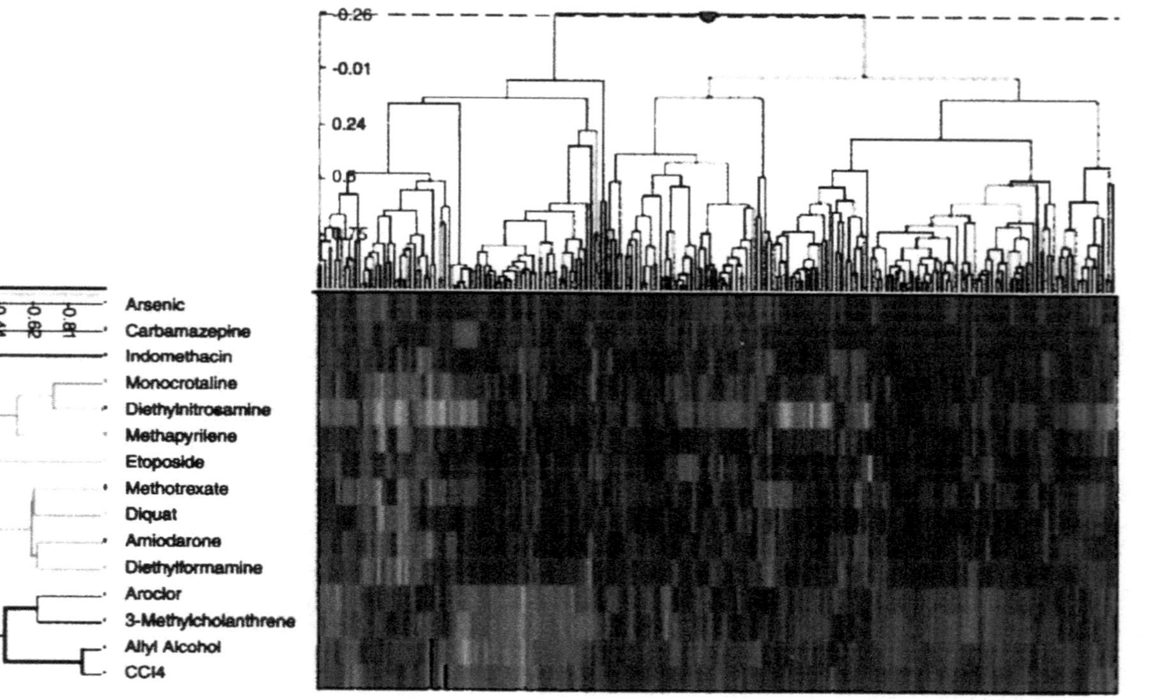


**FIGURE 13.3** (A) Two-dimensional graph showing the gene changes occuring in livers from rats treated with the 15 known hepatotoxins. A total of 231 genes with a $p$-value of 0.01 were shown to be regulated ±twofold or more by at least three compounds. The hepatotoxins ($y$-axis) and genes ($x$-axis) were clustered using agglomerative clustering. Increases in mRNA levels are represented as shades of dark gray and decreases in mRNA levels are represented by shades of light gray. (B) Two-dimensional graph showing the gene changes occuring in livers from rats treated with the 15 known hepatotoxins. A total of 231 genes with a $p$-value of 0.01 were shown to be regulated ±twofold or more by at least three compounds. The hepatotoxins ($y$-axis) and genes ($x$-axis) were clustered using divisive clustering. (C) Graph showing gene expression profile groupings using $k$-means clustering from pools of 15 hepatotoxins. The conditions are the same as in parts A and B.

B.

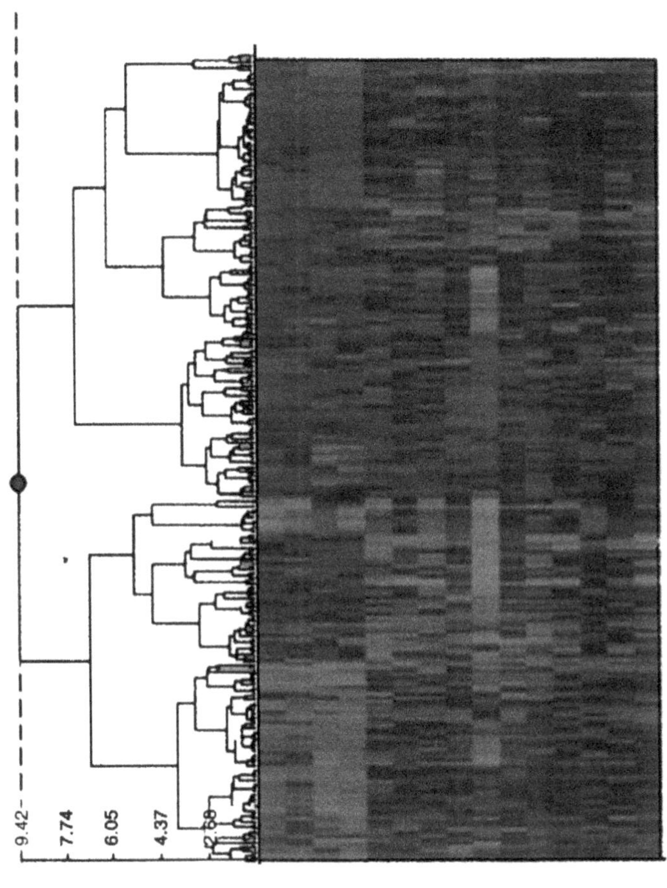


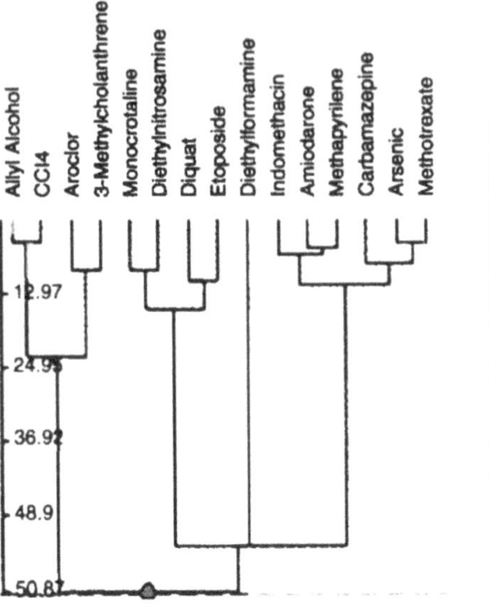


-1 1

FIGURE 13.3 (B)

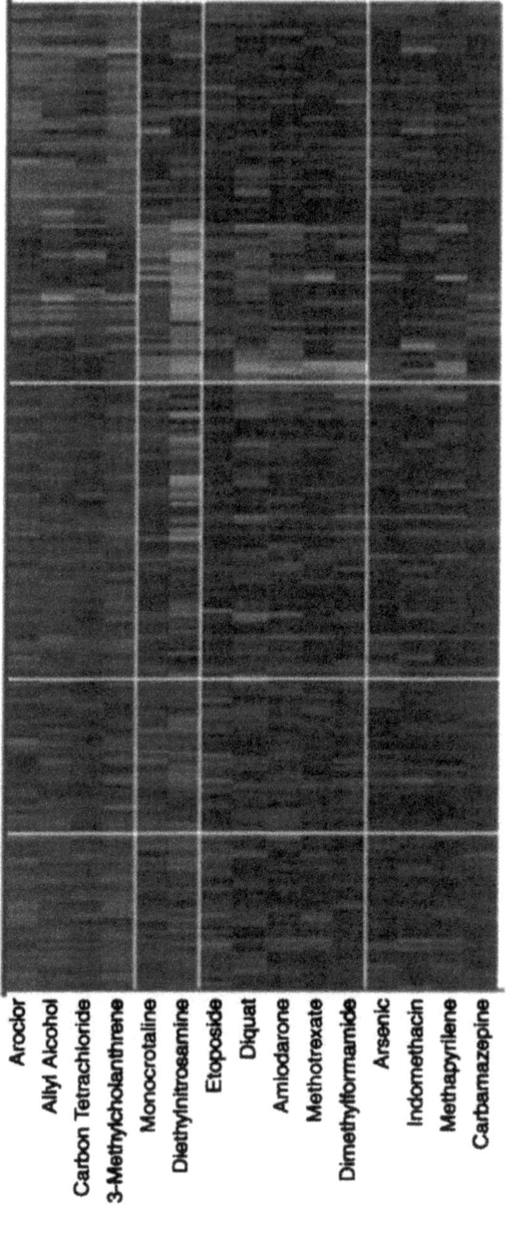
C.
Aroclor
Allyl Alcohol
Carbon Tetrachloride
3-Methylcholanthrene
Monocrotaline
Diethylnitrosamine
Etoposide
Diquat
Amiodarone
Methotrexate
Dimethylformamide
Arsenic
Indomethacin
Methapyrilene
Carbamazepine
-1
1

FIGURE 13.3 (C)

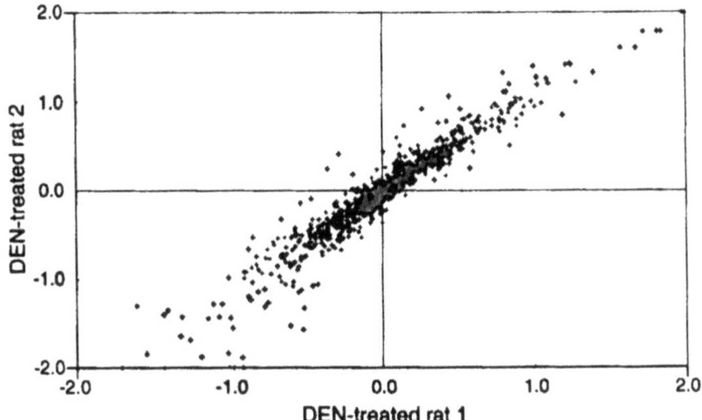


**FIGURE 13.4** Comparison plot showing the regulation of genes by two different rats treated with diethylnitrosamine. Genes that were regulated at a $p$-value of 0.01 are represented by gray dots. Genes commonly regulated in both samples are represented by black dots. The correlation coefficient of genes regulated in common in diethylnitrosamine treated rats was 0.979 ± 0.005.

of pooling. To address this issue, we performed microarray analysis on samples from the three individual animals treated with Aroclor 1254 and diethylnitrosamine and two of the animals treated with carbon tetrachloride. The microarray results showed strong correlation between the individual animals from the different treatment groups. Figure 13.4 shows the result of comparison plots of the expression analysis from two of the animals treated with diethylnitrosamine. The genes regulated in common between the two different animals are shown in red. While there were some gene changes between the animals, the correlation coefficient of genes regulated in a similar manner by animals from the same treatment group was 0.971 for animals treated with diethylnitrosamine, 0.972 for animals treated with Aroclor, and 0.878 for animals treated with carbon tetrachloride.

The expression profiles from the individual animals were clustered with the signature profiles from the pooled samples. Figure 13.5 shows the results of the cluster analysis using divisive and $k$-means clustering. Both cluster methods show that the signature profiles from the individual Aroclor-, diethylnitrosamine-, and carbon-tetrachloride-treated animals cluster with the signature profiles from the pooled samples. In addition, the signature profiles from the Aroclor-treated animals continue to cluster with 3-methylcholanthrene and the signature profiles from carbon-tetrachloride-treated animals continue to cluster with allyl alcohol.

Figure 13.6 shows a close-up view of some of the regions of the $k$-means cluster analysis from Figure 13.5B. Figure 13.6A shows a cluster of genes that are upregulated with diethylnitrosamine treatment. Many of these genes are also upregulated by monocrotaline, etoposide, allyl alcohol, and carbon tetrachloride. Figure 13.6B shows genes that are downregulated by diethylnitrosamine, and Figure 13.6C shows genes that are upregulated by Aroclor and 3-methylcholanthrene treatment.

### 13.3.3 Association with Clinical Chemistry Changes

Using supervised learning, an attempt was made to correlate alterations in some of the clinical chemistry parameters with changes in gene expression (see methods). Several genes were identified that were consistently either upregulated or downregulated in association with changes in a clinical chemistry parameter. A list of some of these genes and their corresponding correlation coefficients is provided in Table 13.1.

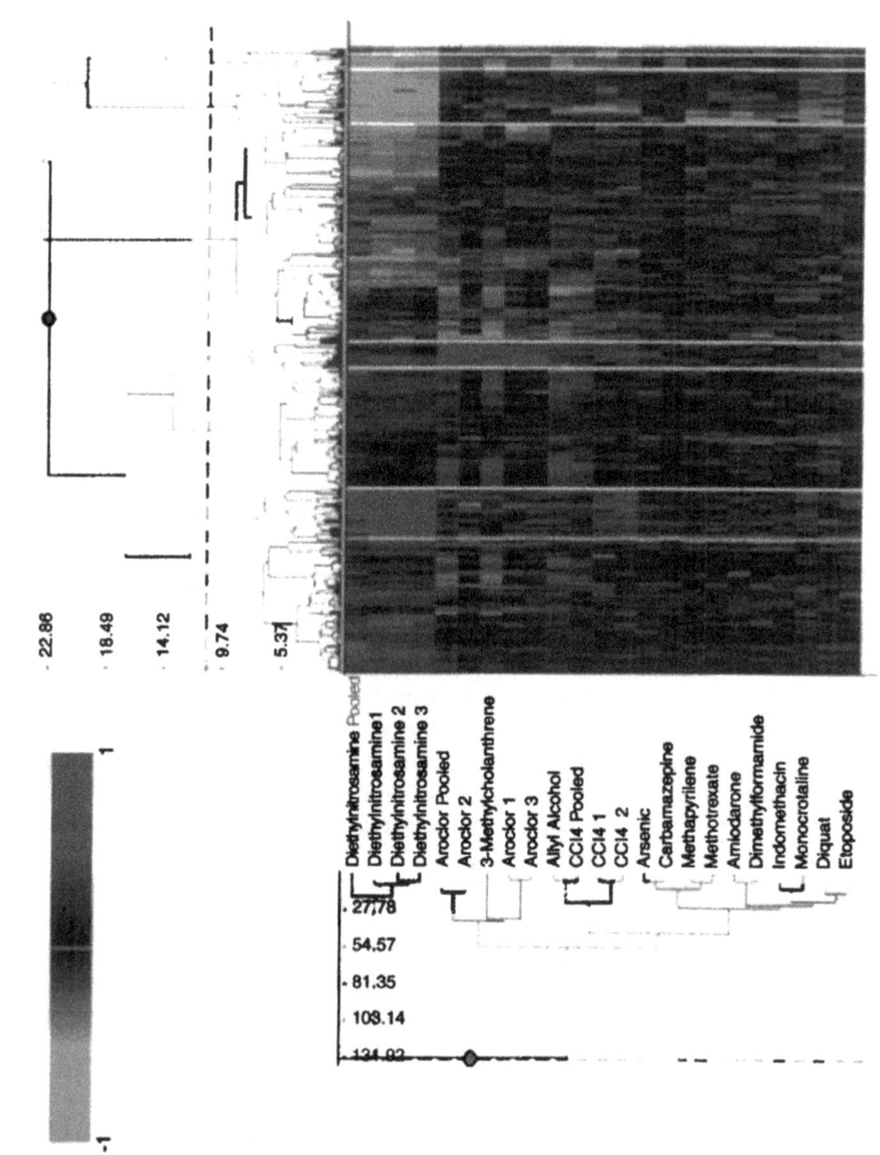


**FIGURE 13.5** (A) Two-dimensional graph showing the expression profiles of pooled samples from 15 hepatotoxins and expression profiles from individual animals treated with Aroclor 1254, diethylnitrosamine, and carbon tetrachloride grouped by divisive clustering. (B) Gene expression profile clusterings from pooled and individual samples. The profiles were grouped using *k*-means clustering. (See color insert.)

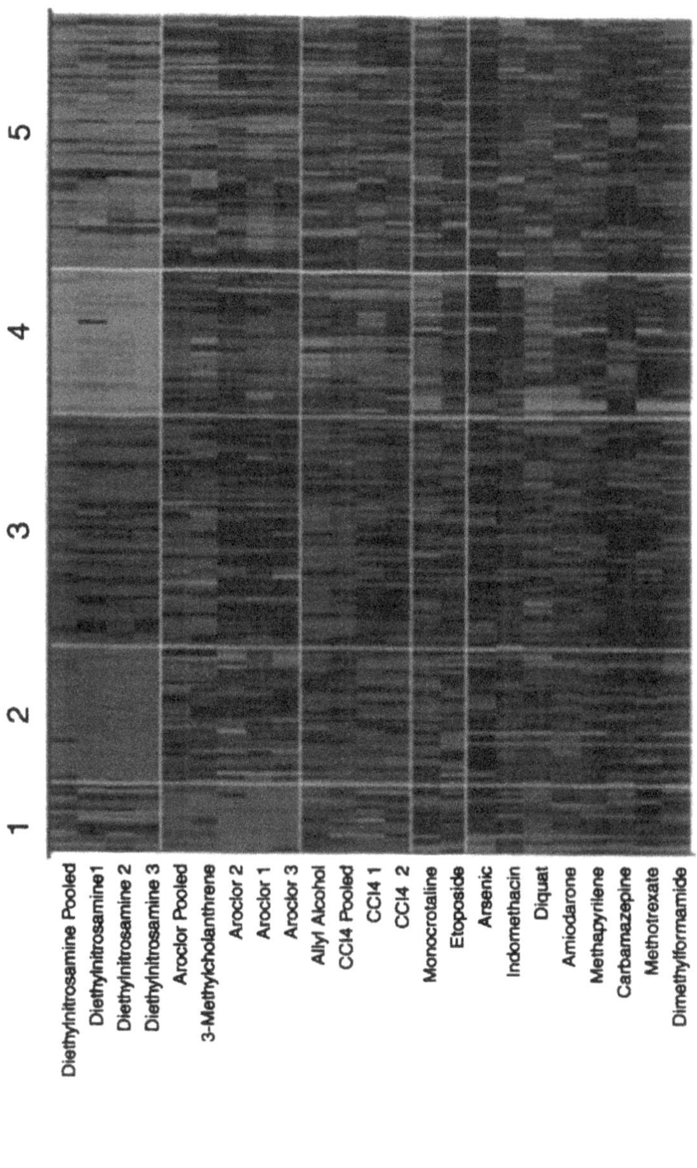
B.
1
2
3
4
5
Diethylnitrosamine Pooled
Diethylnitrosamine1
Diethylnitrosamine 2
Diethylnitrosamine 3
Aroclor Pooled
3-Methylcholanthrene
Aroclor 2
Aroclor 1
Aroclor 3
Allyl Alcohol
CCl4 Pooled
CCl4 1
CCl4 2
Monocrotaline
Etoposide
Arsenic
Indomethacin
Diquat
Amiodarone
Methapyrilene
Carbamazepine
Methotrexate
Dimethylformamide
-1
1

FIGURE 13.5 (B)

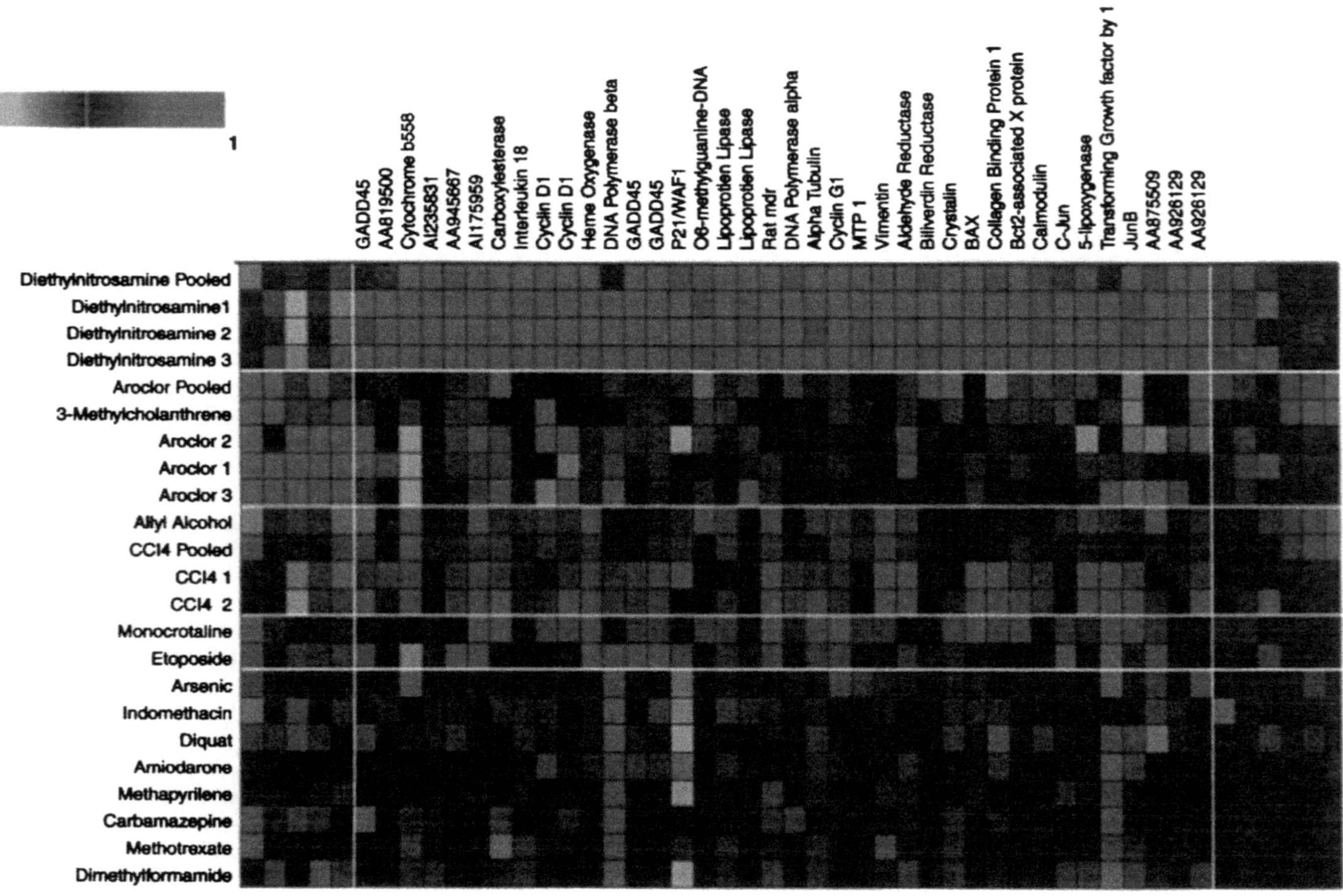


**FIGURE 13.6** (A) Close-up view of group 2 from Figure 13.5B. The individual gene names belonging to group 2 are shown above the graph. (B) Close-up view of group 4 from Figure 13.5B. (C) Close-up view of group 1 from Figure 13.5B. (See color insert.)

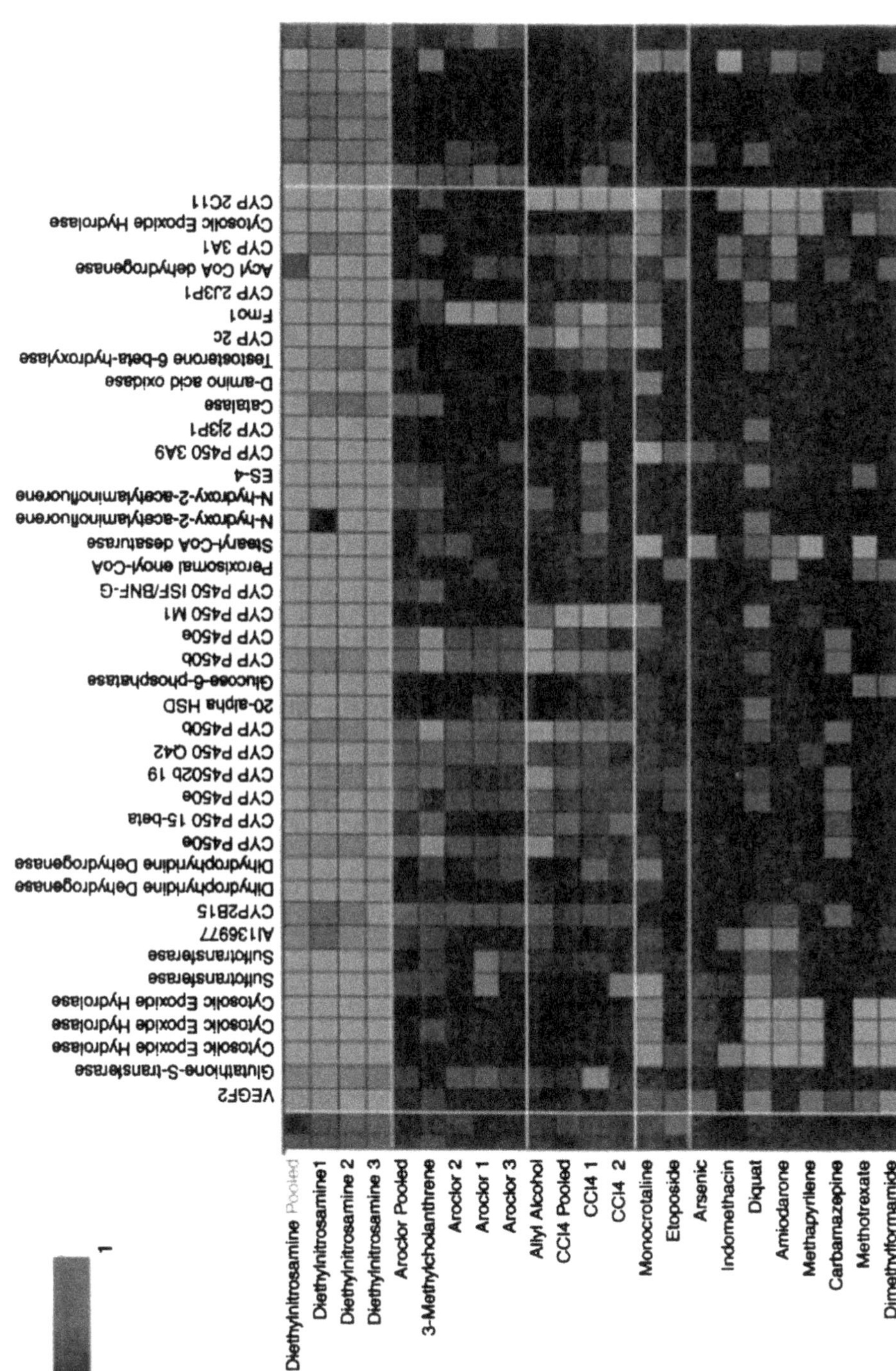
B.
-1
1
Diethylnitrosamine Pooled
Diethylnitrosamine1
Diethylnitrosamine 2
Diethylnitrosamine 3
Aroclor Pooled
3-Methylcholanthrene
Aroclor 2
Aroclor 1
Aroclor 3
Allyl Alcohol
CCl4 Pooled
CCl4 1
CCl4 2
Monocrotaline
Etoposide
Arsenic
Indomethacin
Diquat
Amiodarone
Methapyrilene
Carbamazepine
Methotrexate
Dimethylformamide
CYP 2C11
Cytosolic Epoxide Hydrolase
CYP 3A1
Acyl CoA dehydrogenase
CYP 2J3P1
Fmo1
CYP 2c
Testosterone 6-beta-hydroxylase
D-amino acid oxidase
Catalase
CYP 2j3P1
CYP P450 3A9
ES-4
N-hydroxy-2-acetylaminofluorene
N-hydroxy-2-acetylaminofluorene
Stearyl-CoA desaturase
Peroxisomal enoyl-CoA
CYP P450 ISF/BNF-G
CYP P450 M1
CYP P450e
CYP P450b
Glucose-6-phosphatase
20-alpha HSD
CYP P450b
CYP P450 Q42
CYP P4502b 19
CYP P450e
CYP P450 15-beta
CYP P450e
Dihydropyridine Dehydrogenase
Dihydropyridine Dehydrogenase
CYP2B15
AI136977
Sulfotransferase
Sulfotransferase
Cytosolic Epoxide Hydrolase
Cytosolic Epoxide Hydrolase
Cytosolic Epoxide Hydrolase
Glutathione-S-transferase
VEGF2

FIGURE 13.6 (B)

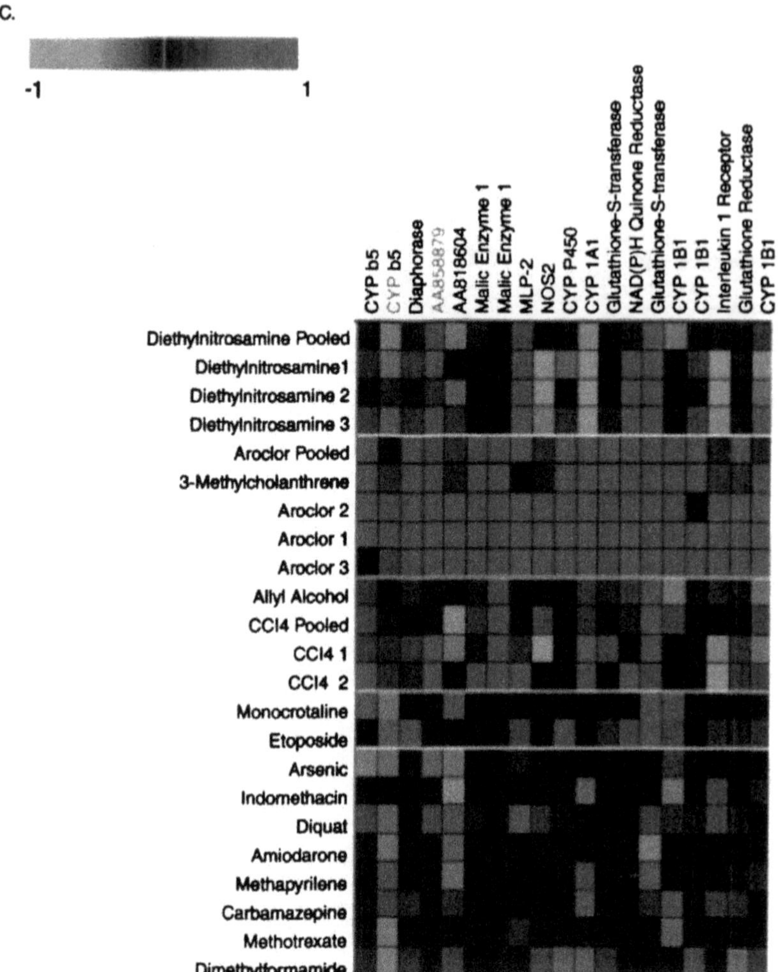


FIGURE 13.6 (C)

## 13.4 DISCUSSION

By using microarray analysis on livers from a 3-day rat study, we have determined gene expression profiles for 15 known hepatotoxins: allyl alcohol, amiodarone, Aroclor 1254, arsenic, carbamazepine, carbon tetrachloride, diethylnitrosamine, dimethylformamide, diquat, etoposide, indomethacin, methapyrilene, methotrexate, monocrotaline, and 3-methylcholanthrene. These compounds elicit a variety of patterns of microscopic injury and alterations in liver-related serum analytes. The main goal of our study was to determine if, using gene expression analysis, these compounds would cluster based on their patterns or mechanisms of toxicity and if gene expression profiles would be useful predictors of toxic responses.

We decided to use 3 days of exposure to the hepatotoxins because we believed that within 3 days significant gene changes corresponding to mechanisms of toxicity would occur. Compound treatment duration for toxicology studies can vary considerably, ranging from single-dose acute to chronic lifetime. We selected a 3-day time point for a variety of reasons. This time point would provide for a complete gene induction response within the liver without the complication of a significant secondary inflammatory response (cellular infiltration) or fibrosis likely to be encountered at later time points. Our histology evaluation confirms we achieved this. Also, an earlier time

**TABLE 13.1**
**Genes Correlated with Changes in Clinical Chemistry Parameters**

| Probe Set | Correlation | Gene Name |
|---|---|---|
| | **Alanine Amino Transferase (ALT)** | |
| S76511_s_at | 0.98 | bax = apoptosis inducer |
| S66618_at | 0.96 | mdr1a = multidrug-resistance transporter P-glycoprotein |
| D63761_g_at | 0.91 | Adrenodoxin reductase |
| M69246_at | 0.91 | Rat collagen-binding protein (gp46) |
| U39207_at | 0.89 | Rattus norvegicus cytochrome P450 4F5 (CYP4F5) |
| D90048exon_g_at | 0.89 | Rat $Na^+$, $K^+$-ATPase |
| X77117exon_1_3_at | 0.87 | NADH-cytochrome b5 reductase |
| M86389cds_s_at | 0.78 | Rat heat shock protein (Hsp27) |
| X67654_at | 0.76 | Gluthathione-S-transferase |
| J04791_s_at | 0.72 | Ornithine decarboxylase (ODC) |
| M24604_g_at | 0.66 | Rat proliferating cell nuclear antigen (PCNA/cyclin) |
| D26307cds_at | 0.62 | Rat jun-D |
| | **Aspartate Amino Transferase (AST)** | |
| Z24721_at | 0.95 | Rattus norvegicus SOD gene |
| U40064_at | 0.93 | PPAR delta protein (PPAR delta) |
| D26112_s_at | 0.84 | Fas antigen |
| U34963_s_at | 0.84 | Programmed cell death repressor BCL-X-Long |
| J04791_s_at | 0.82 | Ornithine decarboxylase (ODC) |
| U47921_s_at | 0.80 | Rattus norvegicus alpha A (insert)-crystallin |
| L06482_at | 0.78 | Rat retinoid X receptor alpha |
| U09540_at | 0.78 | Cytochrome P450 (CYP1B1) |
| U59809_s_at | 0.78 | Mannose 6-phosphate/insulin-like growth factor II receptor |
| D26307cds_at | 0.74 | Rat jun-D |
| X13933_s_at | 0.62 | Calmodulin (pRCM1) |
| Z75029_s_at | 0.61 | Heat shock protein 70 (Hsp70) |
| | **Alkaline Phosphatase (ALP)** | |
| U73174_g_at | 0.99 | Glutathione reductase |
| L23088_at | 0.94 | P-selectin |
| J04791_s_at | 0.83 | Ornithine decarboxylase (ODC) |
| U17260_s_at | 0.80 | Arylamine *N*-acetyltransferase-1 (AT-1) |
| X68041cds_s_at | 0.76 | Epididymal secretory superoxide dismutase |
| U39207_at | 0.73 | Cytochrome P450 4F5 (CYP4F5) |
| L16764_s_at | 0.72 | Heat shock protein 70 (Hsp70) |
| M86389cds_s_at | 0.70 | Rat heat shock protein (Hsp27) |
| AJ224120_at | 0.66 | Peroxisomal membrane protein Pmp26p (Peroxin-11) |
| D26112_s_at | 0.66 | Fas antigen |
| U37464_s_at | 0.66 | MEK5alpha-2 (MEK5) |
| AB004096_at | 0.61 | Lanosterol 14-demethylase |

*(continued)*

**TABLE 13.2 (CONTINUED)**
**Genes Correlated with Changes in Clinical Chemistry Parameters**

| Probe Set | Correlation | Gene Name |
|---|---|---|
| | | **Cholesterol** |
| U39207_at | 0.90 | Cytochrome P450 4F5 (CYP4F5) |
| U35245_g_at | 0.88 | Rat vacuolar protein sorting homolog r-vps33b |
| M63122_at | 0.88 | Rat tumor necrosis factor receptor (TNF receptor) |
| AB004096_at | 0.87 | Lanosterol 14-demethylase |
| U61405_at | 0.87 | Aryl hydrocarbon receptor nuclear translocator 2 (ARNT2) |
| U73174_g_at | 0.82 | Glutathione reductase |
| D26112_s_at | 0.79 | Fas antigen |
| X68041cds_s_at | 0.79 | Epididymal secretory superoxide dismutase |
| U17260_s_at | 0.78 | Arylamine *N*-acetyltransferase-1 (AT-1) |
| AJ224120_at | 0.76 | Peroxisomal membrane protein Pmp26p (Peroxin-11) |
| U37464_s_at | 0.69 | MEK5alpha-2 (MEK5) |
| L23088_at | 0.61 | P-selectin |
| | | **Total Bilirubin** |
| M64301_g_at | 0.90 | Rat extracellular signal-related kinase (ERK3) |
| U73174_g_at | 0.87 | Rattus norvegicus glutathione reductase |
| U89695_at | 0.85 | Non-histone chromosomal architectural protein HMGI-C |
| U35245_g_at | 0.83 | Rat vacuolar protein sorting homolog r-vps33b |
| X07365_s_at | 0.81 | Glutathione peroxidase |
| M21060_s_at | 0.79 | Rat copper-zinc containing superoxide dismutase |
| U17260_s_at | 0.78 | Arylamine *N*-acetyltransferase-1 (AT-1) |
| U47921_s_at | 0.77 | Rattus norvegicus alpha A (insert)-crystallin |
| D38380_at | 0.75 | Rattus norvegicus mRNA for transferrin |
| U25650_f_at | 0.72 | Low-affinity nerve growth factor receptor precursor (LNGFR) |
| X06769cds_at | 0.72 | Rat c-*fos* |
| D63761_at | 0.70 | Adrenodoxin reductase |
| | | **Glucose** |
| M15427_s_at | 0.98 | Rat c-raf protooncogene |
| X05137_at | 0.96 | Fast nerve growth factor receptor (NGFR) |
| L23088_at | 0.93 | P-selectin |
| AF025670_g_at | 0.90 | Caspase 6 (Mch2) |
| Z24721_at | 0.88 | R.norvegicus SOD gene |
| S79263_s_at | 0.87 | Interleukin-3 receptor beta |
| U35245_g_at | 0.87 | Vacuolar protein sorting homolog r-vps33b |
| X64403_at | 0.86 | c/ebp gamma |
| L06482_at | 0.83 | Rat retinoid X receptor alpha |
| U25650_f_at | 0.81 | Low-affinity nerve growth factor receptor precursor (LNGFR) |
| X94185cds_s_at | 0.81 | Dual-specificity phosphatase, MKP-3 |
| U40836mRNA_s_at | 0.79 | Cytochrome oxidase subunit VIII |

*(continued)*

**TABLE 13.3 (CONTINUED)**
**Genes Correlated with Changes in Clinical Chemistry Parameters**

| Probe Set | Correlation | Gene Name |
|---|---|---|
| | | **γ-Glutamyl Transferase (GGT)** |
| D90048exon_g_at | 0.91 | Rat $Na^+$, $K^+$-ATPase |
| M86389cds_s_at | 0.90 | Rat heat shock protein (Hsp27) |
| X67654_at | 0.88 | Gluthathione-S-transferase |
| S66618_at | 0.87 | mdr1a = multidrug-resistance transporter P-glycoprotein |
| U39207_at | 0.79 | Cytochrome P450 4F5 (CYP4F5) |
| D44591_s_at | 0.78 | Inducible nitric oxide synthase |
| Y09332cds_s_at | 0.75 | Cytosolic peroxisome proliferator-induced acyl-CoA thioesterase |
| D83792_at | 0.75 | Rattus norvegicus mRNA for p27 |
| M58040_at | 0.74 | Rat transferrin receptor |
| U34963_s_at | 0.72 | Programmed cell death repressor BCL-X-Long |
| M24604_g_at | 0.72 | Rat proliferating cell nuclear antigen (PCNA/cyclin) |
| M31931cds_s_at | 0.72 | Rat cytochrome P450 (P450olf1) |
| | | **Blood Urea Nitrogen (BUN)** |
| L38437_at | 0.99 | NADH ubiquinone oxidoreductase subunit (IP13) |
| L23088_at | 0.92 | P-selectin |
| U17260_s_at | 0.92 | Arylamine *N*-acetyltransferase-1 (AT-1) |
| U11681_at | 0.91 | Rapamycin and FKBP12 target-1 protein (rRAFT1) |
| M21210_i_at | 0.90 | Glutathione peroxidase (GSH-PO) |
| J05460_s_at | 0.87 | Cholesterol 7-alpha-hydroxylase |
| M33329_f_at | 0.84 | Hydroxysteroid sulfotransferase a (STa) |
| M32167_at | 0.83 | Rat glioma-derived vascular endothelial cell growth factor |
| X55286_at | 0.82 | HMG-CoA reductase |
| U73174_g_at | 0.81 | Glutathione reductase |
| M76767_s_at | 0.81 | Fatty acid synthase |
| AB004096_at | 0.81 | Lanosterol 14-demethylase |

point would likely have included immediate-early response genes. Finally, while for some compounds it is possible that longer or shorter treatments may have been more optimal, we believe that if gene expression cluster analysis is to be used to predict mechanisms of toxicity for thousands of compounds, it must be robust enough to cluster based on a single time point.

The dose levels for the compounds were selected from the literature or from our own personal experience (see Materials and Methods section). For all of the compounds, the goal was to dose at a level where significant hepatotoxicity would be expected at 7 days of treatment. This meant that, because the rats were necropsied on day 4, none of the expected histological changes were seen with some of the compounds (e.g., arsenic). Despite the absence of histopathologic changes after treatment with arsenic, gene changes were observed that might be indicative of changes in cellular function.

The decision to pool the mRNA from the rat livers was made in order to obtain a representative analysis of gene expression changes across more than one animal. For many of the compounds, the three rats in a treatment series gave very similar histopathologic and clinical chemistry results.

We performed microarray analysis on individual animals for three of the treatment groups: Aroclor, diethylnitrosamine, and carbon tetrachloride. For these treatment groups, relatively little variability between the different animals was observed; comparison plots of the gene expression profiles showed a high degree of similarity, and the individual animal expression profiles all clustered with the pooled profile. It is possible that for other treatment groups there would not have been as high a degree of similarity between animals as there was with Aroclor, diethylnitrosamine, and carbon tetrachloride. However, our results show that for these compounds the decision to pool the RNA samples did give a representative analysis of gene expression changes and suggest that pooling RNA samples from the same treatment group may be a viable method for detecting key gene changes caused by a compound.

We used three different clustering methods in order to determine which compounds gave similar expression profiles. These methods differ in their mathematical approach to form the clusters. Our results show that the major clusters are the same using the different methods, but some of the minor clusters vary. This suggests that it may be helpful to do more than one clustering method in order to determine which clusters are significant.

The results from our study suggest that many of the clusters formed from the gene expression results do correlate with what is known about the mechanisms of toxicity caused by the compounds. In addition, many of the clusters correspond well with observed clinical chemistry or histopathology results. The liver gene expression profiles for 3-methylcholanthrene and Aroclor 1254 formed one cluster using different methods of statistical cluster analysis. This result agrees with the clinical chemistry results, the histology results, and what is known about the compounds from the literature. Both of these agents are aromatic hydrocarbons that have similar mechanisms of toxicity.[26,27] Previous studies have shown that the liver cell hypertrophy caused by both 3-methylcholanthrene and Aroclor 1254 is due to an increase in the hepatocellular content of smooth endoplasmic reticulum.[28] Results from the microarray analysis showed that both compounds caused an upregulation of cytochrome P450 1A1 and 1B1, malic enzyme, and glutathione S-transferase (GST) (Figure 13.6C). All of these genes have been shown previously to be regulated by aromatic hydrocarbons and potentially play a role in the proliferation of the endoplasmic reticulum.[29–33]

The results from the microarray analysis showed a tight cluster formed between carbon tetrachloride and allyl alcohol. Both of these compounds caused necrosis in the liver, with elevated levels of serum alanine amino transferase and aspartate amino transferase. It was somewhat surprising that these compounds clustered tightly together, as our studies and others have shown that carbon tetrachloride generally results in necrosis in the midzonal and centrilobular region, while allyl alcohol causes necrosis in the periportal region of the liver,[34–36] implying that the cell injury was produced by different mechanisms. However, it has also been shown previously that both compounds cause cellular injury by free-radical formation, which could result from the alteration of a number of genes, but especially genes involved in oxidation.[37] An examination of the genes that are regulated in a similar manner between carbon tetrachloride and allyl alcohol reveals that many are involved in oxidation, such as heme oxygenase, p38-activated mitogen kinase, cytochrome b558, and GST (Figures 13.6A and C). Previous studies showed that some of these genes are regulated by these compounds.[38]

Microarray analysis showed diethylnitrosamine clustering alone, while etoposide and monocrotaline clustered close together using *k*-means and unsupervised clustering. All of these compounds are known DNA damaging agents. Figure 13.6A shows a number of genes that are upregulated with diethylnitrosamine treatment. Many of these genes have been shown previously to be upregulated by diethylnitrosamine and are related to growth arrest and DNA damage, such as cyclin G1 and D1, p21/Waf, GADD45, and Bax.[39] Some of these genes are also upregulated by monocrotaline and etoposide, both of which are DNA damaging agents that have been shown to cause an arrest in the $G_2$ phase.[40,41] There are some expressed sequence tag (EST) sequences clustering with these genes, which suggests they may also have a role in cell cycle arrest and DNA damage.

Figure 13.6B shows a number of genes that are downregulated with diethylnitrosamine treatment. One of these genes is sulfotransferase, which has been shown previously to be downregulated in rat livers following diethylnitrosamine treatment.[42] Sulfotransferase was also downregulated by monocrotaline and diquat. The downregulation of sulfotransferase is thought to be a primary pathway in the activation of hepatic carcinogens.[43,44] Several of the hepatotoxins downregulated epoxide hydrolase, including diethylnitrosamine, monocrotaline, diquat, amiodarone, methapyrilene, methotrexate, and dimethylformamide. Many of these agents have been shown to cause oxidative stress, which would result in regulation of epoxide hydrolase.[13,45–49]

Supervised learning analysis was used to correlate gene expression with changes in clinical chemistry parameters to identify genes that might serve as *in vitro* markers for detecting hepatotoxic responses. Also, identification of the genes that are regulated in association with necrosis may lead to a better understanding of the underlying pathogenic mechanism. Elevated serum levels of ALT and AST are associated with necrosis in the liver.[50] Genes identified as consistently up- or downregulated in association with changes in serum ALT or AST, or both, were proliferating cell nuclear antigen (PCNA), Hsp70-3, calmodulin, ornithine decarboxylase, bax, JunD, and GST. Expression of many of these genes has been shown previously to be altered in livers with necrosis. PCNA is involved in cell-cycle regulation and has been shown to be upregulated in livers following cell necrosis.[51,52] Bax is an apoptosis gene shown to be activated following exposure to certain hepatotoxins, including carbon tetrachloride.[53] Calmodulin has also been linked to necrosis in the liver following treatment with carbon tetrachloride, and treatment of rats with the anticalmodulin drug fluphenazine significantly protects the liver against carbon-tetrachloride-induced necrosis.[54] Genes such as Hsp70, JunD, GST, and ornithine decarboxylase are involved in oxidative metabolism, and changes in their expression might promote necrosis.[55–57] Monitoring the regulation of these genes individually would likely not be an accurate predictor, but as a group the genes may serve as markers for hepatocellular necrosis.

The identification of genes for which regulation correlates with changes in clinical chemistry parameters could be a useful measure for other liver disorders as well. Alkaline phosphatase levels can be elevated due to a number of reasons, one of which is damage to the intrahepatic bile duct. This is often accompanied by elevations in cholesterol levels.[58] Interestingly, out of the 12 genes that were correlated with changes in ALP levels, 8 were also correlated with changes in cholesterol. Thus, monitoring the regulation of these 8 genes may prove to be useful markers for intrahepatic bile duct damage.

With a small database, it is difficult to gauge the predictability of gene expression for determining mechanisms of toxicity. Certainly, we saw some compounds cluster together reproducibly that we expected to see, such as Aroclor 1254 with 3-methylcholanthrene. However, it might have been expected that methapyrilene would cluster with allyl alcohol, as both have been shown to result in periportal necrosis in rats.[35,59] Instead, methapyrilene clustered with arsenic. Previous research has shown that treatment with arsenic can result in necrosis in the periportal region, but it also results in other forms of hepatotoxicity as well, such as steatosis.[60] In addition, other clusters varied depending on the type of clustering algorithm used. For instance, diquat clustered with etoposide using divisive clustering and with methapyrilene using agglomerative clustering. Interestingly, however, all of these compounds have been shown to cause DNA damage, suggesting that they may regulate common pathways.[40,61,62]

## 13.5 SUMMARY

The purpose of this study was to begin to answer the questions of whether different mechanisms of toxicity can be determined from gene expression using unsupervised clustering. Our results suggest that, when using a limited database, the compounds clustered together in fairly good agreement with the observed clinical chemistry and histopathology results. In some cases, compounds were clustered based on gene expression that could not have been clustered by either histology or clinical chemistry as no meaningful changes in these categories were observed. The size of a useful database will ultimately depend on the amplitudes of the signature profiles, but most certainly a larger database will be required in order to have complete determinations of mechanisms based on signature profiles.

## REFERENCES

1. Young, R.A., Biomedical discovery with DNA arrays, *Cell*, 102, 9–15, 2000.
2. Golub, T.R., Slonim, D.K., Tamayo, P. et al., Molecular classification of cancer: class discovery and class prediction by gene expression monitoring, *Science*, 286, 531–537, 1999.
3. Ross, D.T., Scherf, U., Eisen, M.B. et al., Systematic variation in gene expression patterns in human cancer cell lines, *Nat. Genet.*, 24, 227–235, 2000.
4. Roberts, C.J., Nelson, B., Marton, M.J. et al., Signaling and circuitry of multiple MAPK pathways revealed by a matrix of global gene expression profiles, *Science*, 287, 873–880, 2000.
5. Scherf, U., Ross, D.T., Waltham, M. et al., A gene expression database for the molecular pharmacology of cancer, *Nat. Genet.*, 24, 236–244, 2000.
6. Ganey, P.E. and Schultze, A.E., Depletion of neutrophils and modulation of Kupffer cell function in allyl alcohol-induced hepatotoxicity, *Toxicology*, 1–2, 99–106, 1995.
7. Reasor, M.J., McCloud, C.M., Beard, T.L. et al., Comparative evaluation of amiodarone-induced phospholipidosis and drug accumulation in Fischer 344 and Sprague–Dawley rats, *Toxicology*, 106, 139–147, 1996.
8. Wolfgang, G.H.I., Donarski, W.J., and Petry, T.W., Effects of novel antioxidants on carbon tetrachloride-induced lipid peroxidation and toxicity in precision-cut rat liver slices, *Toxicol. Appl. Pharmacol.*, 106, 63–70, 1990.
9. Mahaffey, K.R., Capar, S.G., Gladen, B.C. et al., Concurrent exposure to lead, cadmium, and arsenic. Effects on toxicity and tissue metal concentrations in the rat, *J. Lab. Clin. Med.*, 98, 463–481, 1981.
10. Regnaud, L., Sirois, G., and Chakrabarti, S., Effect of four-day treatment with carbamazepine at different dose levels on microsomal enzyme induction, drug metabolism and drug toxicity, *Pharmacol. Toxicol.*, 62, 3–6, 1988.
11. Williams, G.M., Iatropoulos, M.J., Wang, C.X. et al., Diethylnitrosamine exposure-responses for DNA damage, centrilobular cytotoxicity, cell proliferation and carcinogenesis in rat liver exhibit some non-linearities, *Carcinogenesis*, 10, 2253–2258, 1996.
12. Mathew, T., Karunanithy, R., Yee, M.H. et al., Hepatotoxicity of dimethylformamide and dimethylsulfoxide at and above the levels used in some aflatoxin studies, *Lab Invest.*, 42, 257–262, 1980.
13. Wolfgang, G.H., Jolly, R.A., and Petry, T.W., Diquat-induced oxidative damage in hepatic microsomes: effects of antioxidants, *Free Radic. Biol. Med.*, 10, 403–411, 1991.
14. Linden, C.J., Toxicity of interperitoneally administered antitumour drugs in athymic rats, *In Vivo*, 3, 259–262, 1989.
15. Fracasso, M.E., Cuzzolin, L., Soldato, P.D. et al., Multisystem toxicity of indomethacin: effects on kidney, liver and intestine in the rat, *Agents Actions*, 22, 310–313, 1987.
16. Graichen, M.E., Neptun, D.A., Dent, J.G. et al., Effects of methapyrilene on rat hepatic xenobiotic metabolizing enzymes and liver morphology, *Fundam. Appl. Toxicol.*, 1, 165–174, 1985.
17. Custer, R.P., Freeman-Narrod, M., and Narrod, S.A., Hepatotoxicity in Wistar rats following chronic methotrexate administration: a model of human reaction, *J. Natl. Cancer Inst.*, 58, 1011–1017, 1977.
18. Perazzo, J., Eizayaga, F., Romay, S. et al., An experimental model of liver damage and portal hypertension induced by a single dose of monocrotaline, *Hepatogastroenterology*, 46, 432–435, 1999.
19. Kleeberg, U., Barth, A., Roth, J. et al., On the selectivity of aryl hydrocarbon hydroxylase induction after 3-methylcholanthrene pretreatment, *Acta Biol. Med. Ger.*, 710, 1701–1705, 1975.
20. Selinger, D.W., Cheung, K.J., Mei, R. et al., RNA expression analysis using a 30 base pair resolution *Escherichia coli* genome array, *Nat. Biotechnol.*, 18, 1262–1268, 2000.
21. Hughes, T.R., Marton, M.J., Jones, A.R. et al., Functional discovery via a compendium of expression profiles, *Cell*, 102, 109–126, 2000.

22. Alon, U., Barkai, N., Notterman, D.A. et al., Broad patterns of gene expression revealed by clustering analysis of tumor and normal colon tissues probed by oligonucleotide arrays, *PNAS*, 96, 6745–6750, 1999.
23. Hartigan, J.A. and Wong, M.A., A *k*-means clustering algorithm, *Appl. Stat.*, 28, 100–108, 1979.
24. Tamayo, P., Slonim, D., Mesirov, J. et al., Interpreting patterns of gene expression with self-organizing maps: methods and application to hematopoietic differentiation, *PNAS*, 96, 2907–2912, 1999.
25. Sherlock, G., Analysis of large-scale gene expression data, *Curr. Opin. Immunol.*, 12, 201–205, 2000.
26. Poland, A., Knutson, J., and Glover, E., Studies on the mechanism of action of halogenated aromatic hydrocarbons, *Clin. Physiol. Biochem.*, 3, 147–154, 1985.
27. Safe, S., Bandiera, S., Sawyer, T. et al., PCBs: structure-function relationships and mechanism of action, *Environ. Health Perspect.*, 60, 47–56, 1985.
28. Mayes, B.A., McConnell, E.E., Neal, B.H. et al., Comparative carcinogenicity in Sprague–Dawley rats of the polychlorinated biphenyl mixtures Aroclors 1016, 1242, 1254, and 1260, *Toxicol. Sci.*, 41, 62–76, 1998.
29. Borlak, J. and Thum, T., Induction of nuclear transcription factors, cytochrome P450 monooxygenases, and glutathione S-transferase alpha gene expression in Aroclor 1254-treated rat hepatocyte cultures, *Biochem. Pharmacol.*, 61, 145–153, 2001.
30. Mehlman, M.A., Tobin, R.B., Friend, B. et al., The effects of a polychlorinated biphenyl mixture (Aroclor 1254) on liver gluconeogenic enzymes of normal and alloxan-diabetic rats, *Toxicology*, 5, 89–95, 1975.
31. Jauregui, H.O., Ng, S.F., Gann, K.L. et al., Xenobiotic induction of P-450 PB-4 (IIB1) and P-450c (IA1) and associated monooxygenase activities in primary cultures of adult rat hepatocytes, *Xenobiotica*, 21, 1091–1106, 1991.
32. Lubet, R.A., Jones, C.R., Stockus, D.L. et al., Induction of cytochrome P450 and other drug metabolizing enzymes in rat liver following dietary exposure to Aroclor 1254, *Toxicol. Appl. Pharmacol.*, 108, 355–365, 1991.
33. Borlakoglu, J.T., Edwards-Webb, J.D., and Dils, R.R., Evidence for the induction of fatty acid desaturation in proliferating hepatic endoplasmic reticulum in response to treatment with polychlorinated biphenyls. Are fatty acid desaturases cytochrome P-450-dependent monooxygenases?, *Int. J. Biochem.*, 23, 925–931, 1991.
34. Recknagel, R.O., Carbon tetrachloride hepatotoxicity, *Pharmacol. Rev.*, 19, 145–208, 1967.
35. Badr, M.Z., Belinsky, S.A., Kauffman, F.C. et al., Mechanism of hepatotoxicity to periportal regions of the liver lobule due to allyl alcohol: role of oxygen and lipid peroxidation, *J. Pharmacol. Exp. Ther.*, 238, 1138–1142, 1986.
36. Butterworth, K.R., Carpanini, F.M., Dunnington, D. et al., The production of periportal necrosis by allyl alcohol in the rat, *Br. J. Pharmacol.*, 63, 353P–354P, 1978.
37. Comporti, M., Three models of free radical-induced cell injury, *Chem. Biol. Interact.*, 72, 1–56, 1989.
38. Serfas, M.S., Goufman, E., Feuerman, M.H. et al., p53-independent induction of p21WAF1/CIP1 expression in pericentral hepatocytes following carbon tetrachloride intoxication, *Cell Growth Differ.*, 8, 951–961, 1997.
39. Lee, Y.S., Kim, W.H., Yu, E.S. et al., Time course of cell-cycle-related protein expression in diethylnitrosamine-initiated rat liver, *J. Hepatol.*, 29, 464–469, 1998.
40. Fukumi, S., Horiguchi-Yamada, J., Iwase, S. et al., Concentration-dependent variable effects of etoposide on the cell cycle of CML cells, *Anticancer Res.*, 50, 3105–3110, 2000.
41. Wilson, D.W., Lame, M.W., Dunston, S.K. et al., DNA damage cell checkpoint activities are altered in monocrotaline pyrrole-induced cell cycle arrest in human pulmonary artery endothelial cells, *Toxicol. Appl. Pharmacol.*, 166, 69–80, 2000.
42. Werle-Schneider, G., Schwarz, M., and Glatt, H., Development of hydroxysteroid sulfotransferase-deficient lesions during hepatocarcinogenesis in rats, *Carcinogenesis*, 14, 2267–2270, 1993.
43. Higgins, M.J., Ficsor, G., Aaron, C.S. et al., Micronuclei in mice treated with monocrotaline with and without phenobarbital pretreatment, *Environ. Mol. Mutagen.*, 26, 37–43, 1995.
44. Ringer, D.P. and Norton, T.R., Further characterization of the ability of hepatocarcinogens to lower rat liver aryl sulfotransferase activity, *Carcinogenesis*, 8, 1749–1752, 1987.
45. Babiak, R.M., Campello, A.P., Carnieri, E.G. et al., Methotrexate: pentose cycle and oxidative stress, *Cell. Biochem. Funct.*, 16, 283–293, 1998.

46. Gallagher, E.P., Buetler, T.M., Stapleton, P.L. et al., The effects of diquat and ciprofibrate on mRNA expression and catalytic activities of hepatic xenobiotic metabolizing and antioxidant enzymes in rat liver, *Toxicol. Appl. Pharmacol.*, 134, 81–91, 1995.
47. Imazu, K., Fujishiro, K., and Inoue, N., Liver injury and alterations of hepatic microsomal monooxygenase system due to dimethylformamide (DMF) in rats, *Fukuoka Igaku Zasshi*, 85, 147–153, 1994.
48. Leeder, R.G., Brien, J.F., and Massey, T.E., Investigation of the role of oxidative stress in amiodarone-induced pulmonary toxicity in the hamster, *Can. J. Physiol. Pharmacol.*, 72, 613–621, 1994.
49. Sarma, J.S., Pei, H., and Venkataraman, K., Role of oxidative stress in amiodarone-induced toxicity, *J. Cardiovasc. Pharmacol. Ther.*, 2, 53–60, 1997.
50. Goldstein, R.S. and Schnellmann, R.G., *Casarett and Doull's Toxicology: The Basic Science of Poisons*, McGraw-Hill, New York, 1996.
51. Lee, J.H., Ilic, Z., and Sell, S., Cell kinetics of repair after allyl alcohol-induced liver necrosis in mice, *Int. J. Exp. Pathol.*, 77, 63–72, 1996.
52. Koukoulis, G., Rayner, A., Tan, K.C. et al., Immunolocalization of regenerating cells after submassive liver necrosis using PCNA staining, *J. Pathol.*, 166, 359–368, 1992.
53. Horn, T.L., O'Brien, T.D., Schook, L.B. et al., Acute hepatotoxicant exposure induces TNFR-mediated hepatic injury and cytokine/apoptotic gene expression, *Toxicol. Sci.*, 54, 262–173, 2000.
54. deFerreyra, E.C., Bernacchi, A.S., Martin, M.F.S. et al., Late protective effects of the anticalmodulin drug fluphenazine on carbon tetrachloride-induced liver necrosis, *Biomed. Environ. Sci.*, 8, 218–225, 1995.
55. Mendelson, K.G., Contois, L.R., Tevosian, S.G. et al., Independent regulation of JNK/p38 mitogen-activated protein kinases by metabolic oxidative stress in the liver, *Proc. Natl. Acad. Sci. USA*, 93, 12908–12913, 1996.
56. Schiaffonati, L. and Tiberio, L., Gene expression in liver after toxic injury: analysis of heat shock response and oxidative stress-inducible genes, *Liver*, 17, 183–191, 1997.
57. Byus, C.V., Costa, M., Sipes, I.G. et al., Activation of 3′:5′-cyclic AMP-dependent protein kinase and induction of ornithine decarboxylase as early events in induction of mixed-function oxygenases, *Proc. Natl. Acad. Sci. USA*, 73, 1241–1245, 1976.
58. Klaassen, C.D. (1996), *Mechanisms of Toxicity*, McGraw-Hill, New York.
59. Ratra, G.S., Morgan, W.A., Mullervy, J. et al., Methapyrilene hepatotoxicity is associated with oxidative stress, mitochondrial disfunction and is prevented by the $Ca^{2+}$ channel blocker verapamil, *Toxicology*, 130, 79–93, 1998.
60. Zimmerman, H.J. (1999), *Hepatotoxicity*, Lippincott/Williams & Wilkins, Philadelphia.
61. Lorico, A., Toffoli, G., Boiocchi, M. et al., Accumulation of DNA strand breaks in cells exposed to methotrexate or N10-propargyl-5,8-dideazafolic acid, *Cancer Res.*, 48, 2036–2041, 1988.
62. Gupta, S., Husser, R.C., Geske, R.S. et al., Sex differences in diquat-induced hepatic necrosis and DNA fragmentation in Fischer 344 rats, *Toxicol. Sci.*, 54, 203–211, 2000.

# 14 Toxicogenomic Analysis in Clinical Pharmacogenomic Studies

*Michael E. Burczynski*

## CONTENTS



## 14.1 INTRODUCTION

Expression profiling studies using microarray technology have already added tremendous value to many fields of biomedical research in recent years. Mechanistic studies have provided insight into transcriptional networks regulated in conditions as disparate as different disease states, toxic exposures, pharmacological administrations, nutrient depravations, and hundreds of other studied scenarios. These experiments have yielded countless scientific leads for the formulation of new and testable hypotheses. Transcripts that are differentially expressed may represent: (1) candidate target genes/proteins for therapeutic intervention (transcriptional regulation of gene X leads to or contributes to an undesirable phenotype), (2) directly regulated genes (transcriptional regulation of gene Y is a direct result of the treatment/condition/disease state), or (3) indirect response genes (transcriptional regulation of gene Z is a secondary downstream event in response to a treatment, condition, or disease state).

One of the tenets of predictive genomics is that, regardless of whether transcriptional changes are causative or responsive in nature (or fall into some other category), the patterns observed may also serve as diagnostics of the treatment, condition, or disease state in question. Of course, the utility of such expression profiles as diagnostics will ultimately depend upon the robustness and reproducibility of the patterns in question (discussed briefly in Chapter 5). In the field of predictive genomic studies, two main approaches are applied to the problem of class identification: (1) class

0-8493-1334-1/03/$0.00+$1.50

discovery, and (2) class prediction. The focus of this chapter is on class prediction methods in clinical studies, but the next section briefly reviews the distinction between these two approaches.

## 14.2 CLASS DISCOVERY VS. CLASS PREDICTION

### 14.2.1 Class Discovery

In class discovery, unsupervised methods are applied to expression profiles in order to "discover" classes while imposing no preordained stratifications in the data. All profiles are treated equally and blindly, and the underlying structure within the expression data drives the similarity or dissimilarity between samples. In hierarchical clustering approaches, the similarity of profiles is described by positioning the profiles as leaves of a dendrogram, where the distance between the leaves in the tree reflects the difference between the expression profiles of the corresponding samples. Once these relationships are recognized, the novel classifications of samples that arise from these unsupervised methods can then be assessed further as to their significance. Clusters of similar expression profiles are investigated for correlation with other parameters measured in the samples. The underlying basis for the subgroupings may or may not be immediately recognized; however, by simultaneously monitoring the expression patterns of thousands of genes, researchers are able to find unanticipated stratifications in otherwise apparently homogeneous sample sets. Often these stratifications can have clinical relevance.

Transcriptional profiling for the purpose of class discovery has identified novel subclasses of previously indistinguishable tumors[1] in the field of cancer research, and these findings are of immense value. Because patients with various tumor subtypes might demonstrate widely disparate responses to subsequent therapy, the identification of tumor-subclass-specific molecular profiles will likely have an impact on human health in a relatively short time. It is easy to imagine a day in the not-too-distant future in which a transcriptional profile measured in a biopsy will dictate whether a patient receives the standard of care or some alternative form of therapy, depending on the expression pattern observed in the affected tissue.

The same principles of class discovery have recently been applied to questions in the field of toxicology.[2] In Chapter 13 we saw how an unsupervised hierarchical clustering approach was used by Waring et al. to "discover" classes of toxicants with similar gene expression profiles. The relevance and the general applicability of these guilt-by-association approaches are not entirely clear; thus, it is important to test whether toxicants sharing similar gene expression profiles also share the same mechanism of toxicity. Waring et al. were able to assess the performance of these unsupervised methods because the toxicants clustered in their study were not unknowns but were rather toxicants of known toxicity and mechanisms of toxic action. These types of controlled studies with paradigm compounds are quintessential at the outset of applying unsupervised methods to the question of class discovery to demonstrate proof of principle. The results indicated that, in several cases using several different clustering approaches, compounds with similar known molecular mechanisms of action (for instance, Aroclor and 3-methylcholanthrene [3MC] or ally alcohol and carbon tetrachloride) yielded similar expression profiles in rat liver following administration *in vivo*. The inference from these studies provides preliminary evidence for the basic tenet upon which the field of predictive toxicogenomics is based: toxic compounds with similar mechanisms of toxicity will evoke similar patterns of gene expression in the appropriate tissue.

### 14.2.2 Class Prediction

While class discovery is concerned with applying unsupervised methods to "discover" stratifications in samples based upon their gene expression profiles, class prediction approaches essentially work in reverse. Class prediction approaches initially employ supervised methods that study the already-known *classes* in a set of known samples (the training set) and then *determine the genes* most highly correlated with each class. In subsequent studies, the expression patterns of these "predictor" genes are assessed in unknown samples (the test set) and a decision is made as to the identity of the unknown sample on the basis of gene expression. For instance, a researcher employing class prediction methods might divide a group of expression profiles into known classes (for toxicogenomic studies, these classes could constitute profiles measured in rat liver in response to DNA damaging agents, free radical stressors, peroxisomal proliferators, etc.) and then use an algorithm to identify the genes for which the expression in rat liver is most highly correlated with each class. This type of approach was one of the first examples of supervised analysis applied in the field of toxicogenomics.[3] The set of samples initially used to identify the class-specific correlated genes is referred to as the *training set*, because each class-specific gene set is identified using an initial set of samples to train the predictive model.

Many methods are available for determining the most highly correlated marker genes for a given class. In the early study mentioned above, a correlation optimization algorithm was applied to expression data for HepG2 cells treated with toxic doses of DNA damaging agents, antiinflammatory compounds, and nongenotoxic controls (Figure 14.1). Using this algorithm, multiple binary comparisons between these three classes of treatments were performed and the genes that maximized the similarity among members of a single class (calculated using a floating Pearson's correlation coefficient) and minimized the similarity among members of different classes were identified.

More recently our laboratory has identified class-correlated marker genes by calculating a class separation statistic. We have used the software program Genecluster (http://www-genome.wi.mit.edu/cancer/software/genecluster2.html) and employed the signal-to-noise statistic,[4,5] which is defined as:

$$(\mu_{class0} - \mu_{class1}/\sigma_{class0} + \sigma_{class1}).$$

In this calculation, μ represents the mean, and σ represents the standard deviation of the expression of the gene in each class. The genes with the highest relative class separation, as determined by this metric, also represent the genes with the highest predictive potential. The performance of this metric compared favorably with other metrics of similarity (Pearson's correlation coefficient) or distance (Euclidean, Manhattan, Battacharyya) that have been used as traditional measures of class separation.

The likely utility of this classifier gene set for prediction can next be established using any number of methods to estimate the statistical significance of the gene set. One common method used by researchers is to randomly permute the training set of sample labels (leading to nonsensical classifications) and then to recalculate the genes most highly correlated with these meaningless classifications. The rank-ordered correlations of genes rising from the permuted data can then be compared with the most highly correlated genes resulting from the original true class distinction. If the measures of correlation for genes from the original (true) classes are more significant than the measures of correlation for genes from randomly permuted classes, then it is likely that the original "true" classification gene set was indeed significant. The particular method we employ is a neighborhood analysis, which is described in detail in other publications.[1,4,5]

Once the statistical significance of the top correlated genes has been established (by permutation and neighborhood analysis or other techniques), the actual utility of these gene sets for the purposes of prediction can finally be evaluated. The robustness of any predictive model can be tested by cross-validation of the training set, as described in detail in Chapter 12. We employ several types of cross-validation to estimate the likely classification error rate; typically, we most often train a predictive model using approximately 70% of the analyzed samples (the training set) and then test the accuracy of prediction on the remaining 30% of the samples (the test set).

It is important to remember that any number of genes can be used in a predictor, and each of these predictor gene sets is a different predictive model with different accuracy, sensitivity, and overall performance. We often attempt to define the smallest set of genes with utility in prediction, although, depending on the approach used, very large sets of genes can also perform well.[6]

To determine the performance of a predictor gene set, we must first select a method for class assignment; that is, we must first determine how we will use the information (the expression of predictor genes) in the unknown samples to assign an unknown sample to one class or the other. Once the method for class assignment is selected (see below), the identities of the unknown samples

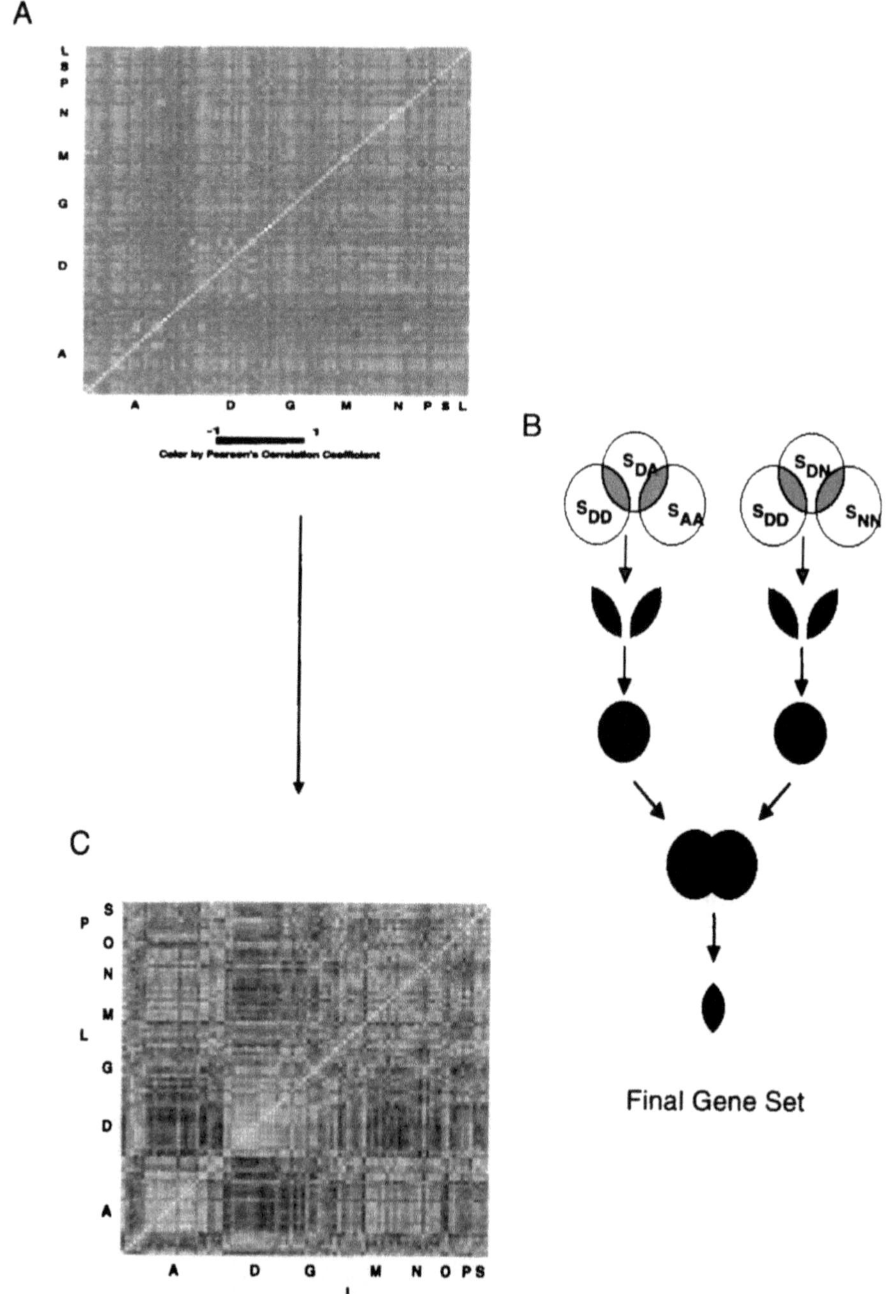


FIGURE 14.1

in a test set can be predicted using the expression patterns of the predictor genes in each unknown sample. The method of class assignment that we currently use is the weighted voting method.[4] In this method, the expression value (for each predictor gene) lying exactly halfway between the mean expression values (of that predictor gene) for the two classes in question defines a decision boundary. Recall that the means of class 0 and class 1 were calculated in the training set, and the identities of unknown samples in the test set are predicted on this basis. This is essentially how the voting method works: Each gene in the predictor set "votes" for class 0 or class 1, depending upon which side of the decision boundary the expression value of that gene in the unknown sample falls. This is too simplistic of a description because the voting method is actually designed to allocate more weight to those genes most strongly correlated with the class distinction vector observed in the original training set. In addition, the resulting numbers of weighted votes for each class are used to define a new variable called the ***prediction strength***, which measures the margin of victory for the predicted class. This variable indicates the degree of confidence associated with the prediction.

If the identities of the samples in the test set are actually known, as they are in initial proof-of-principle studies, then this exercise can provide an estimate of the accuracy and sensitivity of the utility of any given predictor gene set in assigning class membership to unknown expression profiles. For instance, the researcher can define a cut-off value for the prediction strength below which calls will not be made. The total number of correct calls divided by the total number of calls made simply refers to the accuracy of the predictor (i.e., 88 correct calls/100 calls made = 88% accuracy). The total number of calls made divided by the total number of samples in the test set describes the sensitivity of the prediction (e.g., 100 calls made/120 profiles ~ 83%).

**FIGURE 14.1 (CONTINUED)** An early supervised approach to toxicogenomic analysis. (A) HepG2 cells were treated with various cytotoxic compounds and relative changes in gene expression were ascertained using microarray analyses. Comparisons of the gene expression patterns between each pair of two treatments were made by Pearson's correlation coefficient and depicted graphically by ordering in a supervised fashion the toxicant expression profiles on the *x* and *y* axes by the assigned mechanisms of action. The color in each cell of the plot reflects the similarity between the two experiments, resulting in a similarity matrix that is symmetric about the main diagonal (identical experiment compared to itself). The letters designate the following toxicant classes: A, antiinflammatory; D, DNA-damagers; G, gene synthesis inhibitors; L, low dose DNA damaging agents (nontoxic); M, metabolic poisons; N, nongenotoxic controls; O, other; P, peroxisomal proliferators; S, steroids. The initial supervised analysis was performed across the entire database of 100 toxic compounds using all genes on the microarray (~250). (B) The optimization algorithm depicted is a "brute force" type of supervised-learning algorithm that relies on the processing power of the computer to perform calculations on all pairs of experiments in all pairs of two groups. The groups of treatments from which data were used for algorithm optimization were DNA damaging agents, antiinflammatories, and nongenotoxic controls. By focusing on one gene at a time and determining the effect of each gene on the average correlation coefficient, the algorithm selected genes that maximized the average intra-group correlation coefficient and minimized the average inter-group correlation coefficient for all pairs of groups in our analysis. Application of the algorithm resulted in six group/group comparisons and their associated optimized gene sets: Gene $\text{Set}_{\text{DNA damagers, DNA damagers}}$, $S_{DD}$; Gene $\text{Set}_{\text{DNA damagers, Antiinflammatories}}$, $S_{DA}$; Gene $\text{Set}_{\text{DNA damagers, Notoxic controls}}$, $S_{DN}$; Gene $\text{Set}_{\text{Anti-inflammatories, Antiinflammatories}}$, $S_{AA}$; Gene $\text{Set}_{\text{Antiinflammatories, Notoxic controls}}$, $S_{AN}$; Gene $\text{Set}_{\text{Notoxic controls, Notoxic controls}}$, $S_{NN}$. The genes in the resulting six gene sets were reduced further by taking the intersection of the union of the intersection of the initial genes sets as displayed graphically in the Venn diagrams. (C) Supervised analysis of all compounds in the database using genes identified by the algorithm-based optimization gene set for distinguishing cisplatin and diflunisal/flufenamic acid. The distinction between toxic doses of DNA damaging agents and NSAIDs is much clearer, providing evidence that genes identified in this supervised analysis may serve as predictors of toxicity in subsequent comparisons of other members of these toxicant classes. (Adapted from Burczynski, M.E. et al., *Toxicol. Sci.*, 58, 399–415, 2000. With permission.) (See color insert following page 112.)

### 14.2.3 Multiclass Prediction in Toxicogenomics: A Theoretical Example

The examples above have referred to the simpler problem of binary class prediction: whether an unknown belongs to class 0 or class 1. However, recent applications have described more complex multiclass techniques that allow the assignment of unknowns to any one of several classes. In some of these prediction approaches, the problem of multiclass prediction is simply broken down into a series of binary decisions, as in the one-vs.-all approach.[6,7] While more detailed description of these methods is beyond the intended scope of this chapter, a theoretical approach to the multiclass class prediction of toxic modes of action for unknown new chemical entities (NCEs) is presented in Figure 14.2.

## 14.3 Toxicogenomic Analysis in Clinical Pharmacogenomic Studies

### 14.3.1 Technical Issues and the Problem of Confounding Variables

As we will see in the next section, the first objective of clinical pharmacogenomic studies is often the identification of markers of disease.[8] As one would expect, this is accomplished by simply comparing expression profiles of normal and diseased tissues and finding statistically significant markers in the comparisons using any number of supervised approaches. The comparisons, however, are not straightforward and caveats associated with these comparisons are abundant. The same issues that affect the quality of expression profiling studies conducted on a smaller scale in *in vitro* systems or animal models are amplified in large clinical studies focusing on human subjects.

Many sources of both technical and biological variability confound pharmacogenomic analysis in clinical trials. As just one simple example, the length of time associated with any given clinical trial is dependent upon patient accrual; most phase II clinical trials typically require from six months to more than a year to accrue sufficient numbers of patients for pharmacogenomic analysis. In a perfect world, frozen tissue samples could accumulate at a central laboratory storage facility and, at the completion of the study, all RNA samples could be prepared, processed, assessed, and analyzed on gene chips at once, thereby minimizing the contributions of technical procedures to variance in the dataset.

In the imperfect real world, however, because our laboratory is continually collecting and analyzing expression data from numerous clinical trials, tissue samples are processed to RNA and hybridized to gene chips in a continuous workflow. It is therefore absolutely critical that our laboratory implements the types of quality controls that were discussed by our colleagues in Chapter 3. In addition to employing spiked-in standard curves for every sample, we assess RNA quality, chip quality, and chip sensitivity for every sample (along with a list of many other parameters beyond the scope of this chapter). Samples not meeting quality control (QC) are flagged on the basis of failing to possess technical measures within an established range of acceptable values based on historical in-house data (for example, a beta-actin 5′ probe to 3′ probe ratio intensity of 0.4 or greater). Indeed, our first-pass analyses of expression data are not geared toward answering biological questions, but rather on confirming that technical issues do not complicate or invalidate our analyses. We utilize the standard curves in each sample to normalize all expression profiles to a pooled standard curve by the scaled frequency method, as described in Chapter 3. Global differences in normalized expression profiles of all samples are then assessed by scatterplot analysis. Because normalized datasets from clinical trials can often contain hundreds of profiles, we routinely use visual representations of Pearson's correlation coefficients between the samples to assess the overall variability in the data. If gross outliers are evident, we attempt to determine whether the outlier profiles in question are correlated with any technical quality in the sample processing: date of hybridization, chip lot number, etc. These approaches allow us to remove technical outliers prior

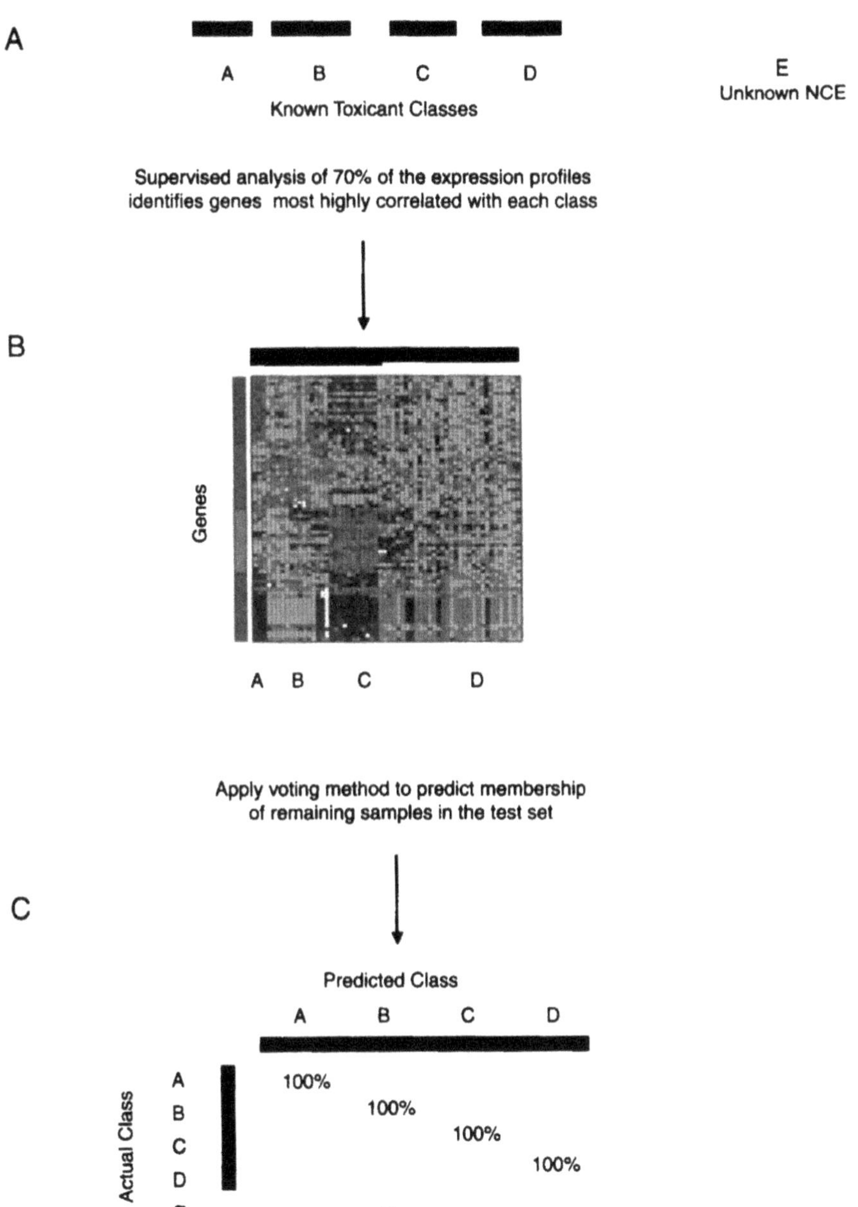


FIGURE 14.2 A multiclass approach to toxicogenomic prediction. (A) *Supervised approach.* As described in the text, a training set is constructed from known toxicant expression profiles. In this example, expression profiles for four classes of toxicants (classes A to D) and a single unknown NCE (class E) have been generated. (B) *Predictor gene selection.* In one version of a supervised approach, 70% of the known samples can be used to identify the genes most highly correlated with (and most likely to be predictive of) each class of toxicant. The significance of the predictor gene sets resulting from this analysis can be estimated using random permutation approaches and neighborhood analysis (not shown). (C) *Results for a voting method.* In this theoretical example, the predictive model based upon the training set above performed remarkably well when assigning membership to the remaining samples in the test set (validation). The NCE in class E was assigned membership in toxicant class B, providing a lead as to a possible mechanism of toxicity. (See color insert.)

to analyzing data to find biologically relevant markers, thus increasing our confidence in the analyses.

Many other sources of artifactual technical variability must be controlled in clinical studies, including site-to-site variability in sample procurement, site-to-site variability in sample processing (if applicable), lot-to-lot variability in sample reagents, lot-to-lot variability in gene chips, and a myriad of other imaginable sources. Even if we suspect that all possible sources of technical variability have been controlled, we have nonetheless instituted further measures to ensure that genes that are ultimately identified as biologically significant in our analyses are unlikely to have originated from technical sources. One of these measures is the ongoing analysis of technical variability.

To understand the technical variability associated with our in-house processes, we have instituted the use of a pool of standard reference RNA that is analyzed along with clinical samples on a routine basis. These technical replicates therefore identify a typical coefficient of variation associated with each probe set on the relevant Affymetrix gene chip design (HgU95A, U133, etc.). Individual coefficients of technical variation are noted for each gene in the biologically relevant gene sets (markers of disease, predictors of outcome, etc.) identified in subsequent analyses. If a gene of interest possesses a large coefficient of technical variation, its significance is assessed accordingly.

Similarly, coefficients of biological variation in a given type of tissue are calculated across a pool of biological replicates (tissue samples from many normal individuals). An understanding of the expected variability of expression in tissue across the normal human population for a given-sized sample set is crucial in order to assess whether genes that appear differentially expressed between normal and diseased tissues simply fall within the range of normal biological variability or appear to be truly significant.

Finally, clinical pharmacogenomic studies are somewhat unique in that they can be impacted by many other factors besides the obvious issues of technical processing and normal biological variability. While this section has only scratched the surface of the possible sources of artificial variance in clinical gene expression data, it is important to note one other important and controllable source of variability that can invalidate clinical comparisons: confounding variables due to patient demographics. Any patient demographic can confound a normal vs. disease comparison (or any other pharmacogenomic analysis in question for that matter): from ethnicity to previous medication history and any other measurable demographic in between. To give a simple example, if the intent of an original normal vs. disease analysis is confounded by an age stratification in the normal vs. disease patient population, age is a confounding variable and it is unclear whether the observed differences in gene expression are age or disease related. It is therefore critical to address these types of issues up front, prior to analyzing clinical data to find markers of interest.

### 14.3.2 Objectives in Clinical Pharmacogenomic Studies and Their Applicability to Toxicity Prediction

The majority, if not all, of the transcriptional profiling approaches summarized in this volume thus far have described results from cell culture model systems or animal models used for the purpose of toxicogenomic analysis. With respect to the pharmaceutical industry, these types of analyses are important during the preclinical phase of drug development. As numerous authors have iterated herein, the most important initial goal and potential benefit of toxicogenomics will be to accelerate the evaluation of toxicity for NCEs. As this textbook duly testifies, the efforts of many determined scientists are currently aimed at discovering whether this goal will be realized in the near future.

Clinical pharmacogenomic (and toxicogenomic) studies, on the other hand, have lagged slightly behind, although they are now beginning to receive greater attention in the pharmaceutical industry. Clinical pharmacogenomic studies are run in parallel with human clinical trials of investigational drugs and are designed to mine expression profiles in tissues from human patients in order to

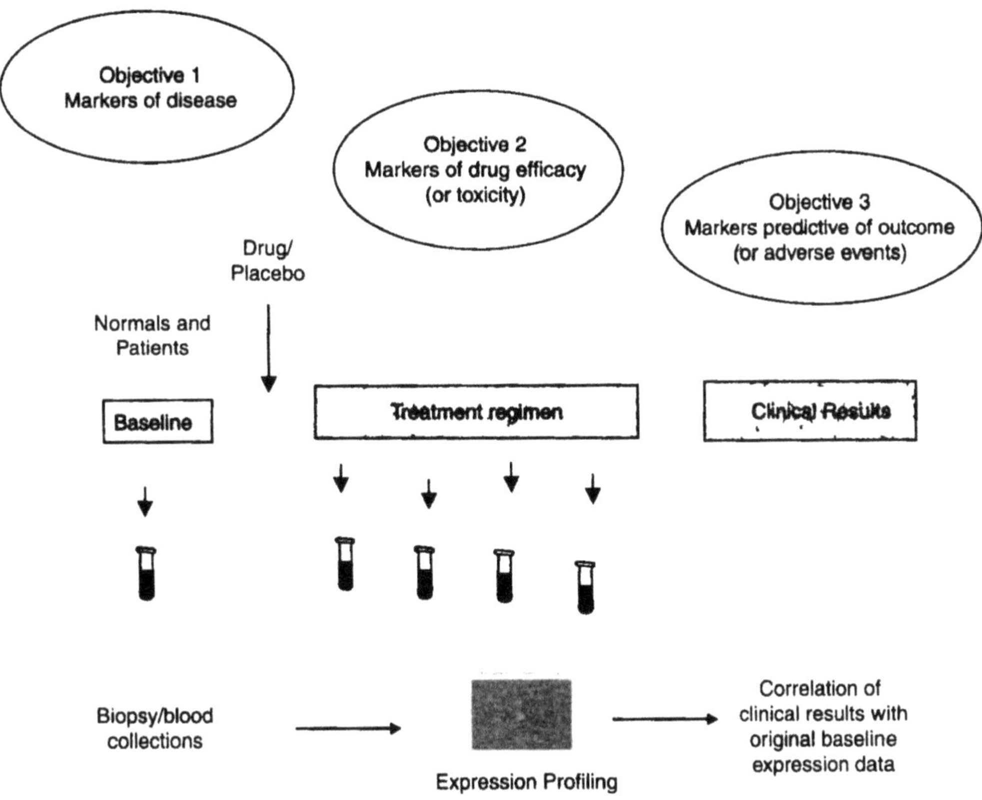


**FIGURE 14.3** Objectives of clinical pharmacogenomic analyses. This schematic depicts the objectives of clinical pharmacogenomic studies. The first objective, identification of markers of disease, is achieved by comparing baseline expression profiles from normal individuals and patients enrolled in the clinical trial. The second objective, identification of markers of drug efficacy, is achieved by comparing expression profiles over time in drug- and placebo-administered patients and correlating them with measures of clinical response. The third objective, identification of markers predictive of outcome, is obtained by comparing baseline and/or longitudinal expression profiles of patients that responded to therapy with those who failed to respond to therapy. As noted in the text, in the event of toxicity during the clinical trial, similar supervised methods can be applied to identify biomarkers of drug toxicity or markers predictive of adverse events. (See color insert.)

identify several types of diagnostics. Depending on the outcome in the clinical trial, these studies have the potential to reveal both markers of beneficial drug activity as well as adverse toxicity. Our laboratory performs clinical pharmacogenomic studies designed to identify three main types of markers: (1) markers of disease, (2) markers of drug efficacy, and (3) markers predictive of outcome (Figure 14.3). While the main focus of our laboratory is the identification of markers correlated with desired clinical pharmacogenomic results, it is obvious that the latter two objectives listed here may ultimately define cognate toxicogenomic markers in the event that toxicity is observed during the study. Thus, in the event of toxicity, markers in groups 2 and 3 become markers of drug toxicity and markers predictive of adverse outcomes, respectively.

The successful identification of each of the potential diagnostics is predicated on several implicit assumptions. Three major ones are that (1) "appropriate" tissues will exhibit sufficiently robust and reproducible alterations in transcription between normal and diseased states, (2) pharmacologic intervention will evoke a transcriptional response in an appropriate tissue, and (3) the transcriptional profiles in the appropriate tissue at baseline (prior to drug therapy) will be related to whether the patient will or can respond to pharmacologic intervention (either beneficially or adversely).

After expression data have been normalized and subjected to tests for quality control, data for patients with confounding variables have been removed, and technical and biological variation for the various genes have been established, we can finally search for pharmacogenomic effects in detail. To identify markers of disease we typically compare tissues from normal individuals with tissues from patients taken at baseline prior to drug therapy. Statistical tests and fold-change analyses are used to identify a disease gene set. However, because these rely on arbitrarily imposed cut-offs in *p*-values and fold changes, we also employ class prediction methods such as the ones described in the previous sections to identify genes that best discriminate between (or among) the classes in question.

Markers of drug efficacy can only be identified once indices of clinical efficacy have been measured. In the case of blinded trials containing placebo arms, definitive analyses of drug-specific effects are delayed until study unblinding. Once the cohorts of patients receiving drug can be classified relative to placebo, class prediction methods can again be applied to identify drug efficacy genes at the time points following drug administration. Multivariate approaches (multiway analysis of variance [ANOVA], etc.) are most often applied to temporal or longitudinal data in which more than two points are compared such that variability over time represents one source of variance and the desired stratification represents another. Any other number of other variables can also be applied in these types of approaches (e.g., dose). Although we normally correlate gene changes with efficacy markers, the same types of approaches could be applied to correlate gene changes with adverse effect markers. Biomarkers measured at early time points that are significantly correlated with either eventual response or eventual toxicity would be of obvious importance in monitoring the progress of patients prior to the end of therapy.

The search for markers predictive of outcome supposes that, in the appropriate tissue, the transcriptional baseline may indicate whether a patient will be capable of responding to the drug. It is easy to imagine a scenario in which, if a certain enzyme is required for the activation of a pro-drug to its efficacious metabolite, then a patient profile in which that enzyme's transcript is missing or extremely low may be highly correlated with poor outcome. Markers predictive of outcome can only be identified once follow-up data for the patients in the trial have been monitored for a sufficient length of time: disease-free survival, time to progression, time to remission, etc. Again, in the event of toxicity during a trial, the baseline expression profiles of patients undergoing adverse events can be compared in supervised analyses to baseline profiles of non-adversely affected patients to determine whether a baseline profile correlated with adverse outcome can be found.

## 14.4 SUMMARY

This brief chapter has attempted to lay a basic foundation for the principles of class prediction in clinical pharmacogenomic studies and how questions regarding toxicity can be answered during the course of these studies. It has highlighted many of the issues facing scientists attempting to perform clinical pharmacogenomic analyses, and it has introduced a general approach to clinical pharmacogenomic studies conducted in concert with clinical trials. Finally, it is readily apparent that toxicogenomic data can be derived in clinical studies if adverse events or toxicities are observed. Identification of gene expression profiles that are predictive of toxicity in human patients may someday lead to the development of useful diagnostic assays that identify toxic reactions to therapies before they become irreversible. If the results of initial studies are confirmed in subsequent larger scale trials, such biomonitoring may one day have a dramatic impact on human health.

## ACKNOWLEDGMENTS

The author gratefully acknowledges Donna Slonim for her comments on the contents of this chapter; Jen Stover and Natalie Twine for their helpful discussions regarding aspects of clinical pharmacogenomic study design; and the pioneering work of Michael McMillian, Joe Cierro, Li Li, James (Brandon) Parker, Bob Dunn, Sam Hicken, Spencer Farr, and Mark Johnson in the field of predictive toxicogenomics.

## REFERENCES


1. Pomeroy, S.L. et al., Prediction of central nervous system embryonal tumour outcome based on gene expression, *Nature*, 415, 436–442, 2002.
2. Waring, J.F., Jolly, R.A., Ciurlionis, R., Lum, P.Y., Praestgaard, J.T., Morfitt, D.C., Buratto, B., Roberts, C., Schadt, E., and Ulrich, R.G., Clustering of hepatotoxins based on mechanism of toxicity using gene expression profiles, *Toxicol. Appl. Pharmacol.*, 175, 28–38, 2001.
3. Burczynski, M.E., McMillian, M., Ciervo, J., Li, L., Parker, J.B., Dunn, R.T., 2nd, Hicken, S., Farr, S., and Johnson, M.D., Toxicogenomics-based discrimination of toxic mechanism in HepG2 human hepatoma cells, *Toxicol. Sci.*, 58, 399–415, 2000.
4. Slonim, D.K., Tamayo, P., Mesirov, J.P., Golub, T.R., and Lander, E.S., Class prediction and discovery using gene expression data, *Proc. Fourth Annu. Conf. Computational Molecular Biology*, 263–272, 2000.
5. Golub, T.R. et al., Molecular classification of cancer: class discovery and class prediction by gene expression monitoring, *Science*, 286, 531–537, 1999.
6. Ramaswamy, S., Tamayo, P., Rifkin, R., Mukherjee, S., Yeang, C.H., Angelo, M., Ladd, C., Reich, M., Latulippe, E., Mesirov, J.P., Poggio, T., Gerald, W., Loda, M., Lander, E.S., and Golub, T.R., Multiclass cancer diagnosis using tumor gene expression signatures, *Proc. Natl. Acad. Sci. USA*, 98, 15149–15154, 2001.
7. Su, A.I., Welsh, J.B., Sapinoso, L.M., Kern, S.G., Dimitrov, P., Lapp, H., Schultz, P.G., Powell, S.M., Moskaluk, C.A., Frierson, H.F., Jr., and Hampton, G.M., Molecular classification of human carcinomas by use of gene expression signatures, *Cancer Res.*, 61, 7388–7393, 2001.
8. Slonim, D.K., Transcriptional profiling in cancer: the path to clinical pharmacogenomics, *Pharmacogenomics*, 2, 123–136, 2001.

# Section 6

## The Future of Toxicogenomics

# 15 The Future of Toxicogenomics

*John C. Rockett*

**CONTENTS**



## 15.1 INTRODUCTION

Toxicology has classically been viewed as the science of poisons. In the modern world, however, it has evolved into a composite of related, but distinct, disciplines that together seek to understand how chemicals of all kinds — both man-made and natural — affect human health and the environment. Indeed, Dr. David Eaton, recent president of the Society of Toxicology, boldly declared during his tenure that toxicology had "come of age" and that genomics would provide the tools to influence its future. Genomics will not be the only tools used, however. Although toxicogenomics can be defined simply as a combination of toxicology and genomics, its future probably lies in a much more encompassing definition, perhaps communicated most clearly by Selkirk and Tenant:[1]

> *Toxicogenomics:* A new scientific field that elucidates how the entire genome is involved in biological responses of organisms exposed to environmental toxicants/stressors. It combines information from studies of genomic-scale mRNA profiling, cell-wide or tissue-wide protein profiling (proteomics), genetic susceptibility, and computational models to understand the roles of gene–environment interactions in disease.

0-8493-1334-1/03/$0.00+$1.50

There can be no doubt that toxicogenomics is still a very young field. The first published use of the term was in 1999,[2] and even now a keyword search for "toxicogenomics" in PubMed (http://www.ncbi.nlm.nih.gov/entrez/query.fcgi) reveals a mere 38 publications (as of July 2002), only 4 of which are data papers. This is typical of an emerging field, and one should not misinterpret this paucity of published material as a sign of disinterest or lack of investment. The fact is, toxicogenomics has stimulated tremendous excitement and interest the world over, particularly among the drug and chemical manufacturers, who publish much less frequently than academics and yet have the resources and infrastructure to quickly pursue and build upon new ideas less constrained by financial and logistical restrictions. Thus, the main body of the toxicogenomic literature, primarily consisting of review papers written by leaders and innovators in the field, conveys the sentiment shared at multiple scientific meetings and in hundreds of laboratories the world over for the last 3 years: namely, toxicogenomics has a big and bright future. But, what exactly does that future hold? That is the proverbial $10 million question. Or, perhaps more aptly, the $10 *billion* question. That sum may be a more realistic estimate of the kind of savings drug and chemical manufacturing companies may realize if toxicogenomics can deliver what it promises: a means to better understand the molecular mechanisms underlying biological responses to toxicant exposure, thus ultimately providing a more informed and therefore more accurate mechanism of conducting toxicity and risk assessments.

Risk assessment, the process that scientists and officials use to estimate the increased risk of health problems in people (and animals) exposed to different amounts of toxic substances, is a most complex area. It includes hazard identification (what problems are caused by the toxicant?), exposure assessment (how many people are exposed and how much of the toxicant do they receive?), dose response (the health problems at different doses of the toxicant), and risk characterization (the extra risk of health problems in the exposed population). Given its complexity, risk assessment is an imprecise science at best. Indeed, some might view it as more of an art form, often based on limited data, subjective interpretations of data, and guesstimates in extrapolating effects from test species to humans. Consequently, risk assessment is a fairly controversial area that could benefit substantially from toxicogenomic data. The key questions that should be addressed for improved risk assessment include: Which animal models are the best predictors for humans? What are the overlaps in human and animal ranges of sensitivity? How much variability exists in a human population? It is hoped that toxicogenomics will provide answers to these questions by identifying appropriate models for extrapolating to human conditions; for facilitating an improved understanding of individual differences in biomarkers of exposure, effect, and susceptibility; and for understanding the effect of gene-environment interactions.

## 15.2 TECHNOLOGY

Many people automatically equate genomic technologies with microarrays and transcriptomics, but this view is far from complete. Microarrays are just one (albeit large) tree in a forest of different technologies. Remarkable progress has been made in recent years in the development of other genomic technologies for studying the structural, functional, and temporal characteristics of nucleic acids, particularly technologies for identifying aberrations of DNA in regions of chromosomes likely to harbor genes important in tumorogenesis or progression. Many toxicants have genotoxic properties that can lead to tumorogenesis, and "toxicogenomicists" of the future should find such techniques useful adjuncts to transcriptomics that provide data to enhance and support risk assessment decisions. A more complete list of the new technologies available includes:

1. *Restriction landmark genome scanning (RLGS).*[3] RLGS is a highly resolving gel-based genome scanning technique in which several thousand fragments of genomic digests are visualized simultaneously and quantitatively analyzed. It allows identification of not only genome copy number differences but also mutations and polymorphisms. It has been

used to identify novel imprinted genes, novel targets of DNA amplification and methylation in human cancer, and deletion, methylation, and gene amplification in mouse models of tumorogenesis.[4] RLGS is uniquely suited for simultaneously assessing the methylation status of thousands of CpG islands. A CpG island is a CpG-rich region associated with the 5′-end of many human genes. Although most CpGs in the human genome are methylated, those within CpG islands are usually methylation-free in somatic cells, regardless of the expression of their associated genes; however, methylation of CpG islands results in silencing of the associated gene. Imprinted genes and many genes on inactivated X chromosomes are inactivated in this way. DNA methylation in many cases appears to be lost in older vs. younger subjects, thus providing a mechanism for the development of susceptible subpopulations. RLGS is also a useful method for integrating methylation analyses with high-resolution gene copy number analyses such as comparative genomic hybridization (CGH).

2. *Comparative genomic hybridization (CGH).*[5] CGH utilizes the hybridization of differentially labeled target and reference DNA to generate a map of DNA copy number changes in target cell genomes. CGH is an ideal tool for analyzing chromosomal imbalances in tumor material and for examining possible correlations between these findings and tumor phenotypes. Conveniently, CGH is not limited to metaphase chromosome-based analysis, being easily adapted to arrays of well-mapped cloned regions.[6]
3. *Spectral karyotyping (SKY).*[7] SKY is based on the simultaneous hybridization of differentially labeled chromosome painting probes (24 in human), followed by spectral imaging that allows the unique display of all human (and other species) chromosomes in different colors. SKY greatly facilitates the characterization of numerical and structural chromosomal aberrations, thus improving karyotype analysis considerably.
4. *Representational difference analysis (RDA).*[8] Array CGH and analysis of loss of heterozygosity (LOH) are limited in that they detect only aberrations in the test loci. Subtraction strategies such as RDA overcome these limitations by detecting and cloning segments of DNA that differ in copy number between two genomes. In this approach, the two genomes to be compared are enzymatically digested and ligated to linker oligonucleotides. These representations of each genome are amplified (using primers complementary to the linker oligonucleotides), and the differences between the two genomes are detected by hybridizing one to an excess of the other.
5. *Quantitative real-time polymerase chain reaction (QRT-PCR).*[9] Real-time PCR methods (e.g., TaqMan, Molecular Beacons) for quantitating the amount of mRNA present in a sample are often used in conjunction with microarray studies as a means to confirm observed gene changes in a limited number of genes. Real-time PCR automates the laborious process of end-point relative quantitative reverse transcription PCR by quantitating reaction products for each sample in every cycle. In this way, a dynamic range in excess of 100-fold is achievable, and, after setup, the process requires no further user intervention or replicates. Data analysis, including standard curve generation and copy number calculation, is performed automatically. As more labs and core facilities acquire the instrumentation required for real-time analysis, this approach will probably become the dominant RT-PCR-based quantitation technique.
6. *Gene expression profiling (GEP).* Most expanded definitions of toxicogenomics contain a reference to gene expression profiling, and indeed this is currently the dominant technique used in toxicogenomic studies. The most extensively used approach in most laboratories is microarray analysis; however, a number of other approaches exist that may be more suitable depending on the model being used or the objective of the research being conducted.[10] For example, one of the greatest limitations of microarrays is that they are closed systems (i.e., they report only on the expression of the genes that they contain). This might be acceptable for species such as human and mouse where tens of

thousands of genes are available for printing on arrays, but, for more exotic species (e.g., amphibians and fishes), where limited sequence data are available, so-called open GEP systems are more useful. These systems, which include differential display (DD), serial analysis of gene expression (SAGE), and suppression subtractive hybridization (SSH), can be used to isolate differentially expressed transcripts between a test and control sample without having any kind of foreknowledge of the genes that might be involved in the model system or their sequences.

7. *Transgenic technology.* Rapid advances in molecular genetics over the past few years have resulted in the ability to manipulate the mammalian genome, allowing scientists to create transgenic animals. Many of these transgenic models are quickly becoming standard research tools facilitating the study of normal gene function and the role of specific genes in human disease. Transgene constructs include regulatory elements that can enhance and direct tissue expression of the transgene. Other technologies capitalize on homologous recombination events to inactivate selected genes, resulting in animals that no longer express their normal gene product. Transgenic models are finding their way into numerous research disciplines, and specific models are being developed for defined applications. In the field of toxicology, for example, various transgenic and knockout models such as the TG.AC mouse and the heterozygote TSG-p53 mouse are being evaluated as alternatives to the standard two year carcinogenicity bioassay.
8. *Single nucleotide polymorphism (SNP) analysis.* SNPs are the most widespread and stable form of genetic variation, occurring about once every 200 to 300 bases. It is estimated that each human has around three million SNPs, with as many as 200 to 300,000 of them located in the protein coding sequences of the human population. Latest estimates are that as many as 10% of these SNPs, so-called cSNPs, have functional implications. In this way, SNPs are thought to explain a large proportion of differences in individual responses to toxicants and drugs, and a large school of thought supports the contention that characterizing these SNPs is the key to deciphering the genetic basis of complex disease. Polymorphism analysis, and particularly SNP analysis, is rapidly developing into one of the most important components of toxicogenomic studies and as such will be discussed in more detail.

With the rapid advances in the Human Genome Project, the role of genetic polymorphisms in drug metabolism will become an important adjunct for the explanation of drug toxicity and interactions. SNP analysis is one of the most important components of toxicogenomics and the technology will develop rapidly. Indeed, this has been recognized for some time, and in order to enhance our knowledge and understanding of the impact of SNPs on human physiological responses (particularly to pharmaceuticals and environmental toxicants), two or three major SNP consortiums have been established. The largest of these is The SNP Consortium (TSC, http://snp.cshl.org), which was established in 1999 through the financial contributions of multiple organizations to advance the field of medicine and aid in the development of genetic-based diagnostics and therapeutics. SNPs are easy to detect and store as digital code, and the initial objective of TSC was to characterize 300,000 SNPs, map 150,000, and maximize public accessibility. In fact, due to progress in sequencing technology, TSC ended up with 1.7 million SNPs, 1.52 million of which were mapped. Of these, 1.4 million are described as "unencumbered" (i.e., have no intellectual property strings attached). Various TSC member laboratories are now in the process of genotyping a subset of those SNPs as a part of the Allele Frequency Project (http://snp.cshl.org/allele_frequency_project/afp_summary_nov2001.shtml), the goal of which is to determine the frequency of 60,000 SNPs in three major world populations (Caucasian, Asian, African American).

Another important SNP database is that established through the NIEHS's Environmental Genome Project (EGP, http://www.niehs.nih.gov/envgenom/home.htm). The EGP database is of particular interest for those in the toxicogenomics field, as the genes it has targeted have been specially selected as being environmentally responsive genes (ERGs), which have been identified as being likely to be important factors in genetic susceptibility to environmentally induced diseases. SNP sequence information for these genes is stored on the GeneSNPs website (http://www.genome.utah.edu/genesnps).

Of course, the use of polymorphism data in risk assessment is the ultimate goal for many organizations, and some well-characterized examples of genetic polymorphisms modifying exposure-related responses are already known. For example, polymorphisms of CYP2D6 (debrisoquine hydroxylase), a cytochrome P450 enzyme, drastically affect its drug metabolizing ability. Another commonly quoted example is that of alcohol intolerance, which is influenced by polymorphisms in the aldehyde dehydrogenase gene. Unfortunately, determining quantitative measures of exposure is difficult in humans, so that combining this with genetic information (such as SNPs) to assess risk is quite problematic and difficult to achieve. Thus, we need to determine functional relationships between genotype and phenotype, remembering that simple polymorphisms may have different effects depending on the chemical and the target organs that are considered.

Most current SNP genotyping methods, based predominantly on sequencing and PCR, are still too slow and expensive for routine use in large association studies with hundreds or more SNPs in a large number of DNA samples. However, SNP genotyping technology is rapidly progressing with the emergence of novel, faster, and less expensive methods, as well as improvements in the existing methods. Sequenom's (http://www.sequenom.com) MassARRAY system, for example, based on matrix-assisted laser desorption/ionization time-of-flight (MALDI-TOF) mass spectrometry, is capable of accurate, cost-efficient analysis of up to 20,000 SNPs per day. Sequencing-based SNP analysis has also improved recently. Pyrosequencing technology (http://www.pyrosequencing.com) can analyze up to 10,000 SNPs per day, and up to 30,000 if triplex analysis is used. This rapid progress in technology development for SNP discovery, in combination with the discovery of millions of SNPs and the development of the human haplotype map, may enable whole genome association studies to be initiated in the near future.

Despite the tremendous interest shown in applying toxicogenomic approaches to evaluating drug and chemical safety, there should be no mistaking that several of the new genomic technologies being used, particularly microarrays, are still far from being suitable for risk assessment. The general consensus within the scientific community is that genomic techniques must be improved so that they can return more sensitive, reproducible, and quantitative data before they can realistically be used in the risk-assessment process.[11–13] A need also exists to standardize and validate the protocols that are developed, and maintain rigorous quality control. The good news is that we anticipate that such technical issues will be overcome in relatively short course. This will be due in no small part to cooperation among stakeholders and the leadership of organizations such as the International Life Sciences Institute (ILSI; http://www.ilsi.org), which has assembled a consortium of representatives from many of the world's major pharmaceutical companies in order to work together to determine the technological limitations of microarray technology. Their ultimate aim is to determine whether this technology really can be applied to mechanism-based risk assessment.

## 15.3 REDUCTION OF ANIMAL USAGE

Pharmaceutical companies and manufacturers of nonpharmaceutical agents such as pesticides use a large number of laboratory animals in testing newly developed compounds. Because there is a general push in the scientific community toward implementing the three Rs (reduce, refine, and replace animal usage) of conscientious animal research, excitement has been generated in regard to the potential of toxicogenomics to facilitate this goal. The expectation is that toxicogenomic data will help improve the prioritization process for determining which chemicals will be included in the expensive, time-consuming, and animal-intensive *in vivo* tests.

For some time now, toxicologists have been calling for modernization of some of the current standard bioassays used to identify chemical carcinogens and assess health risk. The answer to both the call for modernization and for reduced animal usage may lie to some extent in transgenic models designed specifically to assess the *in vivo*, tissue-specific mutagenic potential of test compounds. In this vein, the so-called Big Blue® Transgenic Rodent Mutation Assay (Stratagene; La Jolla, CA) was developed as a rapid, modern alternative to the 2-year animal bioassay proscribed by the National Toxicology Program (NTP). The NTP bioassay involves the dosing of a large number of animals with the endpoint of tumor formation. In contrast, the endpoint in the Big Blue assay is mutation of a test gene, and the assay takes only 3 to 4 months. The Muta™Mouse transgenic model (CRP, Inc.; Denver, PA) was developed at approximately the same time as the Big Blue system and has also enjoyed extensive use in genotoxicity studies. Other transgenic mouse systems have also been developed for toxicity tests, including strains such as TSG-p53®, TSG-p53/Big Blue, TG.AC, and Xpc knockout mice.

Other systems have also been developed that may contribute to the reduction in animal usage. Xenometrix, for example, developed a series of *in vitro* tests called Gene Profile Assays (GPAs). The aim of each assay is to detect a variety of transcriptional responses in human liver cells following exposure to the test compound. The assay utilizes chloramphenicol acetyl transferase (CAT) reporter constructs driven by wild-type promoters or response elements from genes known to play a pivotal role in the response of the liver to foreign substances (e.g., Hsp70, c-*fos*, GRP78). CAT expression from any of the reporters following treatment of the cells with a test compound indicates potential toxicity.

In truth, however, the excitement about the ability of toxicogenomic techniques to reduce animal usage may be premature. We are still very much at the beginning of determining how gene expression changes impact biological processes in whole animals, and it may be a long time before this knowledge can be transferred to *in vitro* approaches to toxicity testing (thus reducing animal usage). Indeed, one widespread school of thought believes that animal usage in toxicogenomic studies actually may well go up for awhile as various time and dose-response experiments are carried out to elucidate biomarkers, gene expression networks, and modes and mechanisms of action of the many classes of toxicants. Nevertheless, in time the knowledge gained from studies in both standard and transgenic strains of animals should permit identification of the best models to use and the most informative time points to employ, thus reducing overall animal usage.

## 15.4 DATA ANALYSIS AND INTERPRETATION

One of the least acknowledged but probably most important reasons that the post-genomic era has been able to develop in such dramatic and conspicuous fashion is the microprocessor and recent advances in computer technology, particularly in nanofabrication techniques and software design. Not only do computers drive the increasingly sophisticated machines that are being developed for genomic studies, but they also are the backbone behind toxicogenomic data capture, storage, analysis, and interpretation. Indeed, it is not too much of an overstatement to say that fields such as toxicogenomics could not exist until contemporary computer power became available to store, analyze, and help interpret and understand the vast quantities of data that they generate. The reader must be aware of the clear distinction between data analysis and data interpretation.

### 15.4.1 Data Analysis

Data analysis, in essence, involves comparing datasets. These might include gene expression profiles of a dose response or time course or genomic sequence information across individuals or species. Data analysis is primarily a number-crunching process, requiring statistical skills and computing power. More complex data analysis includes patterning and clustering techniques to identify biomarkers and potential gene networks. All this is easier said than done. The vast amount of data generated by even a single genomic experiment can be enormous, and making sense of it can be a tremendous task. Many methods of analyzing gene expression data have evolved rapidly over the last few years, including various types of cluster analysis, principle component analysis, self-organizing maps, and discriminant analysis. However, nobody can yet say that one approach is better than the others. Because each type of analysis offers a slightly different view of the data, it is more than likely that several approaches will emerge and persist as standards. The main drawback with such analytical approaches, particularly those that are unsupervised, is that they often have difficulty recognizing subtle differences between tissues with different pathologies.

A supervised learning approach may be more helpful in such cases. In supervised learning, classification of genes is carried out with prior knowledge of the correct answer. This approach tries to learn the characteristics of known classes and is useful in predicting whether a new tissue belongs to a certain category. In addition, it can be used to validate previous classifications based on unsupervised learning and may find outliers that were previously unrecognized. Supervised learning algorithms, such as SVMs (support vector machines) and C4.5 (decision trees),[14] are often able to discern between subtly different biological specimens which provides the ability to distinguish between different pathologies.

It has been suggested that one of the best ways to learn about the human genome is through comparative genomics, and it seems that this field will become increasingly important in toxicogenomic analysis as more genomes are sequenced. Increasingly powerful software is thus being developed for comparative genomic analysis — the assembly, alignment, and comparison of gene sequences. It is often the case that gene or amino acid sequences are highly conserved across species, and the ability to compare sequence data for such unrelated organisms as human, rat, fruit fly, nematode, and shark will prove invaluable in the identification of important structural and functional regions of genes and their protein products. For example, the little skate (*Raja erinacea*) has been used in the study of the bile salt export pump (BSEP), which regulates transportation of bile salts across the canalicular membrane of hepatocytes. BSEP is an important target for induction of drug- and estrogen-induced cessation of bile excretion (cholestasis) in mammalian liver, and inadvertent agonism or antagonism of BSEP has stopped many drugs from getting to market. Mutations in the human BSEP result in a form of liver disease called progressive intrahepatic cholestasis, or PFIC type II. Comparative gene sequence alignments and functional studies have shown that mutations of BSEP in patients with PFIC type II disease are in conserved regions in human and small skate,[15] demonstrating that comparative genomics can be used to identify functionally important regions of genes and provide insights into mechanisms of human disease.

### 15.4.2 Data Interpretation

Analysis of data from most toxicogenomic experiments gives plenty of information, but, as author Caleb Carr's insightful comment reveals: "*It is the greatest truth of our age: information is not knowledge.*" This will undoubtedly draw nods of knowing approval from those familiar with any of the "-omic" fields. Recognizing patterns of gene expression and clustering genes to begin elucidating molecular networks is one thing. Using these to determine the mechanism or mode of action of a toxicant is quite another. Thus, conducting experiments and analyzing data to determine gene changes or gene sequences are the easy steps. Of more concern is how to interpret these vast quantities of complex genomic data. The technology for interpreting genomic data, particularly from transcriptomics, lags behind that for analysis. Without a clear understanding of, for example, gene–gene and gene–environment interactions, differences between species and individual responses, and qualitative and quantitative linkages between toxicity and disease, a real potential exists for disagreement or misinterpretation of data where risk assessment is concerned. For example, what does it mean if a gene changes twofold following exposure to a test substance? Very likely not much if the gene is already expressed constitutively at a high level. For other genes, however, this type of fluctuation could be quite significant. The point is, gene changes must be considered, wherever possible, in light of the normal range of their expression levels. This kind of interpretation currently relies on human input and knowledge but may one day be able to be programmed into "interpretational software."

It may be that convenient genomic biomarkers of toxicity can be discovered in certain species for certain tissues and certain chemicals or families of chemicals; however, the more we discover about cell function, the more complex we realize it is. The truth is, even in the case of commonly used laboratory animals such as rats and mice, we do not know enough about the biology of systems in even a single strain to make wholly encompassing predictions about the meaning of a change in expression of a single gene or the impact of disrupting a particular network of genes. Changes in gene expression and characterization of SNPs simply may not be enough to understand toxic effects, due to the number of environmental and biological variables, such as modifier genes and stochastic effects.

It seems that accurately interpreting toxicogenomic data will require complex computer analysis using algorithms yet to be developed. Such algorithms will interpret toxicogenomic data in light of the genetic networks that are active in a cell, which in turn will depend on species, tissue type, development stage, age, and so forth. In the meantime, the interpretation of such data comes down to the hunches and intuition of researchers based on an incomplete understanding of cell systems. Indeed, toxicogenomic data are often used as hypothesis generators for further mechanistic studies with techniques more appropriate for functional genomic analysis such as gene knock-out or knock-in technology.

### 15.4.3 Data Storage and Mining

The extent to which toxicogenomics will impact our understanding of biological systems and the risk assessment process probably lies in databases and how they are populated and queried. The more details available (e.g., on the changes in gene expression over time and dose in various tissues) or the more sequences across individuals, strains, and species for a particular gene, the stronger the conclusions that we will be able to be make concerning the effect of a toxicant or the structural and functional characteristics of a gene. It will thus be most interesting over the next few years to see how genomic information will be used to create databases for categorizing chemicals according to their mode of action. Indeed, several such toxicogenomic databases have already been produced by private and public entities. For example, an integrated database of gene expression and drug sensitivity profiles was recently constructed and used to identify genes with expression patterns that showed significant correlation to patterns of drug responsiveness.[16] A most promising and exciting development in the pipeline is the NIEHS-funded Comparative Toxicology Database (CTD; http://www.niehs.nih.gov/envgenom/abstract/r2111267.htm) being assembled by investigators at the Marine Desert Island Biological Laboratory (MDIBL; Salisbury Cove, Maine). They are building a comprehensive database to compare gene sequences and functions among humans, laboratory animals such as rats and mice, and aquatic organisms. It will be the first community-based, publicly accessible database devoted to genes of human toxicological significance and will include a plethora of information that includes genomic information (e.g., ontology, gene, and protein sequences), references, toxicants, assays, a repository for reagents and associated contact information, sequence analysis tools, and integration with related databases such as PubMed, GenBank, OMIM, and Toxnet. Microarray and expressed sequence tag (EST) data will also be added at a later juncture. It is anticipated that the database will provide novel insights into the dynamics of gene–environment interaction and human health. Although it is not slated to go online until 2006, limited access may be available to collaborators and registered users as early as 2004.

One of the most intriguing prospects of toxicogenomics is the possibility of data sharing. A typical microarray experiment holds so much data that only a fraction of it is likely to be of interest to any given researcher. The ultimate achievement would be to have a single, publicly accessible database or at least an integrated series of linked databases that can be queried simultaneously. This would permit the same data to be used over and over again by different researchers looking at different genes and gene networks. Of course, strict controls must be in place for data submission, and detailed annotation of the experiments must be made available.

Such a grand vision may be just a pipe dream. Factors such as lack of cross-platform concordance and the disarrayed state of gene nomenclature (see GeneCards, http://bioinformatics.weizmann.ac.il/cards/) and annotation make this vision a rather forlorn hope at present. Nevertheless, a movement is afoot to encourage the standardization of genomic data. In the gene expression analysis field, for example, where such standardization is most needed, the European Bioinformatics Institute (EBI, http://www.ebi.ac.uk/) has set up the Microarray Gene Expression Database Group (http://www.mged.org/), a cross-organization entity whose goal is to "facilitate the establishing of gene expression data repositories, comparability of gene expression data from different sources and interoperability of different gene expression databases and data analysis software." It is to be hoped that such standards, whether derived from EBI or another organization, will be adopted across the majority of the toxicogenomic community. This will not only facilitate public database development but may also prevent much wasted time and effort when buyouts and mergers in pharmaceutical and biotechnology companies bring together former competitors and their extensive repositories of genomic data.

Whatever the future holds for database development, it is clear that toxicogenomicists of the future will be a different breed from most of those around today. Not only will they have to possess traditional toxicology and molecular biology skills, but they must also have an understanding of information technology and applied and\or developmental bioinformatics skills, including the ability to mine extensive datasets and even conduct so-called virtual, *in silico*, or e experiments (e.g., differential gene expression analysis).

### 15.4.4 Gene Pathway Identification

Current analysis of toxicogenomic data has been based largely on the analysis of single genes. Two main approaches have been employed. One is functional analysis, whereby transcriptional changes of individual genes are linked to probable physiological and pathological outcomes. For this approach, only genes with established downstream (protein) functions are generally considered. Second, statistical correlations are made between groups of gene changes to determine gene changes that have a discernible motif (so-called pattern recognition) for identification of chemical exposures. It is becoming increasingly recognized that, in order for toxicogenomic data to be applied to risk assessment, the gene interaction networks and pathways that are being impacted (and in what way) must be determined. This is especially important in cross-species extrapolations where metabolic and toxicity pathways are sometimes quite different. Gene pathway elucidation is clearly a difficult undertaking, but software programs are being developed to facilitate this task. For example, Cytoscape (http://www.cytoscape.org), designed by researchers at The Institute for Systems Biology (http://www.systemsbiology.org/home.html) and The Whitehead Institute for Biomedical Research (http://www.wi.mit.edu/home.html), is a bioinformatics software platform for "visualizing molecular interaction networks and integrating these interactions with gene expression profiles and other state data." The key feature of Cytoscape is an algorithm that finds "active pathways" — subnetworks of genes that jointly show significant differential expression over a set of experimental conditions observed in microarray experiments. The identification of signaling and regulatory pathways will play an important role in risk assessments, as the regulation of a single gene means very little. Genes such as *p53*, for example, are involved in multiple cellular processes that may lead to the life or death of a cell depending on a number of other factors. Thus, a change in *p53* means little unless the pathway in which it is participating can be identified. Thus, interpreting toxicogenomic data requires complex computer analysis using algorithms that will view toxicogenomic data in light of gene expression networks.

## 15.5 USE OF TOXICOGENOMIC DATA IN RISK ASSESSMENT

There exists a strong linkage between risk assessment and management and research needs. Risk assessment is a dynamic process in that the quality of risk analysis will improve only as the quality of the input (scientific data) improves. Traditional hazard identification is evolving into the evaluation of modes and mechanisms through biologically based models. Toxicogenomics will be particularly useful in identifying and demonstrating mechanism and mode-of-action data in such models by indicating specific events and sequences of events leading to toxic effects, and assisting in the identification of measurable key events. These might include receptor–ligand changes, DNA/chromosome effects, increases in cell growth and organ weight, hormonal or other physiological perturbations, and hyperplasia and cellular proliferation.

A proverb widely quoted from the genomic literature tells us that, "Genes load the gun; the environment pulls the trigger," referring of course to how our genetically preprogrammed disease predispositions are activated by environmental factors. Understanding gene–environment interactions is thus a key element of risk assessment. The interaction between genes and the environment in the development and progression of disease pathologies is complex, and the great challenge facing us today is to find relevant biomarkers that predict risk, convey an estimate of exposure, and indicate the predisease state. The question of course is how toxicogenomics information will be used to achieve this.

### 15.5.1 Gene Expression Profiling in Risk Assessment

Early detection of toxic exposures is a developing art, but many groups have already successfully classified chemical exposures based on profiling of mRNA from treated animals.[17–21] This kind of information might be useful for risk assessment in that significant changes in expression in a small set of highly discriminatory genes can together act as a biomarker of toxic mechanisms or endpoints. Indeed, Thomas et al.[22] were able to identify such a diagnostic set of 12 transcripts that provided an estimated 100% predictive accuracy for a set of five different classes of toxicants. Expansion of this approach to additional chemicals of regulatory concern could serve as an important screening step for toxicological testing. With expression data from such diagnostic gene sets in hand, risk assessors could say of a chemical being tested: *"This group of genes changed in such and such a manner, which means that this chemical belongs to 'X' class of toxicants."* It has also been shown that correlations exist between gene expression profiles and histopathology and clinical chemistry.[23] With such data available, risk assessors could expand their conclusions: *"This group of genes changed in such and such a manner, which means that this chemical belongs to 'X' class of toxicants and will likely cause 'Y' outcome."* And, finally, it has been shown that correlations exist between gene expression profiles and mechanisms of toxicity.[24] With all the aforementioned information in hand, the time will come when risk assessors may be able to say: *"This group of genes changed in such and such a manner, which means that this chemical belongs to 'X' class of toxicants and will likely cause 'Y' outcome through 'Z' mechanism."*

Of course, the situation is not as simple as this explanation perhaps suggests. The data on which to base these conclusions must first be generated, reproduced, annotated, and deposited in appropriate integrated databases. There is also the burning question of what dose and time points would be optimal for generating the gene expression profiles for such analyses, which, in turn, depend on species, strain, tissue, age, gender, and a host of other factors. Although in many instances it will be most difficult to pin down a definite answer to this question, it should be remembered that if gene expression cluster analysis is to be used to predict toxicity, it must be robust enough to do so based on data from a single time point; otherwise, those testing the chemicals may find assays becoming too numerous and expensive to perform in a cost-efficient manner.

It has been demonstrated that gene expression profiling can also be used to elucidate gene networks through analysis of coordinately expressed genes,[25] and work is ongoing in this area to refine and improve the algorithms and approaches being used. This is a very exciting application of gene expression data, although admittedly much work is still needed to support and enhance the initial studies.

### 15.5.2 DNA Repair Systems in Risk Assessment

Despite such findings from the gene expression field, some risk assessors have suggested that genomic analysis may be less relevant for risk assessment than measuring functional phenotypes such as enzyme activation/deactivation and DNA repair function. Various alternative approaches have thus been suggested. One such approach for evaluating risk of toxicity of a chemical, at least in terms of its carcinogenic ability, may be to assess its impact on the function of DNA repair genes. The reasoning behind this is that it is not necessarily the amount of DNA damage a cell sustains *per se* that produces a cancerous phenotype, but the amount of damage present at the time of cell division. If the DNA repair systems of a cell are compromised by a chemical, then it is more likely that the chemical will cause toxicity. This idea stems from studies such as those carried out by Wu et al.[26] on a group of patients with hepatocellular carcinoma (HCC) to determine whether constitutional genetic instability, based on the quantification of mutagen-induced chromatid breaks in cultured lymphocytes, modifies an individual's risk of HCC development. The study findings suggested that differences in host factors related to the predisposition to chromosome breakage or the capacity for DNA repair, or both, may be involved in HCC development. The question remains as to whether genotypes associated with reduced repair capacity can be used as biomarkers of increased cancer risk. If so, elucidation of the molecular systems that protect and repair cell function will provide a new generation of surrogate biomarkers for monitoring cell damage, thus aiding risk assessment.

### 15.5.3 mRNA:Protein Ratios in Risk Assessment

Another alternative approach has been suggested based on the fact that RNA levels do not always correlate with levels of their corresponding protein product. In examining the correlation coefficients for mRNA and protein in the same tumors, investigators have found that some show good correlation while others do not, and some are negatively correlated (perhaps as a result of negative feedback). This has led to a proposal that evaluating the cellular response to a toxic challenge should not necessarily be based on changes in gene expression *per se*, but on how the expression relationship changes between an mRNA and its protein.

### 15.5.4 The Impact of Genetic Predisposition on Risk Assessment

The risk of developing disease due to genetic predisposition is a complex area of study. Diseases can be grouped into three main categories according to the degree of environmental influence:

1. *Those that arise due to genetic factors only.* In such diseases, individual risk of disease development is high, population risk is low (affected individuals are relatively few), and environmental exposures are not a factor. Approximately 2800 conditions are known to be caused by defects in just one gene, and about one in ten people has or will develop an inherited genetic disorder. Some single gene disorders are quite common (e.g., the occurrence of cystic fibrosis [CF] in the western hemisphere is about 1/2500), and, in total, diseases that can be traced to single gene defects account for about 5% of all admissions to children's hospitals. Some of these diseases show little or no variation in

their timing and degree of severity, usually indicating a monogenetic origin and no environmental influence. For example, severe combined immunodeficiency X1 (SCID-X1) is an X-linked inherited disorder characterized by an early block in T and natural killer lymphocyte differentiation. The block is caused by recessive mutations of the gene encoding the gamma cytokine receptor subunit of interleukin-2, -4, -7, -9, and -15 receptors.[27] An understandable belief is that diseases having an exclusive monogenetic origin will be the first to be characterized in terms of the molecular mechanisms of their development and progression. This has to some degree already been proven to be true. Given the lack of environmental and other genetic factors influencing the development of such diseases, it should be relatively easy to uncover enough mechanistic and pathophysiological data to enable diagnostic tests and preventative or curative treatment.

2. *Those diseases for which the risk and severity depend on the impact of modifier genes on disease alleles.* The population impact of such genes and the role of exposure are not known. A large percentage of monogenetic disorders demonstrate phenotypic variation that can only be a consequence of environmental factors or other genetic elements. Sibling studies of affected families have demonstrated that, for numerous diseases where a single allele is the source of the disease, the phenotypic outcome is dependent on the action of so-called modifier genes. The body of evidence available suggests, for example, that the variations seen in the CF and neurofibromatosis phenotype and in the development of cancer in women with BRCA-1/-2 mutations, results from complex interactions between numerous gene products.[28–30] In such cases where risk or severity of a disease depends on the presence of modifier genes, risk assessment becomes much more onerous. This is because the identification and characterization of all potential modifier genes are extremely difficult tasks. Large population studies may well pick up some new individual genetic determinants associated with a moderate (2–10×) relative risk of disease, but elucidating the individual contributions of many loci to polygenic traits is much more problematical. Current studies into identifying such loci are usually limited to simple polymorphism association studies, which explore the association of functional single nucleotide polymorphisms in genes implicated in the disease with the severity of the clinical phenotypes.
3. *Those diseases triggered by susceptibility alleles for which the individual risk of disease progression is low, but the population risk is high due to the fact that there are many susceptible alleles.* Here, environmental exposure is an important contributing factor to disease development. Because the number of contributing alleles and environmental exposures are so large (and, in most cases, not fully documented), these diseases will be the most difficult to characterize. Thus, in an age when the majority of monogenic human disease genes have been identified, a particular challenge for the coming generation of human geneticists will be resolving complex polygenic and multifactoral diseases. Given our current level of knowledge and understanding, this appears to be an almost staggering burden. Fortunately, toxicogenomics may provide the tools to help alleviate this burden. It is anticipated that the high throughput nature and powerful analytical properties of many modern genomic technologies, coupled with advances in database building and the algorithms and software used to analyze data and mine the databases, will help in facilitating breakthrough discoveries in this complex frontier of biological systems research.

### 15.5.5 Risk Posed by Inadvertent Exposure to Environmental Toxicants

Of course, people are not just at risk from drugs they knowingly take. Risks are also associated with exposures derived from the environment. Such exposures occur through air, water, and food and are often inadvertent, unknown, or unavoidable. Assessing the risk posed by such exposures is becoming increasingly important as the number and volume of chemicals and chemical mixtures being released, inadvertently or on purpose, into the biosphere is gradually increasing. Toxicogenomics will be particularly useful in identifying and demonstrating the mode of action of any toxic effects through highlighting the gene-expression networks and/or pathways that are impacted in individuals who develop diseases as a result of environmental exposures. Such information will also help identify and measure key events that are useful in risk assessment, including changes following receptor–ligand interaction or changes in DNA and chromosomes, such as DNA strand breaks or base modifications induced by environmental toxicants. Toxicogenomics will also help our understanding of whether toxicology data generated from animal laboratory models and environmentally located "sentinel" species are relevant to human health.

### 15.5.6 Risk of Adverse Response to Prescribed Pharmaceuticals

Pharmacogenomics has been defined as the use of genetic information to predict the safety, toxicity, and/or efficacy of drugs in individual patients or groups of patients.[31] By such definitions, then, we can see that pharmacogenomics is very closely related to toxicogenomics. Indeed, in many cases the two are used in conjunction: Pharmacogenomic studies commonly focus on the discovery of compounds that bind to specific proteins of interest or determining the protein targets to which a potential drug candidate binds. Toxicogenomic studies are often used at the same time to identify biomarkers of toxic endpoints induced by such interactions.

In this way, pharmacogenomic and toxicogenomic studies together will soon begin to reveal risk presented to a patient by a selected treatment regimen and permit selection of possible alternative therapies. Of the 600 or so pharmaceutical compounds on the U.S. market, about 500 have toxic side effects in some individuals. It has been estimated that a staggering 100,000 die each year from the side effects of properly prescribed medicines, and a further two million develop serious side effects.[32] One reason for the side effects is that drugs are rarely 100% specific in their action. They often bind to non-target receptors or combine with other drugs or compounds in the body to induce toxic side effects. Side effects can also be produced by metabolites created when the body breaks down the original parent compound. Another reason for adverse drug or chemical effects is genetic disposition: polymorphisms in certain key drug metabolizing genes give rise to enzyme products that metabolize the drug either too quickly or not quickly enough. This leads to drug resistance or extreme toxicity respectively. There is thus a need to identify early those patients who are highly refractory to current standard-of-care treatments. In the future, such drug efficacy might be investigated through a combination of toxicogenomic and pharmacogenomic studies, whereby prognostic markers predicting responses to chemotherapy are first identified. The paradigm then is to take a biopsy of the diseased tissue from a patient and conduct *ex vivo* pharmacogenomics tests to identify responders and nonresponders to the first choice therapy, as determined by evaluation of the predetermined prognostic markers. The results of such tests may dictate the use of alternative therapies when they are available.

The main reason why toxic side effects are sometimes not seen until after a drug goes to market is that clinical trials of new drugs are very costly and logistically complex to arrange and conduct. Consequently, although they may provide sufficient data for regulatory agency approval, they are usually not large enough to incorporate enough people with all possible sensitivities. As already discussed, some individuals are inherently more sensitive to drugs, with the incidence of adverse reaction being dependent on the penetrance of the genetic factors responsible for the susceptibility, the doses being allocated, environmental factors (diet and other drugs in the system), and certain stochastic factors. The cost of taking a drug from discovery to market has been estimated variously at up to $800 million. A major contributor to this cost is the expense of bringing drugs through trials that ultimately fail. It is not surprising then that pharmaceutical companies are continually striving to find a quick way to screen out overly toxic compounds ("fail fast, fail cheap"). It is possible that, in the future, it may be possible to associate drug toxicity (or ineffectiveness) with particular SNPs, in which case a pharmaceutical could be directed to be prescribed only to people with (or without) a certain specified genotype.

## 15.6 ISSUES IN APPLYING GENOMIC DATA TO RISK ASSESSMENT

There are a number of hurdles that must be overcome before genomic technologies can be applied to risk assessment. These include:

1. *Regulatory agencies.* The impact of toxicogenomics on risk assessment will rely to a large degree on the increased acceptance and involvement of regulatory agencies. It is expected that, in the not too distant future, chemical and pharmaceutical manufacturers may begin submitting toxicogenomic data to the regulatory agencies in support of their approval packages. Given the current state of the science, much research remains to be done before this would become routine or standard procedure. Nevertheless, the regulatory agencies are understandably concerned about when they will be in a position to accept data from toxicogenomic methods in support of approval applications. The prevailing opinion is that this will occur when consensus is reached within the scientific community and respective regulatory agency that a certain method is robust, sensitive, accurate, and informative enough to justify its use. Many scientists and regulators believe that toxigenomic methods such as gene expression profiling hold much promise, but a good deal of work is still needed before such data can be usefully incorporated into risk assessment for environmental exposures. It is widely anticipated that the next 5 to 10 years will indeed see at least some forms of toxicogenomic data proven robust and sensitive. By that time we should also have appropriated an increased understanding of the complexities of gene expression networks and the biological effects associated with certain gene structure and function in different tissues. These two factors together may then permit toxicogenomic data to be used in risk assessment studies.
2. *Biological relevancy.* One thing that must be done is to make sure that the changes seen in toxicogenomic assays are biologically relevant. Although genomic information has real potential to improve risk prediction, changes in genotype or phenotype *per se* may not be relevant to risk, and variability between individuals in exposure and sensitivity must be incorporated into the risk analysis process. For example, some individuals smoke 40 cigarettes a day all their life without developing lung cancer, whereas others who smoke much less may develop cancer at a relatively early age. Many smokers show characteristic genetic or phenotypic changes in their lung epithelia that are generally indicative of increased risk of progressing to a disease state. For some individuals, however, these changes do not represent a significantly increased risk as certain genetic makeups and/or life-style and environmental factors (notably diet) may strongly reduce the possibility of further disease progression.
3. *Intellectual property rights (IPRs).* IPRs are not an insubstantial barrier to genetic studies. Since 1980, the U.S. Patent and Trademark Office (PTO) has granted more than 20,000 patents on genes or other gene-related molecules (for humans and other organisms), and more than 25,000 applications claiming genes or related molecules remain outstanding. A discussion on the ethics of patenting genes could fill a whole other book; however, it is clear from such figures that many companies and institutions are hoping to exert control over the commercial use of genes, and already this control is beginning to generate opposition. Institutions such as the American College of Medical Genetics (ACMG, http://www.acmg.net/) have issued position statements on gene patents and accessibility of gene testing. The ACMG statement decries certain practices related to current patterns of enforcement of patents on genes that are important in the diagnosis, management, and risk assessment of human disease. In their words:[33]

Enforcement has been effected in one or more of these ways: monopolistic licensing that limits a given genetic test to a single laboratory, royalty-based licensing agreements with exorbitant up-front fees and per-test fees, and licensing agreements that seek proportions of reimbursement from testing services. These limit the accessibility of competitively priced genetic testing services and hinder test-specific development of national programs for quality assurance. They also limit the number of knowledgeable individuals who can assist physicians, laboratory geneticists and counselors in the diagnosis, management and care of at-risk patients.

Such concern is echoed widely across the scientific and risk assessment community, and it can only be hoped that legislation or self-regulation by patent-owning institutions can be effected that will not prevent or hinder the use of genomic data in risk assessment situations.

4. *Improvements in risk management systems*. The technological revolution must be accompanied by improvements in both the risk management and risk assessment systems themselves, such that information on human variability and the associated uncertainties can be properly utilized.
5. *Use of data by competition*. Most pharmaceutical and chemical manufacturing companies are understandably recalcitrant to share data on their chemicals in any way. Much of the concern lies in the vagaries of IPRs associated with the mining of databases. For example, a recent European Union directive on database protection has introduced a "fair use" provision, stipulating that "member states may allow lawful users the free extraction of database contents if used for the purposes of illustration for teaching or scientific research," as long as the source is acknowledged and "to the extent justified by the non-commercial purpose to be achieved." Whether such wording is strong enough to both protect originator rights and facilitate the traditional reuse of scientific data is still under debate. Meanwhile, the debate about database legislation and IPRs rages on in the United States and appears stalled in Congress.[34] The lack of agreement on this issue is regrettable, as much useful information remains locked up in the archives of such private institutions, untapped by their curators because of lack of knowledge or interest or a jealous desire to guard their data, however obsolete it is to them, from competitors. It could be argued that it would be in the best interests of medical research for gene expression laboratories to pool data in a central database, perhaps after they have extracted information that interests them. In this event, the benefits to companies in terms of providing a more complete understanding of biological function might outweigh any potential losses to competitors from revealing certain datasets. It may be in the future of toxicogenomics that certain pharmaceutical and biotechnology companies start scrambling to form strategic alliances with one another in order to reduce experimental costs and mine the untapped wealth of one another's data.
6. *Lack of necessary technology*. The current box of tools used to obtain genomic data, while powerful, is still insufficient to derive all the genetic data required to cover the various types of genomic information and regulation, including X-chromosome activation, genomic imprinting, autoimmunity, aging, and exogenous effects.

## 15.7 ETHICAL CONSIDERATIONS

Perhaps the largest obstacles to the use of toxicogenomic data in risk assessment are the complex social, moral, and legal issues relating to the protection of human subjects, the privacy of genetic information, and the possibility of discriminatory use of the data. A number of ethical and social implications associated with post-genome technology are emerging, and the majority of the general public are becoming highly sensitized and wary of almost all forms of genetic prying. This means that the scientific community must be prepared to educate healthcare providers, the public healthcare system, and the general public on genetic literature, as well as dealing with issues such as loss of privacy, risk of discrimination, and loss of control over personal genetic information.

The first question that arises for ethical consideration is, "Will originators of genetic information make the utmost effort to interpret them properly?" A genuine concern is that the vast quantities of data generated by GEP can provide plenty of ground for "misinterpretation" of data where regulatory and other matters are concerned. Whether such "misinterpretations" are in favor of or against approval of a chemical or pharmaceutical or whether they are a "deliberate" or a "genuine" mistake will be impossible to determine. Integrity and appropriate ethical conduct should of course start with the companies producing the chemicals and testing them. Unfortunately, it is the financial impact that often rules business decisions, and so the burden of thoroughly deciphering the data and understanding its implications may rest ultimately with regulatory agencies.

Another significant threat from the widespread introduction of genomic data might come in the form of workplace discrimination, where employers may find it more convenient or cost effective to remove workers rather than eliminating potential workplace hazards. The questions then arise:

1. When does worker selection based on genetic data constitute discrimination? For example, would it be discrimination to dismiss airline pilots because they are found to be susceptible to narcolepsy?
2. What obligations do employers have to protect workers from placing themselves (and others) at risk?
3. What responsibilities do potential employees have in accepting employment in a position that potentially offers risks when matched with their genetic disposition?

In this complex area of "toxicogenethics," so to speak, bioethicists appear set to play an increasingly important role in the design of experiments that characterize the genetic information of human subjects and application of the data generated therein. The Ethical, Legal, and Social Implications (ELSI) program (http://www.nhgri.nih.gov/About_NHGRI/Der/Elsi/) was launched in 1990 by the National Human Genome Research Institute (NHGRI; http://www.genome.gov/) in order to address these complex issues. ELSI research projects have focused on a wide range of issues, including discrimination in insurance and employment based on genetic information, when and how new genetic tests should be integrated into mainstream healthcare services, informed consent in genetic research protocols, and public and professional education about genetics and related ELSI issues. ELSI currently has an annual budget in excess of 14 million dollars for continuing support of basic and applied research that identifies and analyzes the ethical, legal, and social issues surrounding human genetics research. The current goals of ELSI (developed in 1998 for report in 2003) are to:

1. Examine the issues surrounding the completion of the human DNA sequence and the study of human genetic variation.
2. Examine issues raised by the integration of genetic technologies and information into healthcare and public health activities.
3. Examine issues raised by the integration of knowledge about genomics and gene–environment interactions into nonclinical settings.
4. Explore ways in which new genetic knowledge may interact with a variety of philosophical, theological, and ethical perspectives.
5. Explore how socioeconomic factors, gender, and concepts of race and ethnicity influence the use, understanding, and interpretation of genetic information, the utilization of genetic services, and the development of policy.

Because ELSI is funded in large part by the Human Genome Project (HGP; http://www.ornl.gov/hgmis/) budget, it has been a justifiable concern that the words of bioethicists can be paid for, leading to one notable quote declaring they are "paid to be guard dogs, but appear more as show dogs." The main critique is that bioethicists have failed to properly address controversial issues that might interfere with "the pace of the science," that their input increases academic discussion but does little to improve policy or legislation. Where toxicogenomics are concerned, one of the biggest failings is that, in most cases, the discussion of its strengths and limitations is not balanced and can therefore lead non-scientists to misinterpret the significance of toxicogenomic findings.

The emergence of bioethics as a significant component in the generation and application of genomic data shows the seriousness with which scientists and policymakers are treating public skepticism over the control of this powerful technology. It is not unlikely that many institutions that carry out genomics research will soon employ the services, hired or contracted, of bioethics specialists. The scientific community should be mindful that the services of bioethicists may currently be viewed as somewhat of a commodity at this time, but nevertheless their increased involvement in research is likely to occur and should be viewed as a mutually beneficial arrangement that can facilitate identification of ethical issues that would otherwise go unnoticed.

## 15.8 SUMMARY

The potential offered by toxicogenomics is currently unproven in most part but offers a revolutionary change to the risk assessment process. Past precedent certainly exists for new genetic information leading to revolutionary changes in science, including the overturning of the central dogma of molecular biology (one gene, one transcript, one polypeptide chain), and the revolution in microbial taxonomy initiated by nucleotide sequence information. Integrating genomics data into risk assessment faces a number of challenges, including the signal-to-noise problems (when do changes become biologically significant?), the large and complex datasets generated, and the current lack of qualitative and quantitative linkages to toxicity and disease.

A number of key relationships must also be elucidated before genomic data can be incorporated into risk assessment, namely those between endpoint and health, lab models and humans, outcomes in new assays with established methodologies, and reproducibility and assay performances. Despite these hurdles, many are optimistic that, with more attention to route- and situation-specific exposures and sensitive subpopulations, the ultimate goal of *biologically based risk assessment* is eminently achievable.

It is perhaps too early to predict exactly when toxicogenomic data will have a measurable effect on risk assessment procedures. It will certainly be delayed until consensus is reached within the scientific community and the responsible regulatory centers about the suitability of any given approach. It is unlikely that a single genomic technology will meet all assessment needs; rather, different methods will apply under different circumstances. Furthermore, risk assessment will embrace an increasingly multidisciplinary approach requiring the integration of pharmacology, pharmacokinetics, statistics, pathology, toxicology, and more. However, as is written in one of the world's most ancient texts:[35]

> Behold they are one people, and they all have the same language. And this is what they began to do, and now nothing which they purpose to do will be impossible for them.

With so many academic laboratories, government regulatory agencies, and international companies speaking the language of toxicogenomics and investing increasingly large amounts of resources in the field, it should not be too many years before the promises it offers will either be revealed as one of the biggest hoaxes in biological history (second only to Piltdown man!) or a springboard to launch pharmaceutical and chemical safety evaluation to a new level. Like a large proportion of the life science community, I support the latter possibility and predict that in the 21st century the new field of toxicogenomics will greatly improve the accuracy of risk assessment, allowing identification of sensitive subpopulations and, ultimately, enabling personalized risk profiling for each individual based on their genetic composition.

## ACKNOWLEDGMENTS

The information in this document has been subjected to review by the National Health and Environmental Effects Research Laboratory and approved for publication. Approval does not signify that the contents reflect the views of the Agency nor does mention of trade names or commercial products constitute endorsement or recommendation for use. Thanks to Drs. Barbara Abbott (USEPA) and Michael Burczynski (Wyeth Laboratories) for critically reviewing the manuscript prior to submission.

## REFERENCES

1. Selkirk, J. and Tennant, R.W., http://www.grc.uri.e.,du/programs/2003/toxico.htm, 2002.
2. Nuwaysir, E.F. et al., Microarrays and toxicology: the advent of toxicogenomics, *Mol. Carcinog.*, 24, 153–159, 1999.
3. Hatada, I. et al., A genomic scanning method for higher organisms using restriction sites as landmarks, *Proc. Natl. Acad. Sci. USA*, 88, 9523–9527, 1991.
4. Costello, J.F., Smiraglia, D.J., and Plass, C., Restriction landmark genome scanning, *Methods*, 27, 144–149, 2002.
5. Kallioniemi, A. et al., Comparative genomic hybridization for molecular cytogenetic analysis of solid tumors, *Science*, 258, 818–821, 1992.
6. Pinkel, D. et al., High resolution analysis of DNA copy number variation using comparative genomic hybridization to microarrays, *Nat. Genet.*, 20, 207–211, 1998.
7. Schrock, E. et al., Multicolor spectral karyotyping of human chromosomes, *Science*, 273, 494–497, 1996.
8. Lisitsyn, N., Lisitsyn, N., and Wigler, M., Cloning the differences between two complex genomes, *Science*, 259, 946–951, 1993.
9. Gibson, U.E., Heid, C.A., and Williams, P.M., A novel method for real time quantitative RT-PCR, *Genome Res.*, 6, 995–1001, 1996.
10. Rockett, J.C., Esdaile, D.J., and Gibson, G.G., Differential gene expression in drug metabolism: practicalities, problems and potential, *Xenobiotica*, 29, 655–691, 1999.
11. Rockett, J.C., Use of genomic data in risk assessment, *Genome Biol.*, 3, reports4011.1–4011.3, 2002.
12. Simmons, P.T. and Portier, C.J., Toxicogenomics: the new frontier in risk analysis, *Carcinogenesis*, 23, 903–905, 2002.
13. Aardema, M.J. and MacGregor, J.T., Toxicology and genetic toxicology in the new era of "toxicogenomics": impact of "-omics" technologies, *Mutat. Res.*, 499, 13–25, 2002.
14. Cai, J. et al., Supervised Machine Learning Algorithms for Classification of Cancer Tissue Types Using Microarray Gene Expression Data, http://www.cpmc.columbia.edu/homepages/jic7001/cs4995/project1.htm, 2002.
15. Cai, S.Y. et al., Bile salt export pump is highly conserved during vertebrate evolution and its expression is inhibited by PFIC type II mutations, *Am. J. Physiol. Gastrointest. Liver Physiol.*, 281, G316–G322, 2001.
16. Dan, S. et al., An integrated database of chemosensitivity to 55 anticancer drugs and gene expression profiles of 39 human cancer cell lines, *Cancer Res.*, 62, 1139–1147, 2002.
17. Burczynski, M.E. et al., Toxicogenomics-based discrimination of toxic mechanism in HepG2 human hepatoma cells, *Toxicol. Sci.*, 58, 399–415, 2000.
18. Bartosiewicz, M., Penn, S., and Buckpitt, A., Applications of gene arrays in environmental toxicology: fingerprints of gene regulation associated with cadmium chloride, benzo(*a*)pyrene, and trichloroethylene, *Environ. Health Perspect.*, 109, 71–74, 2001.
19. Bartosiewicz, M.J. et al., Unique gene expression patterns in liver and kidney associated with exposure to chemical toxicants, *J. Pharmacol. Exp. Ther.*, 297, 895–905, 2001.

20. Hamadeh, H.K. et al., Prediction of compound signature using high density gene expression profiling, *Toxicol. Sci.*, 67, 232–240, 2002.
21. Hamadeh, H.K. et al., Gene expression analysis reveals chemical-specific profiles, *Toxicol. Sci.*, 67, 219–231, 2002.
22. Thomas, R.S. et al., Identification of toxicologically predictive gene sets using cDNA microarrays, *Mol. Pharmacol.*, 60, 1189–1194, 2001.
23. Waring, J.F. et al., Clustering of hepatotoxins based on mechanism of toxicity using gene expression profiles, *Toxicol. Appl. Pharmacol.*, 175, 28–42, 2001.
24. Waring, J.F. et al., Microarray analysis of hepatotoxins *in vitro* reveals a correlation between gene expression profiles and mechanisms of toxicity. *Toxicol. Lett.*, 120, 359–368, 2001.
25. Banerjee, N. and Zhang, M.Q., Functional genomics as applied to mapping transcription regulatory networks, *Curr. Opin. Microbiol.*, 5, 313–317, 2002.
26. Wu, X. et al., Mutagen sensitivity as a susceptibility marker for human hepatocellular carcinoma, *Cancer Epidemiol. Biomarkers Prev.*, 7, 567–570, 1998.
27. Noguchi, M. et al., Interleukin-2 receptor gamma chain mutation results in X-linked severe combined immunodeficiency in humans, *Cell*, 73, 147–157, 1993.
28. The BioSpace Glossary, http://www.biospace.com/gls_detail.cfm?t_id = 1728, 2002.
29. Acton, J.D. and Wilmott, R.W., Phenotype of CF and the effects of possible modifier genes, *Paediatr. Respir. Rev.*, 2, 332–339, 2001.
30. Bruder, C.E. et al., Severe phenotype of neurofibromatosis type 2 in a patient with a 7.4-MB constitutional deletion on chromosome 22: possible localization of a neurofibromatosis type 2 modifier gene?, *Genes Chromosomes Cancer*, 25, 184–190, 1999.
31. Weber, B.L. and Nathanson, K.L., Low penetrance genes associated with increased risk for breast cancer, *Eur. J. Cancer*, 36, 1193–1199, 2000.
32. Lazarou, J., Pomeranz, B., and Corey, P., Incidence of adverse drug reactions in hospitalized patients a meta-analysis of prospective studies, *JAMA*, 279, 1200–1205, 1998.
33. American College of Medical Genetics Position Statement on Gene Patents and Accessibility of Gene Testing, http://www.acmg.net/Pages/ACMG_Activities/policy_statements_pages/current/Gene_Patents_&_Accessibility_of_Gene_Testing_Position_Statement_on.htm, 2002.
34. Russo, E., EU Database Directive Draws Fire, http://www.the-scientist.com/yr2002/jul/russo_p18_020708.html, 2002.
35. Moses, The Book of Genesis, in *The Holy Bible*, chapter 11, verses 6–7.

# Index

NOTE: Italicized page numbers refer to tables and illustrations.

## D



## E

## H

## T

## U



## V



## W



## X



## Z

Milton Keynes UK
Ingram Content Group UK Ltd.
UKHW050453071024
449327UK00015B/361

9 780367 395308